Population Biology of Passerine Birds

An Integrated Approach

NATO ASI Series

Advanced Science Institutes Series

A series presenting the results of activities sponsored by the NATO Science Committee, which aims at the dissemination of advanced scientific and technological knowledge, with a view to strengthening links between scientific communities.

The Series is published by an international board of publishers in conjunction with the NATO Scientific Affairs Division

A Life Sciences B Physics	Plenum Publishing Corporation London and New York
C Mathematical and Physical Sciences D Behavioural and Social Sciences E Applied Sciences	Kluwer Academic Publishers Dordrecht, Boston and London
F Computer and Systems Sciences G Ecological Sciences H Cell Biology	Springer-Verlag Berlin Heidelberg New York London Paris Tokyo Hong Kong Barcelona

Series G: Ecological Sciences Vol. 24

Population Biology of Passerine Birds

An Integrated Approach

Edited by

Jacques Blondel

Centre National de la Recherche Scientifique,
Centre d'Ecologie Fonctionnelle et Evolutive,
B.P. 5051, 34033 Montpellier Cédex, France

Andrew Gosler

Department of Zoology,
Edward Grey Institute of Field Ornithology,
South Parks Road, Oxford OX1 3PS, Great Britain

Jean-Dominique Lebreton

Centre National de la Recherche Scientifique,
Centre d'Ecologie Fonctionnelle et Evolutive,
B.P. 5051, 34033 Montpellier Cédex, France

Robin McCleery

Department of Zoology,
Edward Grey Institute of Field Ornithology,
South Parks Road, Oxford OX1 3PS, Great Britain

Springer-Verlag Berlin Heidelberg NewYork
London Paris Tokyo Hong Kong Barcelona
Published in cooperation with NATO Scientific Affairs Division

Proceedings of the NATO Advanced Research Workshop on Demographical, Physiological, Behavioural and Genetical Aspects of Population Biology of Passerine Birds held at Evisa, Corsica, France October 3–8, 1989.

ISBN-13: 978-3-540-51759-7 e-ISBN-13: 978-3-642-75110-3
DOI: 10.1007/ 978-3-642-75110-3

Library of Congress Cataloging-in-Publication Data
NATO Advanced Research Workshop on Demographical, Physiological, Behavioural, and Genetical Aspects of Population Biology in Passerine Birds (1989 : Evisa, France) Population biology of passerine birds : an integrated approach / edited by Jacques Blondel ... [et al.]. p. cm.—(NATO ASI series. Series G, Ecological sciences ; vol. 24) "Proceedings of the NATO Advanced Research Workshop on Demographical, Physiological, Behavioural, and Genetical Aspects of Population Biology in Passerine Birds, held at Evisa, Corsica, France, October 3–8, 1989"—T.p. verso.

1. Passeriformes—Congresses. 2. Bird population—Congresses. I. Blondel, J (Jacques). II Title III. Series: NATO ASI series. Series G, Ecological sciences ; no. 24. QL696.P2N38 1989 598.8'045248—dc20
90-10391 CIP

Softcover reprint of the hardcover 1st edition 1990

2131/3140-543210 – Printed on acid-free-paper

PREFACE

There is in Europe a long tradition of research on population biology of passerine birds, especially hole-nesting species such as Tits and Flycatchers. For instance, three studies have continued without a break for several decades. The first, on the Great Tit, was initiated in Holland in the 1930's by H.N. Kluyver. The second concerns the Pied Flycatcher which has been intensively studied since 1941 in Finland by Lars von Haartman. A third long-term programme was started in 1947 on tits by David Lack at the Edward Grey Institute of Field Ornithology (Oxford). These works had and still have a great impact on the elaboration of modern theories of population biology. The success obtained by these pioneer studies triggered the development of many programmes throughout Europe and North America. The amount of knowledge accumulated so far, as well as the emergence of new conceptual issues and impressive methodological improvements gave a stimulating impetus to this field of research. Although interest focussed mainly on the dynamics and the regulation of bird populations during the 1950's, more recent developments have concerned such major issues as genetics, ecological physiology, social behaviour, optimal foraging theory and biometry. These new developments convincingly demonstrated that what is important in nature is less the fate of an average individual in a population than the fate of each singular individual. Thus, population biologists have a growing interest in what an individual is, and why it is as it is within a population in a given set of environmental characteristics. They have also clearly demonstrated that because of a large amount of genetic variation, populations may change within observable periods of time. For a long time populations and species have been considered to have evolved slowly in response to long-term environmental changes. In fact, modern studies reveal that the dynamic nature of environments produces changes in selective pressures on heritable variation over very short periods of time. This emphasizes the importance of studying natural populations in a world where mankind imposes everywhere tremendous changes on the environment.

There have been two previous workshops on this theme, both held in Wageningen (Holland). The first one (1979) focussed on **"The integrated study of bird populations"** and the second one (1985) on **"Causal and evolutionary aspects of the determination of bird numbers with special reference to hole nesting birds"**. The proceedings of these meetings have been published in the journal Ardea in 1980 and 1987 respectively. The aim of this third workshop was to present and discuss in an integrated way the state of the art in this rapidly developing field of research. Of course there are so many aspects in population biology and so many dedicated scientists working on this subject that this book is necessarily selective. It certainly does not claim to be a textbook. Our goal was rather to bring together a panel of specialists working on different topics in order to encompass some major trends of research. Fifty-three scientists from 15 countries of Europe, North America and North Africa participated in the meeting. This book is the publication of most of their contributions. It is divided into 7 parts: first a presidential address given by Lars von Haartman on the theme of the "breeding time" of a population of the Pied Flycatcher which he has been studying for 40 years. Then follow five parts of unequal importance: **Reproductive biology and population regulation** (12 chapters), **Between-individual variation within populations** (6 chapters), **Costs and benefits of reproduction** (7 chapters), **Behavioural aspects of reproduction** (7 chapters), **Recruitment, floaters, immigrants** (6 chapters). The book ends with two conclusions. The first, by C.M. Perrins, discusses the extent to which birds evolve different characteristics in different habitats. The second conclusion, by A. van Noordwijk, addresses the general point on what has been new in the Conference and why and how to understand populations in nature.

We greatly acknowledge the NATO Scientific Affair's Division which gave us the opportunity to organize this workshop. We are indebted to the French Ministère de l'Environnement, the Centre National de la Recherche Scientifique (CNRS) and the Parc Naturel Régional de Corse for funding parts of the workshop. This Conference was held in the Parc Naturel Régional de Corse and we are grateful to its Director, M. Leenhardt, for having given us the wonderful facilities of the Evisa's Paisolu d'Aitone. We are indebted to the Scientific Committee (Dr. H.N. van Balen, Dr. J. Blondel, Dr. J. Clobert, Prof. A. Dhondt, Dr. J.D. Lebreton, Prof. C.M. Perrins and Prof. C. Ferry) for having given us invaluable advice and checked the scientific quality of the talks. Thanks are also given to all individual participants who made this meeting an outstanding event and to the referees of the manuscripts who greatly improved the quality of the written contributions.

This workshop would not have been so successful without the enthusiastic involvement of many people who helped us before, during and after the workshop: A. Dervieux, P. Isenmann, P. Lenck, P. Lepaulmier, M. Maistre, J.L. Martin, Ph. Perret, M. Sophy, J.C. Thibault, J. Vilanova and his colleagues. R. Ferris made a nice drawing of the Corsican Nuthatch. Maryse Gautier and Josiane Trauchessec carried out the prolonged and unrewarding administrative preparation for the workshop and F. Romane gave us computer facilities. We owe special thanks to Mireille Cartan-Son who made a wonderful job of editing the manuscripts. Lastly, we would like to warmly thank the publisher, Springer-Verlag, and we hope that this book will stimulate further research in the field of population biology of birds.

Jacques Blondel
Andrew G. Gosler
Jean-Dominique Lebreton
Robin H. McCleery

PARTICIPANTS

*ALATALO, Rauno V. Department of Biology, University of Jyväskylä, Yliopistonkatu 9, SF-40100 Jyväskylä, Finland

*ASKENMO, Conny. Department of Zoology, University of Göteborg, Box 25059, S-400 31 Göteborg, Sweden

*BLONDEL, Jacques. Centre Louis Emberger, C.N.R.S., BP 5051, 34033 Montpellier cedex, France

*CLOBERT, Jean. Laboratoire d'Ecologie, URA 258, C.N.R.S., Ecole Normale Supérieure, 46, rue d' Ulm, 75230 - Paris Cedex 05, France

*COWIE, Richard J. School of pure and applied Biology, University of Wales, College of Cardiff, P.O. Box 915, Cardiff CF13TL, Great Britain

*DHONDT, André A. Department of Biology, Universiteitsplein 1, University of Antwerp (U.I.A.), B-2610 Wilrijk, Belgium

DIAS, Paula C. Av. M. Conceiçao, 13, 2E, Carcavelos, 2775 Parede, Portugal

*DRENT, Pietr. Instituut voor Oecologisch Onderzoek, Boterhoeksestraat 22, Postbus 40, 6666 GA, Heteren, Holland

*EKMAN, Jan. Department of Zoology, University of Stockholm, Stockholm, S-10691 Sweden

FERRY, Camille. Etaules, 21121, Fontaines les Dijon, France

*FIDALGO, Lilia. Serviço Nacional de Parques, Reservas e conservaçao da Natureza, Reserva Natural das Dunas de S. Jacinto, S. Jacinto, 3800 Aveiro, Portugal

*GOSLER, Andrew G. Department of Zoology, Edward Grey Institute of Field Ornithology, South Park Roads, OX1 3PS, Oxford, Great Britain

*GUSTAFSSON, Lars. Department of Zoology, Uppsala University, Box 561, 751 22 UPPSALA, Sweden

*HAFTORN, Svein. University of Trondheim, The Museum, 7000 Trondheim, Norway

*HENRICH, Sabine. Zoologisches Institut, Rheinsprung 9, Ch-4051 Basel, Switzerland

*HILDEN, Olavi. Department of Zoology, University of Helsinki, P. Rautatiekatu 13, SF-00100 Helsinki 10, Finland

*HOGSTAD, Olav. University of Trondheim, Dept. of Zoology, 7055 Dragvoll, Norway

*ISENMANN, Paul. Centre Louis Emberger, C.N.R.S., B.P. 5051, 34033 Montpellier cedex, France

*KÄLLANDER, Hans. University of Lund, Ecology Building, 223 62 Lund, Sweden

*LAMBRECHTS, Marcel M. University of Antwerp (U.I.A.), Department of Biology, Universiteitsplein I, 2610 Wilrijk, Belgium

*LEBRETON, Jean-Dominique. Centre Louis Emberger, C.N.R.S., B.P. 5051, 34033 - Montpellier Cedex, France

LECOQ, Nuno P. Serviço Nacional de Parques, Reservas e Conservaçao da Natureza, Rua da Lapa, 73, 1200 Lisboa, Portugal

*LESSELLS, Kate M. Department of Animal and Plant Sciences, University of Sheffield, Sheffield S10 2TN, Sheffield, Great Britain

*LUNDBERG, Arne. Department of Zoology, Uppsala University, Box 561, S-751 22 Uppsala, Sweden
*MARTIN, Jean-Louis. Centre Louis Emberger, C.N.R.S., BP 5051, 34033 Montpellier cedex, France
*MATTHYSEN, Eryk. Department of Biology, Universiteitsplein 1, University of Antwerp (U.I.A.), B-2610 Wilrijk, Belgium
*McCALLUM, D. Archibald. Savannah River Ecology Laboratory, Drawer E., Aiken, SC 29802, USA
*McCLEERY, Robin H. Department of Zoology, Edward Grey Institute of Field Ornithology, South Park Roads, OX1 3PS, Oxford, Great Britain
MOALI, A. Laboratoire d'Ecologie des Animaux Terrestres, INES Bilogie, 15000 DZ Tizi Ouzou, Algérie
*MøLLER, Anders P. Department of Zoology, Uppsala University, Box 561, S-751 22, Uppsala, Sweden
*NAGER, Ruedi. Zoologisches Institut, Rheinsprung 9, Ch-4051 Basel, Switzerland
*NILSSON, Jan-Åke. University of Lund, Department of Ecology, Animal Ecology, Ecology Building, S-223 62 Lund, Sweden
*NUR, Nadav. Point Reyes Bird Observatory, 4990 Shoreline Highway, Stinson Beach, CA 94970, USA
*ORELL, Markku. Department of Zoology, University of Oulu, Linnanmaa A, SF-90570 Oulu, Finland
*PERRINS, Christopher M. Department of Zoology, Edward Grey Institute of Field Ornithology, South Park Roads, OX1 3PS, Oxford, Great Britain
*POWER, Harry W. Department of Biological Sciences, Rutgers University, P.O. Box 1059, Piscataway, NJ 08903, U.S.A.
*ROBERTSON, R.J. Queen's University, Department of Biology, Kingston, Ontario, K7L 3N6, Canada
*SLAGSVOLD, Tore. University of Oslo, Zoological Museum, Sars gate 1, N-0562 Oslo 5 Norway
*SMITH, Susan M. Department of Biological Sciences, Mount Holyoke College, South Hadley, MA 01075, U.S.A.
THIBAULT, Jean-Claude. Parc Naturel Régional de Corse, France
*TINBERGEN, Joost M. Instituut voor Oecologisch Onderzoek, Boterhoeksestraat 22, Postbus 40, 6666 GA, Heteren, Holland
*TÖRÖK Janos. Department of Systematic Zoology and Ecology, and Department of Genetics, Eötvös University, 1088 Puskin 3, Budapest, Hungary
TURAN, L. Hacettepe Universitesi Egitim Fakültesi Fen Bilimleri Bölümü Biyoloji Egt. ABD, Beytepe-Ankara, Turkey
*van BALEN, J. Hans. Instituut voor Oecologisch Onderzoek, Boterhoeksestraat 22, Postbus 40, 6666 GA, Heteren, Holland
*van NOORDWIJK, Arie J. Zoologisches Institut, Rheinsprung 9, CH 4051 Basel, Switzerland
VANSTEENWEGEN, Chistian. CRBPO, Muséum National d'Histoire Naturelle, 55 Rue de Buffon, 75005 Paris, France
*VERHULST, Simon. Instituut voor Oecologisch Onderzoek, Boterhoeksestraat 22, Postbus 40, 6666 GA, Heteren, Holland

VIALE, Denise. Université de Corte, Corse, France
VICENTE, Luis. Departamento de Zoologia e Antropologia, Faculdade de Ciencias, Bloco C-2, 3° Piso - Campo Grande, Universidade de Lisboa, 1700 Lisboa, Portugal
VICENTE, A.
*von HAARTMAN, Lars. Lemsjöholm, Via Turku 21999, Finland
*WINDT, W. Arbeitsgruppe für Verhaltensforschung, Ruhr-Universität Bochum, Postfach 102 148, D-4630 Bochum, West Germany
*WINKEL, Wolfgang. Aussenstation Braunschweig für Populationsökologie, "Vogelwarte Helgoland", Weddel, Bauernstr. 14, D-3302 Cremlingen, FR Germany
WINKEL, Doris. Aussenstation Braunschweig für Populationsökologie, "Vogelwarte Helgoland", Weddel, Bauernstr. 14, D-3302 Cremlingen, FR Germany
*ZANDT, Hagen. Zoologisches Institut, Rheinsprung 9, Ch-4051 Basel, Switzerland

* Presenter of spoken paper

CONTENTS

PRESIDENTIAL ADDRESS

REPRODUCTIVE BIOLOGY AND POPULATION REGULATION

BETWEEN-INDIVIDUAL VARIATION WITHIN POPULATIONS

COSTS AND BENEFITS OF REPRODUCTION

BEHAVIOURAL ASPECTS OF REPRODUCTION

RECRUITMENT, FLOATERS, IMMIGRANTS

CONCLUSIONS

SUBJECT INDEX

PRESIDENTIAL ADDRESS

BREEDING TIME OF THE PIED FLYCATCHER *FICEDULA HYPOLEUCA*

Lars von HAARTMAN

Lemsjöholm, via Turku, 21999 Finland

EGG DEVELOPMENT AND ITS CONSEQUENCES

Egg development (Stieve 1919) starts in spring with a slow recovery of the ovary. After arrival and under the influence of the courtship of a territorial male (cf. also Silverin 1979), a restricted number of presumptive egg cells, the oocytes, successively enter a "period of rapid growth", reach maturity, and are laid one after another, early every morning, until the clutch is complete.

This process is illustrated by a dissection of a female (Fig. 1), soon after laying her first egg. One of the oocytes (1) has already entered the oviduct, where it will be fertilized, and the albumen and shell deposited, so that it is ready to be laid next morning. In the ovary 3 oocytes (2-4) of increasing size are discernible. They will evidently be laid 2, 3, and 4 days later - intervals of more than one day being exceptional. A fifth oocyte may quite recently have entered the phase of rapid growth, although no change is yet visible. If this is the case, the oocyte in question will develop into an egg to be laid after 5 days. If number 4 is the last oocyte which has commenced rapid growth, egg development will take one day less. Finally, the ovary contains two small cups, calyces, remnants of the follicles which surrounded oocytes 0-1, which have now left the ovary. The calyces are resorbed within a relatively short time.

If the rapid growth of the oocytes requires 5 days, we may expect that a female, that has lost her clutch, will be able to start a new one only after a minimum of 5 days. Likewise, an increase in temperature will (if it has any effect at all) be followed by laying after no less than 5 days. Finally, the interval between laying dates of the primary and secondary females of polygynous males should be influenced by the length of time required for the growth of the oocytes.

In an earlier investigation (von Haartman 1969), it was found that lost clutches were renewed only after 5 or more days. More recently, a single 4-day interval was observed (Tab. I). Berndt and Sternberg (1972) found 8 cases of 4-day intervals out of 100 (!). In the Great Tit also, the 5-day rule is occasionally broken. These cases, however, can be regarded as exceptions.

A minimum interval of 5 days between clutch loss and replacement could be the rule even without being caused by oocyte growth, as a consequence of the time it would take the Pied Flycatcher female to find a new mate in an often distant territory and to build a nest, which is known usually to take more than 5 days (Stjernberg 1974). Great Tits do not generally change their mate or

NATO ASI Series Vol. G 24
Population Biology of Passerine Birds
Edited by J. Blondel et al.

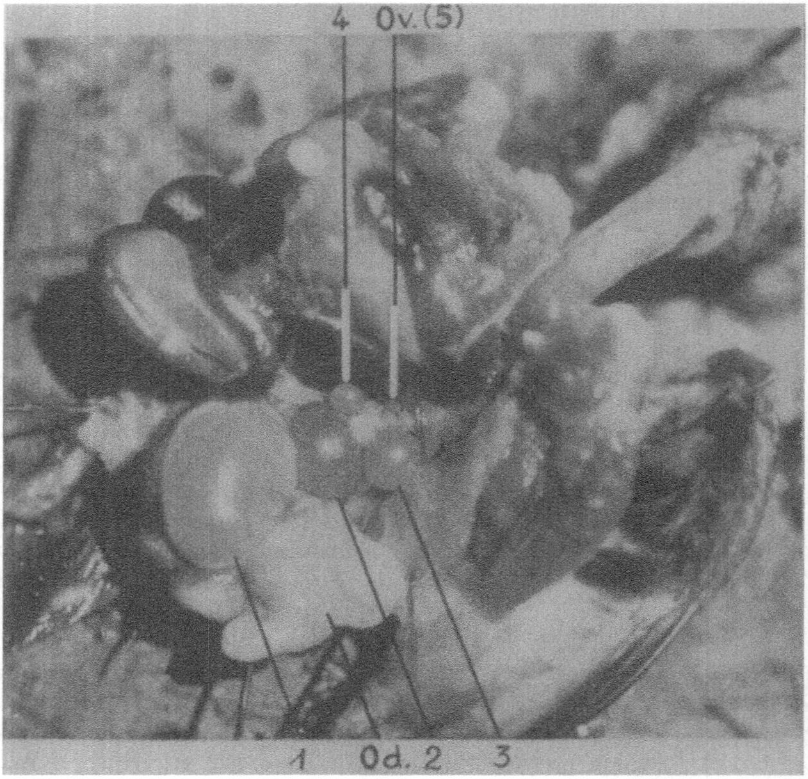

Fig. 1. Dissection of female Pied Flycatcher some 4 hours after laying her first egg. 1-4 : oocytes, which will be laid after 1-4 days. The ovary (ov.) contains some equally large oocytes, one of which may just have begun rapid growth, and will be laid after 5 days. Od., oviduct. Between oocytes 2, 3, 4, and ovary an empty follicle (calyx). From von Haartman 1951 a.

search for a nest-hole outside the original territory, and their replacement nests are often extremely carelessly built, so as to demand a minimum of work. This behavioural difference may be the reason why loss-replacement intervals of 7 days or more were more common in the Pied Flycatcher than in the Great Tit. Kluyver (1951) did not, however, ascribe much importance to the "determinant day", 5 days before laying, in the Great Tit. Unfortunately, there has been no anatomical analysis of egg development in this species.

The situation at the end of laying was not studied satisfactorily. In a Great Tit which had just laid her 9th egg, a 10th and last egg was found in the oviduct, but I found no enlarged oocytes in the ovary. Either there had not been more than 10, or there had been an excess, which had already been resorbed.

Table I. Observed interval between clutch loss and replacement in the Pied Flycatcher (PF) and Great Tit (GT) at Lemlsjöholm, 20 km W of the city of Turku, SW Finland. The length of the intervals was not precisely 5, 6, etc. days, but included a number of surplus hours, as eggs are laid early in the morning, but ringing, the usual cause of nest desertion, is performed during the day.

Days	3	4	5	6	7	8	9	≥10
PF	-	1	6	1	2	-	1	1
GT	1	1	15	6	-	-	-	3

When the relation between rising temperature in May and the appearance of the first egg in the Pied Flycatcher population was calculated for the first 14 study years at Lemsjöholm and Kilpisjärvi, some correlation was found with the temperature 7-5 days earlier (Tab. II). Järvinen (1983) found a similar relation between the fifth day before the start of egg-laying and temperature, but he thinks that the oocytes of the Pied Flycatcher "may be in a more advanced state at Kilpisjärvi...enhancing a relatively quick laying". This assumption would demand a minimum of anatomical work to be prooved or falsified.

Table II. First egg laid in the Pied Flycatcher populations at Lemsjöholm (L, 60°30'N) and Kilpisjärvi (K, 69°N) in relation to temperature 10-0 days before laying ; 0 = day of laying ; 14 years of observation in both areas. The temperature is given as the difference between observed temperature and average temperature of the different days during the years of observation (v. Haartman 1956, Järvinen 1983).

Days before laying	10	9	8	7	6	
Temp. L.	+0.1	+0.3	+0.0	+1.0	+0.9	
Temp. K.	-0.2	-0.5	-0.6	-0.1	-0.1	
Days before laying	**5**	**4**	**3**	**2**	**1**	**0**
Temp. L.	+1.2	+0.0	-0.5	-0.4	-0.5	-0.3
Temp. K.	+0.9	+0.6	+1.3	+1.1	+0.3	+0.5

The successive polygamy of the Pied Flycatcher male begins with his searching for a new territory, disappearing for longer or shorter periods from his old one. After finding a suitable nest-hole, he starts to advertise it, trying to attract a female. As he leaves his primary territory just before or when his primary female starts to lay, the secondary female should be expected usually to start laying a minimum of five days later. Shorter intervals could occur in occasional simultaneous polygamy, perhaps also if the male left his primary female unusually early. The empirical data (Tab. III) seem to confirm these assumptions.

Table III. Intervals (in days) between start of laying of primary and secondary females of polygynous Pied Flycatcher males at Lemsjöholm (from von Haartman 1951).

Intervals	0	1	2	3	4	5	6	7	8	9	≥10
Number of cases	1	-	-	2	-	1	5	-	1	1	5

EGG DEVELOPMENT LARGELY INDEPENDENT OF EXTERNAL FACTORS

Once the rapid growth of the oocytes has started, the number of eggs laid is largely independent of external influences. Daily removal of up to 5 eggs, or addition of extra eggs the day before or on the first day of laying did not cause an increase, or reduction, of the number laid (von Haartman 1967, cf. also Berndt 1943). Nor was the number affected by an extremely cold spell which started on 20 May 1944 and reached its peak on 23 May, when the ground was covered by snow. No flycatchers were heard singing, and few other birds for that matter. In fact the only sign of the presence of the Pied Flycatchers was the regular appearance of a new egg in the nests every morning (Fig. 2). Only after 2-3 days had the birds lost so much energy that laying had to stop, but after 2-3 days of "refuelling" they started anew and continued until normal clutch sizes were obtained.

BEGINNING AND END OF THE BREEDING SEASON

Laying at Lemsjöholm started on 14 May at the earliest, soon reached a peak, became rarer after a relatively short time and decreased during a prolonged period, ending on 29 June at the latest. (Fig. 3).

As in many other species, by no means only altricial passerines, clutch size sharply decreases as the season progresses (von Haartman 1951 b, 1982). One may ask why Pied Flycatchers do not start laying earlier in order to maximize clutch size, and why they continue until the end of June, when reproduction is low.

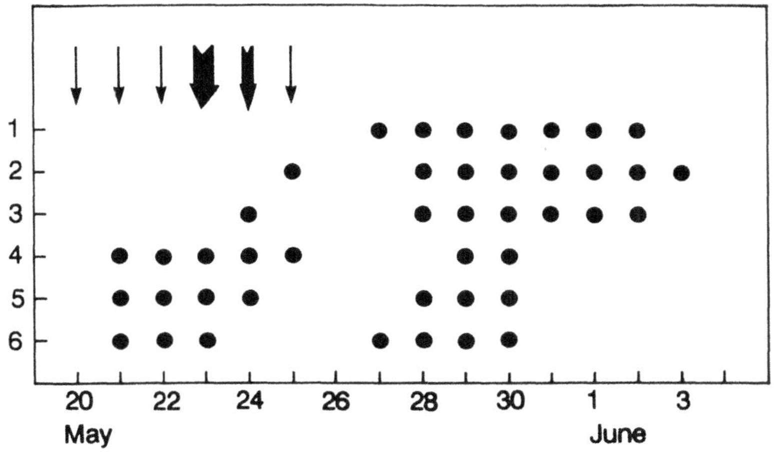

Fig. 2. Laying in Pied Flycatchers at Lemsjöholm during a cold spell in late May 1944. Arrow : cold; thick arrow : snow cover; medium thick arrow : melting snow cover; dots : eggs laid; 1-6: number of females.

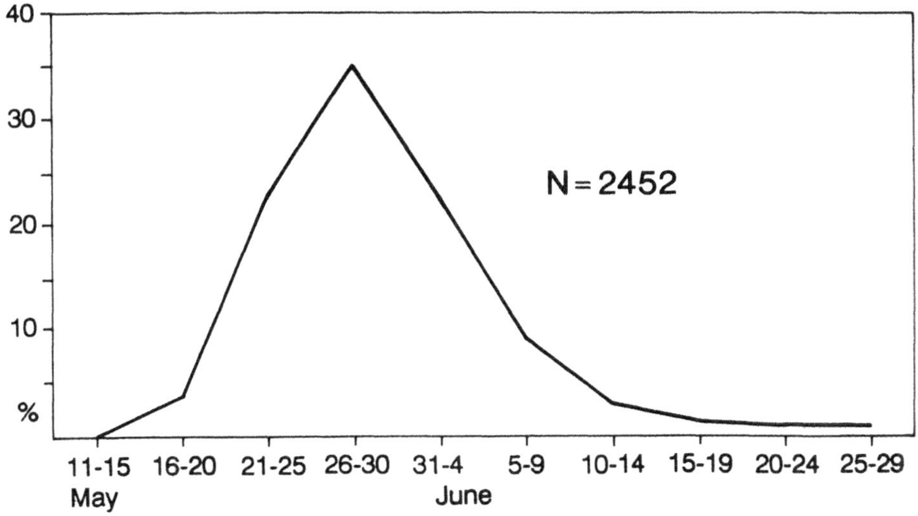

Fig. 3. Percentages of clutches of the Pied Flycatcher at Lemsjöholm, started during 5-day periods. Numbers combined of all dated clutches 1941-89.

The simplest answer to the first question would be that earlier breeding would demand earlier arrival, involving increasing risks of succumbing during cold spells. However, the most severe cold spell witnessed by me, that of late May 1944, did not have other visible effects than a short break in the laying.

In North Finland the breeding time of Pied Flycatchers starts earlier in relation to vegetation development than in South Finland. This is typical of many passerines (von Haartman 1962, 1963). The difference between the Pied Flycatcher's breeding time in South and North Finland is strikingly small (Fig. 4).

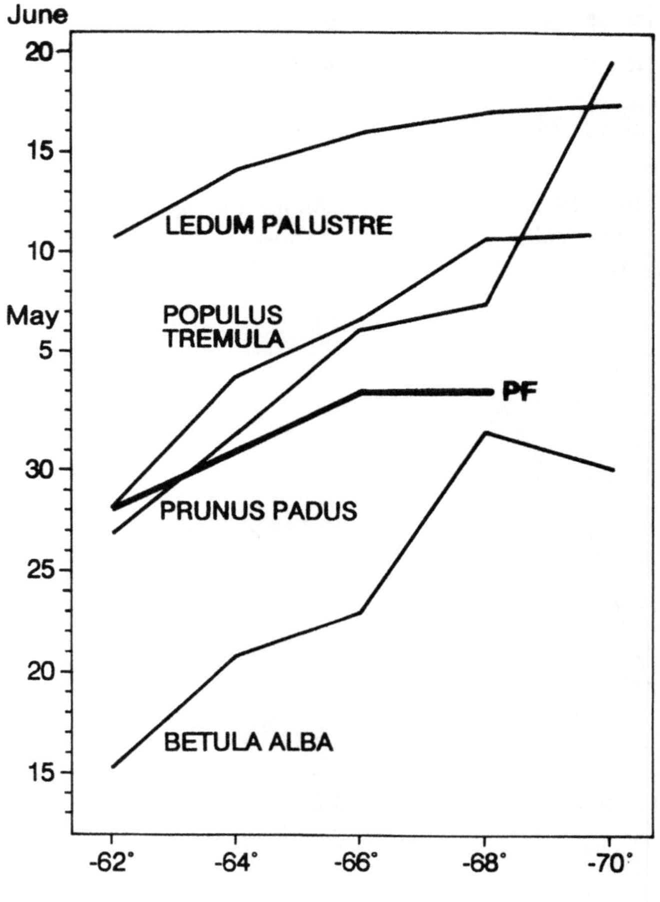

Fig. 4. Average start of breeding in Pied Flycatcher (PF) populations at different latitudes in Finland. Data on ringing Pied Flycatcher broods counting back 31 days (see text). NB the start of rapid oocyte growth is generally 5 days before the start of laying and determines this start. The average leafing of birch and aspen, as well as the start of flowering of bird-cherry and *Ledum*, are given (data from von Haartman 1962, 1963).

Why the Pied Flycatcher starts laying so early in the North is a difficult question (see also Slagsvold 1975a, 1977, Järvinen 1983). Several factors may be responsible. The longer days close to the Arctic Circle is one, and the more rapid development of summer (plant and insect life), when it finally comes, is another. Also, the shortness of the summer may necessitate an early start of breeding in order to give the young time to mature before autumn migration. An extreme example of the last principle is the Honey Buzzard, *Pernis apivorus*, which breeds almost simultaneously over large areas of Central and North Europe.

In the zone of 60°- 62°N the mean ringing date of young Pied Flycatchers was 28 June, and in the zone of 66°- 68°N it was only 6 days later. The preferred age for ringing Pied Flycatcher young is probably about 12 days, the incubation requires c. 14 days, and c. 6 eggs are laid. Accordingly, we have to go back c. 31 days to reach the average laying dates, i.e. to c. 28 May and 3 June, respectively. There is little doubt that the temperature and vegetation in the North lag behind when Pied Flycatchers start to lay.

The biotic factors which might influence the evolution of the breeding times include competition with other hole-nesting birds and with other animals such as wasps, utilizing nest-boxes and natural holes. The main competitors at Lemsjöholm were the Great Tit and the Wryneck, *Jynx torquilla*. The latter, known for its habit of throwing out the contents of nest-holes, has been decreasing in numbers in my area.

Slagsvold's (1975b) summary of the competition between the Great Tit and the Pied Flycatcher shows how little has been known of this matter. I hope to publish some observations later ; for the time being I will restrict myself to a few brief observations on aggressive competition (see also Slagsvold 1978).

Pied Flycatchers were usually killed by wounds in the occipital region. Most victims were found within the nest-boxes, often incorporated in the nest-material of the Great Tit ; a few were lying outside on the ground. Killing mainly took place during the building and laying periods of the tits. All the kills occurred in May, from 13 May on. Killing reached a peak in two years, when the breeding times of the species collided, the laying of the tits being delayed. In all, 13 Pied Flycatchers were found killed (a minimum figure !). In the same period c. 4000 adult Pied Flycatchers were breeding in the study area.

In 4 years, chosen by chance, 4 out of 125 nest-boxes containing Great Tit clutches were taken over by the Pied Flycatcher, the female building her nest over that of the tit. The tits probably produced repeat clutches, which on an average (von Haartman 1969) are 9 % smaller than the original clutches.

The influence on the "Darwinian fitness" in the populations caused by the aggressive competition of the two species seems to be of the same order of magnitude. The heavy but rare loss of fitness inflicted by tits killing flycatchers, and the slight but common loss of fitness caused by flycatchers taking over tit nests seem roughly to balance each other.

The possible effect upon the evolution of the breeding season is, however, difficult to evaluate. It is clear that other factors, such as the amount of food or temperature conditions, may have a much stronger effect as ultimate causes.

The prolonged end of the breeding season affects only a few females : in many years there were no late breeders. Not only are late clutches small, the latest being only two thirds of the earliest clutches (von Haartman 1967, 1982), but the death rate in the offspring is also higher. The number of fledglings is reduced, especially when the male does not participate at feeding (Alatalo et al., 1981, Winkel and Winkel, 1984), and the percentage of returns of ringed young from late broods is low, indicating severe mortality (von Haartman 1967, also Campbell's data in Lack 1966).

In most cases late breeding females are probably repeating lost clutches. The conflict between refraining from renesting and the drudgery of raising a reduced number of young with a reduced chance of living to a mature age may be an ultimate factor shortening the end of the breeding season. The breeding season of the Pied Flycatcher is shorter than that of the majority of passerines in the area, and true second broods have never been found.

INFLUENCE OF HABITAT

A one-legged Pied Flycatcher female and a similar Great Tit bred in my area. As I ring the left leg, but they had lost the right one, I knew that their disability was not my fault. Both bred late and laid a small clutch. An analogous effect can be expected if the individual is fit, but the habitat poor.

Since the classic study by Palmgren (1932), it has been known that birches produce c. 2.5 times more invertebrates per unit branch weight than spruce and pine. A considerable number of my nest-boxes were placed in birch or pine forest. These forests were old ; during the later part of the investigation a forester would have characterized them as economically unproductive. They were not monocultures. The breeding times and clutch sizes are shown in table IV.

The result came as a surprise to me. There is a statistically significant difference in breeding time in favour of the pine forest (X^2-test, $p=0.0074$), and the clutches were somewhat larger. As birches come into leaf relatively late, the pines may have offered more food at the beginning of the breeding season. I checked the possibility that late clutches would show a reverse trend with respect to clutch size, but this was not the case. Late clutches were smaller in both habitats, but the difference remained.

I can offer no explanation for this result which is at variance with several earlier findings (Berndt and Winkel 1967, Lundberg et al. 1981, Gezelius et al. 1984). One may, however, agree that the differences generally were small. It may be noted that Pied Flycatcher broods leave the territory and disappear immediately after fledging. They are not bound to the habitat in which they were born, and are not influenced by it when they later choose their own breeding sites (von Haartman 1949, Curio 1959). If, as seems plausible, the most critical period with respect to feeding the brood occurs after fledging, then the supply of food around the nest should play a smaller role than among the majority of birds, in which the fledged broods remain in the territory.

Table IV. Beginning of laying in Pied Flycatcher females at Lemsjöholm breeding in birch or pine forests in 1951-89. "Late clutches" indicate clutches started between 31 May and 29 June. N = sample size; M = mean.

Pentads	May				June							May	Average	Average
beginning	11	16	21	26	31	5	10	15	20	25	N	M	clutch	late clutches
Birch	-	2	88	144	121	55	17	7	8	4	446	31.1	6.14	5.77
Pine	1	17	117	203	107	45	18	7	7	4	525	29.6	6.25	5.88

Table V. Time of starting clutches in Pied Flycatcher females ringed at unknown (x) age. Observations at Lemsjöholm. Last column : 34.25 = 3.25 June. N = sample size; M = mean.

Pentads	May				June							
beginning	11	16	21	26	31	5	10	15	20	25	N	M
Age												
x	2	75	521	758	466	193	75	27	20	15	2152	29.62
x+1	0	8	28	69	40	11	4	5	1	0	166	29.66
x+2	0	4	15	31	14	8	1	1	0	2	76	29.71
x+3	0	0	6	7	4	2	2	0	1	0	22	30.95
Older	0	0	1	1	3	2	0	1	0	0	8	34.25

PARALLEL ANNUAL VARIATION IN BREEDING TIMES OF PIED FLYCATCHERS AND GREAT TITS

The start of breeding might be expected to be in some way correlated with food supply. As this parameter is difficult to measure with a bird so versatile in its choice of food as the Pied Flycatcher, indirect evidence must be sought. The Great Tit, though probably preferring larger prey, has a food choice not so very different from that of the Pied Flycatcher - which is definitely less of a flycatcher than the name would imply (von Haartman 1954, Meidell 1961). One would, therefore, expect that early breeding in the Great Tit would be parallelled by early breeding in the Pied Flycatcher, etc. As the Pied Flycatcher on an average starts to breed a good fortnight later than the tit, the factor(s) responsible for early or late breeding in the Great Tit should still to some extent be prevailing when the Pied Flycatchers start to lay. Such a correlation exists (Fig. 5), although, as might be expected, it is not very strong.

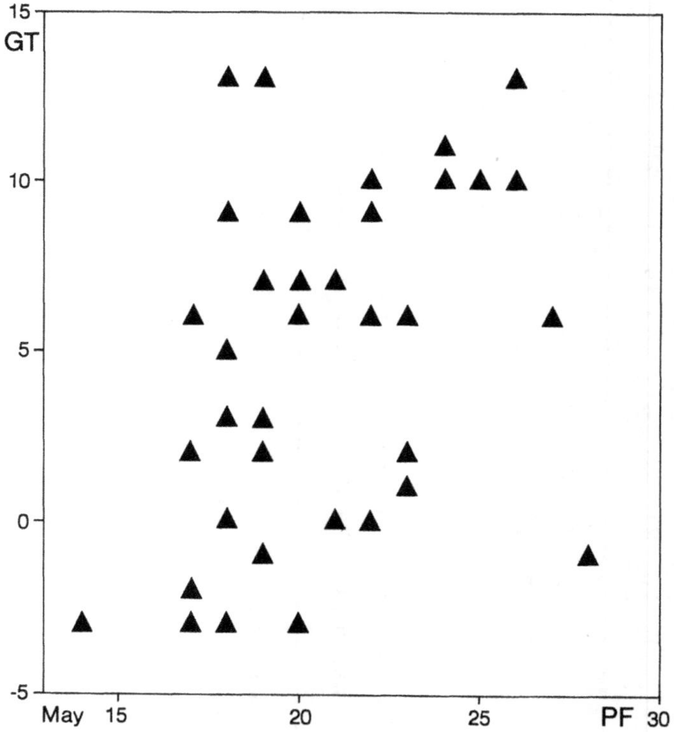

Fig. 5. Laying of the first egg of the season in the Great Tit and Pied Flycatcher populations at Lemsjöholm in 1953-89; 1958 omitted, since no data are available on the Great Tit. $r = 0.36$, $p = 0.031$.

AGE AND BREEDING TIME

The notorious lack of site tenacity in young and adult female Pied Flycatchers in North Europe effectively prevents a detailed analysis of clutch size in relation to age. The data available on females ringed as nestlings indicated, though not significantly, a difference between l-yr-old and (2-3)-yr-old females. A much larger set of data is available on females ringed at unknown age (x) and recovered later. The x-yr-old females, which must include relatively many 1-yr-old birds (von Haartman 1988), though not all 1-yr-olds breed, might be expected to start breeding later than the (x + 1) and (x + 2)-yr-old females. In fact, there is not the slightest indication that this is so (Tab. V). Comparison of simultaneously laid clutches shows that 1-yr-old females lay smaller clutches than older females (von Haartman 1967).

ANNUAL VARIATIONS IN BREEDING TIME 1941-1989

The above indicated variations are shown in Fig. 6. The average of all the clutches was not used as criterion because late, probably repeat clutches, would exercise an undesirable influence upon the result, their number varying strongly. Instead, the following methods were used :

(1) Exclusion of late, probably repeat clutches, from the calculation of averages. Admittedly there is no infallible way to recognize repeat clutches, as they are usually laid by unringed females, whose past is not known. The method amends a cardinal source of error to some degree, but is subjective.

(2) Use of the date of laying the first egg in the population. This method is objective, but to some extent influenced by population size. Further, mere chance will play a greater role when a single date is used than with data from a population.

(3) Use of the median of the laying dates in the population. This includes late, usually repeat clutches, but minimizes their influence. Medians may well offer the best way to visualize the annual variations in breeding time.

In the histograms showing the median and the date of laying the first egg in the population, floating averages were introduced. The smoothed curves so obtained facilitate the apprehension of trends.

The annual variations in breeding time correlate obviously over wide areas. Thus, Virolainen's (1984) averages for 13 years strongly correlate with mine, though Lemsjöholm is at a distance of 150 km WNW from Virolainen's study area, and his averages included repeat clutches.

The annual variations may shed some light upon long-term changes typical of the climate in the northern parts of Europe, Asia, and North America, but probably occurring everywhere. Recently, two predictions have been made with respect to climate. The first, in vogue about 20 years ago, was based on long-term analysis of annually recurrent phenomena influenced by the temperature. Examples are the layers of new ice formed yearly on the Greenland ice sheet, their amount of oxygen isotopes being correlated to the

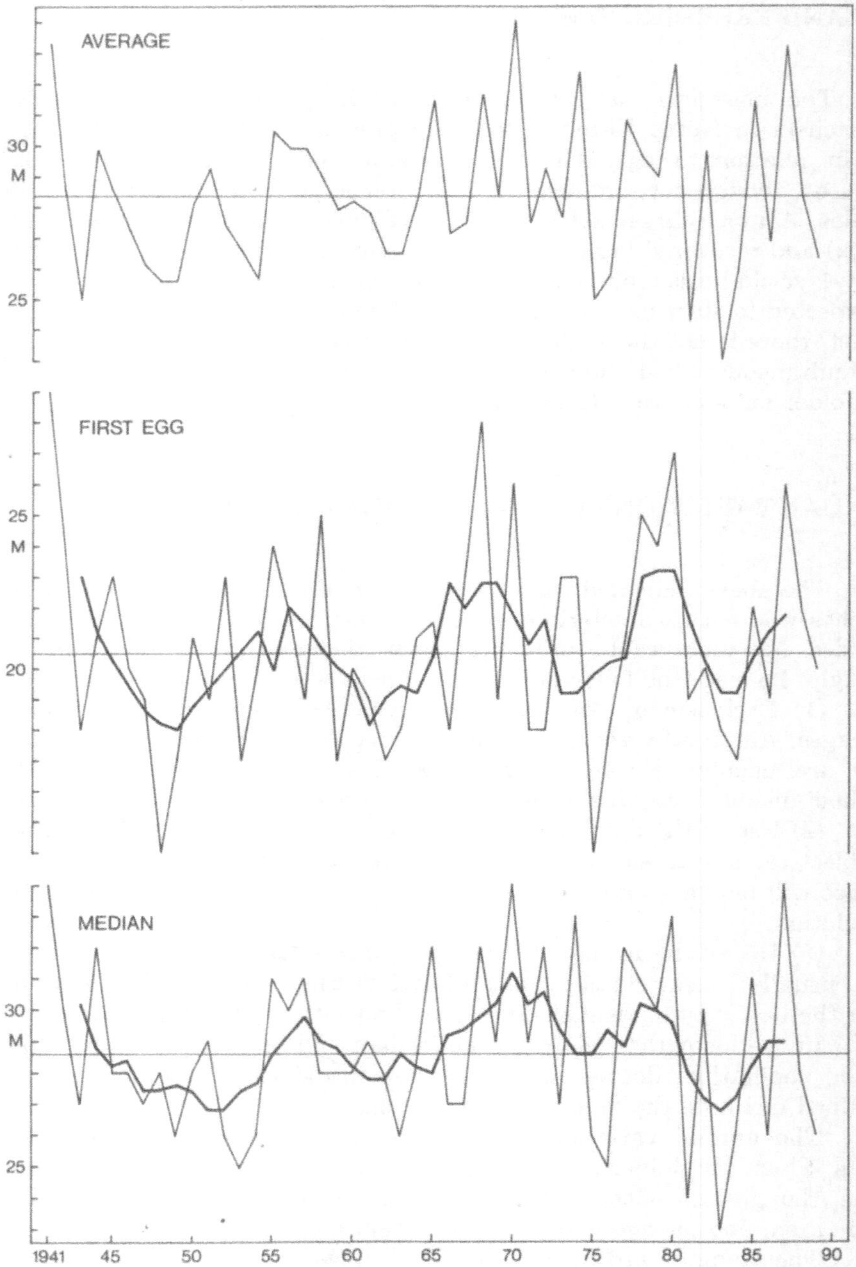

Fig. 6. Annual variation in breeding time of the Pied Flycatcher at Lemsjöholm 1941-1989. Concerning the methods used to visualize the variation, see the text. Thick lines : floating averages, using 5 day periods.

temperature (Johnsen et al. 1970), and the annual rings of the slowly growing pines of Lapland (Sirén 1961). By the former method it has been possible to reconstruct variations in the climate during the last 100,000 years, and by the latter during the last 800 years.

When the empirical curves are extrapolated, the two methods give consistent forecasts for the future climate. A deep decline in temperature could be expected, culminating in the 1980s (Fig. 7). Occasionally, such periods of severe cold have been referred to as "small ice-ages".

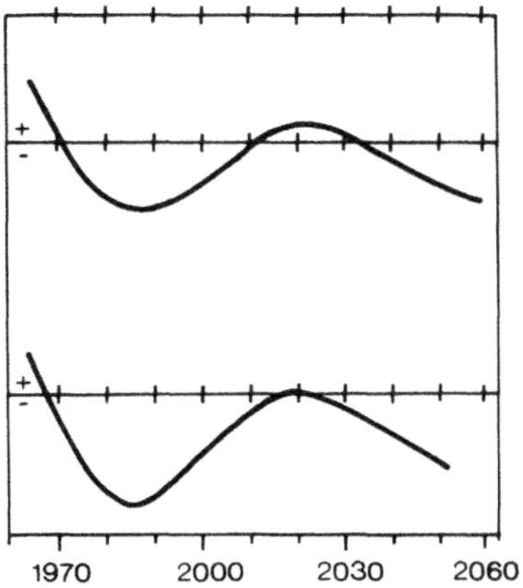

Fig. 7. Forecast of climatic change at Finnish latitudes based on growth rings of pines in Lapland (Sirén 1961) and analysis of the annual ice layers on the ice sheet of Greenland (Johnsen et al. 1970). Redrawn from Elomaa 1979.

Lately, another hypothesis has come to the fore, that of the "green-house effect". In consequence of human industry, the amount of some gases, mainly carbon dioxide, forming a layer in the atmosphere, is increasing with accelerated speed. The layer acts like the glass roof on a green-house ; it lets more energy in than out. A minority of students see in the green-house effect a promise of an almost paradisic situation, but the majority fears that the water from melting ice-caps in the polar regions will raise the level of the oceans, inundating valuable land, cities on coasts and estuaries, etc. A final choice between the two hypotheses is not possible at present, but the breeding times of the Pied Flycatcher were not delayed in the 80s (Tab. VI) as demanded by the "small ice-age" hypothesis. Laying in the Great Tit has seemingly changed a trifle, but the change is too small to be of any importance. The "green-house effect", on the other hand, can hardly have had time to express itself yet.

There may also be a third alternative. The trends for a sinking and for a rising temperature may balance each other, so that no dramatic changes occur.

Looking at the histograms in Fig. 6, one can hardly fail to observe a marked change taking place in the mid-60s. From then on, the averages, medians, and dates of first egg laying in the population, fluctuate on a scale not seen during the first part of the observation period (Tab. VII).

Table VI. Average of laying medians and dates of first egg in the Pied Flycatcher (PF) at Lemsjöholm, and average of dates of first egg in the Great Tit (GT). The average for the tit for the 50s concerns only the years 1953-57 and 1959.

	1941-49	1950-59	1960-69	1970-79	1980-89
PF median	28.8	28.1	28.6	29.6	28.2
PF 1st egg	20.6	20.5	20.5	21.0	21.0
GT 1st egg	-	(4.2)	4.3	4.2	6.0

Table VII. Median laying date in the Pied Flycatcher population at Lemsjöholm during the first and second half of the observation period. M = mean, S.D. = standard deviation.

	May								June					
Median	23	24	25	26	27	28	29	30	31	1	2	3	M	S.D.
1941-64	0	0	1	4	3	8	2	2	2	1	0	1	28.3	2.140
1965-89	1	1	1	4	3	1	4	1	0	5	2	2	29.0	3.253

To the best of my knowledge, no corresponding change has yet been reported by climatologists. Elderly people, of course, sometimes complain that nowadays the summers are either too hot or too cold. In order to shed light on this matter, the May-June temperatures since 1941 should be analysed. It would be a surprise if a small song-bird was the first herald of an unknown change in the climate, a matter which concerns all of us so deeply.

REFERENCES

Alatalo RV, Carlson A, Lundberg A and Ulfstrand S (1981) The conflict between male polygamy and female monogamy : the case of the Pied Flycatcher *Ficedula hypoleuca*. Amer Naturalist 117: 738-753

Berndt R (1943) Wie reagiert der Trauerfliegenschnäpper, *Muscicapa h. hypoleuca* (Pall.) auf die Fortnahme seines Geleges während der Gelegeperiode ? Beitr Fortpfl Biol Vögel 19: 77-78

Berndt R and Sternberg H (1972) Über Ort, Zeit und Grösse von Ersatzbruten beim Trauerschnäpper, *Ficedula hypoleuca*. Beitr Vogelkunde 18: 3-18

Berndt R and Winkel W (1967) Die Gelegegrösse des Trauerschnäppers *Ficedula hypoleuca* in Beziehung zu Ort, Zeit, Biotop und Alter. Vogelwelt 88: 97-136

Curio E (1959) Verhaltensstudien am Trauerschnäpper. Zeitschr Tierpsychologie, Beiheft 3: 1-118

Elomaa E. (1979) Ilmaston vaihteluista vuoteen 2060 mennessä. Ilmatieteen laitos UDK 551.52/57: 60-66

Gezelius L, Grahn M, Källander H and Karlsson J (1984) Habitat-related differences in clutch size of the Pied Flycatcher *Ficedula hypoleuca*. Ann Zool Fennici 21: 209-212

von Haartman L. (1951a) Successive polygamy. Behaviour 3: 256-275

von Haartman L (1951b) Der Trauerfliegenschnäpper. II. Populationsprobleme. Acta Zool Fennica 67: 1-60

von Haartman L (1956) Territory in the Pied Flycatcher *Muscicapa hypoleuca*. Ibis 98: 460-475

von Haartman L (1962) Faglarnas häckningstider i statistisk belysning. Zoologisk Revy 1962: 5-16

von Haartman L (1963) The nesting times of Finnish birds. Proc XIIIth Internat Ornithol Congr 611-619

von Haartman L (1967) Clutch-size in the Pied Flycatcher. Proc XIVth Internat Ornithol Congr 155-164

von Haartman L (1969) The nesting habits of Finnish birds. I. Passeriformes. Comment Biol Soc Scient Fennica 32: 1-187

von Haartman L (1982) Two modes of clutch size determination in passerine birds. Journ Yamashina Inst Ornithol 14: 214-219

von Haartman L (1988) The biological significance of the nuptial plumage of the male Pied Flycatcher. Proc XVIIIth Internat Ornithol Congr, Second print, corrected according to the author's manuscript, Helsinki: 1-21

Järvinen A (1983) Breeding strategies of hole-nesting passerines in northern Lapland. Ann Zool Fennici 20: 129-149

Järvinen A and Lindén H (1980) Timing of breeding and the clutch size in the Pied Flycatcher *Ficedula hypoleuca* in Finnish Lapland. Ornis Fennica 57: 112-116

Johnsen SJ, Dansgaard W and Clausen HB (1970) Climatic oscillations 1200 - 2000 A.D. Nature 227: 482-483

Kluijver HN (1951) The population ecology of the Great Tit, *Parus m. major* L.. Ardea 39:1-135

Lack D (1966) Population studies of birds. Clarendon Press, Oxford

Lundberg AR, Alatalo V, Carlson A and Ulfstrand S (1981) Biometry, habitat distribution and breeding success in the Pied Flycatcher *Ficedula hypoleuca*. Ornis Scandinavica 12: 68-79

Meidell O (1961) Life history of the Pied Flycatcher and Redstart in a Norwegian mountain area. Nytt magasin for Zoologi 10: 5-48

Palmgren P (1932) Zur Biologie von *Regulus r. regulus* (L.) und *Parus atricapillus borealis* Selys. Eine vergleichend-ökologische Untersuchung. Acta Zool Fennica 14: 1-113

Silverin B (1979) Seasonal development of reproductive organs in the female Pied Flycatcher *Ficedula hypoleuca* and experimental studies in her renesting ability. Gonadal endocrine activities and interrelations related to the breeding behaviour of the Pied Flycatcher *Ficedula hypoleuca*. Gotenburg, Ph. D. thesis

Slagsvold T (1975a) Breeding time of birds in relation to latitude. Norwegian Journ Zool 23: 213-218

Slagsvold T (1975b) Competition between the Great Tit *Parus major* and the Pied Flycatcher *Ficedula hypoleuca* in the breeding season. Ornis Scand 6: 179-190

Slagsvold T (1978) Competition between the Great Tit *Parus major* and the Pied Flycatcher *Ficedula hypoleuca*: an experiment. Ornis Scand 9: 46-50

Sirén G (1961) Skogsgränstallen som indikator för klimatfluktuationerna i norra Fennoscandien under historisk tid. Metsäntutkimuslaitoksen julkaisuja 54:2: 1-66

Stieve H (1919) Die Entwicklung des Eierstockeies der Dohle (*Coloeus monedula*). Archiv f mikroskopische Anatomie 92: 388-405

Stjernberg M (1974) Nest-building by the Pied Flycatcher *Ficedula hypoleuca*. Ornis Fennica 51: 85-109

Virolainen M (1984) Breeding biology of the Pied Flycatcher *Ficedula hypoleuca* in relation to population density. Ann Zool Fennici 21: 187-197

Winkel W and Winkel D (1984) Polygynie des Trauerschnäppers (*Ficedula hypoleuca*) am Westrand seines Areals in Mitteleuropa. J Orn 125: 1-14

REPRODUCTIVE BIOLOGY

AND POPULATION REGULATION

COMPARATIVE REPRODUCTIVE BIOLOGY OF FOUR BLUE TIT POPULATIONS IN THE NETHERLANDS

Johan H. van BALEN and Roel P.J. POTTING

Institute for Ecological Research, Boterhoeksestraat 22, PO BOX 40, 6666 GA Heteren, The Netherlands

After a long period in which research on the population biology of hole-nesting birds was mainly conducted on the Great Tit (*Parus major*), recently more attention is being given to its counterpart, the Blue Tit (*Parus caeruleus*). This is partly due to the increasing number of studies from the mediterranean region (Blondel et al. 1987), where the Blue Tit is often the most numerous tit species. As this species seems to be a promising subject for a comparative demographical study over a large area of Europe, an analysis of data collected in several populations from the Netherlands is of interest. Correlative studies in contrasting habitats can give us clues about the factors that influence parameters such as date of egglaying, clutch size and breeding success, but they need experimental verification.

In this paper we will first compare the values obtained for each parameter in the various areas. Next we will examine if there are trends in the course of the study period, and analyse the correlations of the annual figures for the different areas. This will show us whether the factors that influence the parameter under study are general or rather local factors. Thirdly the relationships with environmental variables and with population characteristics will be considered, and the presumed causes of the established relationships will be discussed. If appropriate, a comparison will be made with the Great Tit, or Great Tit parameters will be used in the analyses.

STUDY AREA AND METHODS

Data were collected in four areas, distributed throughout the country, two are mature deciduous oak forests and the others are a mixture of deciduous and coniferous forest plots:

Vlieland: all the wooded areas (292 ha) on the island (53.17 N, 5.03 E) are included. The forest was planted about 1900, and is mainly Pine (*Pinus* ssp.) with local patches of Birch (*Betula* ssp.), Oak (*Quercus robur*), and other deciduous trees. Since the early sixties selective felling has increased the proportion of deciduous trees. The number of nestboxes has gradually increased from 125 (1956) to 200 (1966) and to 350 (1986), to keep in pace with the increased suitability of the habitat. For some purposes the total area is divided into Vlieland-East (the 214 ha forest north of the village) and Vlieland-West (the remaining four small forest plots). The vegetation of the two areas is

NATO ASI Series, Vol. G 24
Population Biology of Passerine Birds
Edited by J. Blondel et al.

similar, but the population in East is closer to the village, and therefore could profit more from food provision by man in winter.

Hoge Veluwe: an area situated in the southern part of the National Park (52.03 E, 5.51 N). From 1960 to 1972 the study area was about 320 ha and the plantations were 60% pine and 40% mixed coniferous-deciduous woodland, with between 215 and 325 nestboxes. From 1973 onward the area was reduced to 165 ha, with an increased proportion of deciduous trees, and about 370 nestboxes. Extra seed food was provided in the winters of 1966/1967 to 1975/1976.

Oosterhout: a small (11.4 ha) deciduous wood with predominantly mature oaks and dense undergrowth. It is (at 51.52 N, 5.50 E) isolated from other woodland by agricultural land. Number of nestboxes increased from 29 (1959) to 78 (1965) and finally about 135 (1966 to 1986). Data for 1982 are not available as Blue Tits were experimentally removed in spring.

Liesbos: study area (18 ha) is part of 190 ha medium-aged to mature oak forest in the southern part of the country (51.35 N, 4.42 E). Since 1957 there have been about 97 nestboxes.

In three areas nestboxes were inspected at least once a week during the breeding season, but in Liesbos in some years only data on laying date and clutch size were collected. Nestlings were ringed in Vlieland (all years), Hoge Veluwe (from 1960) and Oosterhout (most years). Blue Tit parents were identified in Vlieland (all years), Hoge Veluwe (up to 1970) and Oosterhout (mainly in 1981 and 1987). For the analysis, data were used from the period 1965-1986.

BREEDING DENSITY

Blue Tit densities are highest in the two deciduous woods, Oosterhout (mean 17.0, S.D. 4.1 first clutches/10 ha) and Liesbos (mean 12.1, S.D. 3.1), and much lower in the mixed forests (Hoge Veluwe mean 1.9, S.D. 1.0, Vlieland-West mean 2.3, S.D. 1.0, Vlieland-East mean 0.5, S.D. 0.3). Differences between areas are more extreme than in the Great Tit that has the same overall pattern although at a higher level. The annual density figures in the deciduous habitats are less variable (coefficient of variation = 24-26%) than those in the mixed habitats (C.V.=44-60%), suggesting again that the former are the more favourable habitats.

The two parts of Vlieland show a fourfold difference in Blue Tit breeding density in favour of the part with the lowest Great Tit density, suggesting that interspecific competition may play a role. Here, too, for both species the part with the highest density is the one where density is least variable.

Trends in Blue Tit density throughout the study period differ strikingly between areas (Fig. 1). In the two oak forests there is neither increase nor decrease, but in the other areas there is a distinct increase. The Blue Tit population of the island of Vlieland is fairly new, the first pair in nestboxes being recorded in 1957. Population increase started in 1965, coinciding with a sharp increase in Great Tit numbers which was ascribed to a sudden

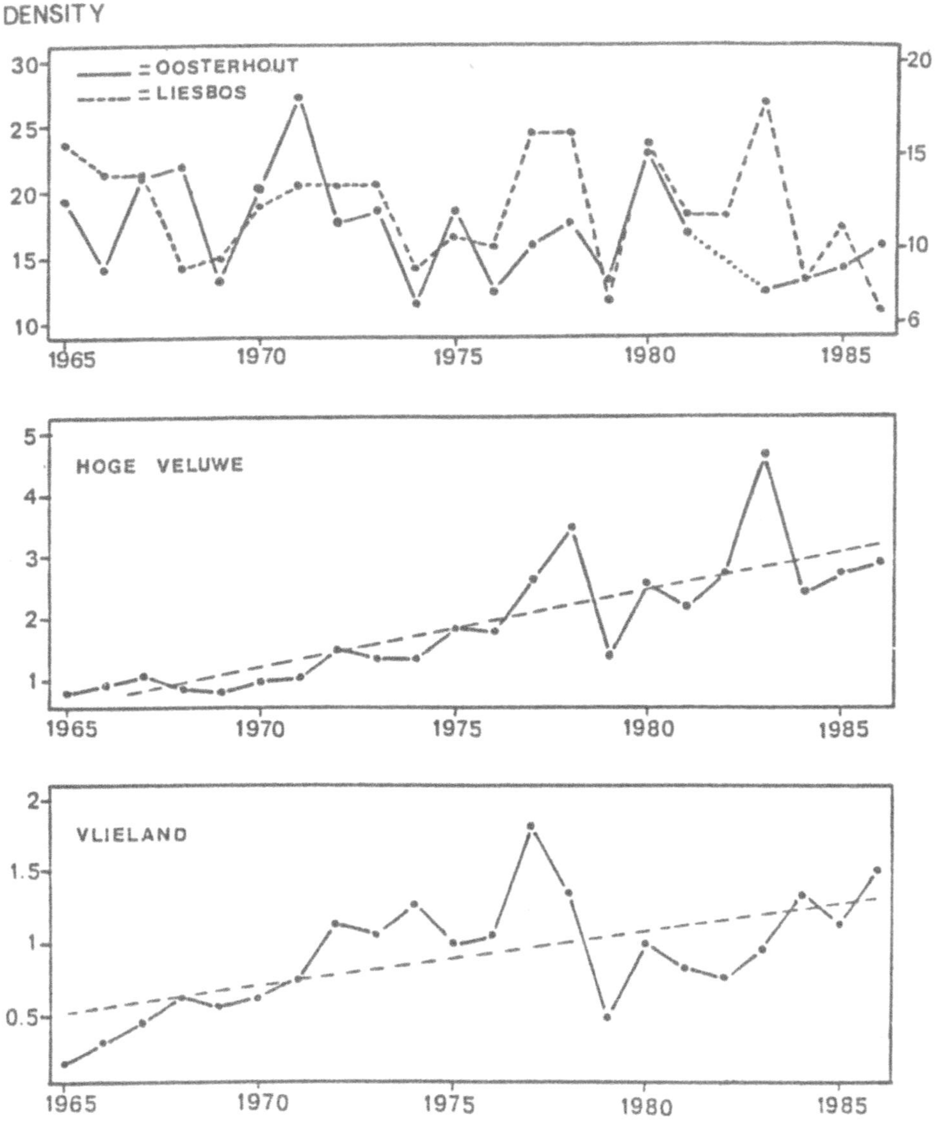

Fig. 1. Fluctuations of Blue Tit density (no. of first broods per 10 ha) from 1965 to 1986 in four areas, with significant regression lines. Upper panel: left axis for Oosterhout, right axis for Liesbos.

improvement of the habitat through selective felling (Kluyver 1971). In addition, presumably feeding by man in winter has increased over the years, possibly leading to increased survival in winter. Both factors may have contributed to the increase in tit density in Vlieland, and, to a lesser extent, in Hoge Veluwe.

The fluctuations in density in the two oak forests are more or less parallel. In general, the annual density figures are positively correlated for most pairs of areas, but this is partly due to the general trends described above. The correlations between the density changes from one year to the next are given in Table I. The two parts of Vlieland are considered separately, as their densities are so markedly different. It appears that the density changes in Oosterhout, Liesbos and Hoge Veluwe are positively correlated. The correlation coefficients for Vlieland-West are somewhat lower, but significant in two out of three cases. Density changes in the low-density area Vlieland-East are not correlated with changes in the other areas, not even with the neighbouring area Vlieland-West. We conclude that density changes are to a large extent determined by general factors, but that local factors must play a greater role in Vlieland-East.

Table I. Correlation coefficients for density changes (absolute difference in density from one year to the next), comparison of three areas and the two subareas of Vlieland.

	Liesbos	Hoge Veluwe	Vlieland-West	Vlieland-East
Oosterhout	0.595**	0.522*	0.301	−0.033
Liesbos	x	0.845***	0.513*	0.028
Hoge Veluwe		x	0.533*	0.193
Vlieland-West			x	0.135

Note: *: $p<.05$, **: $p<.01$, ***: $p<.001$.

The effect of environmental factors on breeding density was analysed by multiple regression analysis. As independent variables were included: year number, Hellmann figure (*i.e.* the absolute sum of all daily mean temperatures below 0°C, measured over the months November-March at De Bilt), and a figure for the Beech Crop Index (BCI). Moreover, we included Great Tit density figures to check if there are indications for interspecific competition. Table II shows that no significant relationships were found in the two oakwoods, though in Liesbos the effect of winter severity was nearly significant. In the mixed woods, there was a very significant increase in density with year number, confirming earlier results, and density was also lower after severe winters. A similar analysis for density changes from one year to the next did not give significant results for the variables year number, Hellmann figure and BCI. For the two oakwoods it showed that Blue and Great Tit densities changed in the same direction (cf.Perrins 1979).

In summary, there are large differences in density between areas, probably connected with food availability. Density has increased markedly over the years in the two low-density areas, for as yet unknown reasons. In these areas, density fluctuates more than in the high-density areas, and it is influenced by winter severity. In the latter areas, fluctuations of Blue and Great Tit run in parallel.

Table II. Results of multiple regression analysis of Blue Tit breeding density in four areas. Data from 1965 - 1986. P-values and r^2, obtained after elimination of the non-significant variables, are given.

Area	year	Hellmann figure	BCI	GT density	r^2
Oosterhout	n.s.	n.s.	n.s.	n.s.	.1468
Liesbos	n.s.	.0512	n.s.	n.s.	.1771
Hoge Veluwe	.0000	.0368	n.s.	n.s.	.7324
Vlieland	.0002	.0187	n.s.	n.s.	.5466
sign	pos	neg	–	–	–

DATE OF LAYING

The mean laying dates of the first clutches are presented in Table III. Differences between areas were analysed with a paired t-test on the annual means.

Table III. Mean laying date of first clutches of the Blue Tit (mean of annual means) in four areas and the two parts of Vlieland. Dates counted from 1 April = 1.

	mean date	S.D.	C.V.(%)
Oosterhout	19.63	4.97	25.3
Liesbos	21.20	4.79	22.6
Hoge Veluwe	26.20	4.11	15.7
Vlieland	26.20	2.83	10.8
Vlieland-West	24.96	3.09	12.4
Vlieland-East	27.94	3.30	11.8

Mean laying dates of the Blue Tit differ between areas by as much as 8.3 days, 4.4 days more than for the Great Tit. Differences between areas were in most cases highly significant, except for the comparison Hoge Veluwe - Vlieland (either West or East). Blue Tits in oak forests lay about six days earlier than those in mixed forests, and show more variability between years, as seen from the coefficients of variation.

Mean laying dates did not change significantly in the course of the study period.

The most likely factors that could explain between-area differences in mean laying date are differences in ambient temperature in spring, and differences in food availability (Clamens and Cramm, in Blondel et al. 1987). As there are appreciable differences in temperature in spring between the Northern and the Southern parts of the country (up to 3°C for the maximum and the minimum temperatures in March and April), this factor must be taken into account. Figure 2 gives the calculated regression lines between mean temperature in the period from 10 March and 20 April and mean laying date.

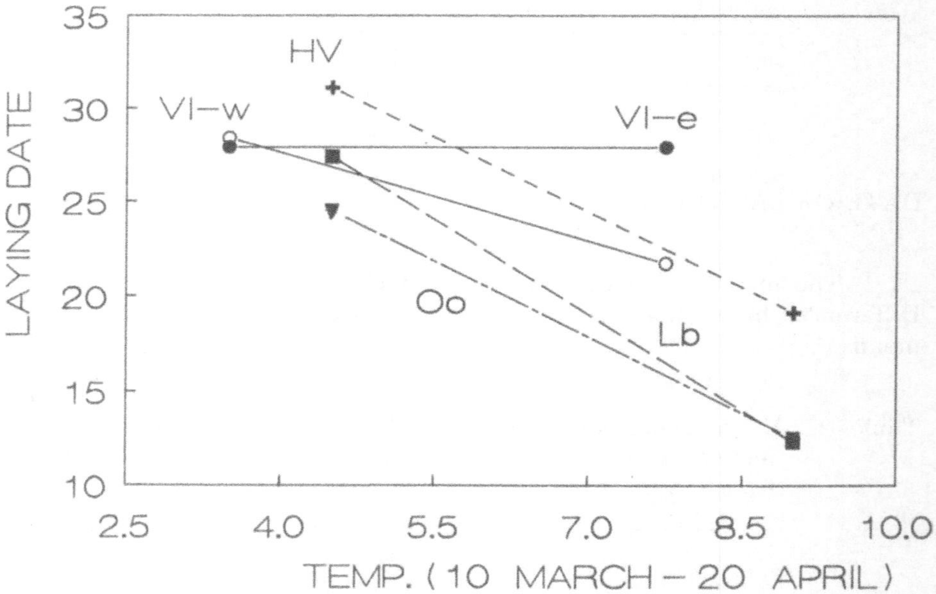

Fig. 2. Regression lines of annual mean laying date of the Blue Tit against ambient temperature in the period from 10 March up to 20 April. Temperature data were taken from De Kooy (52.55 N, 4.48 E) for Vlieland, and from De Bilt (52.06 N, 5.10 E) for the other study areas. Regression equations: for Oosterhout (Oo): Y= 36.42 - .664 X (p=.0015), for Liesbos (Lb): Y= 42.42 - .836 X (p=.0000), for Hoge Veluwe (HV): Y= 43.09 - .666 X (p=.0000), for Vlieland-West (VL-w): Y= 33.96 - .395 X (p=.0110), Vlieland-East (VL-e): p= .4218, n.s.

They were significant for all areas except Vlieland-East. Part of the between-area differences, e.g. the late laying in Vlieland, can be ascribed to the low prevailing temperatures, but this is not the whole story. Over a large range of temperatures the laying dates in the two oak forests are earlier than those in the mixed forests, indicating that other factors play an additional role, e.g. food availability. Other indications that factors other than ambient temperature play a role are the significant difference in laying date between the two parts of Vlieland, and the difference between Oosterhout and Liesbos. On the basis of temperature alone we would not expect the Blue Tits in Oosterhout to lay earlier, Liesbos being the more southerly location.

The annual mean laying dates in the various areas are mutually correlated (Table IV), as expected from the overriding effect of temperature, but Vlieland-East is an exception: the only significant correlation is with Vlieland-West. This corresponds with the absence of a correlation with temperature in the former area.

Table IV. Correlation coefficients for annual mean laying dates in three areas and the two parts of Vlieland.

	Liesbos	Hoge Veluwe	Vlieland-West	Vlieland-East
Oosterhout	0.820***	0.676***	0.519*	0.278
Liesbos	x	0.768***	0.441*	−0.035
Hoge Veluwe		x	0.341	0.122
Vlieland-West			x	0.563**

Note: *: p<.05, **: p<.01, ***: p<.001.

The annual mean laying dates of Blue and Great Tit are strongly correlated within all areas. This is not surprising as the laying dates of both species depend to a large extent on ambient temperature in March and April (Fig. 2, see also van Balen 1973 and Schmidt 1984).

In mixed forest the laying dates of the two species are not different, but in oak forest Blue Tits lay on average 3 to 4 days earlier than Great Tits. Such a difference was also found by Dhondt et al. (1984). Oakwood is the habitat with the highest breeding density for both species. If this difference is related to density one would expect a similar correlation between years within one area. For the three areas and the two parts of Vlieland, correlations were computed between the difference in laying date of the two species and their respective densities. In two of the ten cases a significant correlation was obtained. The data for Hoge Veluwe and Vlieland-West do indeed show the expected relationship: in high density years the Blue Tits lay earlier than the Great Tits but in low-density years the difference is reversed (Fig. 3).

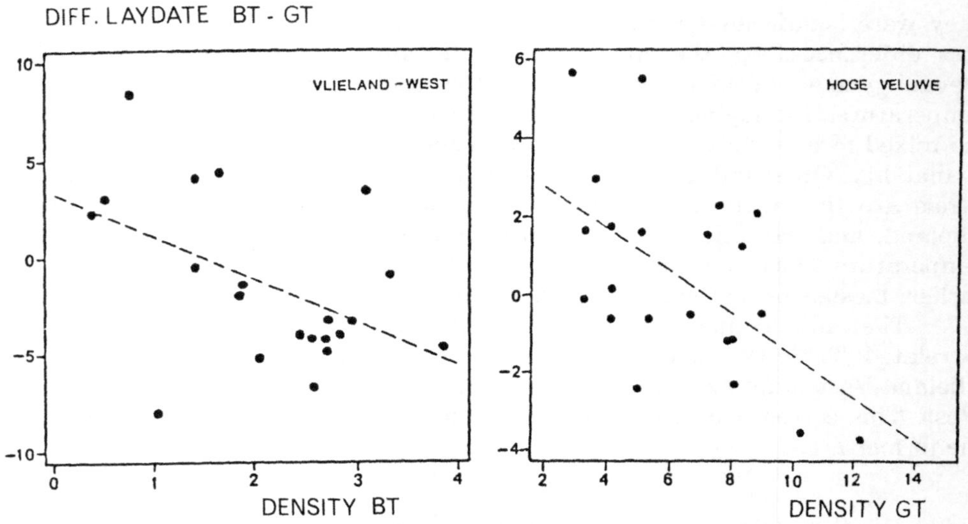

Fig. 3. Correlation between the difference in laying date between Blue and Great Tit (negative = Blue Tit earlier) and breeding density.

We have shown earlier that Blue Tit laying dates in the two parts of Vlieland differ significantly. It appears that at least part of this difference is caused by the differential age composition of the two sub-populations. In Vlieland first-year females start egglaying on average 2.1 days later than older females ($t=3.65$, $n=377$, $p<.001$). Not only are most of the females in Vlieland-East yearlings, and lay their eggs correspondingly later than the older females, but also the difference in laying date between East and West is larger in years when the age difference is greater (Fig. 4).

CLUTCH SIZE

The mean size of the first clutches of the Blue Tit in the four areas and in the two subareas of Vlieland is given in Table V. Differences between the areas were tested with a paired t-test on the annual means, excluding years with fewer than five clutches.

Clutches are largest in the two oak forests, significantly smaller in Vlieland, and smaller still in Hoge Veluwe. All differences are significant except those between Oosterhout and Liesbos, and between Vlieland-West and East.

There is no consistent increasing or decreasing trend in clutch size in the course of the study period, despite the steadily increasing breeding density in Hoge Veluwe and Vlieland.

The annual variation in clutch size is large (Table V). The annual values for the various areas are in most cases not correlated, not even the figures for Vlieland-West and East. Presumably clutch size is to a large extent determined

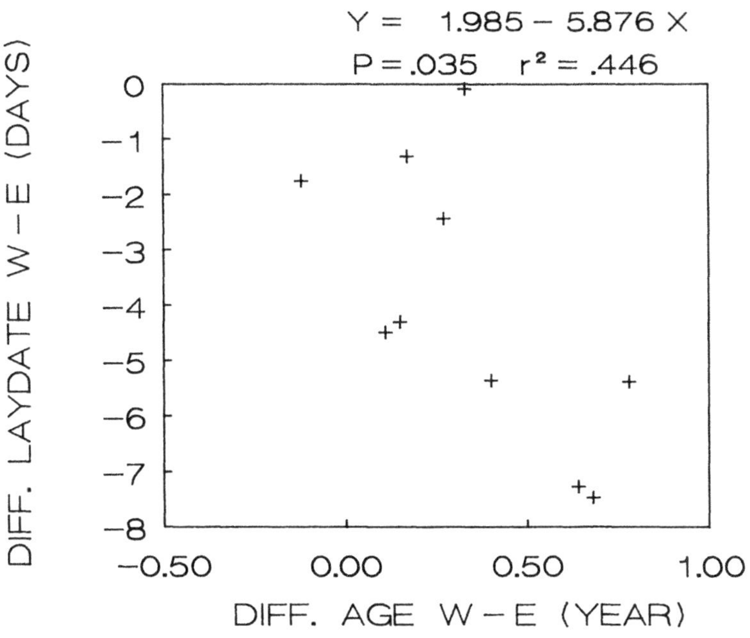

Fig. 4. Relationship between the difference in laying date of the Blue Tit in Vlieland-West and -East, and the age difference of the females in these two subareas. Data from 1977-1986.

Table V. Size of first clutches of the Blue Tit in four areas and in the two parts of Vlieland. Mean of annual means, and highest and lowest value.

Area	mean	S.D.	C.V.(%)	min.	max.
Oosterhout	11.57	0.64	5.5	10.4	12.8
Liesbos	11.94	0.64	5.4	10.8	13.3
Hoge Veluwe	10.61	0.77	7.3	9.0	11.9
Vlieland	11.20	0.38	3.4	10.7	12.2
Vlieland-West	11.22	0.42	3.7	10.5	12.2
Vlieland-East	11.21	0.72	6.4	9.5	12.6

by local factors. The exception is formed by the two oak forests ($r=.529$, $p<.05$), where conditions are presumably more alike than those in the other areas.

Factors that might influence annual and local variation in clutch size include breeding density, date of laying and food availability. There is no information on the annual variation in food availability in the laying season. For each of the four areas, variations in the annual mean figures for clutch size were analysed by regression. Density figures for both Blue and Great Tit were taken as independent variables, together with the mean laying date and the year number. In none of the areas is there a correlation with the average date of laying in the year concerned: clutch size is similar in early and late years. In Oosterhout clutch size increased through the years, but the relationship was hardly significant ($p=.04$). In the other areas there were no trends with time, and clutch size was also not related to Blue or Great Tit density. A similar analysis was performed with the change in clutch size from one year to the next as the dependent variable, and the corresponding change in breeding density (plus year number and laying date) as independent variables. Here, it appeared that the number of eggs is in some subtle way affected by density changes: in Oosterhout and Liesbos there is a significant negative relationship between change in Blue Tit clutch size and the corresponding change in Great Tit breeding density (for Oosterhout: $p=.0017$, $r^2=.450$; for Liesbos: $p=.0006$, $r^2=.506$).A similar result was obtained by Clamens and Cramm (in Blondel et al. 1987) in a Holm Oak wood in southern France.

A correlation such as this could be partly based on differences in age composition that go hand in hand with density changes. In dense populations there is a higher proportion of yearling females, and yearlings (in Vlieland) lay on average 0.5 eggs less than older birds($t=3.70$, $p<.001$).

Clutch size decreases strongly in the course of the laying season. This also applies to first clutches (Fig. 5). In all four areas there is a significant (Oosterhout, Vlieland) or highly significant (Liesbos, Hoge Veluwe) decrease with relative laying date. The decrease is strongest in Hoge Veluwe, corresponding with earlier findings on the Great Tit (van Balen 1973). In Vlieland the decrease is very small, and is not significant in all years except one. Assuming that the seasonal decline in clutch size is ultimately linked to the survival chances of the nestlings, we expect that survival chances in Vlieland should not deteriorate as quickly as they do in the other areas. This will be confirmed in the next section.

So far we have seen that there are differences in clutch size between the areas, that cannot be explained by factors like laying date and breeding density. Furthermore, annual variation in clutch size is large, is not correlated between areas and cannot be explained by variation in laying date and breeding density, except that increases in Great Tit density reduce the number of Blue Tit eggs in high-density areas. Clutch size depends more on local than on general factors.

BREEDING SUCCESS

Failure of eggs to produce fledglings falls into two categories: total and partial losses. Total losses are often caused by predation or unfavourable weather, and are therefore erratic in time and space. Partial losses are often connected with (quantitative or qualitative) food shortage and indicate the factors that can influence brood size.

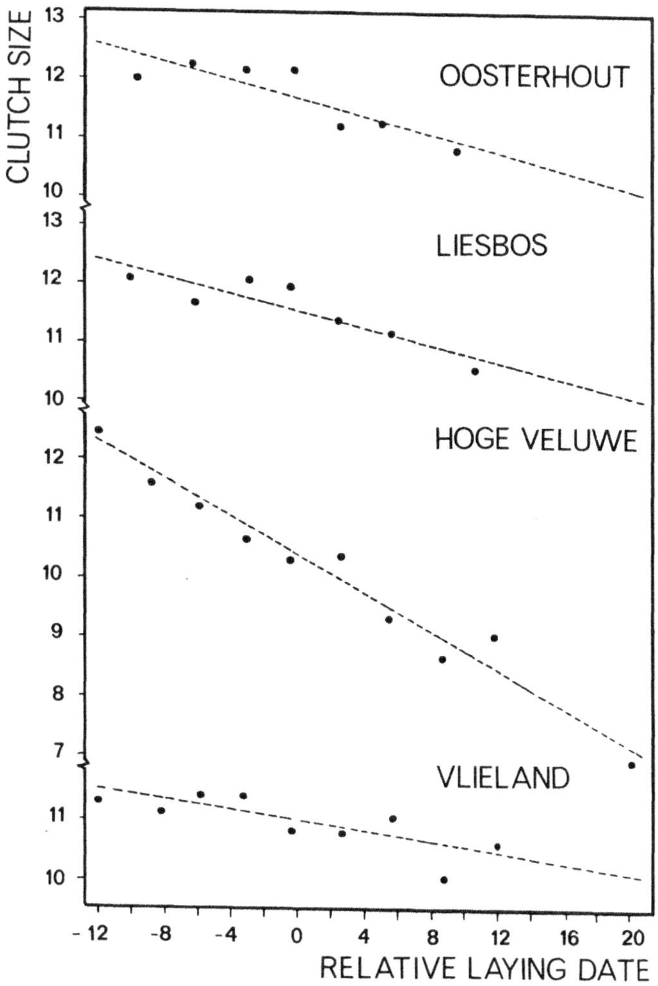

Fig. 5. Decrease in size of first clutches with relative laying date in four areas. (0 = mean laying date in the year concerned). Data from 1974 - 1986.

Table VI gives the percentage of successful clutches for the three areas for which adequate data are available. The mean percentages for the three areas do not differ. In Oosterhout, success seems to be rather variable, due to two years in which there was heavy predation and some damage by people. The annual figures for percentage success are not correlated to the corresponding figures for partial success that will be discussed hereafter.

For all clutches that produced at least one fledgling we computed three measures of reproductive success: hatching success (proportion of eggs hatched), fledging success (proportion of young fledged) and breeding success, being the product of hatching and fledging success.

Table VI. Proportion of successful clutches in three areas in the years 1965-1986. Successful = at least one fledgling.

Area	mean	S.D.	n	min.	max.
Oosterhout	.853	.209	21	.214	1.000
Hoge Veluwe	.875	.063	22	.693	.979
Vlieland	.821	.104	22	.571	1.000

Table VII gives the average values of success together with the mean number of fledged young in successful first broods. It shows clearly that reproductive success is highest in the oak forest Oosterhout, that hatching success is lower in the two mixed forests and that fledging success is lowest in Vlieland. Breeding success in Vlieland is only 74% of that in Oosterhout and the corresponding value for the number of fledged young per successful brood is 72%. Not only do the mean values decrease from Oosterhout to Vlieland, but there is also an increase in variability, as measured by the coefficients of variation. Both findings demonstrate that conditions for incubating eggs and rearing nestlings are best in Oosterhout and worst in Vlieland.

Table VII. Reproductive success in three areas. All clutches with at least one fledgling are included. N = 277 (Oosterhout), 600 (Hoge Veluwe), 425 (Vlieland).

	Oosterhout			Hoge Veluwe			Vlieland		
	mean	S.D.	C.V.	mean	S.D	C.V	mean	S.D.	C.V.
Hatching success	0.92	0.11	12.0	0.84	0.18	21.4	0.81	0.20	24.7
Fledging success	0.96	0.07	7.3	0.92	0.14	15.3	0.82	0.20	24.4
Breeding success	0.89	0.14	15.7	0.77	0.21	27.3	0.66	0.20	30.3
Fledglings per successful clutch	10.2	2.12	20.8	8.4	2.95	35.1	7.3	2.73	37.4

Fledging success did not show a decreasing or increasing trend through the years, but hatching success (and consequently breeding success) decreased significantly in Hoge Veluwe and Vlieland (Fig. 6). The presumed cause of this phenomenon will be treated in the Discussion.

When the annual values for the measures of breeding success are compared between the three areas we do not find any significant correlations. Within areas, however, hatching and fledging success are usually correlated: for

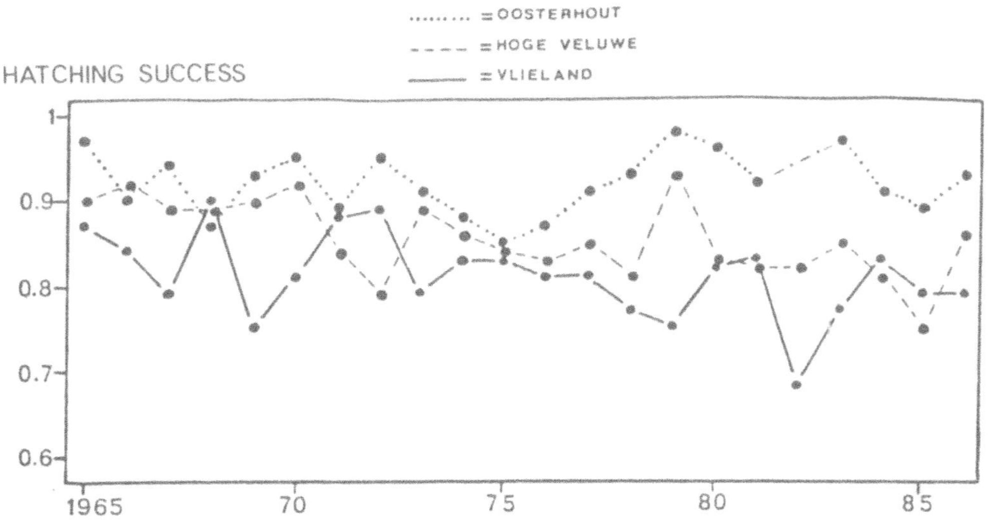

Fig. 6. Trends of hatching success in the course of the years. Those in Hoge Veluwe (Y = 0.9047 - 0.0004 X, p=0.0026) and in Vlieland (Y = 0.8513 - 0.0004 X, p=0.0344) decrease significantly.

Oosterhout r=0.497 (p=.022), for Hoge Veluwe r=0.586 (p=.003), but for Vlieland: r=0.297 (p=.180). These findings indicate that breeding success mainly depends on local rather than on general conditions.

Breeding success varies strongly during the season. Figure 7 shows the two components of breeding success in relation to relative hatching date. Dates over 25 refer to repeat and second clutches. In Oosterhout hatching success declines slightly and fledging success more strongly during the first-brood period. In Hoge Veluwe hatching success shows a steep decline during the first month and increases greatly in the late broods, and fledging success declines throughout the season. Results from Vlieland deviate from this pattern: during the period of the first broods both hatching and fledging success increase. The most consistent differences between the areas are found for fledging success. Probably, changes in food availability for the nestlings during the breeding season do not vary in parallel in the three areas.

A regression analysis of the breeding success of individual first broods, with relative hatching date and clutch size as independent variables, confirmed that breeding success was negatively related to date in Oosterhout and Hoge Veluwe, and positively in Vlieland.

In order to analyze the relationships of breeding success with breeding density and clutch or brood size, multiple regressions were performed with the annual means, including the number of the year and the mean date of laying as independent variables. The results (Table VIII) differ for the three areas. In the oak forest there is a positive correlation between hatching success and mean clutch size: hatching success (and hence also breeding success) is greater in years when clutches are large. This correlation also holds when all non-

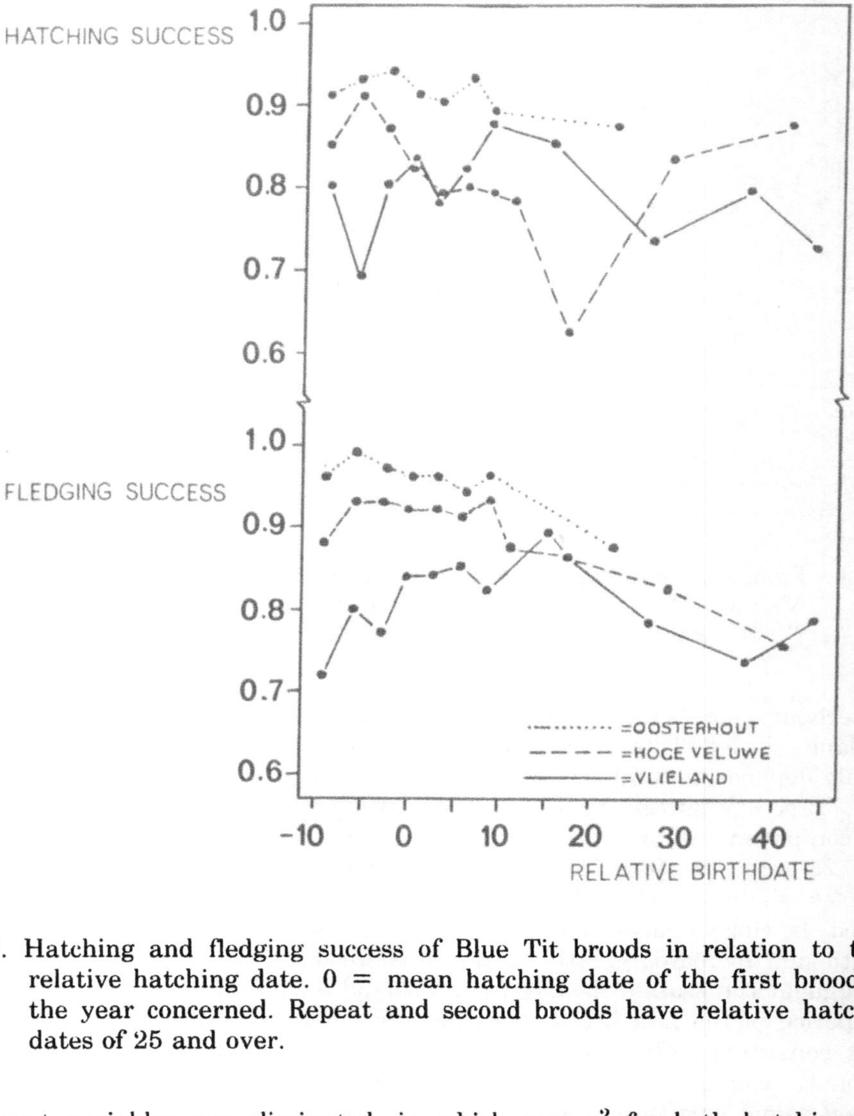

Fig. 7. Hatching and fledging success of Blue Tit broods in relation to their relative hatching date. 0 = mean hatching date of the first broods in the year concerned. Repeat and second broods have relative hatching dates of 25 and over.

significant variables are eliminated, in which case r^2 for both hatching and breeding success is highly significant. In the mixed forest at Hoge Veluwe the same correlation holds, and in addition there is a positive correlation with mean laying date: hatching success is higher in late years. When breeding density is left out, there is also a relationship between hatching success and year number: hatching success decreased in the course of the study period, as already shown. This relationship was also found in the Vlieland data. Moreover, in this area, fledging success appeared to be negatively related to clutch size, suggesting that in years with large broods the tits are not so successful in rearing their young to fledging as in years when broods are smaller.

Table VIII. Results of multiple regression analysis of hatching success (HS), fledging success (FS) and breeding success (BS) in three areas. Data from 1965 - 1986, except years with <5 broods. Regression coefficients and r^2 are given with their significance.
(*): p<.10, *: p<.05, **: p<.01, ***: p<.001.

Area		year	laying date	clutch size	density	r^2
Oosterhout	HS	−.0015	−.0004	+.0397*	+.0001	.3982
	FS	−.0011	+.0001	+.0152	−.0013	.2218
	BS	−.0026	−.0002	+.0517**	−.0006	.4357(*)
Hoge Veluwe	HS	−.0025	+.0041**	+.0301**	−.0056	.7429***
	FS	+.0002	+.0005	+.0148	−.0060	.2001
	BS	−.0014	+.0041	+.0410**	−.0135	.4868**
Vlieland	HS	−.0053*	+.0030	−.0181	+.0480	.3133
	FS	+.0019	−.0022	−.0939*	−.0409	.2872
	BS	−.0035	+.0006	−.0880	+.0100	.1745

There were no significant relationships between breeding density and the various measures of breeding success.

Apart from the effect of predation and disturbance on total losses, we have seen that breeding success differs in the three areas, and that it depends on local rather than general conditions. The differences in the seasonal changes in breeding success, declining in Oosterhout and Hoge Veluwe and increasing in Vlieland, agree with this view. These conditions may be similar in the egg-laying and the incubation period, as clutch size and hatching success are positively correlated. Food availability may be a prominent factor.

SECOND BROODS

A summary of the occurrence of second broods in the areas for which we have sufficient data is given in Table IX. The percentage of second broods increases from Oosterhout to Hoge Veluwe, and is very high in Vlieland. The interval between the mean laying dates of first and second broods increases from 38.8 days in Oosterhout to 41.8 days in Hoge Veluwe and to 44.8 days in Vlieland (Table X). Hence, for the average Blue Tit, the breeding season lasts much longer in Vlieland than in the other areas.

Earlier we have seen that the decrease in clutch size within the first broods is much less steep in Vlieland than in the other areas. In this respect it is striking that the number of eggs in repeat and in second clutches is larger in Vlieland than in the other areas (Table X). Presumably conditions do not deteriorate as quickly as in the other areas.

Table IX. Percentage of second broods in three Blue Tit populations. Data from 1965 - 1986.

Area	no. of successful first broods	no. of second broods	percentage second broods mean	min.	max.
Oosterhout	343	12	3.5	0	20.0
Hoge Veluwe	668	64	9.6	0	25.6
Vlieland	457	170	37.2	10.0	100.0

Table X. Mean laying date and clutch size of first, repeat and second clutches of the Blue Tit in three areas. Data from 1965 - 1986.

	first			repeat			second		
Laying date	mean	S.D.	N	mean	S.D.	N	mean	S.D.	N
Oosterhout	19.6	6.4	386	44.3	12.2	20	58.4	5.6	12
Hoge Veluwe	26.2	6.8	749	51.0	11.5	54	68.0	6.4	62
Vlieland	26.3	6.0	553	54.9	8.9	44	71.1	8.2	155
Clutch size	mean	S.D.	N	mean	S.D.	N	mean	S.D.	N
Oosterhout	11.5	1.56	386	7.7	2.00	20	6.8	0.97	12
Hoge Veluwe	10.5	2.24	749	6.9	2.19	54	6.9	1.42	62
Vlieland	11.1	1.53	553	8.7	2.62	44	7.8	1.50	155

DISCUSSION

Ten years ago Perrins (1979) stated that the Blue Tit is basically a bird of broad-leaved deciduous woodland, especially oakwoods. That this species is well-adapted to oakwoods is confirmed by the present study: the breeding density is high and relatively stable, clutches are large and eggs are laid early in the season, and breeding success is higher and less variable than in the other habitats. The fact that egg-laying dates in oakwoods are not only earlier but also more variable between years, means that the factors to which Blue Tits respond are more variable in oakwoods than in mixed woods.

These findings on the differences in breeding density, laying date, clutch size and breeding success in different habitats confirm the results obtained by other authors (Gibb 1950, Lack 1955, 1958, Dhondt et al. 1984).

Having established that the Blue Tit is well-adapted to conditions in deciduous oak woodland, it becomes most interesting to study in which ways populations in other, contrasting habitats, such as conifer plantations or evergreen oak woodland, differ from the populations in deciduous oak woodland.

This is especially interesting where there are evident differences in timing of the breeding season (cf. Blondel et al. 1987).

In the mixed forests studied by us the Blue Tits usually nest in or close to deciduous plots or patches of deciduous trees that are surrounded by conifers. On the one hand this suggests that there is a preference for deciduous habitat, but on the other hand this does not preclude the fact that breeding success is much lower than in oakwood.

The island population of Vlieland deviates from the mainland populations in several respects: it is a fairly young population (<50 years old), with a low but sharply increasing density, hardly any decline of clutch size (in first clutches) with laying date, low breeding success, and an increase of breeding success (in first broods) with date. Moreover, many birds produce a second clutch (cf. the findings by Winkel (1975) in a larch forest in Germany). Some of these phenomena are similar to those found for Great Tits in pine plantations. The latter birds time their breeding season in a way similar to their conspecifics in oakwood, but the peak in food (caterpillar) availability is several weeks later in the pine than in the oakwood. This results in heavy nestling mortality and low fledging weights (van Balen 1973). Indications that such a situation also applies to the Blue Tits in Vlieland are twofold: in 1988 the peak in caterpillar abundance in the oaks in Vlieland occurred about 9 days later than in the oaks in Hoge Veluwe, and the distribution of the small patches of deciduous trees (where most of the Blue Tits nest) in Vlieland is such that the tits' nests are close to large blocks of conifers. Many of the peculiarities of the Blue Tits can be understood if one assumes that much of the food is collected in the conifer plantations. Observations on foraging behaviour are, however, lacking.

Breeding success, calculated as the number of fledglings per egg laid, is invariably high in deciduous oak woods (Table VII and Gibb 1950, Lack 1955, 1958, Källander 1983), and considerably lower in mixed forest, evergreen oaks, conifers, and gardens (e.g. Blondel et al. 1987, Cowie and Hinsley 1987).

The relationship between breeding success and hatching date differed strikingly in the three study areas: in Oosterhout and Hoge Veluwe breeding success declines with date, as has been found before by Perrins (1965, 1970) for the Great Tit in Wytham and by van Balen (1973) for the same species in Oosterhout and Liesbos. In Vlieland, however, breeding success was largest for late first broods. In two Dutch pine plantations the seasonal trend in Great Tit fledging success varied from year to year in correspondence with the timing of the peak of food availability, and success often peaked in the second half of the breeding season (van Balen 1973). Moreover, van Noordwijk et al. (1981) found that selection for laying date in the Great Tit in Hoge Veluwe and Vlieland was variable from year to year, sometimes favouring early, and sometimes middle or late first broods. In Vlieland late first broods were favoured in many years. It is evident that the relationship between reproductive success and date does not follow general rules, but that it is to a large extent determined by local conditions, e.g. the seasonal trend in food availability. In the Blue Tit, parental survival and recruitment need to be studied before we can draw definite conclusions about timing and parental fitness.

One could ask if the population in Vlieland has shown any changes through the years indicating an increasing degree of adaptation to the new environment. However, apart from the breeding density, that increased sharply until 1977 and stabilised thereafter (Fig. 1), there were no apparent changes, either in laying date, or in clutch size, breeding success or proportion of second broods.

Earlier it was shown that hatching success declined significantly over the years in Hoge Veluwe and Vlieland. This can be associated with the presumed occurrence of calcium shortage in the eggshells of Great Tits and several other passerines, that has been recently observed in Buunderkamp, another study area with coniferous and mixed deciduous forest on poor sandy soils. Here, the percentage of females without eggs or with eggs of inferior quality has increased sharply in recent years (Drent and Woldendorp 1989). Observations made in 1989 have shown that egg volume and eggshell weight in the Great Tits from Hoge Veluwe are slightly larger than the values for Buunderkamp, but lower than in Oosterhout and Liesbos (Graveland, pers. comm.). It seems fairly certain that the decline in hatching success observed in Hoge Veluwe and Vlieland is caused by calcium shortage, which is presumably related to acid precipitation.

We concluded earlier that short-term density changes are often correlated between areas, and must be determined to a large extent by general, rather than local factors, among which winter severity probably is prominent. Combining this conclusion with the finding that density is more variable in the low-density than in the high-density areas, it seems that Blue Tits in less favourable areas suffer more from cold in winter than birds in the favourable deciduous oak woodland.

By comparing the annual values for different areas it was found that changes in breeding density and laying date are influenced by general rather than local factors, whereas the reverse applies to clutch size and breeding success. This implies that in studies of clutch size or breeding success differences between areas must be taken into account: patterns like annual and seasonal variation in these parameters do indeed vary between areas.

The fact that breeding success is determined by local factors, but changes in breeding density are influenced by general factors implies that most of the local variation in productivity is removed by factors operating after fledging. For the Great Tit the beechmast crop and winter severity play an important role because of their effect on juvenile survival (van Balen 1980). The size of the beech crop probably has little effect on Blue Tit numbers, although Källander and Karlsson (1981) have found significant correlations between population size and beechmast crop in two out of three areas. Winter severity has an evident effect as was also found by Källander and Karlsson (l.c.). Further study is needed to find out which feeding resources are important for juvenile survival of this species.

There are several indications that competition between Great and Blue Tits affects density or breeding parameters of the Blue Tit. The opposing density figures of the two species in Vlieland-West and -East agree with findings by Dhondt and Eyckerman (1980), who concluded that Blue Tit breeding numbers were influenced by competition with Great Tits for roosting sites in winter. The correlation between Great Tit or Blue Tit density and the difference in laying date between the two species, that was found in two areas, could mean that either the Blue Tit lays earlier or the Great Tit later in years with a high density of competitors, this leads to a lower intensity of competition. Further research is needed to find out whether the Blue Tit or the Great Tit or both respond to this.

Dhondt et al. (1984) argued that differences in laying date are probably caused by differences in the timing of food availability. Regarding the difference in timing between Great and Blue Tit, as related to density, one could argue that Blue Tits would take smaller prey in the egg-laying period, as they do later (Török 1986). Blue Tits would prey on the early stages of caterpillars, leaving less prey of adequate size for Great Tits. This could result in delayed egg-laying by Great Tits at high Blue Tit density.

ACKNOWLEDGEMENTS

This study could not have been made without the dedicated efforts of many workers in the field, notably H.M. van Eck, J. Visser and D. Westra. We thank A.J. Cavé, P.J. Drent, J.M. Tinbergen and S. Verhulst for helpful comments on the manuscript.

REFERENCES

Balen JH van (1973) A comparative study of the breeding ecology of the Great Tit *Parus major* in different habitats. Ardea 61: 1-93

Balen JH van (1980) Population fluctuations of the Great Tit and feeding conditions in winter. Ardea 68: 143-164

Berndt R, Winkel W, Zang H (1983) Über Legebeginn und Gelegestärke von Kohl- und Blaumeise (*Parus major*, *P. caeruleus*) in Beziehung zur geografische Lage des Brutortes. Vogelwarte 32: 46-56

Blondel J, Clamens A, Cramm P, Gaubert H, Isenmann P (1987) Population studies on tits in the mediterranean region. Ardea 75: 21-34

Cowie RJ , Hinsley SA (1987) Breeding success of Blue and Great Tits in suburban gardens. Ardea 75: 81-90

Dhondt AA, Eyckerman R (1980) Competition between the Great and the Blue Tit outside the breeding season in field experiments. Ecology 61: 1291-1296

Dhondt AA, Eyckerman R, Moermans R, Hublé J (1984) Habitat and laying date of Great and Blue Tit *Parus major* and *P. caeruleus*. Ibis 126: 388-397

Drent PJ, Woldendorp JW (1989) Acid rain and eggshells. Nature 339: 431

Gibb J (1950) The breeding biology of the Great and Blue Titmice. Ibis 92: 507-539

Kluyver HN (1951) The population ecology of the Great Tit, *Parus m. major*. Ardea 39: 1-135

Kluyver HN (1971) Regulation of numbers in populations of Great Tits (*Parus m. major*). Proc Adv Study Inst Dynamics Numbers Popul (Oosterbeek 1970): 507-523

Källander H (1983) Aspects of the breeding biology, migratory movements, winter survival, and population fluctuations in the Great Tit *Parus major* and the Blue Tit *P. caeruleus*. Thesis, Univ of Lund

Källander H, Karlsson J (1981) Population fluctuations of some North European bird species in relation to winter temperatures. Proc Second Nordic Congr Ornithol 1979: 111-117

Lack D (1955) British tits (*Parus* spp.) in nesting boxes. Ardea 43: 50-84

Lack D (1958) A quantitative breeding study of British tits. Ardea 46: 91-124

Noordwijk AJ van, van Balen JH, Scharloo W (1981) Genetic variation in the timing of reproduction in the Great Tit. Oecologia 49: 158-166

Perrins CM (1965) Population fluctuations and clutch size in the Great Tit, *Parus major*. J Anim Ecol 34: 601-647

Perrins CM (1970) The timing of birds' breeding seasons. Ibis 112: 242-255

Perrins CM (1979) British Tits. Collins, London

Schmidt K-H (1984) Frühjahrstemperaturen und Legebeginn bei Meisen (*Parus*). J Orn 125: 321-331

Schmidt K-H, Steinbach J (1985) Jahreszeitliche Änderung der Gelegegrösze bei der Kohlmeise (*Parus major*)? J Orn 126: 163-173

Török J (1986) Food segregation in three hole-nesting bird species during the breeding season. Ardea 74: 129-136

Winkel W (1975) Vergleichend-brutbiologische Untersuchungen an fünf Meisen-Arten (*Parus* spp.) in einem niedersächsischen Aufforstungsgebiet mit Japanischer Lärche *Larix leptolepis*. Vogelwelt 96: 41-63, 104-114

POPULATION DYNAMICS AND REGULATION OF A HIGH DENSITY BLUE TIT POPULATION

André A. DHONDT, Erik MATTHYSEN, Frank ADRIAENSEN and Marcel M. LAMBRECHTS

Department of Biology, University of Antwerp (U.I.A), B-2610 Wilrijk, Belgium

In Belgian woodlands Blue Tit *Parus caeruleus* breeding numbers can be increased by providing nestboxes. When all nestboxes have an entrance hole which permits Great Tits also to use them, the breeding density of Blue Tits is around 1.2 - 1.5 pairs per ha in optimal habitat (Dhondt et al. 1982, Dhondt 1985a).

Blue Tit breeding density can be further increased by the provision of small-holed (diameter 26mm) nestboxes that cannot be used by Great Tits (Dhondt and Eyckerman 1980, Dhondt 1989). In this new situation numbers tend to stabilise, suggesting that the population is regulated at a higher level than before (Dhondt 1985a) whereby intraspecific competition is probably important (Dhondt et al. 1982).

In order to identify factors contributing to this stabilisation in numbers we have been studying, since 1979, the dynamics of Blue Tit populations around Antwerp, N. Belgium. Plot C, on which we will report here, is one of the study sites. It is an isolated study area of 17 ha in which 100 small-holed nestboxes have been present since January 1979. In this paper we will try to identify some of the factors contributing to variation and to the regulation of numbers. In the study period there have been four cold winters. We will try to determine what the effects of density and winter cold are in an isolated high density population of Blue Tits in N. Belgium, at the population level (effects of density and winter condition on population processes) and at the individual level (which individuals are more influenced than others).

MATERIAL AND METHODS

The study plot (Plot C) is the wooded part of an isolated private estate "Calixberg", surrounded by houses and meadows. In 17 ha of this estate, covered with mixed deciduous trees, 100 wooden nestboxes were put up in December 1978. The boxes have inside dimensions of about 23 cm high x 9 cm wide x 12 cm deep, and have a circular opening with a diameter of 26 mm. After breeding all old nests are removed from the boxes.

During each winter at least two visits were made during the night to trap birds roosting in the nestboxes. During the breeding season nestboxes were

NATO ASI Series, Vol. G 24
Population Biology of Passerine Birds
Edited by J. Blondel et al.

routinely visited once a week and their contents noted. Supplementary visits were made to trap the adults and weigh and measure the young at 15 days. A high proportion of the breeding birds was identified. At capture, birds were ringed, weighed and measured (wing and tarsus length). All nestlings were ringed, sexed, weighed and measured (tarsus length) at 15 days of age.

Considering all the birds that were identified breeding over the period 1980-1989 and counting each individual once, 40 % of 220 males and 13% of 252 females were known to have been born locally. The proportion of autochtonous yearling recruits, however, varied significantly between the years, as the partial effect of year in a 3-way G-test (effect of sex and year on origin of juvenile recruits) is significant ($X^2=17.496$, 9 df, $P<0.05$). This analysis confirms that the proportion of individuals locally-born was higher in males than in females (partial effect of sex : $X^2=35.4$, 1 df, $P<0.001$).

Although a large proportion of the breeding population is not ringed as a nestling in the study area, the data suggest that the breeding birds do not usually come from far away. Three ringed males and 13/14 ringed females came from adjacent villages; the remaining female was born at 17 km distance. We therefore think that most unringed birds also come from the immediate surroundings of the study plot. Some immigrants are "grandchidren" of breeding birds.

RESULTS

Reproduction

Over 11 years we recorded 465 first broods of which 396 (85.2%) produced fledglings. Only nineteen failed breeders produced a repeat clutch and only nine true second broods were observed in 11 years. An additional seven "extra clutches" were recorded; these are clutches laid much later than first clutches, but which cannot be attributed to a pair that started a first brood. They are included as breeding pairs, but not included in the analysis of first broods. The proportion of eggs laid yielding fledglings is 70.2 % of 5,291 eggs for first broods, 7.3% of 55 eggs for second broods, 42.3 % of 142 eggs for repeat clutches and 45.8 % of 59 eggs for extra pairs. The very low success of second broods and the low success of other non-first broods probably explains why so few Blue Tits, in this area, start a repeat or second brood. 97.61 % of the 3,712 fledglings produced over the 11 year study period fledged from first broods. 99.4 % of 161 young known to survive at least 3 months stemmed from first broods. The contribution to the population of non-first broods is therefore quite small, and one may wonder why Blue Tits, in this study area, ever bother to start a new clutch after the initial one has failed.

More detailed information on reproduction is given in Table I.

Table I. Reproduction of Blue Tits in Plot C (1st broods).

year	n.pairs	n.1st clutches	n.succ. clutches	mean c/s	mean n.fldg	mean laydate
1979	28	26	21	11.48	9.24	108.8
1980	39	39	36	11.33	9.39	106.1
1981	40	39	33	11.06	9.15	99.3
1982	41	40	35	11.89	9.43	104.7
1983	51	49	47	11.62	9.74	109.5
1984	45	44	43	11.09	9.51	114.6
1985	47	47	42	11.93	9.43	110.5
1986	44	44	41	11.51	9.39	117.4
1987	39	39	32	12.34	10.97	108.3
1988	58	58	39	11.03	7.74	107.2
1989	40	40	27	11.52	9.15	93.8
sum	472	465	396	11.52	9.37	107.95

Notes:
N.pairs: number of breeding pairs = first + extra clutches
N.succ. clutches: at least 1 fledgling = sample size for following colunms
laydate: 1 = 1 January, 91(92) = 1 April in normal(leap) years.

K-factor analysis

We have carried out a k-factor analysis until the birds are 1-year old for the cohorts 1979 to 1988. The following terms were used:

N_0: potential number of eggs laid = number of females * 12.2
N_1: numbers of eggs laid
N_2: number of eggs hatched
N_3: number of pulli fledged
N_4: number of yearlings present in next breeding season, both breeding and non-breeding
N_5: number of yearlings breeding
k_0: eggs not laid : $\ln N_0 - \ln N_1$ (calculated)
k_1: eggs not hatched: $\ln N_1 - \ln N_2$ (measured directly)
k_2: young not fledged: $\ln N_2 - \ln N_3$ (measured directly)
k_{012}: breeding losses: $k_0 + k_1 + k_2$
k_3: overwinter losses: $\ln N_3 - \ln N_4$ (calculated)
k_4: non-breeding yearlings: $\ln N_4 - \ln N_5$ (calculated)
K: total K: $k_0 + k_1 + k_2 + k_3 + k_4$

Figure 1 shows that k_3, overwinter loss is clearly the key-factor, since it varies in parallel with total K (correlation coefficient between k_3 and K is $r=0.91$. k_{012} and k_4 are numerically about equal. Two k-values seem to be density dependent: k_{01}, potential eggs not hatched and k_3, overwinter losses (see fig. 2, 3). The latter relationship is quite interesting, if true, because k_3 would be an example of a density dependent key-factor. When fewer young fledge, i.e. when population density is lower, a higher proportion of the cohort survives. It must be underlined, however, that the numbers used combine in fact the actual survival of the juveniles and the net dispersion (immigrants - emigrants). k_3 therefore combines two processes i.e. survival and dispersion. The suggested density dependence does, therefore, not provide information on what is actually happening in the population. Direct measurements of overwinter losses are required to determine whether this apparent density dependence is a statistical artifact or not.

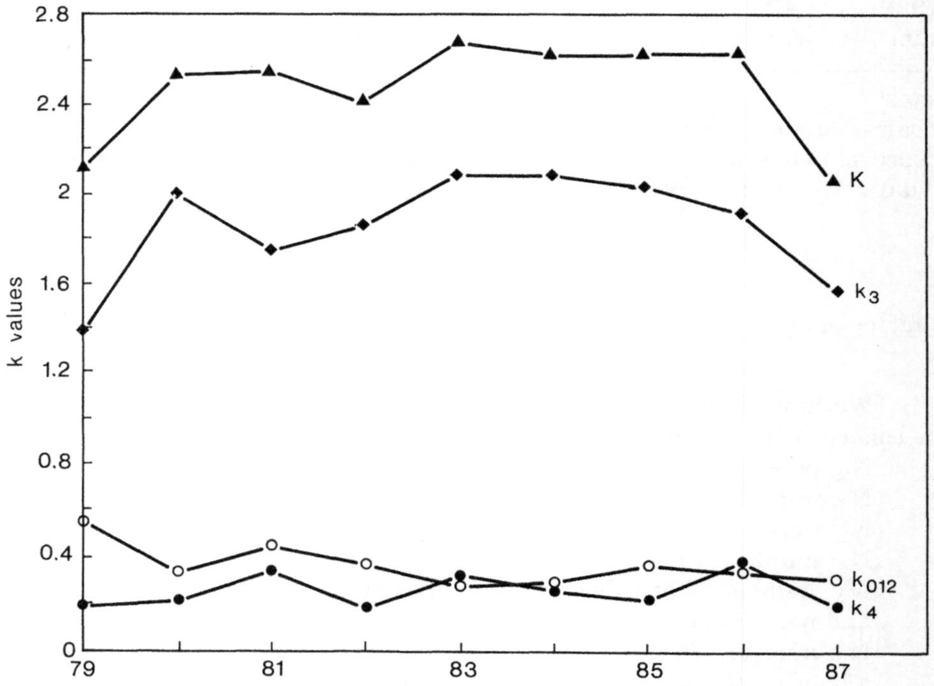

Fig. 1. K-factor analysis of Blue Tit population in Plot C (for more details see text). k_{012}: breeding losses; k_3: overwinter loss; k_4: non-breeding yearlings. K: total cohort losses. Note that the k-values due to breeding losses and due to non-breeding of yearlings are about equally important. Note also that k_3 is the key-factor causing population fluctuations.

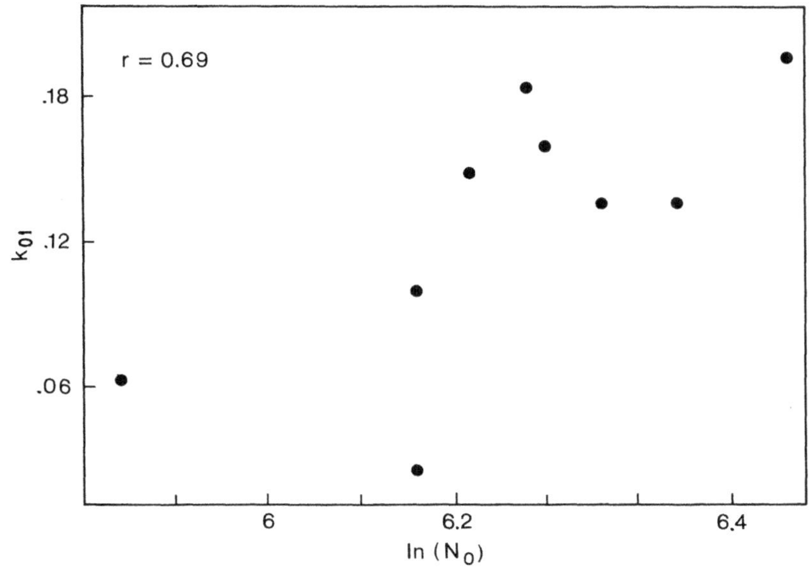

Fig. 2. K-factor analysis of Blue Tit population in Plot C. Losses due to non-laying (k_0) plus non-hatching of eggs (k_1) plotted against natural logarithm of population size. Note the high correlation, suggesting density dependence.

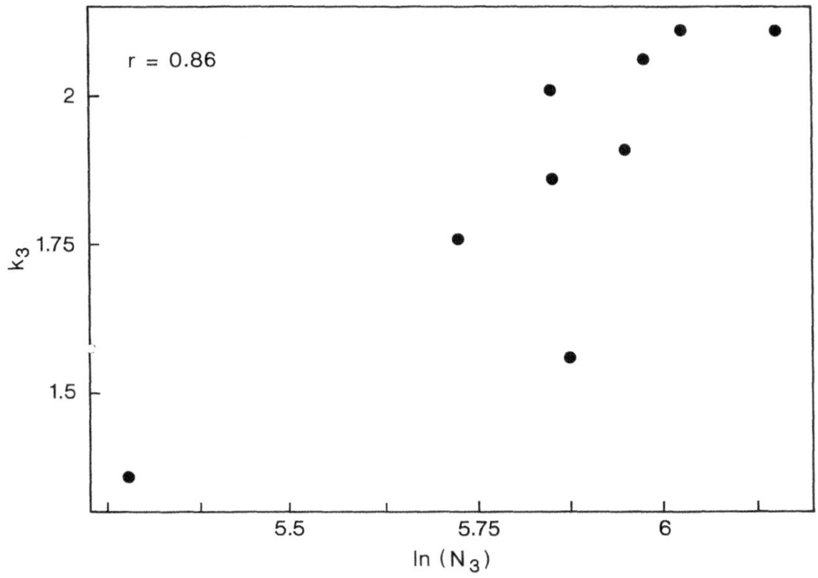

Fig. 3. K-factor analysis of Blue Tit population in Plot C. Overwinter loss plotted against the natural logarithm of number of fledglings. Note the high correlation, suggesting density dependence of k_3 that is also the key-factor (see Fig. 1).

Adult survival: comparison of males and females

We have used the program SURGE (see Clobert et al. 1987, 1988) as it permits the comparison of different models and also takes into account differences in recapture rates. In an initial analysis we included both sexes and we compared the models shown in Table II.

The conclusion from this analysis is that males and females have the same survival rate and that the value of 47.7 for females and 49.4 for males calculated in model 4 do not differ significantly. Similarly the capture rates of 89.9% for females and 82.3% for males are not significantly different. The overall averages calculated in model 1 are an annual survival rate of 48.5 % and a recapture rate of 86.3 %.

Table II. Comparison of survival and recapture rates between sexes of Blue Tits in Plot C (1979-1989) using SURGE.

Model	Survival Rate	Capture Rate	-Deviance	Df
1.	M=F=ct	M=F=ct	1216.08	2
2.	M=F=ct	M=ct; F=ct	1214.29	3
(not better than 1: X^2 = 1.79 , 1 df, ns)				
3.	M=ct; F=ct	M=F=ct	1216.08	3
(not better than 1)				
4.	M=ct; F=ct	M=ct; F=ct	1214.08	4
(not better than 1: X^2 = 2.00, 1 df, P>0.05)				

*: Df = parameters estimated
M,F = annual adult survival rate of males, females.
In all models adult survival and capture rates are assumed to be age and time independent (=ct).
In Model 1 male and female survival rate and are not different.
In Model 2 survival rates are not, but capture rates are different.
In Model 3 survival rates are, but capture rates are not different.
In Model 4 both survival rates and captures rates are different.
None of the more complex models (2-4) is signigficantly better that model 1, so that the latter is accepted.

Adult survival by sex: effect of age, density, winter type

Since it is not evident that the same factors would influence the survivorship of the two sexes we have then analysed the data by sex. In the case of males, all the more complex models are rejected, and the simplest model with one survival rate (49.4 %) and one recapture rate (82.3 %) is accepted.

In the case of females, the data suggest that there could be an effect of the size of the breeding population and of cold winters (Table III). To test this we used models in which the annual survival rates depend on population size and winter type (cold or warm). The results are very clear: both partial effects are significant (see Table IV).

Table III. Comparison of different survival models by sex

Model	Surv.	Capt. rate	df	- Deviance males	females
s, p:	ct	ct	2	587.48	626.60
s_t, p:	time dep.	ct	11	581.13	618.41
s_{wi}, p:	winter cold	ct	3	587.11	622.88 *
s_{dens}, p:	dens.dep	ct	3	586.31	623.91
$s_{1,2}$, p:	age dep.	ct	3	586.98	626.31
$s_{1,2,3}$, p:	age dep.	ct	4	586.92	613.64 **

s: annual survival rate; p: capture rate.
In all models capture rate is constant (=ct).
model s_t: survival is time dependent, e.g. varies between years.
model s_{wi}: survival correlated with severity of the winter.
model s_{dens}: survival correlated with breeding density.
$s_{1,2}$: survival of yearlings different from older birds.
$s_{1,2,3}$: survival of yearlings different from 2 year olds, different from older birds.

Testing different age-dependent models shows no improvement for the males, whereas for females the model with 3 survival rates has a significantly better fit than models with one, two or more than 3 survival rates. The age-specific annual survival-rates are:

yearling females: 0.485; 2-year-old: 0.589; older : 0.295.

This shows that once females have passed the age of 3 years their survival decreases sharply. Several "old-age" effects in female Blue Tits, aged four or more, have been shown by Dhondt (1989). A similar effect cannot be established for males since the $s_{1,2,3}$, p model is not better than the s,p model ($X^2 = 0.5$, 1 df).

Although the average annual survival of males and females is similar, the factors influencing this process differ between the sexes. For the females both winter cold and breeding density contribute significantly to variation in survival (Table IV). In males the effect of density on survival is at best marginal (r=0.44, P= 0.20) and that of winter cold totally absent. This suggests that selection pressures on adults of the two sexes differ.

Table IV. Adult annual survival rates as obtained by SURGE using an [s(t), p] model for each sex separately, and the results of a model constraining the female survival rate to be a function of breeding density and winter type.

Year	female	male	number of breeding females	cold winter
1979	63.1	55.5	28	no (=0)
1980	54.8	46.0	39	no
1981	44.4	62.0	40	yes (=1)
1982	56.0	57.4	41	no
1983	44.2	44.1	51	no
1984	48.7	52.5	45	yes
1985	37.5	53.7	47	yes
1986	38.1	39.3	44	yes
1987	54.8	44.0	39	no
1988	46.4	42.8	58	no

The estimated equation linking female survival to breeding numbers and winter type is:
$\log(s/(1\text{-}s)) = 1.163\ (\pm .657) - 0.0243\ (\pm 0.0144)$ breeding numbers $- 0.414\ (\pm .212)$ winter type
Both slopes are significantly lower than 0 (one-sided likelihood ratio tests: $P=0.0457$ and $P=0.0255$ respectively).
80.1% of the variation in female annual survival is explained by variation in winter type and breeding population size.

Reproduction and female survival

For females, we have also studied the possible effect of reproduction on overwinter survival. We have used clutch-size as the measure of the investment a female is willing to make. In this analysis we have included all females identified with a successful (i.e. young fledged) first brood. Since female survival and clutch-size vary with age, we have studied the relation bewteen female survival and clutch-size by age class.

An overall analysis (Table V) shows no effect of clutch-size on survival. However, if one compares the females laying very large clutches (13 or more) with all others, then the latter survive significantly better ($G=4.549$, $P<0.05$). In contrast to 2-year-old females (see below) no effects of breeding density or winter type can be established.

For the 2-year-old females we have analysed the survival data according to clutch-size (large or smaller than 11 eggs) and according to winter-type (warm, cold) (Table VI). A 3-way G-test yields a complete interaction between

Table V. Survival of yearling females and clutch size (c/s).

c/s	survived no	survived yes	sum	% survived
$\leq$10	22	20	42	48%
11	27	24	51	47%
12	20	18	38	47%
$\geq$13	14	29	43	67%
Total	83	91	174	52%

Table VI. Survival of 2 year old females and clutch size.

	warm winter				cold winter				
c/s	survived no	survived yes	sum	% survived	survived no	survived yes	sum	% survived	
$\leq$11	21	14	35	40	7	7	14	50	G=0.310
$\geq$12	10	33	43	77	15	12	27	44	G=7.320
Sum	31	47	78	60	22	19	41	46	
	G=10.849				G=0.110				

the three factors, which shows that survival varies both with clutch-size and with winter-type, whereby the latter two interact. We have therefore performed 2-way G- tests, as shown in the table. This indicates that :

- in warm winters females laying larger clutches survived better than those laying small clutches (77 % v. 40 %, P<0.001)
- this is not true in cold winters (44% v. 50 %, NS)
- survival of females laying small clutches is not different in cold or warm winters (50 % v. 40 %, NS)
- survival of females laying large clutches is significantly poorer in cold winters than in warm winters (44 % v. 77 %, P<0.01). A similar, although not significant, trend exists in 3-year-old females: 28.6 % of 35 females laying 11 or fewer eggs survived compared to 44.4% of 18 females laying first clutches of 12 eggs or more (G= 1.266, NS).

With reference to the questions posed at the beginning, the answers are:

- adult male survival does not vary between years and is thus not dependent on density, winter coldness or reproductive effort. Male survival also does not change with age. Furthermore, although not shown here, male survival does not vary with clutch-size, nor with the number of fledglings for 167 yearling males as well as for 92 second year males.

- adult female survival is density dependent, but is also influenced by age. Furthermore females laying large clutches survive much better when winters are warm, but not better than females laying smaller clutches when winters are cold.

The higher survival of females laying large clutches cannot be due to these females living in better quality territories, since this effect is not found in their partners.

Structure of the breeding population: age structure

Adult survival and the age structure of the breeding population should be directly related. If adult survival averages 48.5 % over a 10 year period and if the population has not shown a clear trend of increase or decrease in that period, then about 51.5 % of the breeding birds should be yearlings. Assuming that all birds present breed in the nest-boxes, the age composition of the breeding population can be calculated. Over 11 years 46.6% of 388 males and 46.9% of 439 females were yearling birds. This is somewhat less than expected from the previous calculations on adult survival.

Although the data suggest that the proportion yearlings varies between years with a very low value in 1989 (26% out of 57 breeding birds) and high values in 1980 and in 1988, a 3 way G-test, analysing the effects of sex and year on age composition does not confirm this (partial effect of year: $G = 14.086$, 10 df, NS).

Birds present but not breeding

The data thus suggest that some yearling birds are missing from the sample trapped in the nestboxes. This then raises the question: In how far does the age composition of the birds caught breeding in nestboxes reflect the actual age composition of birds present? It was shown elswhere that in Great Tit a some old birds fail to breed (Dhondt 1985b). No such data are available for the Blue Tit, but the number of such birds would be very small anyway. From the population point of view, they can thus be disregarded. More important, however, both from the population and from the evolutionary point of view are non-breeding yearlings. Dhondt (1989) has shown that birds which do not breed as yearlings do not compensate for this later in life by having more young or higher survival; they have effectively lost a year of their lifetime reproduction. Furthermore a large proportion of birds which do not breed as yearlings die before the next season and therefore never reproduce. If the number of non-breeders is large, this results in a gross under-estimation of population size. If the proportion of non-breeding birds varies strongly between years or areas we could make mistakes when estimating possible effects of competition.

Not all birds, however, breed as yearlings. We have calculated above that around 85% of the adults that bred once were trapped on the nest in subsequent years. Over the period 1980-1989, 420 (94.6%) females were trapped out of 444 breeding pairs in nestboxes, compared with the calculated capture rate for females of 89.9%, using the recapture histories. The 5% difference can be caused by one or a combination of the following: (i) some adult females do not breed, (ii) some adult females breed in natural cavities, (iii) some adult females desert before egg laying. The result nevertheless implies that (nearly) all Blue Tits breed in nestboxes in this population and that we have managed to capture a high proportion of the breeding population each year.

We will estimate the proportion of nonbreeding yearlings in different ways, in particular to compare sexes and years.

An overall estimation based on the age of local-born recruits

In Table VII the age at first breeding for local-born nestlings is given. Blue Tits in this population are non-migratory and movements are dispersive. We assume therefore that an individual which is born in the plot and later breeds there has not left the area for extended periods, in particular we assume it would have been present in breeding seasons in which it was not identified on the nest. 79% of the males and 76% of the females were first trapped at age 1. Taking an annual adult survival rate of 50 % a bird first trapped at age 2 is the survivor of 2 alive at age 1, one first trapped at age 3 the survivor of 4 present at age 1 and one first trapped at age 4 the survivor of 8 present at age 1. The number of birds presumed present were calculated in this way. The data show that only 57% of the birds presumed present were actually trapped breeding as yearling. Note the very similar values obtained for males and females.

Table VII. Age at first breeding for Blue Tits born in Plot C and subsequently breeding there (breeding seasons 1980-89).

	Number Trapped		Presumed present	
Age	Males	Females	Males	Females
1	70	25	70 (57%)	25 (56%)
2	14	6	28 (23%)	12 (27%)
3	4	2	16	8
4	1		4	
Sum	89	33	122	45

An overall estimation based on the age of non-local-born recruits

A relatively large number of birds are first caught breeding as "adult", i.e. when at least 2 years old. When these birds arrived in the study area is, however, usually not known.

In Table VIII these birds are used to estimate the proportion Blue Tits first breeding as yearlings assuming that (1) the birds did not breed in a different area in a previous year, and (2) all these birds are 2 years old, not older. Although both assumptions are probably incorrect, they bias the results in opposite ways, and it is not possible to make more accurate assumptions. The values obtained in this sample are similar to those in the previous one. In fact a 3-way G-test comparing the age at first breeding (1 year versus older) between birds of different origin (local-born, not local born) and sexes shows no effects of sex or origin (partial effect sex: $G=1.98$, 1 df, NS; partial effect origin $G=2.42$, 1df, NS).

Table VIII. Age at first breeding of Blue Tits not born in Plot C (breeding seasons 1980-89).

	Number Trapped		Presumed present	
Age	Males	Females	Males	Females
1	104	164	104 (76%)	164 (67%)
≥ 2	16	40	32	80
sum	120	204	136	244

Characteristics of birds not known to breed as a yearling

We have compared the measurements (wing-length, tarsus length, weight on nest) between birds breeding both at age 1 and 2, with those only found breeding at age 2 and could not find any significant differences. We have also compared the reproduction of these two groups of females at age 2. Although there was a tendency for females only breeding at age 2 to lay later, to lay a smaller clutch, to fledge fewer young and to recruit fewer young a year later, none of these effects was statistically significant. It is, therefore, clearly true that females that skip their first breeding season do not compensate by better reproduction in the second season, nor later (cf. Dhondt 1989).

Annual variation in percentage of non-breeding yearlings

Since the proportion of non-breeding yearlings does not vary significantly with sex or place of origin, we have lumped the data by year to study possible annual variation. In the calculation we have assumed that all non-yearling recruits were 2 years old, including the birds ringed as nestlings that were known to be older. This will lead to an underestimation of the percentage of non-breeders. We have also assumed that all immigrant non-yearling recruits were present in the previous season without breeding, which will lead to an overestimation. We can only hope both errors will tend to compensate one another. We further assumed the survival rate of breeders and non- breeders would be similar and have used 50% as an easy approximation.

In this way, for each breeding season we produced totals of yearling breeders, and of non-yearling first-time breeders. From the above assumptions it follows that each non-yearling first-time breeder in one year represents 2 non-breeding yearlings in the previous season. That number is given in Table IX under the heading non-breeding yearlings.

Table IX. Estimation of the proportion of non-breeding yearlings per year. All birds have been used (both sexes, both origins) (see text for further details).

year	yearlings breeding	yearlings non-breeding	total	% Non-Breeding
1980	37	10	47	21
1981	33	10	43	23
1982	34	24	58	41
1983	41	10	51	20
1984	40	24	64	38
1984	36	14	50	28
1986	38	12	50	24
1987	35	30	65	46
1988	54	14	68	21
Total	348	148	496	29.8

How can we explain this high proportion of birds not breeding in nestboxes as yearlings ?

- not all birds breed in nestboxes. If this is true, regardless of age, one should find an important difference between the direct estimate of percentage birds trapped (based on known captures and known breeding pairs) and that estimated by the SURGE calculations. This is not so.
- not all yearling birds breed in nestboxes. If this is true one would expect more birds to first breed as adults in years following peak densities and fewer birds in years following low density years.
- not all birds breed.

In order to start the formulation of an answer to this question we have tried to correlate the proportion non-breeding yearlings to factors such as winter coldness, breeding number, breeding number in the previous year, number of juveniles and year. A multiple regression analysis shows that the proportion non-breeding yearlings is positively correlated with the number of juveniles (the more juveniles, the higher proportion non-breeding ones), but that breeding density has a significant negative effect (the proportion of non-breeding yearlings is lower in years of high density).

Percentage non-breeding yearlings = 33.61 ($\pm$ 13.793, P=0.05)
+ 0.94 ($\pm$ 0.209, P=0.005) NUMBER OF YEARLINGS
- 1.25 ($\pm$ 0.293 , P=0.006) BREEDING POPULATION SIZE

The first effect is not very surprising and suggests that intra-specific competition among yearlings exists. The second relationship suggests that the question was asked the wrong way round, and could mean that in years in which a high proportion of the yearlings can or do breed, the size of the breeding population will be high. What influences the ability of yearlings to breed is at present unclear.

CONCLUSIONS

In this incomplete analysis of a high density Blue Tit population we have been able to show that reproduction and female adult survival are density dependent, and that juvenile survival and immigration may also be density dependent. The analysis suggests that a large proportion of yearlings does not breed. Although cold winters have no effect on the size of the following breeding population they certainly have an effect on its composition as we have shown differential survival of females according to clutch-size and the severity of winter. Elsewhere (Dhondt 1988) we have shown that after cold winter nestlings are larger, which is caused by differential recruitment of larger yearlings (unpublished).

This study confirms the extreme complexity and variability of selective factors acting upon individuals in populations.

ACKNOWLEDGEMENTS

We am very grateful to the Bracht family, who has kindly given the permission for this study on their estate. The work was supported by grants from the Belgian National Foundation for Scientific Research. Frank Fierens gave invaluable assistance both in the field and in preparing the computer data base. Chris Perrins commented on an earlier draft, and J.D. Lebreton helped with some calculations with SURGE.

REFERENCES

Clobert J, Lebreton JD, Allainé D (1987) A general approach to survival rate estimation by recaptures or resightings of marked birds. Ardea 75: 133-142

Clobert J, Perrins CM, McCleery RH, Gosler AG (1988) Survival rate in the Great Tit *Parus major* in relation to sex, age, and immigration status. J Anim Ecol 57: 297-306

Dhondt AA, Eyckerman R (1980) Competition between Great Tit and the Blue Tit outside the breeding season in field experiments. Ecology 61: 1291-1296

Dhondt AA, Schillemans J, De Laet J (1982) Blue Tit territories at different density levels. Ardea 70: 185-188

Dhondt AA (1985) Do old Great Tits forego breeding? The Auk 102: 870-872

Dhondt AA (1985) The effect of interspecific competition on numbers in bird populations. Acta XVIII Congressus Internationalis Ornithologicus, p 792-796

Dhondt AA (1988) The necessity of population genetics for understanding evolution: an ecologist's view. in: de Jong G (ed) Population Genetics and Evolution. Springer-Verlag, Berlin Heidelberg, p 14-18

Dhondt AA (1989) Blue Tit. in: Newton I (ed) Lifetime Reproduction in Birds. Academic Press, London, p 15-33

BREEDING PATTERNS OF THE COAL TIT *(PARUS ATER)* AND THE GREAT TIT *(PARUS MAJOR)* IN PINE STANDS IN PORTUGAL

Lilia FIDALGO

Reserva Natural das Dunas de San Jacinto, 3800 Aveiro, Portugal

Many studies of tit populations (*Parus spp.*) have been carried out over many years in Europe. In these studies different aspects of tit biology have been analysed in an ecological and evolutionary perspective. The extension to Portugal of such studies which were developed long ago in the atlantic and mediterranean contexts may bring new and interesting insight to the analysis of existing data.

Our country clearly belongs to the area of mediterranean influence which is characterized by a profoundly man-modified landscape, which results in a diversification of habitats, and by a well defined climate with hot summers and mild winters.

The present study was initiated in 1987 in the San Jacinto Dunes Nature Reserve (Fig. 1). In this paper I present basic data on the breeding biology of three species breeding in nest-boxes in the study area: the Coal Tit *Parus ater*, the Great Tit *Parus major*, and the Crested Tit *Parus cristatus*. The low occupation rate of the latter made it impossible to analyze the results for this species.

The data collected for the Great Tit and the Coal Tit will be analyzed and discussed by comparison with results obtained by other authors in northern Europe and the Mediterranean area. Emphasis has been given to the Great Tit because the biology of this species is well documented regarding the influence of habitat variation on its breeding biology, clutch size and breeding success.

STUDY AREA AND METHODS

The study area is the San Jacinto Dunes Nature Reserve, near Aveiro, (40°29'N; 8°44'W), a 666 ha pinewood planted on sand dunes between the Atlantic Ocean and a lagune system (Fig. 1). The area is a uniform plantation of Cluster Pine, *Pinus pinaster*, 100 years old, with an undergrowth dominated by *Myrica faya* and *Acacia longifolia*. Being in a nature reserve, this undergrowth is well developed. This is infrequent in other pinewood plantations severely exploited for wood in Portugal.

In this geographical situation, the climate benefits from both mediterranean and atlantic influences, presenting low annual thermic amplitudes and high humidity values for the whole year (maximum temperature: 18°4 C, minimum temperature: 9°9 C).

NATO ASI Series, Vol. G 24
Population Biology of Passerine Birds
Edited by J. Blondel et al.

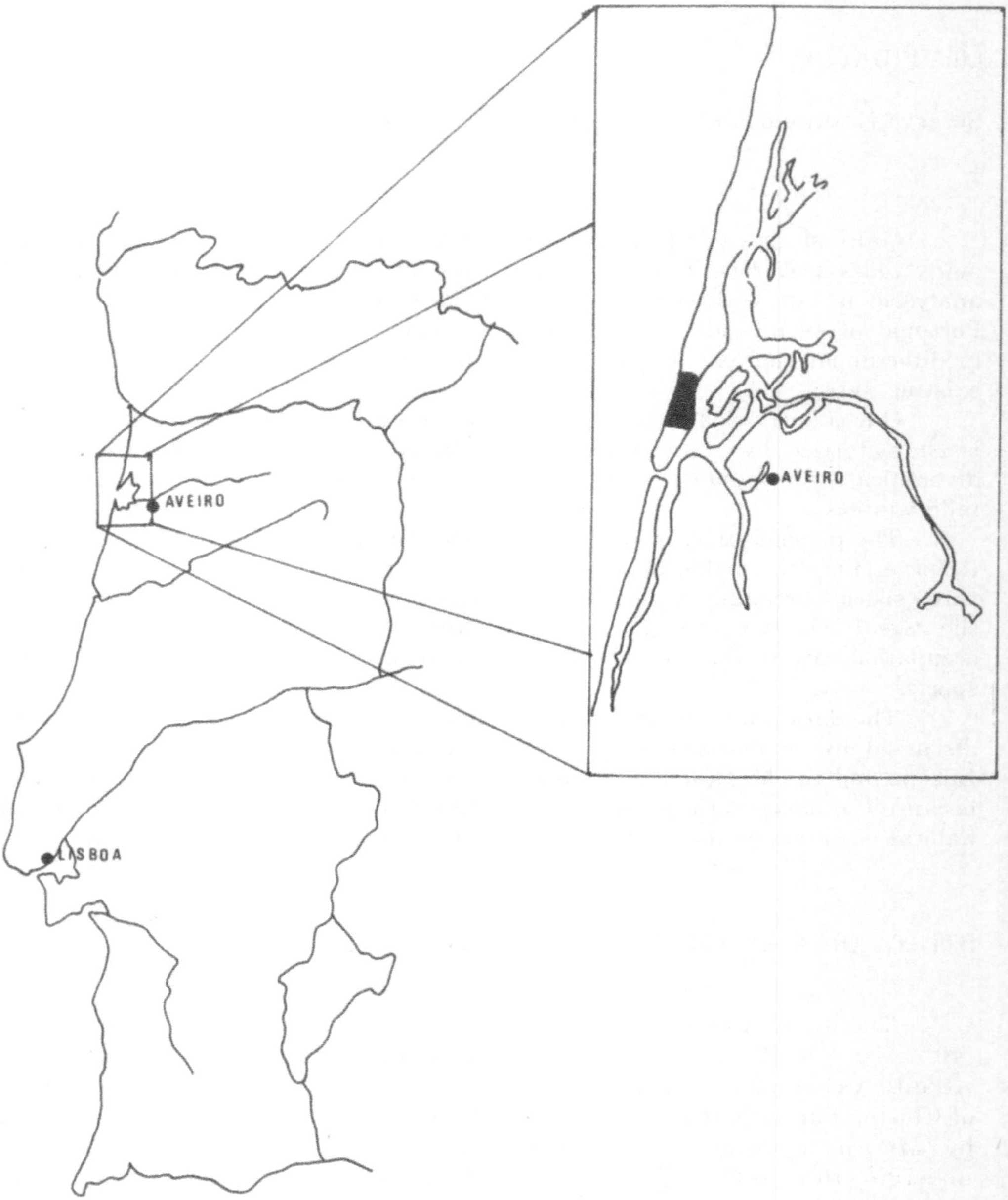

Fig. 1. Location of the study area (black) between lagoons and the sea.

Breeding density

The breeding density of tits in the study area is higher than that in other Cluster Pinewoods in our country (Pina 1982), and also higher than the values reported in other pine plantations in Europe. Rufino et al. (in press), using the line transect method proposed by Järvinen and Väsinänen (1977), found in San Jacinto 7.3 pairs ha^{-10} for the Coal Tit. This is a higher value than those recorded by Lack (1964) in pine plantations in Great Britain and by Blondel (1985) in Holm oak (*Quercus ilex*) woods in Southern France. The same authors (Rufino et al., op. cit.) recorded 4.3 pairs ha^{-10} for the Great Tit, which is a low value when compared with those recorded for deciduous woods, but also higher than the values recorded in coniferous woods in northern Europe (Cowie and Hinsley 1987, Lack 1964, Leclercq 1976, van Balen 1973). Breeding density of the Great Tit in San Jacinto is the same as in the best area (mixed coniferous-deciduous forest) in Northern Finland (Oulu)(Orell pers. com.)

Nestboxes and the procedure for data collection

Since 1975, 425 nest-boxes, with a 3 cm entrance hole, have been distributed in the wood with no other initial aim than to increase the tit populations. The boxes, made of pinewood are 37.5 cm high and 16.5 cm deep, and they were placed 1 to 3 m above the ground, with the entrance hole oriented to south or south-east. The distance between them was not uniform, ranging from 5 to 70 m. In 1988 this distance was made uniform and the boxes are now 50 m apart. The total area covered with nest-boxes is approximately 100 ha.

From the end of February to the end of July nest-box inspections were made weekly, and sometimes even more frequently when they were occupied. This method provided us with data on occupation rates, breeding phenology, clutch size and numbers of young hatched and fledged. The boxes were cleaned after the young had fledged to detect the presence of unfertilized eggs or dead young trampled by the other chicks.

All young tits and as many adults as possible were ringed, on or after the 15th day after hatching, and their weights recorded.

RESULTS

Laying date

Table I summarizes the results concerning laying date, clutch size, fledging weight and breeding success for the 1st clutches (ratio of the number of

Table I. Breeding data for first clutches of Coal Tit, Great Tit and Crested Tit in San Jacinto in 1987.
m = mean, S.D. = standard deviation, n = sample size.

	Coal Tit	Great Tit	Crested Tit
laying date of 1st egg			
m	16 March	18 April	24 March
S.D.	7.37	8.32	7.36
n	24	30	6
clutch size			
m	6.17	6.92	5.67
S.D.	0.88	1.10	0.52
n	17	24	6
breeding success			
%	84.78	38.55	55.88
n	15	24	6
fledging weight (g)			
m	8.2	14.3	10.7
S.D.	0.58	2.17	1.26
n	81	76	23

young flegded to the total of eggs laid by the female). The earliest laying date was 16 March (Coal Tit), which is the earliest annual mean laying date ever registered for tit populations studied in Europe (Blondel et al. 1987, Dhondt et al. 1984, den Boer Hazewinkel 1987, Perrins 1965, van Balen 1973) although single clutches may be laid earlier, even in Central Europe, e.g. 25 February 1959 (Glutz von Blotzheim 1962). The breeding season of this species also ends earlier than recorded for other tit populations in the rest of Europe, by the end of May. In contrast, the breeding phenology of the Great Tit generally agrees with the data reported for different habitats by the authors cited above. However data for Finland (Orell and Ojanen 1983) and Corsica (Blondel 1985) are exceptions.

I hypothesize that the factors which determine the onset of laying in both species in San Jacinto are not the same. In the Great Tit these factors seem to be similar to those which determine the onset of breeding in other populations submitted to different environmental conditions (namely temperature).

I recorded only three second clutches of Coal Tit (12% of the pairs which laid a clutch) and these were produced by the pairs that laid their first clutches earlier. Seventeen Great Tit pairs attempted a second clutch (78.3% of the birds which laid a clutch). Their first one had a low fledging success and I think these 2nd broods are almost all repeats.

Clutch size

The Great Tit lays the largest clutch but the variation in clutch size is very high (4 to 9 eggs). Coal Tit and Crested Tit clutches varied between 5 to 8 and 5 to 6 eggs, respectively. The average clutch size of second broods is lower for both the Great Tit and the Coal Tit (Table II).

Table II. Breeding data for second clutches of Coal Tit and Great Tit in San Jacinto in 1987.
m = mean, S.D. = standard deviation, n = sample size.

	Coal Tit	Great Tit
clutch size		
m	5.33	5.69
S.D.	1.15	0.87
n	3	16
breeding success		
%	75.00	49.45
n	3	16
fledging weight (g)		
m	7.8	15.9
S.D.	0.41	2.58
n	10	47

When plotting laying date against clutch size, combining first and second clutches, there is a significant negative correlation between these two parameters (Fig. 2 and 3) The same is not true when considering only first broods. In this case there is no significant correlation. Thus, there is no "calendar effect" on clutch size in this Pine plantation, at least in this particular year.

Breeding success

The high breeding failure recorded for Great Tit (Table I) was almost entirely due to the mortality of young (59.3%, n=24). Coal Tit mortality was lower and failures were mainly due to unfertilized eggs. Great Tit mortality affected almost all nests, whatever their location in the study area.

The average fledging weight of young Great Tits in first clutches (14.3g)

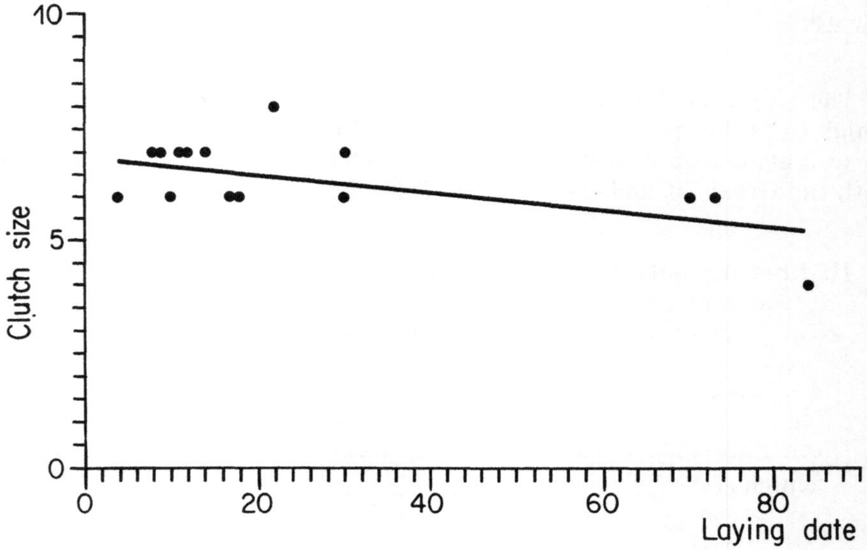

Fig. 2. Relation between laying date and clutch size in 1st and 2nd clutches of the Coal Tit. Value 1 of laying date= 1st March, r=-0.582 (P<0.05), N= 16.

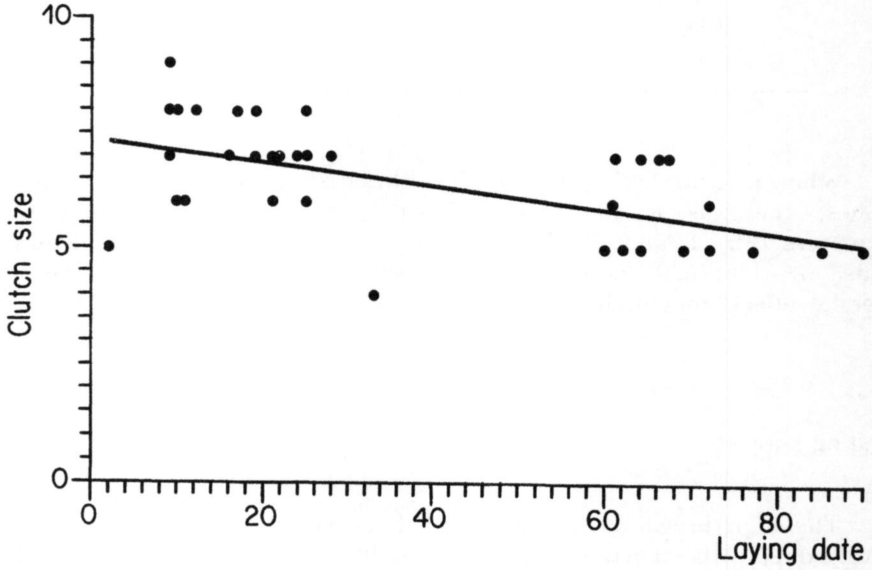

Fig 3. Relation between laying date and clutch size in 1st and 2nd clutches of Great Tit. Value 1 of laying date= 1st April, r=-0.589 (P<0.05), N=39.

is especially low when compared with the average adult weight (16.6g). I suggest that the low breeding success of the Great Tit is related to a very poor food supply for the young. Actually, the tits which died before fledging presented the lowest weight on the 15th day (8-13g). The breeding success was higher in second broods of Great Tit (49.4%, n=16) because of a lower mortality rate of young in the nest (36.6%). The average weight of nestlings was also higher in late broods (1st and 2nd broods combined). There was a significant correlation between weight and laying date (Fig. 4). Although late clutches were smaller than early ones I did not record any relation between nestling weight and clutch size.

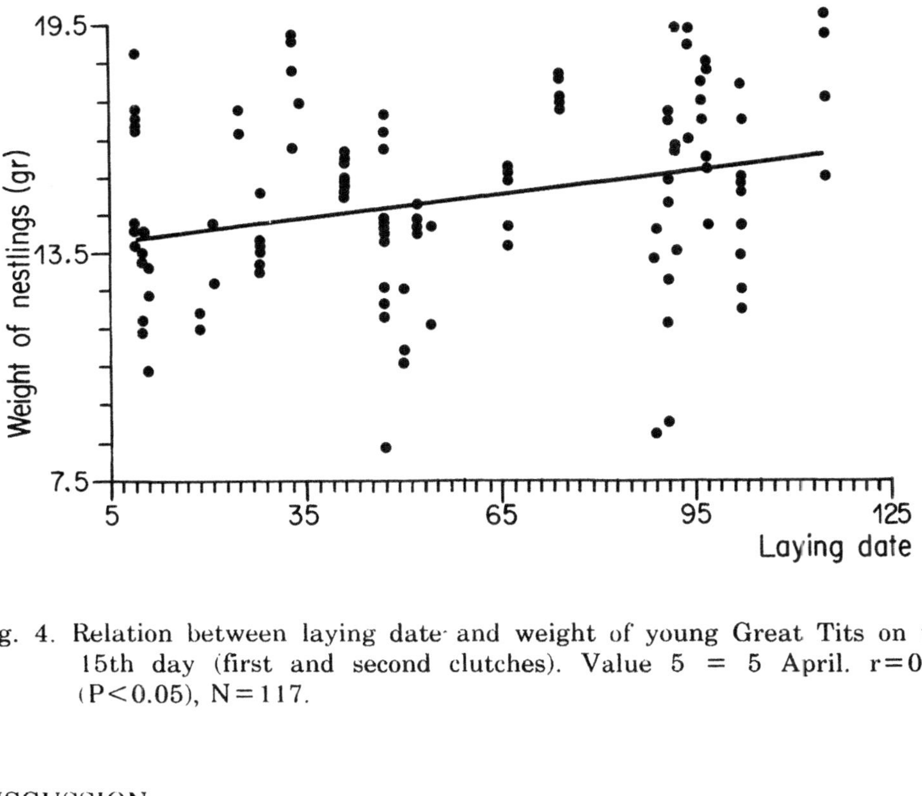

Fig. 4. Relation between laying date and weight of young Great Tits on the 15th day (first and second clutches). Value 5 = 5 April. $r=0.29$ ($P<0.05$), $N=117$.

DISCUSSION

These results point out striking differences in breeding patterns between the Coal Tit and the Great Tit in this study area. Breeding patterns (laying date and clutch size) of the Coal Tit seems to be better adapted to the environment than those of the Great Tit. This results in a better breeding success of the former species.

Since Great Tits have a low breeding success, especially in earlier broods, we should expect a larger investment in second and repeat broods. The latter are apparently determined by the failure of the first ones, as the birds started laying immediately after fledging the first brood.

I may suppose that the mechanism which determines the start of the breeding season of the Great Tit in my study area does not present a good adjustment to the peak of the food supply, a feature similar to that which has been described by van Balen (1973) in Pine woods in the Netherlands. Great Tits seem to adjust their breeding time to feeding conditions which prevail in deciduous woods. Indeed, several studies have shown that the food supply is much poorer and occurs much later in coniferous plantations than in oakwoods (Clamens et al. 1989, Leclercq 1977, Perrins 1965). Although I have no data on the variation of the food supply over the breeding season, it seems that the same situation may occur in our habitat.

As stated above, the timing of the breeding season of the Great Tit in San Jacinto does not differ significantly from that which has been reported in other habitats in Europe, either deciduous or coniferous. The difference in breeding success between first and second clutches also supports the hypothesis of a later and lower occurence of the food supply in pinewoods, which is in accordance with the higher fledging weights recorded in late broods.

Regarding the temperature effect on the laying date, our data do not allow us to look for relations between these two parameters. However, the mild climate with low thermic amplitudes where temperature is buffered by the effect of the ocean should result in a longer breeding season. Accordingly, the very early laying date of the Coal Tit should allow this species to produce second clutches as recorded by Blondel (1985) in Southern France. However, this is not the case in San Jacinto where few second clutches of Coal Tit are produced.

Two non-mutually exclusive hypotheses may account for this low proportion of second clutches in the Coal Tit:

1) The high breeding density of Coal Tit may increase competition for food resources.

2) Food shortage later in the season could limit the number of second clutches. However such a hypothesis does not explain the high success of late broods in the Great Tit.

On the average, I recorded in my study area small first clutches for both species, when compared with values recorded in northern Europe and the Mediterranean (Blondel et al. 1987, Orell and Ojanen 1983). This is partly because clutch size decreases with latitude. However, these low values of clutch size could also result from trade-offs between life-history traits, i.e. between fecundity and survival in poor food habitats, as suggested by many authors (Lack 1964, Blondel 1985, Orell and Ojanen 1983).

The decrease in clutch size as the season advances has been recorded in many tit populations. Usually this "calendar effect" on second clutches results in a lower breeding success, which is likely to be an adaptive response of tit populations, especially in deciduous habitats (Lack 1966, Perrins 1965, van Balen 1973). In San Jacinto the tendency for the clutch size to decrease during the breeding season seems to be counteracted by increasing food resources as noted by Isenmann (1983).

CONCLUSION

Many features of the breeding biology of tits in San Jacinto differ from those recorded elsewhere in deciduous oak woods in Europe. Two hypotheses may be put forward to explain these differences:

1) The food supply is very poor and occurs later in the season than in deciduous habitats,

2) Birds adjust their breeding time and eventually their clutch size as if they were in deciduous woods.

Future studies will focus on: 1) breeding patterns in oak woods, in order to compare the two kinds of habitats, and 2) measurements of food availability and its variation in deciduous woods and pine woods.

REFERENCES

Blondel J(1985) Breeding strategies of the Blue Tit and Coal Tit (*Parus*) in mainland and island mediterranean habitats: a comparison. Journal of Animal Ecology, 54: 531-556

Blondel J, Clamens A, Cramm P, Gaubert H and Isenmann P (1987) Population studies on tits in the mediterranean region. Ardea 75: 21-24

Clamens A and Isenmann P (1989) Effects of supplemental food on the breeding of Blue and Great Tit in Mediterranean habitats. Ornis Scandinavica 20: 36-42

Cowie RJ and Hinsley SA (1987) Breeding success of Blue Tit and Great Tit in suburban gardens. Ardea 75: 81-90

Dhondt AA, Eyckerman R, Moermans R and Huble J (1984) Habitat and laying date of Great and Blue Tit - *Parus major* and *Parus caeruleus*. Ibis 26: 388-397

Glutz von Blotzheim U (1962) Die Brutvögel der Schweiz

den Boer Hazewinkel J (1987) On the costs of reproduction: parental survival and production of second clutches in the Great Tit. Ardea 75: 123-126

Isenmann P (1983) A case of higher clutch size in second broods in Great Tit (*Parus major*) breeding in a mediterranean evergreen habitat. Le Gerfaut 73: 121-126

Järvinen O and Vaisänen R (1977) Line transect method: a standard for field work. Polish Ecological Studies 3(4): 11-15

Lack D (1964) A long term study of the Great Tit (*Parus major*). Journal of Animal Ecology, 33(supl): 159-173

Lack D (1966) Population Studies of Birds. Clarendon, Oxford

Leclercq B (1976) Etude expérimentale des facteurs limitant la densité des mésanges en forêt. Alauda 44: 301-318

Leclercq B (1977) Etude phénologique des paramètres liés à la reproduction des mésanges en futaie de chênes. La Terre et la Vie 31: 599-619

Orell M and Ojanen M (1983) Effects of habitat laying and density on clutch size of the Great Tit *Parus major* in northern Finland. Holarctic Ecology 6: 413-423

Perrins CM (1965) Population fluctuations and clutch size in the Great Tit, *Parus major* L. Journal of Animal Ecology 34(3): 601-647

Pina JP (1982) Avifauna nidificante de povoamentos artificiais de *P. pynaster* AITON e *Eucalytus globulus* LABIEL. Relatorio de actividades do curso de Eng Silvicultor ISA Lisboa

Rufino R, Neves R and Araujo A (*in press)* Passeriformes nidificantes na mata da Reserva Natural das Dunas de S.Jacinto. CEMPA Serv Nac Parques, Reservas e Conservação da Natureza, Lisboa

Van Balen JH (1973) A comparative study of the breeding ecology of the Great Tit *Parus major* in different habitats. Ardea 61: 1-91

LONG-TERM STUDY OF A NORTHERN POPULATION OF THE BLUE TIT *PARUS CAERULEUS*

Olavi HILDÉN

Department of Zoology, University of Helsinki, P. Rautatiekatu 13, SF-00100 Helsinki 10, Finland

Population studies of the Blue Tit *Parus caeruleus* have been made chiefly in Western and Southern Europe (e.g. Källander 1976, Perrins 1979, 1990, Blondel 1985, van Balen and Potting 1990, Dhont 1989, Dervieux et al. 1990). It may therefore be interesting to complete the picture by comparative data from the northernmost part of the species' present range, where the environmental conditions differ in many respects. The significance of new data from northern breeding areas is enhanced by the fact that the Blue Tit is a fairly new species there, after a rapid successful expansion. In southern Finland, where this study has been carried out, the species was an extremely scarce breeding bird at the beginning of the century (e.g. Palmgren 1913, Sundström 1927, Hildén 1986). Now it is almost as abundant as the Great Tit *Parus major* in the best habitats. The question rises, how well the Blue Tit has become adapted to this relatively recently colonized breeding area.

In this paper I summarize the main results of my long-term Blue Tit study. It is a preliminary report focused on presentation of a few basic reproductive and age structure parameters in the population, without any detailed general discussion of the results.

STUDY AREA AND METHODS

The study area lies on the southern coast of Finland, about 30 km west of Helsinki. In its present extent, the study has been in progress since 1972, i.e. during 18 years. There are ca. 250 wooden nest-boxes placed along routes in varying habitats, ranging from barren coniferous to rich deciduous forests. The diameter of the total nest-box area is ca. 5 km. In 1966-71, the study area was smaller and the number of boxes only 100-130; this material will be used chiefly for monitoring the population size.

Most of the nestboxes have an entrance hole of 32mm but in about 25 % of them, the entrance is only 28 mm, which prevents the Great Tit from occupying the boxes. Since 1972, the number and location of the nest-boxes have remained roughly the same, except for minor shifts caused by forest cutting, building of new houses and other environmental disturbances. Each year some old or damaged boxes have been replaced by new ones to keep the quality of the boxes constant.

The nest-boxes were checked 4-5 times each breeding season in order to record (1) the date of the first egg, (2) the size of the completed clutch, (3) the

NATO ASI Series, Vol. G 24
Population Biology of Passerine Birds
Edited by J. Blondel et al.

number of young hatched, and (4) the number of young fledged. In addition, most females and young, and part of the males, have been ringed. The annual population indices have been calculated by comparison of paired samples, i.e. the same nest-boxes in two consecutive years. For practical reasons, the data on replacement and second clutches are somewhat incomplete.

POPULATION TRENDS

In the late 1960s, the Blue Tit population was considerably lower than now (Fig. 1). During the early 1970s, characterized by exceptionally mild winters and warm summers, the population grew rapidly and reached its peak in 1973 (42 pairs). This peak was followed by a crash in 1974 (14 pairs), probably caused by the exceptionally heavy emigration that took place in the preceding autumn, the largest Blue Tit exodus ever recorded in Finland (Hildén, 1974 a, b). After this drop the population showed a steady increase over seven years, reaching in 1981 about the same level as in 1973 (41 pairs). A large drop again occurred in 1982, this time caused mainly by a very low breeding success in the cold and rainy 1981 summer (Hildén et al. 1982). The population recovered in three years, and the extremely cold 1984/85 winter did not slow down the increase. After another extremely cold winter in 1986/87, however, the population dropped to a third of its previous value (12 pairs), but was quickly recovered during the two following years (cf. Hildén 1989).

Although the population peaks in my area have remained at the same level since 1973, the total Finnish population has shown a continuous steep increase since the late 1960s (Fig. 2). This suggests that the population in southernmost Finland, at least in the best habitats, has already reached a saturation level, whereas the populations breeding further north are still increasing in numbers. This is supported by the rapid extension of the breeding range to the north in the 1970s and 1980s.

THE BREEDING SEASON

Egg-laying in the study area starts in late April or early May, a few days earlier than in the Great Tit. The mean laying date of the population has varied annually by more than two weeks during the 18 year period of study, ranging from 29 April in 1989 to 14 May in 1985, with an average of 7 May. Timing of the breeding season is clearly correlated with the ambient temperature: laying starts early in springs with a warm spell in late April - early May and is delayed when cold weather prevails during the period.

As a rule, a replacement clutch is laid if the first clutch is destroyed. Depending on the date of nest loss, replacement clutches are laid from mid-May to late June. The proportion of pairs laying a genuine second clutch varies much between years. In 1972 and 1977, no second clutches were recorded, in 1979

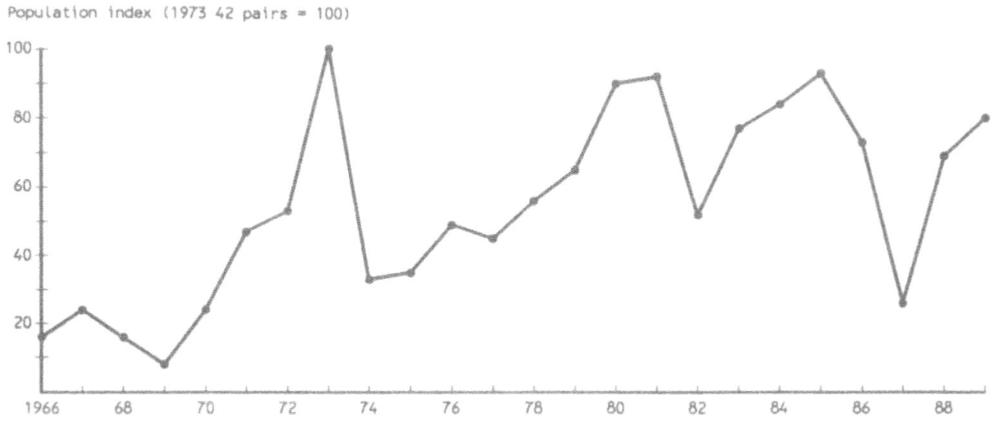

Fig. 1. Development of the Blue Tit population in an area of southern Finland from 1966 to 1989.

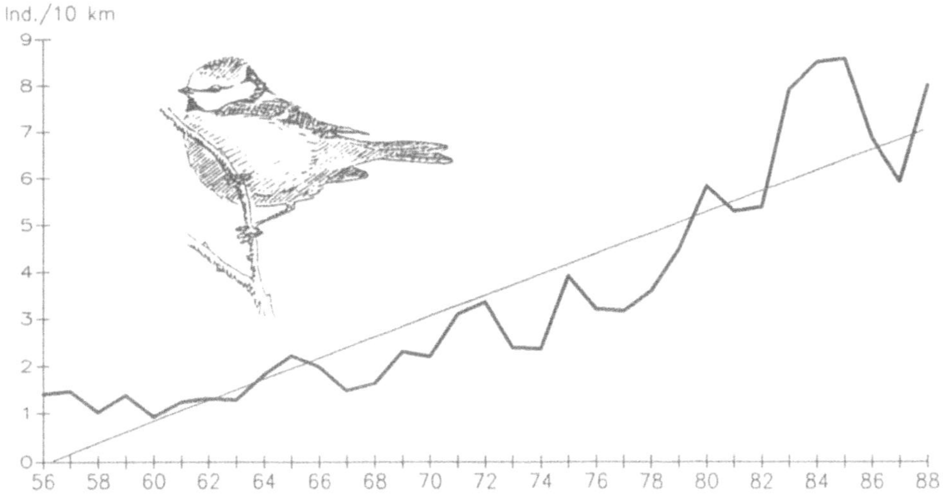

Fig. 2. Long-term population trend for the Blue Tit in Finland according to the national mid-winter censuses. Completed from Hildén (1988).

their proportion rose to about a quarter of the population. On average, about 10 % of the pairs lay a second clutch, which is a slightly higher proportion than in the Great Tit, contrary to the situation in Western and Central Europe (Perrins 1979, de Laet and Dhont 1989).

The earliest second clutches are laid in mid-June, immediately after the first broods have left the nest. In extreme cases laying of the second clutch may even start when the young of the first brood are still in the nest. The latest clutches have been initiated during the first half of July, in which case the young will leave the nest in mid-August.

CLUTCH SIZE

The mean size of first clutches within my study area has varied annually in rather narrow limits, from 10.71 (1986) to 11.81 (1989), averaging 11.23 ± 1.21 (Fig. 3). The annual variation is much smaller than that reported from population studies in western Europe (e.g. 8.4 to 13.4 in Oxford, Perrins 1990). The reason could be that the tits in northern forests feed on more diverse and thus less fluctuating food supply than the populations further south, where a few caterpillar species, widely fluctuating in numbers, seem to form the main part of the food brought to the young.

One-year-old females lay on average significantly smaller clutches than the older ones (Table I). Among individuals whose age is exactly known, 2-year-old females lay significantly larger clutches than yearlings, then the clutch size remains at the same level among 3-year-old birds but begins to decline among the females that are at least 4 years old.

The result is confirmed by an analysis of the clutch sizes of the same individually known females over successive years. In 39 such females, the mean clutch was significantly larger in the second breeding year (11.82 ± 1.30) than in the first one (11.18 ± 1.59); in 21 of the cases, the clutch was 1-4 eggs larger, in 9 cases identical, and in 9 cases one egg smaller. In 11 females checked in their second and third breeding year, no significant difference in the clutch size was found.

My results agree with those of Dhont (1989) from Belgium : in his area the clutch size increases by about 0.5 eggs between the ages of 1 and 2, does

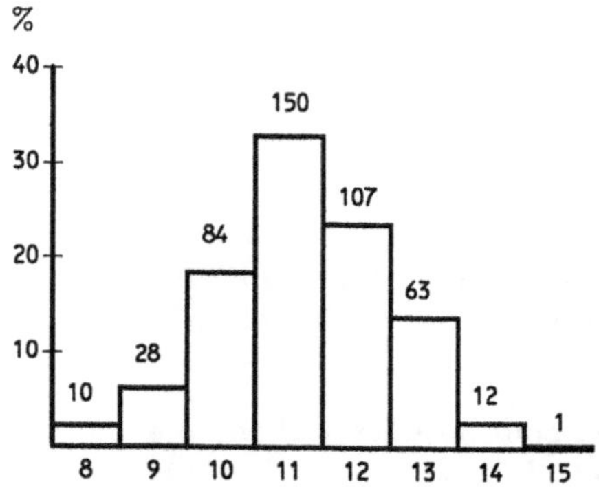

Fig. 3. Clutch size distribution of the Blue Tit in a population of southern Finland in 1965-1989 (first clutches only).

Table I. Clutch size of the Blue Tit in relation to the age of the female in southern Finland in 1972-89 (first clutches).

Age of the female	8	9	10	11	12	13	14	15	Mean±SD	n
			Clutch size							
1 yr	8	20	49	89	56	35	2	-	11.08±1.29	251[1]
≥2 yrs	2	6	31	59	43	23	9	1	11.41±1.26	174
2 yrs	-	1	9	11	15	5	4	1	11.65±1.34	46[2]
3 yrs	-	1	2	5	5	3	1	-	11.59±1.28	17
≥4 yrs	1	-	6	5	5	1	-	-	10.89±1.18	18[3]

1 The difference between these two age classes : $t=2.639$, $P<0.01^{**}$
2 The difference compared to 1-year-old females : $t=2.778$, $P<0.01^{**}$
3 The difference compared to 2-3 yrs old females : $t=2.216$, $P<0.05^{*}$

not change significantly between the ages 2 and 3, but then decreases significantly again once the females are 4 years old. In the Great Tit population of my study area, a significant decline in the clutch size occurs later, at the age of 6 years (Hildén 1981), but Perrins and Moss (1974) and Dhont (1989) could not show any significant decrease in clutch size with age in this species.

Figure 4 shows the relationship between mean clutch size and mean

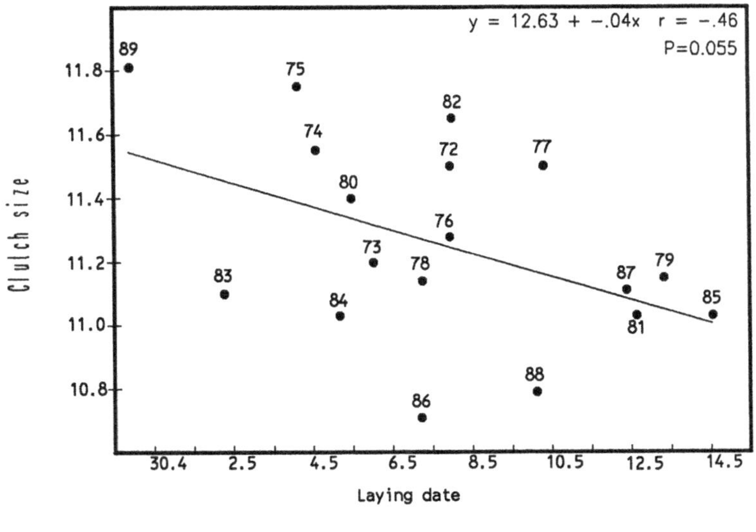

Fig. 4. Relationship between mean clutch size (first clutches) and mean laying date in a Blue Tit population of southern Finland 1972-1989. Numbers refer to years.

laying date in the population during 18 years. There is a slight tendency in the clutch size to be smaller in years when laying is delayed, but the significance of this trend is mainly due to one single year, 1989, when the breeding was exceptionally early and the clutches were large. Several years (e.g. 1977, 1982, 1983, 1984 and 1986) deviated strikingly from the trend. Hence, no clear "calendar effect" between the clutch size and laying date is found, in contrast to the Pied Flycatcher *Ficedula hypoleuca*, for example (von Haartman 1967, Järvinen 1983, Källander et al. 1987, Svensson 1987, Hildén and von Haartman 1987).

Within the breeding season, however, there is a marked decline in clutch size (Fig. 5). During the first 10 days of laying the mean clutch size is high, above 11.2, but declines thereafter steadily and rather steeply towards the end of the season, so that the replacement and second clutches laid from late June to early July usually contain 6 to 9 eggs only. The seasonal decline in clutch size is found in most other Blue Tit studies as well (e.g. Perrins 1979, van Balen 1989).

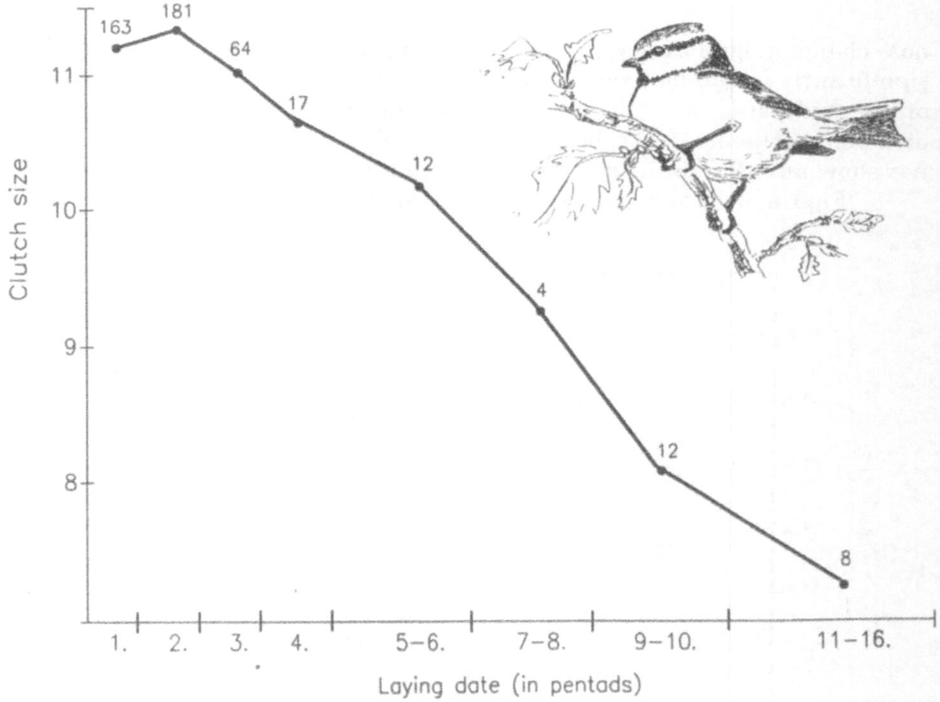

Fig. 5. Seasonal changes in clutch-size of the Blue Tit in a population of southern Finland. Data have been presented in 5-days averages, starting each year from the earliest clutch found and combining all the years 1972-1989. Numbers of clutches for each pentad are shown.

BREEDING SUCCESS

The total losses of eggs or young in first clutches have remained usually between 11 and 35 %, average 22.3 % (n=489). For replacement and second clutches the frequency of total losses has been the same, 22.4 % (n=85). The causes of nest losses in first clutches during the whole 18 year period are summarized in Table II.

The Great Spotted Woodpecker *Dendrocopos major* is by far the most common cause of nest failure. Its significance as a nest-robber is, however, totally dependent on the quality of the boxes used. Woodpeckers destroy mainly boxes that are old, decayed or have too thin walls (less than 2 cm). If boxes of poor quality were used, the predation would be even higher. Some woodpeckers are real specialists of nest-robbing, destroying within their territory all nest-boxes they are able to penetrate. On the other hand, by using woodpecker-proof boxes (made of cement and sawdust, for instance), the predation can be completely prevented.

In successful nests, from which at least one young is fledged, some eggs or young are often lost (Table III). More than 99 % of the egg-losses concern unhatched eggs, then only in four nests part of the eggs, 29 in total, disappeared during incubation. In replacement and second clutches egg losses are about the same but chick mortality is significantly higher than in the first clutches. The figures in Table III are likely to be slightly underestimated, because some unhatched eggs may disappear or remain unnoticed, and some large chicks may be taken by predators when peeking out of the hole.

The fledging success in my Blue Tit population during the 18 year study period can be summarized as follows. First clutches produce 7.37 fledglings per pair, on average. Annually the fledging success has varied from 3.1 (1987) to 9.4 (1975), mainly due to the predation rate of the Great Spotted Woodpecker and the weather conditions during May-June. For replacement clutches the mean number of fledglings is 5.14, for second clutches 4.0. Considering the average frequency of replacement (22 %) and second clutches (10 %) in the population, the final number of offspring produced per pair is 8.9 per year, on average.

SURVIVAL AND MORTALITY RATES

A simple way to estimate the average annual mortality of a bird species is to determine the proportion of first-breeders in a nesting population over several successive years. As the juveniles take the place of the adults dying each year, the percentage of individuals breeding for the first time should tend to equal the mortality of adults. For the Blue Tit in my study area, this method is applicable as the following conditions are met: (1) the first-breeders are easily distinguished from older birds by the colour of their primary coverts, (2) the adults are highly sedentary, (3) in spite of wide annual fluctuations there has

Table II. Causes of total losses of eggs or young in 109 unsuccessful Blue Tit first clutches in southern Finland in 1972-89.

Cause of the nest failure	No.	%
Destroyed by Great Spotted Woodpecker *Dendrocopos major.*	46	42
Deserted (mostly due to taking off the incubating female)	20	18
Destroyed by mammalian predators	15	14
All young dead (mostly starved)	12	11
Occupied by Pied Flycatcher *Ficedula hypoleuca*	7	6
All eggs unhatched	4	4
Occupied by Great Tit *Parus major*	1	1
Occupied by Wryneck *Jynx torquilla*	1	1
Occupied by bumblebees (*Bombus*)	1	1
Female died in the nest	1	1
Female "incubating" an empty nest cup	1	1
Total	109	100

Table III. Proportion of egg and chick losses in successful Blue Tit clutches in southern Finland in 1972-89.

	No. of nests	% with losses	No. of eggs/young	% lost
Eggs				
First clutches	381	32.8	4 279	5.7
Replacement and second clutches	68	32.4	552	6.7
Young				
First clutches	334	43.7	3 526	10.4
Replacement and second clutches	27	81.5	212	23.6

not been any definite long-term trend in the population, and (4) most pairs breed in optimal habitats (see Hildén 1982). However, the average mortality value so obtained should be considered only as an approximation.

In my Blue Tit population during 1972-89, 60 % of the breeding females (n=473) and 57 % of the males (n=289) were juveniles, on average. These figures estimate roughly, as mentioned above, the average annual mortality of adults. Although not statistically significant, the slight difference in mortality between the sexes is most likely real. Higher annual mortality of females, compared with that of males, has been found in most species so far studied, e.g. in the Great Tit (Kluyver 1951, Bulmer and Perrins 1973, Greenwood et al. 1979, Orell and Ojanen 1979, Hildén 1982).

An annual mortality rate of roughly 60 % fits well with the proportions of older age classes in the population (Table IV). The oldest breeding birds recorded so far (2 females) have been at least 7 years old. They both showed signs of senescence : one had only one chick and 10 unhatched eggs, the other "incubated" an empty nest cup. The longevity record of Blue Tits ringed in Finland is so far 8 years (Saurola 1988).

As mentioned above, each Blue Tit pair produces annually 8.9 fledglings, on average. Hence in a population of 100 pairs, 890 fledglings are produced. If 40 % of the 200 breeding birds survive to the next season and if the population remains stable, then 80 adults are left and 120 are replaced by first-year breeders. So 770 of the initial cohort of 890 juveniles die, which corresponds to a first-year mortality of more than 85 %, i.e. much higher than during later years. It is possible, however, that the mortality among yearlings is somewhat lower for two reasons. First, a fraction of the juveniles may be forced to emigrate to new breeding areas, and the emigration is not balanced by

Table IV. Proportions of different age classes among breeding adult Blue Tits in a population in southern Finland in 1976-89; recorded numbers compared with expected numbers with an average annual survival of 40 %.

Ringed as pulli or yearlings			Ringed as adults ($\geq$2 yrs old)		
Age	Recorded	Expected	Age	Recorded	Expected
2 yrs	66	..			
3 yrs	31	26	$\geq$3 yrs	22	..
4 yrs	6	10	$\geq$4 yrs	10	9
5 yrs	3	4	$\geq$5 yrs	3	4
6 yrs	-	2	$\geq$6 yrs	1	2
7 yrs	-	1	$\geq$7 yrs	2	1

a corresponding immigration. This would explain the rapid growth of the Finnish population and its expansion to the north. Second, as the population in southermost Finland is close to a saturation level, some juveniles may be excluded from breeding. Proof of this is lacking, however.

At what time of the year does the main first-year mortality occur? One could speculate that inexperienced young birds would have difficulty in coping with the severe winter conditions in the north and that the main mortality would take place during the coldest winter months, especially when considering that the Blue Tit is a southern species which only recently has colonized Finland. But this seems not to be the case. Results from the Finnish winter bird censuses show convincingly that Blue Tits survive with relatively small losses through the winter (Hildén 1988, 1989). Hence the species seems to have become well adapted to the harsh winters in the northern parts of its range.

Most likely the juvenile mortality is heaviest soon after the broods have left the nest, as in most birds, and again during the dispersal period in late September-October. Young Blue Tits in the northern parts of the range are very mobile, and in autumns of high population densities their movements grow to large-scale eruptions. In springs following heavy emigrations, lively migration occurs again in March-April (Hildén 1974 a,b). These migratory movements seem to play a central role in the population regulation of the Blue Tit in the north, contrary to the situation in more southern breeding areas where the species is mainly sedentary.

REFERENCES

Balen JH van and Potting R (1990) Comparative reproductive biology of the Blue Tit *Parus caeruleus* in four habitats in the Netherlands. This volume

Blondel J (1985) Breeding strategies of the Blue Tit and Coal Tit *(Parus spp.)* in mainland and island Mediterranean habitats: a comparison. J Anim Ecol 54: 531-556

Bulmer MG and Perrins CM (1973) Mortality in the Great Tit *Parus major.* Ibis 115: 277-281

Dervieux A, Isenmann P, Clamens A and Cramm P (1990) Breeding time and breeding performances of the Blue Tit *Parus caeruleus* in two Mediterranean habitats. This volume

Dhondt AA (1989) The effect of old age on the reproduction of Great Tits *Parus major* and Blue Tits P. *caeruleus*. Ibis 131: 268-280

Greenwood PJ, Harvey PH and Perrins CM (1979) Mate selection in the Great Tit *Parus major* in relation to age, status and natal dispersal. Ornis Fennica 56: 75-86

Haartman L von (1967) Clutch-size in the Pied Flycatcher. Proc XIV Int Ornithol Congr Oxford: 155-164

Hildén O (1974a) Finnish bird stations, their activities and aims. Ornis Fennica 51: 10-35

Hildén O (1974b) Occurrence of irregular migrants at Finnish bird stations in 1971-73 (Finnish with English summary). Lintumies 9: 97-111

Hildén O (1981) Nesting time and clutch size of the Great Tit *Parus major* in relation to the age of the female (Swedish with English summary). Proc Second Nordic Congr Ornithol Stavanger: 75-78

Hildén O (1982) Proportion of juveniles as a measure of adult mortality in a breeding population of Great Tits *Parus major* and Blue Tits *Parus caeruleus*. Ornis Fennica 59: 149-151

Hildén O (1986) Long-term trends in the Finnish bird fauna: methods of study and some results. Vår Fågelvärld, Suppl 11: 61-70

Hildén O (1988) Thirty years of Finnish winter bird censuses. Sitta 2: 21-57

Hildén O (1989) The effects of severe winters on the bird fauna of Finland. Memoranda Soc Fauna Flora Fennica 65: 59-66

Hildén O and Haartman L von (1987) A decade of population studies in box-nesting small Passerines in Finland (Finnish with English summary). Lintumies 22 : 70-75

Hildén O, Järvinen A, Lehtonen L and Soikkeli M (1982) Breeding success of Finnish birds in the bad summer of 1981. Ornis Fennica 59: 20-31

Järvinen A (1983) Breeding strategies of hole-nesting passerines in northern Lapland. Ann Zool Fennici 20: 129-149

Källander H (1976) Data on the breeding biology of the Blue Tit *Parus caeruleus* and the Marsh Tit *P. palustris* in southwest Scania (Swedish with English summary). Vår Fågelvärld 35: 1-7

Källander H, Karlsson J, Rosenlund N and Svensson S (1987) Clutch-size and breeding success of the Pied Flycatcher *Ficedula hypoleuca* in two habitats in South Sweden. Proc 5th Nordic Ornithol Congr Göteborg : 76-83

Kluyver HN (1951) The population ecology of the Great Tit *Parus major*. Ardea 39: 1-135

de Laet JF and Dhont AA (1989) Weight loss of the female during the first brood as a factor influencing second brood initiation in Great Tits *Parus major* and Blue Tits *P. caeruleus*. Ibis 131: 281-289

Orell M and Ojanen M (1979) Mortality rates of the Great Tit *Parus major* in a northern population. Ardea 67: 130-133

Palmgren R (1913) Helsingfors-traktens fagelfauna. Acta Soc Fauna Flora Fennica 38/2: 1-224

Perrins CM (1979) British Tits. Collins, London

Perrins CM (1990) An analysis of clutch-size in Blue and Great Tits. This volume

Perrins CM and Moss D (1974) Survival of young Great Tits in relation to age of female parent. Ibis 116: 220-224

Saurola P (1988) Bird ringing in Finland 1913-1987 (Finnish with English summary). Lintumies 23: 234-248

Sundström KE (1927) Ökologisch-geographische Studien über die Vogelfauna der Gegend von Ekenäs. Acta Zool Fennica 3: 1-170

Svensson S (1987) Nesting density and breeding performance of the Pied Flycatcher *Ficedula hypoleuca* near the tree line in Swedish Lapland. Proc Fifth Nordic Ornithol Congr Göteborg: 84-94

BREEDING TIME AND BREEDING PERFORMANCE OF THE BLUE TIT (*PARUS CAERULEUS*) IN TWO MEDITERRANEAN HABITATS

Alain DERVIEUX, Paul ISENMANN, Alex CLAMENS and Patrice CRAMM

Centre d'Ecologie Fonctionnelle et Evolutive L. Emberger (CNRS), BP 5051, F-34033 Montpellier-cédex, France

In a population of birds the date of egg laying and clutch size largely depend on the age structure, the genetical structure of the population in a given year, and the quality and variability of the habitat (van Noordwijk et al. 1981 a and b, Martin 1987). To investigate the last point, we studied a Blue Tit (*Parus caeruleus*) population breeding in two neighbouring habitats in mediterranean France (see Blondel et al. 1987, Isenmann et al. 1987 for details). The first habitat is dominated by evergreen Holm Oaks (*Quercus ilex*), the second by deciduous Downy Oaks (*Quercus pubescens*). The vegetation of the former habitat reflects the climatic constraints of the mediterranean zone where evergreen plants are widespread and dominant, whereas the latter habitat is more representative of temperate European habitats. The peak of food availability in spring occurs later, is smaller and lasts longer in the evergreen oak habitat than in the deciduous habitat (Cramm 1982, Isenmann et al. 1987, Clamens 1988). Blue Tits in the evergreen habitat actually lay fewer eggs per clutch and start to breed later than those in the deciduous habitat. In this paper, we analyse the differences in clutch size and breeding success between and within years as a clue of the explanations previously suggested to account for the between habitat differences in Blue Tit reproductive biology (Blondel et al. 1987, Clamens and Isenmann 1989).

MATERIAL AND METHODS

Both study sites are situated near Quissac (Gard, southern France, 43°54'N, 04°00'E), 100 m a.s.l.. Being 4 km apart, they have a similar climate. The first habitat is a 20 ha coppice of Holm Oaks with a few isolated Downy Oaks. A variable number (55-129) of Schwegler nestboxes has been erected since 1979. The second site is a wood of full grown Downy Oaks covering 10 ha, with 50 nestboxes erected since 1982. Two tit species breed there, the Blue Tit and the Great Tit (*Parus major*). We report here only the data collected on the former species.

All the nestboxes were visited at least once a week during the whole breeding season (March to June) in order to follow the laying period, clutch size, and breeding success of each breeding pair. Only the values from first clutches are used (192 clutches in the Holm Oak habitat from 1979 to 1988, 117 clutches in the Downy Oak habitat from 1982 to 1988), since second clutches

NATO ASI Series, Vol. G 24
Population Biology of Passerine Birds
Edited by J. Blondel et al.

were rare. Standard analysis of covariance F-tests were used to compare the regression lines of clutch size against laying date (Biomeco software, Lebreton et al. 1989). For the variation of the breeding success against laying date, we used program GLIM to fit a logistic regression, accounting for overdispersion (McCullagh and Nelder 1983, Payne 1985). Dates were noted in days from January 1st.

RESULTS

Annual means

Annual means are summarized on Tab. I and Fig. 1. The overall mean laying date in the evergreen Holm Oak site was 7.3 days later than that in the deciduous Downy Oak. The 10 annual values in the Holm Oak habitat were less clumped than the 7 values in the Downy Oak site. In Holm Oak, 17 days separated the earliest from the latest year (10-27 April), while in the Downy Oak this value was 12 days (5-17 April). In the Holm Oak site, Blue Tits laid on the average 2.3 eggs less than those in the Downy oak site. The overall breeding success (young fledged from eggs laid) did not differ significantly between sites.

Table I. Overall means of breeding parameters in the two habitats (± one standard deviation).

	Laying date	Clutch size	Breeding success	Number of years
Holm Oak	108.2±5.09	8.43±1.51	0.61±0.38	10
Downy Oak	100.9±4.16	10.75±1.71	0.62±0.41	7

Clutch size variation in relation to laying date

Table II shows a strong annual variability in the relationship between clutch size and laying date. In the Holm Oak site, the slopes of the regression lines are negative, except during two years (1979, 1983), and are not significantly different from each other. However, the year effect (time-lag in relation to the annual mean date of egg laying during the 10 years) was highly significant ($P<0.001$): clutch size decreased from early to late years (Fig. 2).

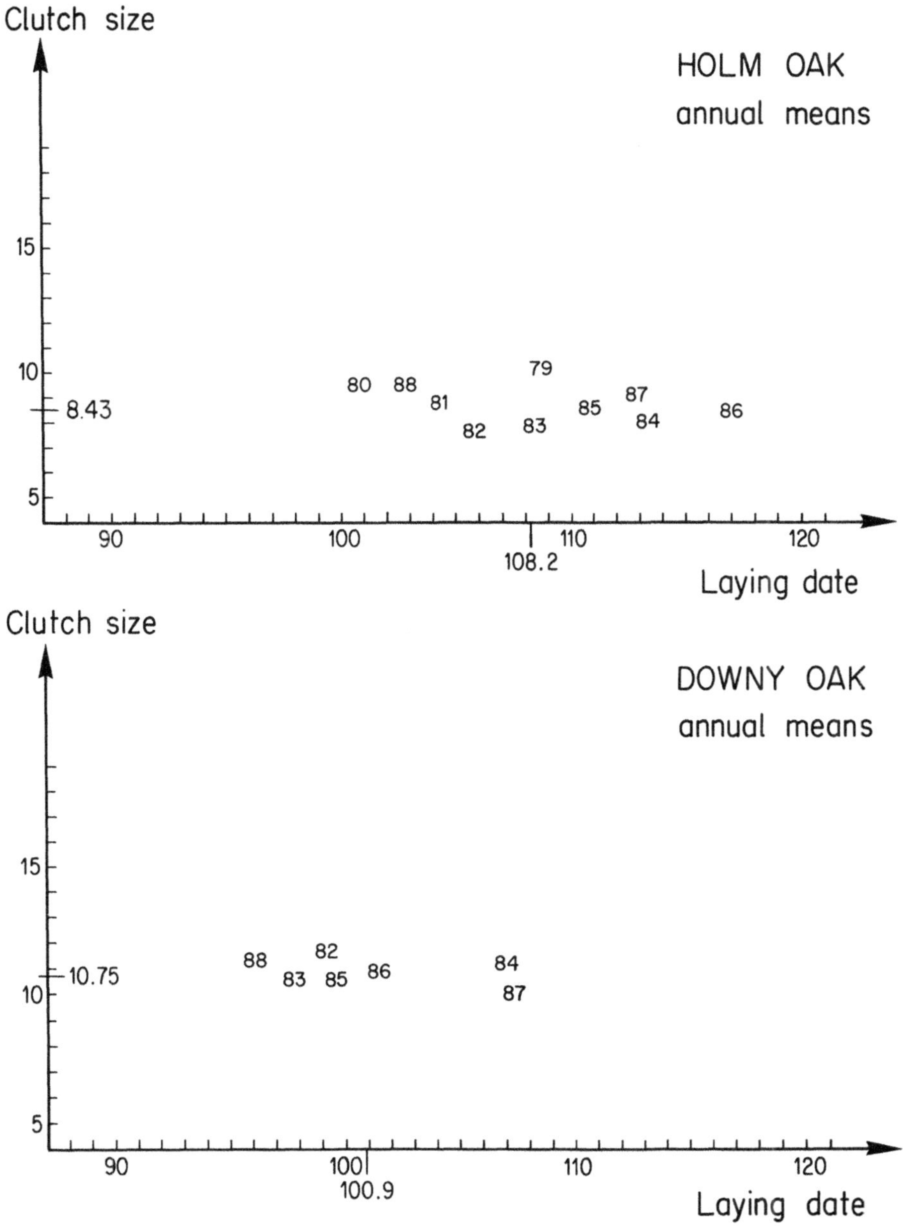

Fig. 1. Annual means of laying dates in the two study sites. Numbers refer to years. For laying date, 1 = 1st January.

Table II. Estimated regression lines of the relationship between clutch size and laying date in the two habitats (years are arranged from early to late ones).

	Intercept	Slope	N	Significance level
Holm Oak				
1980	29.98	-0.2055	9	**
1988	17.59	-0.0815	32	*
1981	11.88	-0.0313	16	NS
1982	10.51	-0.0287	25	NS
1983	5.31	+0.0222	21	NS
1979	-14.48	+0.2263	6	NS
1985	17.38	-0.0814	18	NS
1987	19.44	-0.0933	17	NS
1984	18.25	-0.0917	21	NS
1986	13.85	-0.0480	27	NS
Downy Oak				
1988	21.41	-0.1073	22	**
1983	30.95	-0.2104	19	NS
1982	26.16	-0.1485	12	***
1985	25.31	-0.1500	20	NS
1986	11.09	-0.0038	24	NS
1984	35.00	-0.2252	20	**
1987	47.87	-0.3559	13	**

In the Downy Oak habitat, the slopes were again negative (non-significant values occurred in three of the seven years) (Tab. II). The between year differences were significant ($P<0.05$), reflecting a more pronounced relationship between clutch size and laying date than in the Holm Oak site. The slopes of the two regression lines in Fig. 2 differred significantly ($P<0.01$). This indicated that in the Downy Oak habitat, clutch size decreased more rapidly in years with late laying date than in the Holm Oak habitat.

A seasonal reduction in clutch size was particularly obvious in all years, except 1979 and 1983, in the Holm Oak site (Table II). Mean laying dates in these two years (108.0 and 108.2) were very close to the overall mean (108.2 over 10 years). The same phenomenon occurred in the Downy Oak site in 1986 (with a mean laying date of 101.1 compared to the overall mean of 100.9 over 7 years). The within year effect was also studied in a more general analysis considering a "standard" year which included the whole set of laying dates.

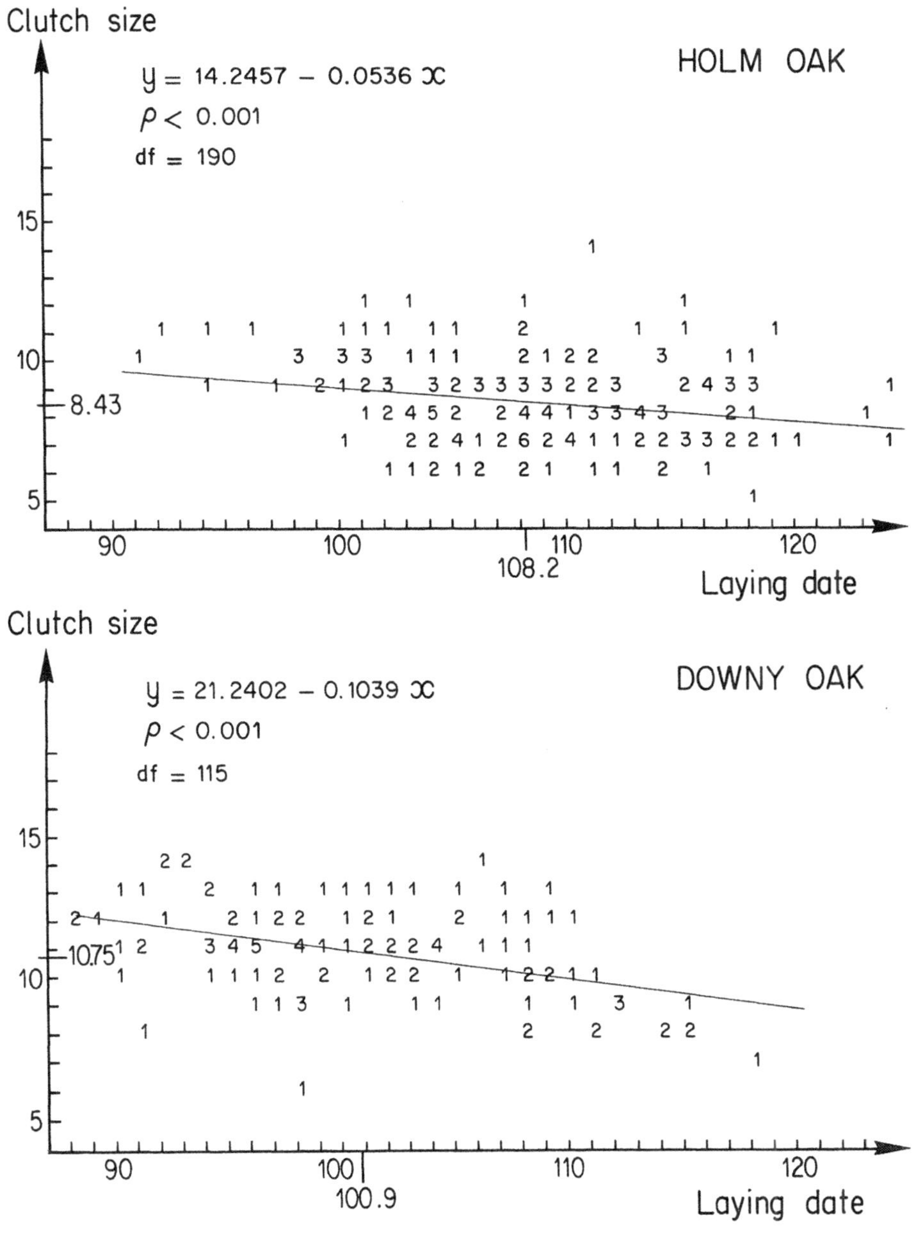

'ig. 2. Regression lines of the variation of clutch size in relation to laying date in the two study sites (8.43 and 10.75 are the overall mean clutch size and 108.2 and 100.9 the laying date in the two habitats respectively; numbers = multiple points). For laying date, 1 = 1st January.

Each laying date could then be compared to the mean laying date of the "standard" year. The slope of the regression line in the Holm Oak site was -0.0633 ($P<0.01$, $df=190$) and that in the Downy Oak site was -0.1392 ($P<0.001$, $df=115$). Again, clutch size decreased faster during the breeding season in Downy Oaks than in Holm Oaks.

Breeding success

To measure variation in breeding success, we used a logistic regression with an overdispersion factor of 6 to 8 (according to site). This resulted from a large variation in breeding success due to predation and desertion which induce a dependance of fate between chicks of a same clutch. This means that the variance in the ratio of fledged clutches to eggs laid was 6 to 8 times that of a binomial. The between year variation in breeding success differred between the two habitats (Fig. 3). In the Holm Oak site, the birds had a lower breeding success in early years than in late years. The slope of the logistic regression was positive, on the verge of significance: $F(1,181) = 3.78$ ($F = 3.90$ at $P = 0.05$). In the Downy Oak site, the slope of the logistic regression was negative: lower breeding success occurred in years when egg laying was late in the season. However, this trend was not significant: $F(1,109) = 1.428$ ($P>0.05$). These opposite trends at each site were obscured by a strong year effect (regression lines are parallel but shifted). The difference between the two sites was significant ($P=0.028$).

The within year pattern of breeding success is such that in the Holm Oak habitat, there is a significant seasonal increase of breeding success over time whereas in Downy Oaks, there is a non significant seasonal decrease. Thus, the within year variation is similar to the between year variation.

Extreme years

Table III illustrates the range of variation between extreme years in Holm Oaks and in Downy Oaks; the values fit with the trends mentioned above. In the Holm Oak habitat (Fig. 4), the trend of decrease in clutch size was significant in 1988 (a year with an early laying date), but not in 1986 (a late year). In the Downy Oak site (Tab. III, Fig. 4), the two slopes were significantly negative and the drop was steeper in 1987 (late laying date) than in 1988 (early laying date). The differences in slopes between the early and the late year are significant. For example in the Holm Oak site, the difference between the extreme years was 14.2 days and the mean clutch size decrease was 1.03 eggs. In Downy Oaks a difference of 11.1 days induced a drop of 1.31 eggs per clutch (Tab. III). The mean breeding success was lower in the early than in the late year in the Holm Oak site, whereas it was the opposite in Downy Oaks (Tab. III).

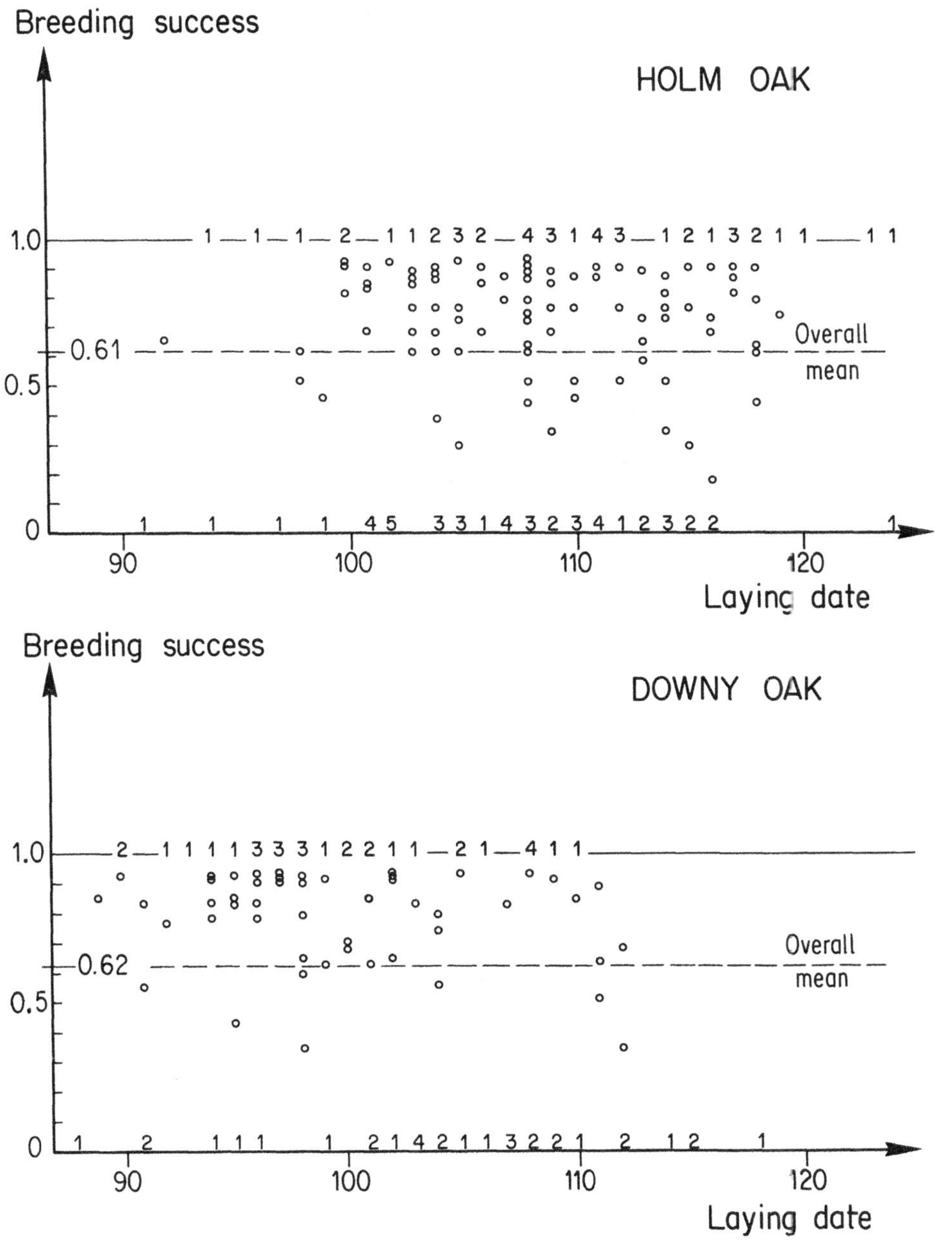

Fig. 3. Variation of breeding success in relation to laying date (numbers = multiple points). For laying date, 1 = 1st January.

Table III. Range of variation of extreme years in the two habitats.

		Laying date	Clutch size	Breeding success	N
Holm Oak	1988	102.4	9.29	0.67	32
	1986	116.6	8.26	0.82	27
Downy Oak	1988	95.7	11.15	0.65	22
	1987	106.8	9.84	0.51	13

DISCUSSION

Kluijver (1951) in The Netherlands, Lack (1966) and Perrins (1965, 1970) in Great Britain, and Neub (1977) in Germany have shown that clutch size and breeding success in Great Tit and Blue Tit decrease with the date of egg laying. Since females lay eggs when they can afford the energetic cost of egg formation (Perrins 1970), this decrease suggests that the synchronization between the nestling period and that of the peak in spring food availability is not necessarily optimal. This feature has its origin in the too rapid growth of feeding resources in spring which does not allow the birds to lay early enough under contrasted temperate climates (Perrins 1970).

Our data show the existence, in these two mediterranean habitats, of a seasonal reduction in clutch size of the Blue Tit, like that existing in northern countries. This reduction is, however, less marked in the typical mediterranean habitat of Holm Oaks. In the Downy Oak habitat, clutch size reduction is more important in years with the latest laying dates, suggesting that females could "forecast" the rapid decrease in feeding resources that immediately follows the food peak. However, the exact mechanism through which females could do so remains unknown (see also the model of Perrins 1970). In the Holm Oak site, by contrast, the slower decrease of feeding resources leads to a less marked reduction in clutch size (Cramm 1982, and in Blondel et al. 1987). Moreover, it is less marked in years with late laying dates. Late laying birds seem to be better synchronized to the availability of food resources. This supports the results of Cramm (1982, and in Blondel et al. 1987) who showed that Blue Tits lay slightly too late in Downy Oaks and slightly too early in Holm Oaks if the spring food peak and the maximum food requirements of the young were synchronized.

The mediterranean populations do not show a reduction in breeding success in relation to the date of egg laying. This suggests that there is a better concordance between the nestling period and the spring food peak in these habitats than in non-mediterranean deciduous habitats. This might be explained by a wider spread of feeding resources in spring and a probably higher level in

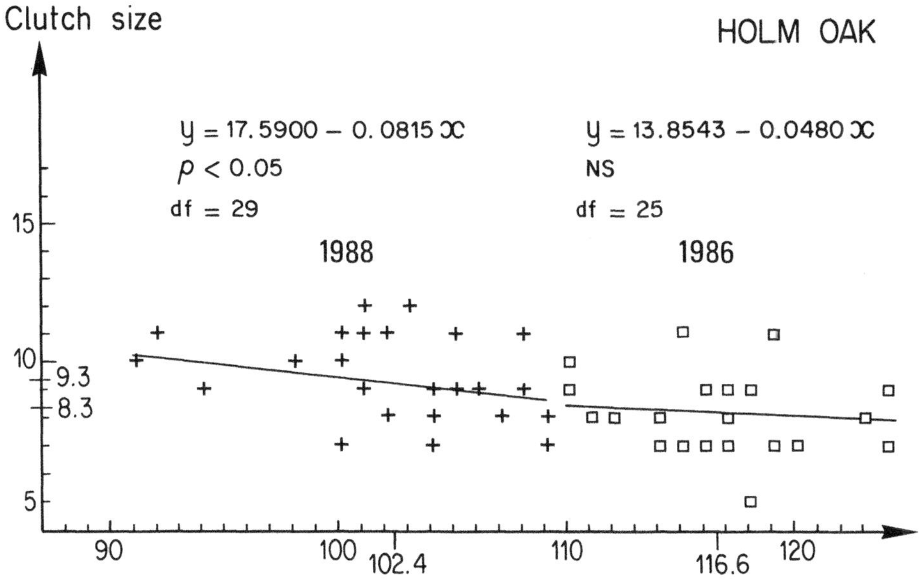

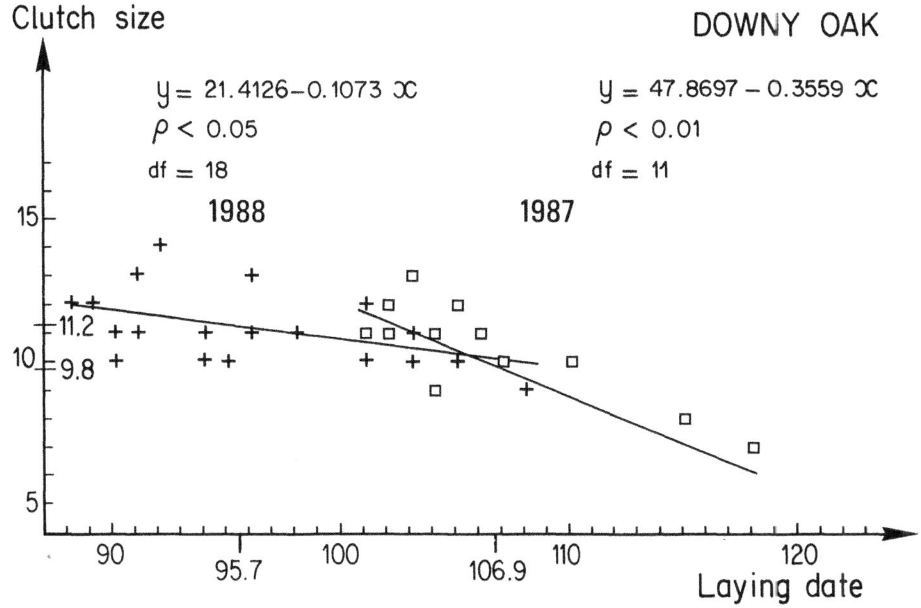

Fig. 4. Regression lines of clutch size in relation to laying date in the two extreme years recorded respectively in the two habitats. For laying date, 1 = 1st January.

feeding resources in winter than in non-mediterranean habitats (Blondel 1969), which allows the females to lay early enough (Clamens 1988). In the Holm Oak site, the breeding success becomes even better in years with late laying dates, because late layers feed their young when food is still abundant enough, because the food resources decrease slowly. Moreover, they have a lower number of nestlings to raise, because of their lower clutch size. The data available suggest the existence of a north-south gradient in the breeding period of european tits. Towards the north, the nestling period is generally slightly too late in relation to the food peak, which appears too fast. In the south, the synchronization seems to be better realized. This gradient is linked to more and more contrasted climates and feeding resources in relation to latitude.

ACKNOWLEDGEMENTS

We thank J. Thompson who did a linguistic revision of our manuscript as well as J. Blondel, J.L. Martin and two referees who made useful comments. J.D. Lebreton also improved our manuscript and helped us in the statistical analysis.

REFERENCES

Blondel J (1969) Sédentarité et migration des oiseaux dans une garrigue méditerranéenne. Terre et Vie 23:269-314

Blondel J, Clamens A, Cramm P, Gaubert H, Isenmann P (1987) Population studies on Tits in the mediterranean region. Ardea 75:21-34

Clamens A (1988) Déterminants écologiques de la date de ponte et de la grandeur de la ponte de la Mésange bleue (*Parus caeruleus*): le cas de la chênaie de Chênes verts. Thèse Univ Montpellier

Clamens A, Isenmann P (1989) Effect of supplemental food on the breeding of Blue and Great Tits in mediterranean habitats. Ornis Scand 20:36-42

Cramm P (1982) La reproduction des mésanges dans une chênaie verte du Languedoc. L'Oiseau et RFO 52:347-360

Isenmann P, Cramm P, Clamens A (1987) Etude comparée de l'adaptation des mésanges du genre *Parus* aux différentes essences forestières du Bassin méditerranéen occidental. Rev Ecol, Suppl 4: 17-25

Kluijver HN (1951) The population ecology of the Great Tit *Parus major*. Ardea 39:1-135

Lack D (1966) Population studies of birds. Clarendon Press, Oxford

Lebreton JD, Roux M, Bacou AM, Banco G (1989) BIOMECO (BIOMétrie-ECOlogie), version 3.3, logiciel pour compatible PC. CNRS, Montpellier

McCullagh P, Nelder JA (1983) Generalized linear models. Chapman and Hall, New York

Martin TE (1987) Food as a limit on breeding birds: a life history perspective. Annual Rev Ecol Syst 18:453-487

Neub M (1977) Evolutionsökologische Aspekte zur Brutbiologie von Kohlmeise (*Parus major*) und Blaumeise (*Parus caeruleus*). Thesis Univ Freiburg-im-Breisgau (W Germany)

van Noordwijk A, van Balen JH, Scharloo W (1981a) Genetic and environmental variation in clutch size of the Great Tit. Netherlands J Zool 31:342-372

van Noordwijk A (1981b) Genetic variation in the timing of reproduction in the Great Tit. Oecologia (Berlin) 49:158-166

Payne CD (ed) (1985) The GLIM System Release 3.77. Numerical Algorithms Group, Oxford

Perrins CM (1965) Population fluctuations and clutch-size in the Great Tit (*Parus major*). J Anim Ecology 34:601-647

Perrins CM (1970) The timing of birds' breeding seasons. Ibis 112:242-255

MODELLING DENSITY DEPENDENCE, ENVIRONMENTAL VARIABILITY, AND DEMOGRAPHIC STOCHASTICITY FROM POPULATION COUNTS : AN EXAMPLE USING WYTHAM WOOD GREAT TITS

Jean-Dominique LEBRETON

Centre d'Ecologie Fonctionnelle et Evolutive, CNRS, BP 5051,
34033 Montpellier-cédex, France

The relative role of density-dependence and of environmental variables in the fluctuations of animal populations was the subject of a well-known debate in the 1950's (Lack, 1954; Andrewartha and Birch, 1954). The present consensus (e.g., Royama, 1977), which applies also to plant populations (e.g., Harper, 1967, p.21), can be summarized roughly as follows :

-(1) long term stability in numbers can only result from density-dependence, which acts as a negative feedback (in cyberneticians' language);

-(2) large short term variations can be induced by environmental variability, which thus may mask the effects of density-dependence;

-(3) population numbers result from the interplay of these two mechanisms in a process which is intrinsically multiplicative over time. As a consequence population numbers rarely follow simple patterns. Great Tit *Parus major* counts in Wytham wood (Oxfordshire) provide a striking example of this complexity (Fig.1).

Spatial subdivision of populations (see e.g. Chesson, 1978) and differences between various types of individuals (e.g., immigrants vs residents, Clobert et al. 1988) lead to further interactions between density-dependent and density-independent mechanisms. However, density-dependence and environmental variability have rarely been considered simultaneously in population models. The influence of variability in nonlinear dynamic models has been largely neglected (Koehl, 1989, p.47, but see Roughgarden, 1975). Even the ad hoc regression approaches to detection of density-dependence from a series of counts (e.g. Morris, in Salt, 1966) do not include environmental variables (see review by Lebreton, 1989).

The goal of this paper is to propose models considering simultaneously density-dependence and dependence on environmental variability (environment-dependence), and to check their ability to predict population changes. We use as few assumptions as possible, in an attempt to provide minimal models for density-dependence and environment-dependence. These models are applied to counts and environmental data of the Great Tit population at Wytham wood (Oxfordshire) (Perrins, 1965, 1971, 1979). The possible use of a series of counts in population biology is discussed. A large part of the results presented here were produced for an unpublished thesis (Lebreton, 1981) (see also Lebreton, 1982a).

NATO ASI Series, Vol. G 24
Population Biology of Passerine Birds
Edited by J. Blondel et al.

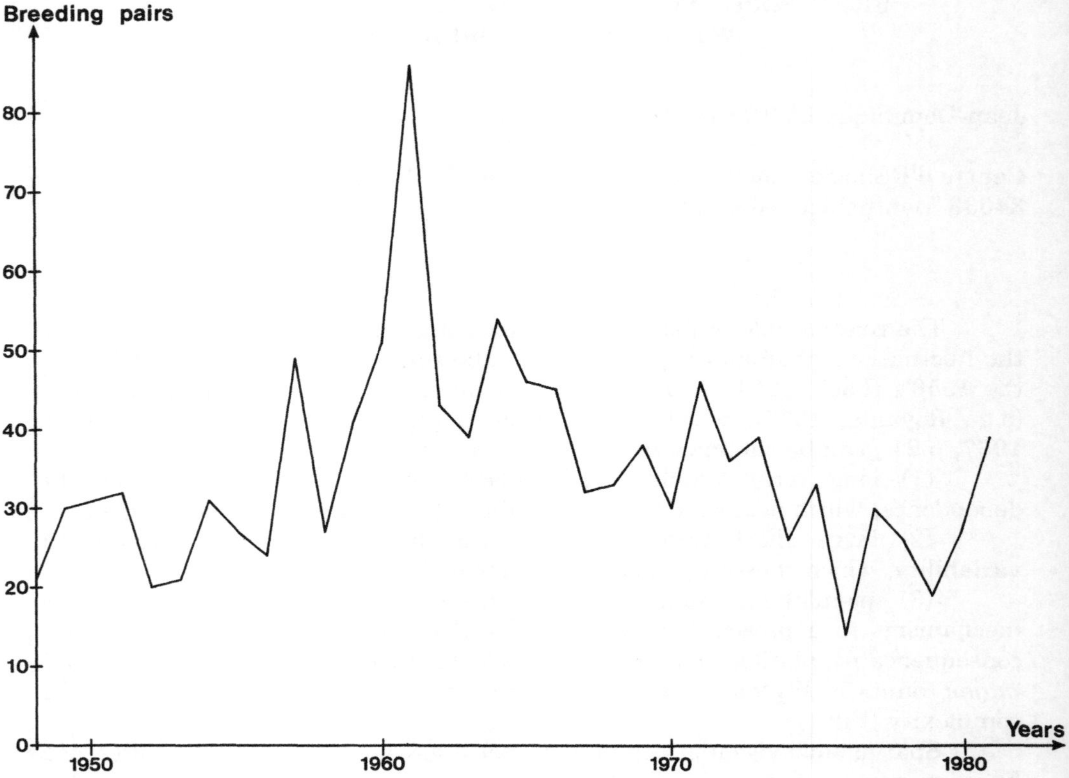

Fig. 1. Number of breeding pairs of Great Tit *Parus major* at Wytham wood (Oxforshire), 1948-1981 (Perrins, 1979 and pers. comm.).

MODELS FOR DISCRETE TIME POPULATION GROWTH

One of the simplest models of population growth is the (discrete time) exponential growth model $N_t = a^t N_0 = \exp(rt) N_0$. The multiplication rate a or equivalently the growth rate $r = \mathrm{Log}(a)$ synthesize the balance between death and recruitment. The recurrence relationship $N_{t+1} = a N_t$, or $N_{t+1} = \exp(r) N_t$ is inherent in this model.

Many density-dependent models can be obtained by making a dependent on N_t (May, 1975, p.9). Among those, the model defined by the recurrence relationship $N_{t+1} = \exp(r_m + bN_t) N_t$, with $b<0$, is a discrete time generalization of the logistic growth model (model B of May, 1976, p.10). In such models, an equilibrium level can be obtained only if the curve which links N_t to N_{t+1} crosses the first diagonal $y=x$ (Fig.2). Such models are particularly noteworthy for the variety of behaviour they present with high values of r_m

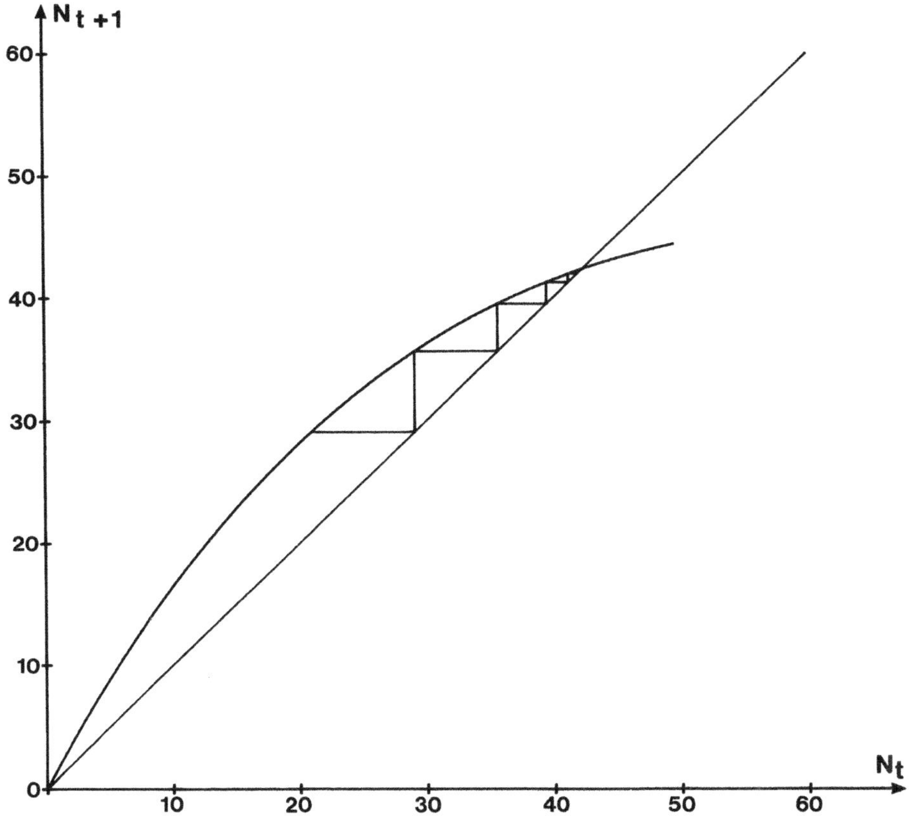

Fig. 2. The density-dependent growth model defined by the recurrence equation $N_{t+1} = N_t \exp(.6485 - .01530\ N_t)$ (horizontal axis: N_t; vertical axis: N_{t+1}). The successive steps between the line $y=x$ and the line $y=x \exp(.6485 - .01530\ x)$ give the successive populations sizes N_0, N_1, N_2,...(here with $N_0=21$), which converge to the equilibrium density $N_\infty \simeq 42.23$.

(periodic behaviour, chaotic behaviour) (May, 1975, p.12-13), which, however, is not of concern here because of the low r_m value observed in vertebrate populations (Lebreton, 1982B).

Year-to-year environmental variability fits easily into this framework. Each time step can be considered as characterized by its own r_m and b values, i.e. by a particular curve linking N_t to N_{t+1} under the environmental conditions prevailing for this particular time step. The particular case of alternating good and poor years is illustrated in Fig.3. If environmental variability is entirely

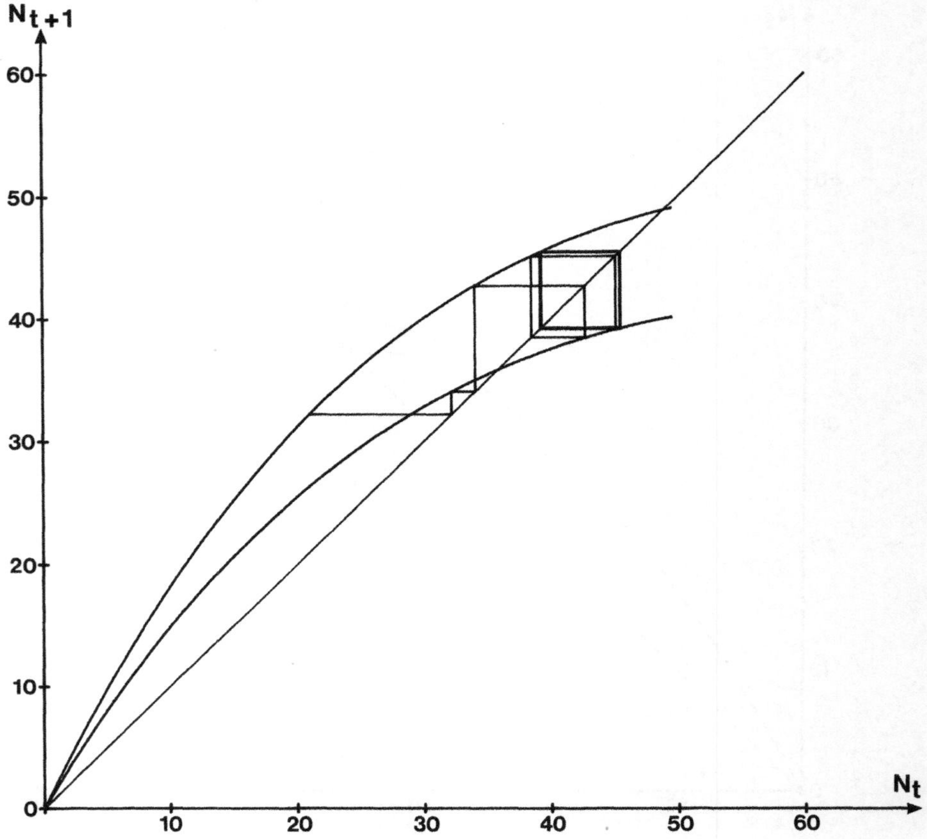

Fig. 3. The density-dependent growth model defined by the alternance of two recurrence equations :$N_{t+1} = N_t \exp(.7485 - .01530\ N_t)$, and $N_{t+1} = N_t \exp(.5485 - .01530\ N_t)$ (horizontal axis: N_t; vertical axis: N_{t+1}). The successive steps give the successive populations sizes N_0, N_1, N_2,..., here with $N_0=21$. Note the convergence to a cycle of period 2. This is a model for the alternance of a good and of a poor year.

dependent on a measurable variable Z_t, r_m and b can be written as simple functions of Z_t. With linear functions, this leads to the model

$$N_{t+1} = N_t \exp(a_0 + a_1 Z_t + b_0 N_t + b_1 Z_t N_t).$$

This model includes an interaction term $b_1 Z_t N_t$ between environmental variability and density-dependence, which represents a variation in the intensity of density-dependence across various environmental conditions. A simpler model considers additive effects :

$$N_{t+1} = N_t \exp(a_0 + a_1 Z_t + b_0 N_t),$$

with, as a straightforward multivariate generalization to p variables Z_{1t}, Z_{2t},..., Z_{pt} :

$$N_{t+1} = N_t \exp(a_0 + \sum a_i Z_{it} + b_0 N_t)$$

DEMOGRAPHIC STOCHASTICITY

However, even for a given population size N_t and a given environment $Z_{1t}, Z_{2t},..., Z_{pt}$, and even in the absence of measurement error, one cannot expect N_{t+1} to be fully determined. At most, the model will predict the expected population size, conditional on the population size N_t and the environment $Z_{1t}, Z_{2t},..., Z_{pt}$ prevailing at time t (denoted by F_t) :

$$E(N_{t+1}/F_t) = N_t \exp(a_0 + \sum a_i Z_{it} + b_0 N_t).$$

There remains for instance a stochastic component in the death process, which ultimately has a binomial structure similar to that of coin tossing. This variation is called demographic stochasticity. As noted by Chesson (1978), it can be thought of as a residual variability in population dynamics models, since it would affect ideal replicates of populations under the same internal (density) and external (environment) conditions. In practice this residual variability partly reflects the influence of systematic sources of variability not included in the model of interest, e.g. environmental variability resulting from unrecorded variables, or differences in demographic parameters across types of individuals.

The demographic stochasticity can be modelled in a straightforward way by considering that each individual i is replaced next year by a random variable X(i), independently of other individuals. N_{t+1}, conditional on F_t is thus the sum of N_t independent and identically distributed random variables :

$$N_{t+1} = X(1) + ... + X(i) + ... + X(N_t)$$

This is called the branching property and models using it are called branching processes. They have been developed mainly as models of density-independent population growth (see e.g. Jagers, 1975). The distribution of X(i), the number of individuals replacing individual i, is called the offspring distribution. This vocabulary is used only for convenience, the individual i being among the individuals which could possibly "replace" itself. We consider here the case where the offspring distribution at time t depends on the state of the population-environment system at time t, i.e. on N_t and $Z_{1t}, Z_{2t},..., Z_{pt}$. Similar models are briefly considered by Zeger and Qaqish (1988, p.1022).

For the sake of simplicity, we will restrict our attention to the case where X(i) is Poisson. N_{t+1}, as a sum of independent Poisson distributions, is also distributed as Poisson (Johnson and Kotz, 1969 p.93). The models we consider are thus :

$$N_{t+1} = \text{Poisson}\ (N_t \exp(a_0 + \sum a_i Z_{it} + b_0 N_t)\ ),$$

conditional on F_t. These are among the simplest models considering simultaneously density-dependence, environment-dependence and demographic stochasticity. Although the interactions among individuals are reduced to the effect of N_t on N_{t+1}, population regulation is modelled in a stochastic context that considers two main sources of biological variability, rather than as a deterministic process with noise added to it. Like other branching processes the models we consider can be extended to several classes of individuals, e.g. age classes, as a stochastic generalization of the density-dependent Leslie matrix models considered e.g. by Beddington (1974).

SOME PROPERTIES OF DENSITY-DEPENDENT BRANCHING PROCESSES

The mathematical study of the previous models is made difficult by their density-dependent component. Few theoretical results are available. This presentation intends to be an informal outline of the main results.

The first striking feature of these processes is that density-dependence, if stabilizing (i.e. if $b_0<0$), implies the certainty of extinction (Lebreton, 1981). However, at density N_t, the probability of extinction during the next time step is

$$P(N_{t+1}=0/F_t)=\exp(-N_t \exp(a_0 + \sum a_i Z_{it} + b_0 N_t)),$$

which most of the time will be negligible. However, density-dependence, by preventing population explosion, i.e. by keeping N_t finite, makes the *cumulative* probability of extinction over time tend to one. Despite the certainty of extinction, N_{t+1} tends to a stationary distribution conditional on nonextinction, called a quasistationary distribution. Thus the fluctuations around the equilibrium tend to stabilize as long as the population is not extinct. This can be proved formally for the simple model

$$N_{t+1} = \text{Poisson}\ (N_t \exp(a_0 + b_0 N_t))$$

without environmental variability (Lebreton, 1981), and by approximation methods in the general model (Lebreton, unpubl.).

FITTING POPULATION GROWTH MODELS

The relationship

$$\text{Log}(N_{t+1})-\text{Log}(N_t)=a_0 + \sum a_i Z_{it} + b_0 N_t,$$

deduced from the deterministic model, is not amenable to standard linear regression which neglects interdependence over time, even in the absence of measurement error (Ito, 1972; Lebreton, 1989). A first possibility is to use time-series approximations (Bulmer, 1975). However, the estimation of parameters may directly rely on maximum likelihood estimation in the density-dependent branching process.

The dependence over time makes the study of the likelihood less straightforward than for independent observations. However, the existence of a quasistationary distribution, which is thus a key result, leads to the same properties in both cases. A more complete account, including technical references and proofs is given by Lebreton (1981, ch.4). Maximum likelihood estimators of a_0, a_1,..., a_p, b_0 benefit from usual properties (see e.g. Zeger and Qaqish, 1988), as was confirmed by simulation (Lebreton, 1981).

Fitting a model by the maximum likelihood method also gives a goodness-of-fit measure called the deviance. For a valid model M1 the deviance Dev(M1) is asymptotically distributed as X^2. This provides a goodness-of-fit test of model M1. Let us consider a model M2 nested in M1, i.e. with q *further* variables. Under the hypothesis that M1 is the valid model, the difference in deviance Dev(M1) - Dev(M2) is distributed as $X^2(q)$. This provides a test of the contribution of the q extra variables to population changes (likelihood ratio test). Such a test on variable N_t, i.e. on b_0, is a test of density-dependence allowing for simultaneous environment-dependence.

Fitting the models I considered by the maximum likelihood method can be done with GLIM (Payne, 1985). The procedure is similar to regression analysis through the origin for Poisson distributed data, (e.g. Frome et al., 1973). This requires a Poisson Error term , Log Link, $YVAR=N_{t+1}$, and $OFFSET=Log(N_t)$.

MATERIAL

Population counts of Great Tits in Oxford are among the few series of counts which at the same time cover a long period of time and are relatively precise. Counts at Wytham wood (Oxfordshire) from 1948 to 1981 were used here, together with various environmental variables listed in Table I. A Poisson distribution for the offspring distribution of each individual (including possible survival of the individual itself) seemed a reasonable approximation (Lebreton, 1981).

Table I. Variables used in the predictive models of Great Tit breeding numbers at Wytham Wood (Oxforshire) (data from Perrins, pers.comm.).

$GTM=N_t$	Great Tit breeding pairs, previous year
BTM	Blue Tit breeding pairs, previous year
BE	Beech mast crop index
BEM	Beech mast crop index, previous year
FEB	average temperature, February
MAR	average temperature, March
APR	average temperature, April
MAY	average temperature, May, previous year
JUN	average temperature, June, previous year
WS	sum of temperatures, January 1 - April 30
FS	sum of temperatures, February 1 - February 28
W16	sum of temperatures, April 16 - April 20
W21	sum of temperatures, April 21 - April 25
W26	sum of temperatures, April 26 - April 30
MS	sum of maximum temperatures, January 1 - April 30
FM	sum of maximum temperatures, February 1 - February 28

There is considerable redundancy in the environmental variables, by definition (e.g. temperature sums for overlapping periods) or as a result of statistical correlation. For instance Beech mast production tends to alternate over years; as a result BE (abbreviated names for variables are given in Table I) has a strong negative correlation with BEM. The redundancy is attested by a Principal Component Analysis on the standardized variables. The correlation circle for the first two axes (Fig.4) shows two groups of positively intercorrelated variables, respectively (FM, WS, MS) and (BE, MAY, JUN, FEB, -BEM). In particular, BEM being opposed to BE, MAY, JUN and FEB, -BEM is positively correlated with them. However, there is still much independent information in the 16 variables since these two first axes account only for 35.4 % of the total variability.

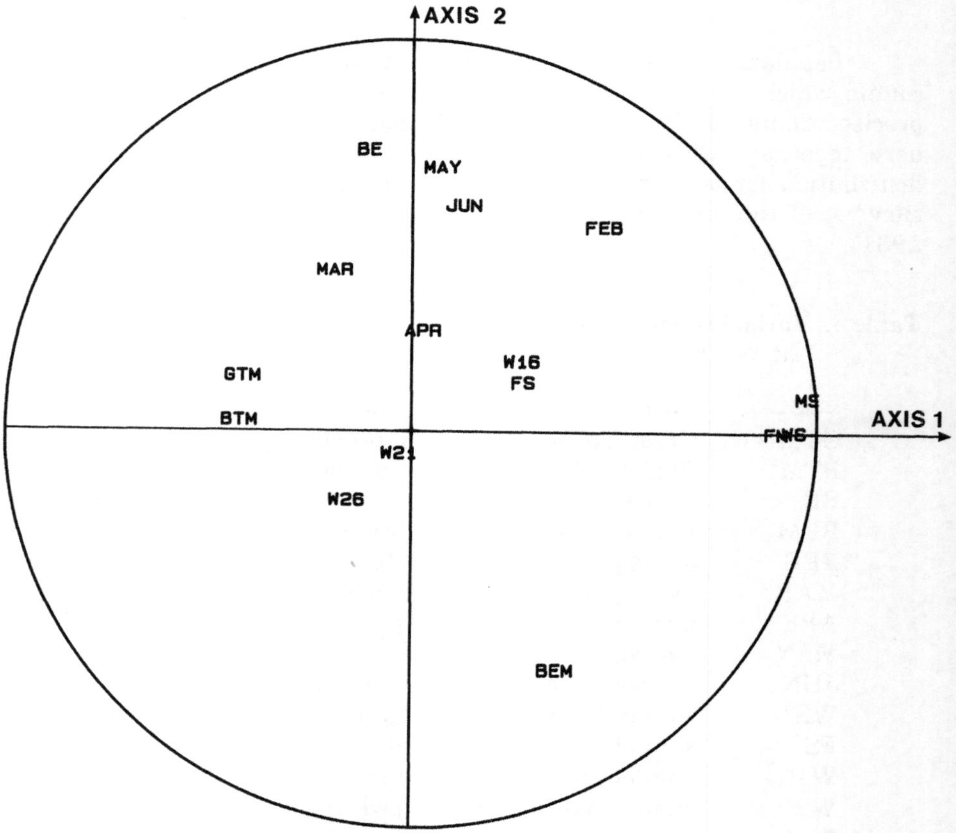

Fig. 4. Principal component analysis of the 16 standardized environmental variables at Wytham wood (Oxforshire) (data from Perrins, pers. comm.). Correlation circle of the variables with the first two axes. Definition and abbreviated names of the variables are given in Table I. The variables FM, MS, and WS form a first group of strongly correlated variables (on axis 1); BE, MAY, JUN and -BEM form a second group, on axis 2.

I introduced variables progressively in the population models on the basis of the largest decrease in deviance. In practice, each variable entered the model as a representative of its group of intercorrelated variables. This decreases the chance that other variables in the same group will contribute significantly to further reductions in deviance. Which variable in a group entered the model first depended partly on chance : another similar series of counts might have led to another order. Any kind of causal interpretation should thus be very cautious. As usual with observational rather than experimental data, there was no replication and many factors were confounded. For these reasons, tailoring a model to fit the data too narrowly would have been meaningless. The models were thus fitted using the first 25 years of data (1948-1972), the last 9 years (1973-1981) being used for cross validation. In all models, I used the number of breeding pairs, to model the female part of the population.

RESULTS

The results of the progressive incorporation of variables in the model are given in Table II. After having incorporated N_t (i.e. the observed population size to account for density-dependence), MS, BEM, W21, and FS, no noticeable improvement was obtained by incorporating further variables. Various results of this model as summarized in table III. The estimated b_0 (slope of N_t) is negative, as expected for density-dependence. The estimated slopes for the four environmental variables were positive, as expected for a production variable (BEM) and three temperature variables (MS, W21, FS).

However, this model was still not acceptable ($X^2(19)=35.73$, $P=.011$). Many reasons could have induced this lack of fit :

1) The functional relationship in our models is restrictive;

2) Variables not considered in the model can influence population growth;

3) The population of Wytham wood is not closed (Clobert et al., 1988), and a purely multiplicative model should thus behave poorly;

4) Lastly, differences between individuals and microhabitats tend to increase the residual dispersion.

Various checks with transformed variables and with interactions between variables did not improve the fit noticeably and the first cause of lack of fit was not further considered. Except perhaps for the problem of immigration, we were thus left with fluctuations which can largely be considered as random from year to year. Their effect may thus be added to the Poisson demographic stochasticity. We considered that the conditional distribution of N_{t+1} was affected by overdispersion and could no longer be considered Poisson. Its variance, instead of being equal to its mean, was proportional to it by a factor c^2. An estimate of c^2 is then the residual deviance divided by the degrees of freedom, i.e. $35.73/19=1.881$. This overdispersion factor can then be used as a denominator for F-tests as a residual variance with 19 degrees of freedom (McCullagh and Nelder, 1983, p.107; Zeger and Qaqish, 1988). In particular, if one accepts the model with overdispersion as valid, one can in turn compare each of the parameter estimates to 0. Such tests take account of overdispersion

Table II. Successive models of Great Tit breeding numbers at Wytham Wood (Oxforshire) (data from Perrins, pers.comm.).
Names of variables are given in Table I.
All models are branching processes N_{t+1} = Poisson (N_t exp(a_0 + $\sum a_i Z_{it}$ + $b_0 N_t$), with various choices of variables.
DEV = deviance; NP = number of parameters. The X^2 probabilities are overestimated because of the rough selection strategy chosen and of the lack-of-fit of models (see text). Hence no noticeable improvement over model (6) seemed possible.

MODEL FORMULA	DEV	NP	TESTS
(1) N_t exp(a_0) (exponential growth)	140.74	24	
			model (2) vs (1) $X^2(1)$= 51.82, P<10-5
(2) $a_0 + b_0 N_t$ (density-dependence without environment-dependence)	88.92	23	
			model (3) vs (2) $X^2(1)$= 25.19, P<10-5
(3) $a_0 + b_0 N_t + a_1$MS	63.73	22	
			model (4) vs (3) $X^2(1)$= 15.34, P=10-4
(4) $a_0 + b_0 N_t + a_1$MS + a_2BEM	48.39	21	
			model (5) vs (4) $X^2(1)$= 7.65, P=.006
(5) $a_0 + b_0 N_t + a_1$MS + a_2BEM + a_3W21	40.73	20	
			model (6) vs (5) $X^2(1)$= 5.00, P=.025
(6) $a_0 + b_0 N_t + a_1$MS + a_2BEM + a_3W21 + a_4FS	35.73	19	
			model (7) vs (6) $X^2(1)$= 1.97, P=.16
(7) $a_0 + b_0 N_t + a_1$MS + a_2BEM + a_3W21 + a_4FS	33.76	18	

Table III. Final model of Great Tit breeding numbers at Wytham Wood (Oxfordshire) (data from Perrins, pers. comm.). The names of variables are given in Table I. The F-Tests to compare the estimates to 0 (with 1 and 19 df) take account of overdispersion.

$$N_{t+1} = \text{Poisson}(N_t \exp(a_0 + b_0 N_t + a_1 MS + a_2 BEM + a_3 W21 + a_4 FS))$$

Coeff.	Variable	Estimate	Standard Error	F-test
a_0		-4.163	.9182	
b_0	Nt	-.01217	.002519	12.89 P<.01
a_1	MS	.00102	.0003269	5.22 P<.05
a_2	BEM	.1166	.03805	5.00 P<.05
a_3	W21	.006616	.002604	3.45 P<.10
a_4	FS	.000785	.0003561	2.66 P>.10

and of the presence of the other variables (Table III, last column). The last variable incorporated (MS) appears then as non significant, even in a one-tailed test, and might be removed from further models.

A confidence interval of level for a prediction of N_{t+1} is then (M-A, M+A), with

$$M = N_t \exp(a_0 + b_0 N_t + a_1 MS + a_2 BEM + a_3 W21 + a_4 FS)$$

(using the parameter estimates in Table III), and A=t (19) 1.881 M. These confidence intervals, based each year on the population size in the previous year and on the prevailing environmental conditions, are given in Figure 5.

The predicted population size for each of the last 9 years can be used as a cross validation of the model. The deviance for these 9 points (15.633) has 9 df and is conditional on the model and on the parameter estimates obtained from the 25 first points. When comparing this deviance to the residual deviance of the model, one obtains F(9,19) = .924 (P .5), which indicates a good ability of the model to predict population changes.

DISCUSSION

Three main factors of change in Great Tit numbers at Wytham wood can be detected with branching models : winter and early spring temperature (MS, FS, W21), Beech production during the previous year (BEM), and density-dependence (N_t). Direct experimental evidence for these factors was available (see e.g. Perrins, 1971), with many more details coming from studies at the demographic parameter level (e.g. Clobert et al., 1988).

Branching models make it possible to test for density-dependence after having accounted for the possible role of environmental variables. This is

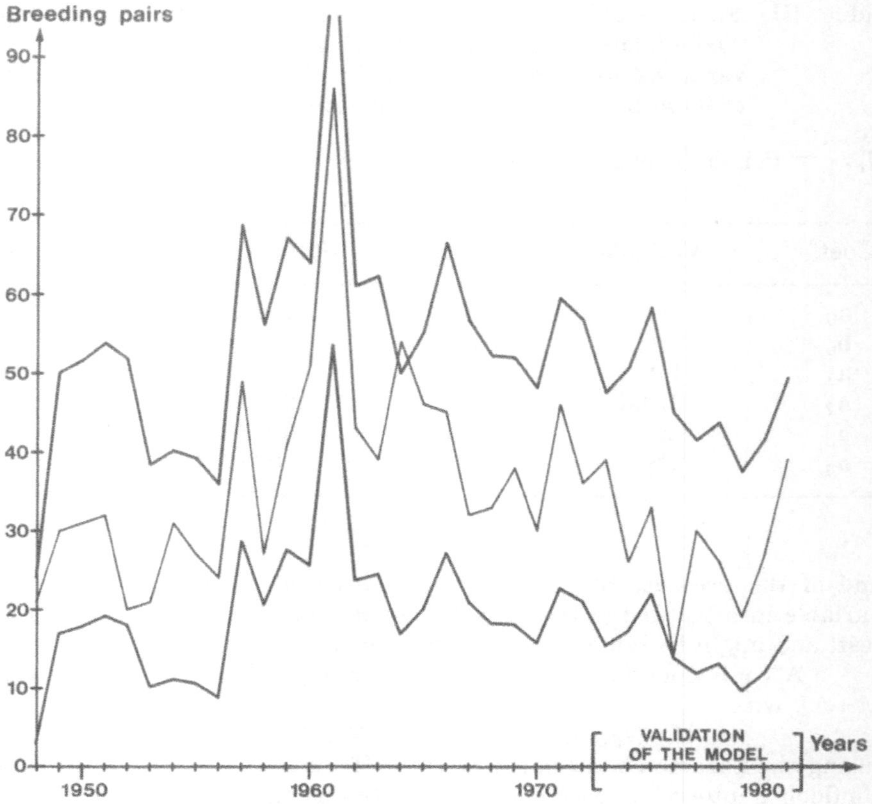

Fig. 5. Confidence interval (at level 0.05) of Great Tit *Parus major* numbers at Wytham wood, as predicted from the final branching model with overdispersion. The observed data (thin line) reach the limits only twice over 34 points (6 %). The model was fitted on the first 25 points, so that the last 9 predictions provide a cross validation of the model. The model does not provide an overall prediction starting from N_0 but a collection of local predictions conditional on the state of the population-environment system at the previous time-step. Details on the final model are given in Table III.

particularly important since negatively autocorrelated variables (like BEM here) tend to induce oscillations in numbers, which would appear as confounded with a negative feedback in a superficial analysis. When there is no measurement error, the tests proposed here do not suffer from the bias of ad hoc regressions on population counts. Branching models can also be approximated by time-series autoregressive models. Bulmer (1975) proposed an autoregressive model allowing to test for density-dependence when there is measurement error and no environment-dependence. Extending his approach to include external variables would lead to combine the advantages of the two approaches (Lebreton, 1989).

However, the potential of analyses of population counts needs to be re-assessed. In the example treated, a great deal of variation could not be removed with reasonably parsimonious models as long as a strict demographic stochasticity was used as residual variability. We considered that the sources of unexplained variability could be summarized in an overdispersion factor inflating demographic stochasticity. The estimate of this overdispersion factor was small enough (1.881) to lead to a reasonably narrow confidence interval for population size next year. The final model was validated with independent data (the last nine years). Less parsimonious models might have been tailored more narrowly to the data ; however they would have been rejected at the validation stage. The final model is thus a compromise between precision and bias, as a consequence of the parsimony principle. Any discussion of causal mechanisms would be restricted by the absence of replication and the confounding between variables : for instance Blue Tit *Parus caeruleus* density (BTM), a strong candidate for selection among the first few variables, became negligible once BEM was introduced.

The precision of predictions and the power on tests would be decreased in models incorporating measurement error. Bulmer's (1975) time series approach has been criticized for its lack of power. The interest of such analyses for a single series of counts without replication is thus quite limited. More direct analyses of demographic mechanisms in such studies usually provide many more results. The largest potential seems to me to be in the analysis of large-scale monitoring programs with replicated counts like the Common Bird Census in Britain (Marchant, 1978). While such counts are usually only looked at in terms of year-to-year population changes they probably deserve a deeper analysis accounting for various kinds of enviromental variables.

On the other hand branching models considering several classes establish a link with demographic models, like LESLIE matrix models (Lebreton, 1982a). In the long run the information from counts and the information on demographic parameters could be confronted in a same model much more deeply than is presently possible.

ACKNOWLEDGEMENTS

I thank C.M. Perrins and his colleagues (Edward Grey Institute) for having kindly made the Great Tit data available to me for this modelling exercise.

REFERENCES

Andrewartha HG, Birch LC (1954) The distribution and abundance of animals. Chicago Univ Press, Chicago

Beddington JR (1974) Age distribution and the stability of simple discrete time population models. J Theor Biol 47: 65-74

Bulmer M (1975) The statistical analysis of density-dependence. Biometrics 31: 901-911

Chesson P (1978) Predator-prey theory and variability. Annu Rev Ecol Syst 9: 23-347

Clobert J, Perrins CM, McCleery RH, Gosler AG (1988) Survival rate in the Great Tit *Parus major* in relation to sex, age, and immigration status. J Anim Ecol 57: 287-306

Frome EL, Kutner MH, Beauchamp JJ (1973) Regression analysis of Poisson-distributed data. J Am Stat Ass 68: 935-940

Harper JL (1967) A darwinian approach to ecology. J Ecol 55: 242-270

Ito Y (1972) On the methods for determining density-dependence by means of regression. Oecologia (Berl) 10: 347-372

Jagers P(1975) Branching processes with biological applications. Wiley, New-York

Johnson NL, Kotz S (1969) Distributions in statistics : discrete distributions. Houghton Mifflin, Boston

Koehl MAR (1989) Discussion : from individuals to population. In: Roughgarden J, May RM, Levin SA (eds) Perspectives in ecological theory. Princeton University Press, Princeton, p 39-53

Lebreton JD (1981)- Contribution à la dynamique des populations d'oiseaux. Modèles mathématiques en temps discret. Thesis, Lyon I Univ, Villeurbanne, France

Lebreton JD (1982a) Applications of discrete time branching processes to bird population modelling. Anais 10a Conf Int Biometria, Embrapa, Brasilia, p 115-133

Lebreton JD (1982b) Modèles dynamiques déterministes définis par des équations de récurrence. In Lebreton JD, Millier C (eds) Modèles dynamiques déterministes en biologie. Masson, Paris, p 58-97

Lebreton JD (1989) Statistical methodology for the study of animal populations. Bull Int Stat Inst 53(1): 267-282

May RM (1976) Models for single populations. In May RM (ed) Theoretical ecology, principles and applications. Blackwell, Oxford, p 4-25

McCullagh P, Nelder JA (1983) Generalized linear models. Chapman and Hall, London

Payne CD (1985) The GLIM System Release 3.77. Numerical Algorithm group, Oxford

Perrins CM (1965) Population fluctuations and clutch size in the Great Tit *Parus major* L. J Anim Ecol 34: 601-647

Perrins CM (1971) Population studies of the Great Tit *Parus major*. Proc Adv Inst Dyn Numb Popul (Oosterbeek, 1970): 524-531

Perrins CM (1979) British Tits. Collins, London

Royama T (1977) Population persistence and density-dependence Ecol Monog 47: 1-35

Salt GW (1966) An examination of logarithmic regression as a measure of population density response. Ecology 47: 1035-1039

Zeger SL, Qaqish B (1988) Markov regression models for time-series : a quasi-likelihood approach. Biometrics 44: 1019-1031

VARIABLE CONE CROPS, MIGRATION, AND DYNAMICS OF A POPULATION OF MOUNTAIN CHICKADEES (*PARUS GAMBELI*)

D. Archibald McCALLUM

Savannah River Ecology Laboratory, Drawer E, Aiken SC 29802 USA

Food is a required resource for individual survival and reproduction, and hence for population stability and persistence (Lack 1954). Wynne-Edwards (1962) suggested that social behaviour functions to regulate population densities within levels sustainable by resources, but his proposal that such behaviour evolves by group selection led to rejection of the entire idea. More recently, Gauthreaux (1978) has argued that dominance-mediated access to local resources can influence which individuals in a population are most likely to emigrate. Murray (1979), Pulliam (1987, 1988), and Łomnicki (1988) have integrated this notion into models showing that social behaviour characterized by individually-selected, rank-related resource access and emigration can enhance population stability. The amount of resource variation in the local environment and the success of the social system at mitigating its effects on population density naturally influence the type of social system that evolves.

Most members of the avian family *Paridae* have rather complex social systems, characterized by coherent, stable, territorial groups. Social rank determines access to resources within the group and territoriality limits access to resources outside the group (Ekman 1989, Matthysen in press). The Mountain Chickadee (*Parus gambeli*) has such a social system (Dixon 1965, Minock 1971, McCallum 1988). The effectiveness of socially mediated emigration at regulating population density should be strenuously tested in the relatively arid conifer forests of the southern Rocky Mountains of the USA, where these birds are often common year-round residents. Not only are the winters cold (temperatures of -25° C are not uncommon), but conifer seeds, an important source of food for many overwintering vertebrates (pers. obs.), are as variable in these mountains (Lanner 1981) as they are in the far north (Bock and Lepthien 1976).

The subject I address in this paper is how the Mountain Chickadee responds to variation in what appears to be its preferred winter food, conifer seeds. Does population density vary with the food supply, and if so, are changes in density mediated through migration, or through survival and/or reproductive rates? Does the social system thereby mitigate environmental variation, and how might its success at this have influenced the social system itself?

Seeds of the Colorado Piñon (*Pinus edulis*) are the food resource that I focused on in this study. The large and nutritious piñon seeds are exploited heavily by Mountain Chickadees and most other winter occupants of the lower elevations of the southern Rocky Mountains. In my study area chickadees harvested piñon seeds for several hours each day from the date cones opened in late August or September for a period of weeks or months, depending upon the size of the crop. In August 1987 chickadees even harvested 1986 seeds from

NATO ASI Series, Vol. G 24
Population Biology of Passerine Birds
Edited by J. Blondel et al.

atop the cotyledons of seedlings. Mountain Chickadees in the study area cached piñon seeds individually in foliage and in the ground throughout the year, especially in autumn, and removed seeds from the ground as late as the next June. Mountain Chickadees in other areas respond to Ponderosa pine and Spruce seeds similarly (Haftorn 1974, pers. obs.). Therefore, although I did not study dietary preferences formally, piñon seeds appeared to be the winter food of preference in my study population.

METHODS

I studied Mountain Chickadees from 1977 to 1988 (intensively 1982-1987) in piñon - juniper (*Juniperus spp.*) - Ponderosa pine (*P. ponderosa*) woodland at an elevation of 2200 m in western New Mexico, USA (McCallum 1988). From 1982 onwards 1400 nestlings from 479 nests were uniquely color-banded, as were some 400 additional adults and immigrant juveniles. Over 20,000 point observations of individual birds, by me and 14 volunteer field assistants, together with nesting data, form the basis of this report.

Statistical analyses in this report are based upon 267 territories in the 277 ha core zone of the study area. Because all territories and virtually all nests were found, estimates of breeding densities are not subject to errors introduced by incomplete recapture. Assuming that territories were successfully enumerated, the only source of error in density estimates is the estimate of the area occupied by the territorial birds. This was approximated by the total area of the core study area, which increased from 1980 through 1985, and was calculated from an aerial photograph using a GTCO digitizing table and ERDAS planimeter software. Year-round resightings coupled with study of 209 nests in areas peripheral to the core zone prevented confusion of local emigration with mortality.

Mountain Chickadees typically fledge in June or July and carry all their juvenal rectrices until the next June (pers. obs.). The distinctive shapes of juvenal rectrices (pers. obs., cf. Laaksonen and Lehikoinen 1976, Meigs et al. 1983) makes age determination (in the hand) possible through their first breeding season. Because almost all breeders were caught before or during their first breeding attempt, it was possible to construct the age structure of the breeding population in each of the years 1983-1987. Previously unmarked breeders were safely identifiable as yearlings in almost all cases. Natural food was augmented with suet and sunflower seeds in selected portions of the study area each year to assess the impact of food on the incidence of migration, overwinter survival, and laying date. A central control area was not provisioned after September 1984, and nest boxes were also absent from this area.

I used cone counts as indices of piñon seed availability. Piñon cone crops were assessed by counting retrospectively (beginning in 1984) cones or cone scars on the apical shoot of one branch facing in each of the eight cardinal directions from the trunk of 88 trees. Cone scars could be detected accurately up to 5 yr after the cone had fallen. Each tree was a randomly selected representative of those in a 2.5 ha grid block. The population-wide cone indices reported here are means of the total counts of cones per year from the 37 trees on which eight healthy and unshaded branches persisted through the entire study period.

All tests were performed using SAS (SAS Institute 1985).

RESULTS

Piñon cones

Cone crops that matured in the late summers of 1983 and 1986 were significantly denser than in other years (McCallum unpubl.). In the analyses that follow these crops will be referred to as "bumper," and others will be referred to as "poor." Such a boom-bust oscillation in reproductive output (Table I) is typical of piñons (Lanner 1981), and was also found in an index of seed production of all conifers, based on USDA Forest Service data for the entire state of New Mexico (McCallum unpubl.).

Mountain Chickadee adults

Breeding densities

The density of breeders in the study population increased rapidly following a population crash in the winter of 1978-79 to a fairly stable level of about 25 pairs / 100 ha by 1983, then crashed again in 1987-88 (Table I, unpubl. data). Piñon cone data are available for the years 1981-1987, and during that period population density was not linearly related to winter food supply (linear regression of growth increment on cone index, $F=0.04$, $df=1,5$, $p=0.42$ [all such tests in this section are one-tailed]; other more complicated regression models were also not significant). In the latter years of the study (1984-1987), after rapid population growth had ceased, the size of the breeding population was influenced by the cone crop the previous autumn. The growth increment was linearly dependent upon the cone crop in these four years ($Y=2.6X-7.9$, $F=11.8$, $df=1,2$, $p=0.04$), and breeding densities were significantly higher following the two bumper crops of 1983 and 1986 than in the other two years ($F=24.2$, $df=1,2$, $p=0.02$).

Age structure of the breeding population

The adult (birds > 1 yr old) and yearling subsets of the breeding population (Fig. 1) differed in their responses to variation in the cone crop during this period of relative stability of population density. The annual change in density of adult breeders was not linearly dependent upon the previous cone

Table I. Annual variation in demographic and migration parameters of a population of Mountain Chickadees.

	YEAR							
	1980	1981	1982	1983	1984	1985	1986	1987
Cone Crop (cones/8 shoots, n=37)	0.81	2.38	0.71	4.00	1.14	2.85	4.40	1.70
Breeding Terr. (/100 ha)	8.3	12.0	15.8	25.2	26.2	22.0	20.2	25.5
Study Area (ha)	48	50	95	159	183	227	227	227
Reproductive Success Rate			0.73	0.68	0.56	0.76	0.61	0.76
Fledglings per Occupied Terr.			3.80	3.55	3.00	3.78	3.35	3.71
Adult Female[1] Survival Rate	1.00	0.60	0.67	0.63	0.42	0.45	0.64	
(n)	4	6	15	40	48	50	46	
Provisioned	1.00	0.60	0.83	0.64	0.44	0.50	0.64	
Control	-	-	0.63	0.62	0.38	0.33	0.64	
Adult Male[1] Survival Rate	0.75	0.57	0.71	0.69	0.57	0.55	0.54	
(n)	4	6	15	40	48	50	46	
Provisioned	0.75	0.57	0.71	0.75	0.65	0.66	0.31	
Control	-	-	0.56	0.68	0.38	0.33	0.65	
Incidence of migration			0.85	0.33	0.91	0.89	0.35	0.99
Yearling Female Breeders (/100 ha)								
Local			-	-	1.0	0.4	0.4	1.7
Summer imm.			3.1	8.8	6.5	0.8	1.7	5.7
Spring Imm.			5.2	8.1	6.5	11.0	7.0	5.2
Yearling Male Breeders (/100 ha)								
Local			-	-	4.9	0.8	1.7	3.5
Summer imm.			4.2	7.5	4.9	0.8	2.2	6.6
Spring Imm.			2.1	9.4	3.2	5.20	2.6	4.4

[1]Survival of the winter of 1978-1979 for 9 color-banded adult males was 0.1 and for 9 color-banded adult females was 0.00.

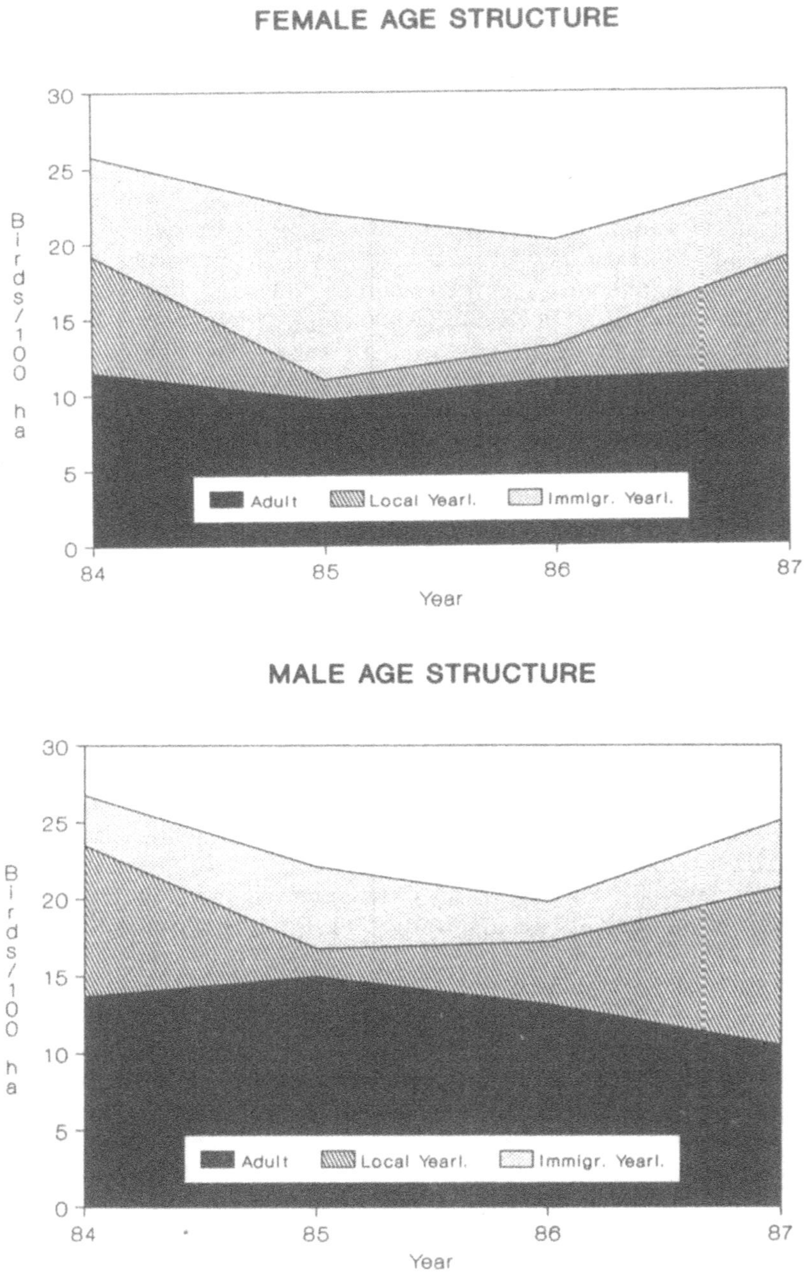

Fig. 1. Variation in population size and age structure over the 4-yr period during which cone crop appeared to influence population dynamics. Locals are yearlings that settled in the study area during the preceding summer, following natal dispersal. Immigrants are yearlings assumed to have settled the following spring.

crop (df=1,2, F=2.24, p=0.27 for females; df=1,2, F=0.003, p=0.48 for males), and, unlike the total population of breeders, adults of neither sex differed significantly in density following bumper and poor crops (F=2.83, p=0.12 for females; F=2.52, p=0.13 for males). The cone-related variation in total density therefore was necessarily attributable to the yearling age class, but this relationship was complex.

The yearling cohort as a whole varied independently of cone crop (linear regression of annual growth increment on previous cone crop, df=1,2, F=0.61, p=0.26 for females; df=1,2, F=1.98, p=0.15 for males), although yearling males were significantly more numerous following bumper crops (ANOVA, df=1,2, F=88.3, p=0.005). But, when the yearlings were divided into a group that had settled on the study area in their first summer, within weeks of fledging ("locals"), and a group that may not have arrived until their first spring ("immigrants", so classified because they were not banded before their first breeding season), an interesting pattern emerged (Fig. 1). Although the numbers of local yearling females did not change in response to cone abundance (df=1,2, F=5.33, p=0.08), the densities of local yearling females were higher following bumper crops (ANOVA: df=1,2, F=168.16, p=0.003) and both tests were significant for local yearling males (linear regression: Y=4.0X-11.6, df=1,2, F=17.87, p=0.03; ANOVA: F=41.06, p=0.01).

As a result, adults + local yearlings (Fig. 1) varied in concert with cone crop (Y=3.9X-11.5 , df=1,2, F=16.46, p=0.03 for females; Y=3.9X-10.7, df=1,2, F=11.20, p=0.04 for males), and, as with the population at large, their numbers were significantly greater after bumper crops of cones (ANOVA: df=1,2, F=39.21, p=0.02 for females; df=1,2, F=13.23, p=0.03 for males).

Thus local yearlings tended to be responsible for much of the annual variation in population density during this period, with immigrants tending to damp this variation (Fig. 1). Numbers of local yearlings were in fact significantly more variable from year to year than numbers of adults for females (df=3,3, F=16.17, p=0.05) and marginally more variable than numbers of immigrants for males (df=3,3, F=12.84, p=0.06).

Survival of breeders

Over the entire period for which cone data are available (1981-1987) annual survival was not linearly related to cone crop alone, or to a combination of cone crop and total population density in July (estimated as the number of breeders + offspring produced on the study area) (Table I). After the population density reached an apparent plateau at about 25 pairs/100 ha, however, survival of unprovisioned adults was higher in the two years with a bumper crop of cones than in the two piñon-poor years (ANOVA of arcsine of annual survival rate: df=1,2, F=115.17, p=0.004 for females; F=118.82, p=0.004 for males). Provisioned females were similarly affected by poor cone crops (df=1,2, F=36.32, p=0.01), and they did not experience higher survival rates than control females in any of the four winters (1983-1986) when this phenomenon was studied. Male survival was improved by provisioning in one low-piñon year (1985, X^2=4.309, p=0.04), with a marginally significant difference in the other (1984, X^2=2.534, p=0.11). [There was no difference in one mast year (1983), but in the other (1986) survival of provisioned males was significantly lower ($X^2=4.179$, $p=0.04$).] As a result, provisioned males did not suffer lower survival rates in piñon-poor years than in piñon-rich years (df=1,2, $F=0.25$, $p=0.33$). The influence of food supply on adult survival is therefore shown on both temporal (between years with bumper and poor crops) and spatial scales (among provisioned and unprovisioned areas).

Density of the population in July apparently did not affect the influence of winter food supply on survival. Adding total population density in July to the models of food-dependent survival described above rendered all results nonsignificant. Survival of provisioned males, however, bore a strong positive relationship to July density when cones were not considered ($Y=1.7X-1.52$, df=1,2, $F=65.25$, $p=0.008$).

Reproduction

Although breeding population density was significantly influenced by the size of the preceeding cone crop, neither reproductive success rate (the proportion of occupied territories that produced a successful nest, arcsine transformed) nor the highly correlated ($r=0.96$, $p=0.002$) per capita fledgling production (mean number of fledglings produced by all territories, successful or not) varied in concert with population density, as measured by the density of occupied territories (1983-1987) (df=1,4, $F=0.29$, $p=0.62$; and df=1,4, $F=1.12$, $p=0.35$, respectively). The only apparent density dependent effect on reproduction occurred in 1984, following a large piñon crop, when densities were high, but many pairs did not nest, resulting in a low success rate. A similarly large breeding population in 1987, also following a bumper piñon crop, had the highest success rate recorded during the study (Table I).

Mountain Chickadee juveniles

Summer dispersal

All juvenile Mountain Chickadees, like the juveniles of almost all other temperate zone parids (Ekman 1989, Matthysen in press) participated in natal dispersal movements within 2-4 weeks after fledging. Because all fledglings left their natal territories, the act of dispersal itself was related to neither the size of the subsequent cone crop nor the presence or absence of local provisioning. These movements were reciprocal, with space vacated by emigrants being filled almost immediately by immigrants. In the three years (1985-87) when my study area was largest (227 ha) the proportion of banded nestlings still present in the core study area at the end of summer dispersal (10 weeks post fledging) was 0.22 - 0.24. These three years spanned the entire range of variation of fledgling density (67 to 94 birds/100 ha), and 60% of the variation in subsequent cone density (1.7 to 4.4 cones/8 shoots). It therefore seems reasonable to conclude that summer dispersal was affected neither by numbers nor by prospective winter food levels in this population. The distance dispersed may be a different matter, but will not be treated here.

Fall migration

On the other hand, movements of juveniles in September (cf. Dixon and Gilbert 1964) were related to the size of the cone crop (linear regression of incidence of migration (arcsine-transformed) on cone index: $Y = -0.24X + 1.48$, $df = 1,4$, $F = 7.61$, $p = 0.05$), but not to total density of breeders and their young ($F = 0.20$, $p = 0.68$), nor to a linear combination of the two ($F = 5.35$, $p = 0.10$). Migration rate was estimated as the proportion of banded juveniles (both local birds banded as nestlings and previously unbanded birds that had settled on the study area) present on 15 August that disappeared between that date and 1 November. This is a maximum estimate, because some mortality must occur, but this period was the most climatically benign part of the year, and insect food should have been available. Mortality during this period would not be expected to vary so greatly. Moreover, independent evidence that Mountain Chickadees respond to variation in cone crops with partial eruptive migrations comes from negative correlations between numbers of chickadees detected on Christmas Bird Counts in the lowlands, where this species normally does not breed, and regional cone crops, for both New Mexico (McCallum unpubl.) and northern Utah (K. L. Dixon, pers. comm.). Although about 5 adults were known to have migrated and returned during the study, migration was overwhelmingly a phenomenon of the juvenile cohort (see McCallum 1988 for justification of this conclusion).

Return of migrants and survival of overwintering juveniles combined to yield the density of local yearlings in the next breeding season. These numbers closely tracked the incidence of migration the previous autumn (Fig. 2.) (regression of density of local yearlings on the arcsine of the incidence of juvenile migration: $Y = -7.5X + 10.2$, $df = 1,2$, $F = 473.8$, $p = 0.001$ for females, $Y = -9.3X + 13.2$, $df = 1,2$, $F = 60.5$, $p = 0.01$), and hence piñon crop controlled much of the variation in breeding densities on the population plateau by controlling autumn emigration of local juveniles.

Spring immigration

Spring immigration was defined as the movement onto the study area of birds that had not previously occupied it. The number of spring immigrants did not vary with the incidence of migration, or with the number of local yearlings present in the spring. The proportion of all yearling females that were putative immigrants (arcsine-transformed) did, however, bear a positive relationship to the incidence of migration the previous autumn ($Y = 0.69X + 0.20$, $df = 1,2$, $F = 24.6$, $p = 0.02$). A similar analysis of males was suggestive, but not significant ($F = 3.0$, $p = 0.11$). Thus the losses of local yearlings resulting from migration were somewhat, but not completely, compensated by the immigration of presumably foreign yearlings. Had this immigration been completely compensatory, the cone-related differences in total density reported above would not have existed.

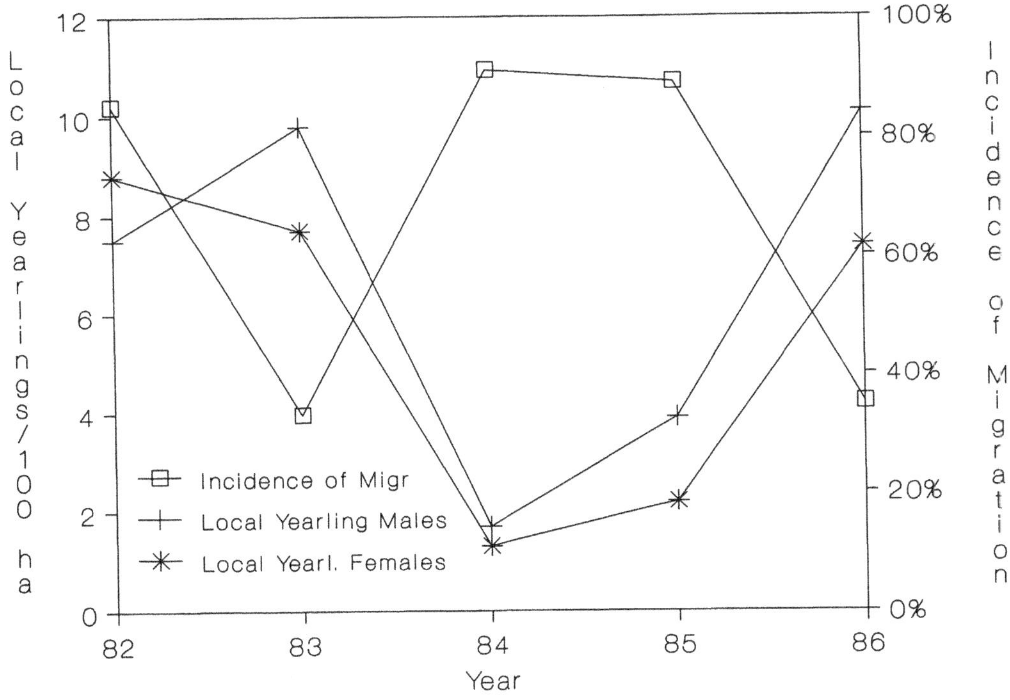

Fig. 2. Covariation of the incidence of migration in a given autumn and the density of local yearling breeders the subsequent breeding season.

The numbers of spring immigrants were estimated from the number of unbanded breeders in the subsequent breeding population. These individuals turned out to be yearlings in almost every case. This estimate was a maximum, because migrants that were not banded before they departed or winter residents that eluded capture were erroneously considered immigrants. The tests reported above are probably conservative, though, because numerous birds eluded capture during piñon-rich winters (many of these were merely visitors, but some probably had settled during the summer) but not in the piñon-poor winters. Moreover, the numbers of returning banded migrants following piñon-poor winters were low, and it is unlikely that the larger numbers of unbanded birds that appeared in the late spring of those years, after the return of the migrants, were predominantly unbanded migrants (McCallum 1988:83). It seems more likely that migration is a dangerous undertaking and that the spring immigrants wintered nearby in areas with better winter food supplies. Support for this interpretation comes from spring emigration of juveniles in 1984 and 1987 following good cone crops. Not only did temporary winter visitors leave, some birds that had been present since the original mid-summer settlement period also did.

DISCUSSION

During the course of 10 yr the study population suffered two crashes, and exhibited sigmoid growth in between. These are not the characteristics of a "well-regulated" population. Crashes are exactly the consequence that must be averted to maintain stability and persistence. Socially mediated resource access, resulting in emigration and immigration of juveniles, was not completely successful at buffering population fluctuations. The crashes occurred. The question this study may shed some light on is whether such a mechanism operated between the crashes. In other words, did socially mediated responses to variation in the food supply tend to reduce the frequency of such crashes?

To answer this question it is necessary to separate the period of rapid growth from the period during which numbers varied around a plateau of about 25 pairs/100 ha. It was this latter period in which cone crop seemed to influence dynamics. Although the short span of 4 yr over which these effects were found limits the degrees of freedom of statistical tests, the short duration of the plateau was biologically real; it was terminated by the almost complete disappearance of previously banded birds. Fewer than 10 were found in brief surveys in 1988.

Food levels influenced population dynamics in two ways. First, low food availability decreased survival of adults. The importance of food to survival was further supported by the enhanced survival of provisioned males in piñon-poor years. Although the adult portion of the population did not decline in numbers following those poor winters (because of the size of the yearling cohorts), food clearly was a limiting resource for adults in those years. The fact that males (which are socially dominant over females during the winter) experienced improved survival with provisioning, while females did not, suggests that high rank tends to mitigate the effect of food limitation. Second, the rate of autumn emigration of juveniles increased in the face of the lower food availability reflected by low cone crops and concomitant high adult mortality. Being both socially subordinate and less experienced at foraging, these individuals presumably would have fared worse than adults if they had stayed on site (Gauthreaux 1978).

If the migrants had returned at a high rate the contribution to population stability would have been patent. As it was, return rates were low, and migration amounted demographically to mortality. Because migrants were essentially lost to the population, the only possible contribution of their emigration to population stability is indirect. If all juveniles remained for the winter, scramble competition would presumably have lowered the population-wide survival rate even further. If adult survival was density independent, because of contest competition for food (as in the model of Lomnicki 1988), the juveniles presumably would have died in as great numbers as they did by migrating.

Spring immigration of presumably foreign yearlings did repopulate the local area somewhat following food-related decreases in population size. This influx did not cover the losses of migrant juveniles. It was, however, important for population stability, because recruited yearlings were the only source for older age classes (immigration of adults was demographically negligible to nonexistent). Without this annual input of foreign yearlings (and no compensatory increase in survival rates owing to smaller population size) the population would have declined gradually to near extinction by 1988, without the crash of 1987.

Migration therefore may have stabilized the population by cutting overwinter mortality, and immigration definitely seems to have contributed to stability by mitigating somewhat the losses of the migrants. The only other return-migration system documented for a parid (Smith and Nilsson 1987) appears to be similar to this one, but spring immigration was not assessed in that case. Spring emigration, which is a necessary counterpart of spring immigration, has been described for Black-capped Chickadees (Smith 1967, Glase 1973), Carolina Chickadees (T. C. Grubb, pers. comm.), and Mountain Chickadees (McCallum 1988).

The availability of a vast expanse of contiguous usable habitat (approximately 3000 km^2) is probably essential to the maintenance of spring immigration and hence the modicum of stability that characterizes the local population. The entire area probably functions as a metapopulation (Murray 1979, Lande 1987) and persistence of Mountain Chickadees in any small area depends upon the relative integrity of the larger patch of habitat.

Although immigration sufficed to replace yearlings lost to cone-related emigration, it apparently cannot offset the loss of adults under the current environmental regime. Nearly complete emigration of juveniles and heavy overwinter losses of adults characterized both crashes. A poor seed crop combined with severely cold weather in December may have been responsible in 1978 (the crash was widespread regionally and taxonomically in the American Southwest, P. B. Stacey, pers. comm.), but weather probably was not a factor in 1987, because most adults were absent by 1 December. Seed production was almost nil that year (pers. obs.), despite the moderate number of cones (Table I). Circumstantial evidence that food scarcity, and not numbers or severe weather, drives the crashes was obtained in January-February 1989, when two heavy snowstorms (> 0.5 m accumulation) had no detectable effect on survival of a large population of resident and visiting chickadees. Piñon seeds were still abundant, and widely harvested, during that period.

In conclusion, the influence of food abundance on the *number* of breeding individuals in the area was mitigated somewhat by age-related individual differences in access to food and the tendency to move from one area to another (Gauthreaux 1978). Thus population dynamics was stabilized somewhat by individual behaviour (Lomnicki 1988). But the influence of variation in the food supply on *which* individuals were present in the area was unmitigated by social behaviour. If cone crops were consistently high, all yearling recruits would be locally produced birds. The resulting exclusion of foreign birds would raise the inbreeding coefficient, reduce the effective population size, increase the potential for local adaptation, and increase the opportunity for cooperative behaviour among related individuals. The social system of chickadees seems well-suited to preventing the overcrowding that would occur under such an environmental regime, as indicated by the spring emigration of winter visitors in 1984 and 1987, following bumper cone crops, and spring emigration in Black-capped and Carolina Chickadees (see above).

The importance of *which* individuals are present becomes clear when one compares the pseudo-stability of this population of Mountain Chickadees with the real stability of populations of cooperative breeders, e.g., the Florida Scrub Jay (*Aphelocoma c. coerulescens*) (Woolfenden and Fitzpatrick 1984). McCallum, Stacey, and Ligon (ms) have argued that habitat improvement through cooperative development or defense of resources was not a viable option for Mountain Chickadees, precisely because the variable environment made waiting for an opportunity to use a cooperatively improved territory unviable. On the other hand, species such as the Florida Scrub Jay and Acorn Woodpecker (*Melanerpes formicivorus*) (Stacey and Ligon 1987) can afford to wait because they live in stable environments or because their cooperative behaviour actually mitigates environmental variation. In these species, the relative constancy of breeding numbers is critically dependent upon contest competition but not upon emigration (Lomnicki 1988). In chickadees emigration and immigration are much more important. Species such as the Willow Tit (*Parus montanus*) (Ekman 1979, 1988) and the Mountain Chickadee (McCallum et al. ms) have probably realized the pinnacle of social complexity attainable by parids in severe and variable environments. Truly cooperative sociality with delayed breeding will be found among parids that inhabit more stable and/or benign environments.

ACKNOWLEDGEMENTS

The Mountain Chickadee data could not have been collected without help in the field from 14 dedicated volunteer field assistants, and financial support from the Josselyn van Tyne Fund of the American Ornithologists' Union, Sigma Xi, the Student Research Allocation Committee (SRAC) and the Biology Graduate Research Allocation Committee (GRAC) of the University of New Mexico, and a University of New Mexico Challenge Assistantship. M. Billings (Cottonwood Gulch Foundation), R. Thompson (USDA Forest Service), and several neighbors graciously allowed access to their lands. This report was prepared while the author was a postdoctoral fellow at the University of Georgia's Savannah River Ecology Laboratory, which is funded by Contract DE-AC09-76SROO-819 between the United States Department of Energy and the University of Georgia. I am grateful to all these people and organizations for their help.

REFERENCES

Bock CE, Lepthien LW (1976) Synchronous eruptions of boreal seed-eating birds. Am Nat 110: 559-571

Dixon KL (1965) Dominance-subordination relationships in Mountain Chickadees. Condor 67: 291-299

Dixon KL, Gilbert JD (1964) Altitudinal migration in the Mountain Chickadee. Condor 66: 61-64

Ekman J (1988) Subordination costs and group territoriality in wintering Willow Tits. Acta XIX Congr Int Ornithol 19: 2373-2381

Ekman J (1989) Ecology of non-breeding social systems of *Parus*. Wilson Bull 101: 263-288

Gauthreaux SA Jr (1978) The ecological significance of behavioural dominance. In: Bateson PPG, Klopfer PH (eds) Perspectives in ethology. Plenum, New York, p 17-54

Glase JC (1973) Ecology of social organization in the Black-capped Chickadee. Living Bird 12: 235-267

Haftorn S (1974) Storage of surplus food by the Boreal Chickadee *Parus hudsonicus* in Alaska, with some records on the Mountain Chickadee *Parus gambeli* in Colorado. Ornis Scand 5: 141-161.

Laaksonen M, Lehikoinen E (1976) Age determination of Willow and Crested Tit *Parus montanus* and *P. cristatus*. Orn Fenn 53: 9-14

Lack D (1954) The natural regulation of animal numbers. Clarendon Press, Oxford

Lande R (1987) Extinction thresholds in demographic models of territorial populations. Am Nat 130: 624-635

Lanner RM (1981) The piñon pine: a natural and cultural history. Univ Nevada Press, Reno

Łomnicki A (1988) Population ecology of individuals. Princeton Univ Press, Princeton

Matthysen E (in press) Nonbreeding social organization in *Parus*. Current Ornithol

McCallum DA (1988) Alternative emigration strategies and the adaptive significance of natal dispersal in a population of mountain chickadees (*Parus gambeli*). PhD dissertation, Univ New Mexico, Albuquerque

Meigs JB, Smith DC, van Buskirk J (1983) Age determination of Black-capped Chickadees. J Fld Ornithol 54: 283-286

Minock ME (1971) Social relationships among mountain chickadees (*Parus gambeli*) Condor 73: 118-120

Murray BG, Jr (1979) Population dynamics: alternative models. Academic Press, New York

Pulliam HR (1987) On the evolution of density-regulating behaviour. In: Bateson PPG, Klopfer PH (eds) Perspectives in ethology. Volume 7, Plenum, New York, p 99-124

Pulliam HR (1988) Sources, sinks, and population regulation. Am Nat 132: 652-661

SAS Institute, Inc (1985) SAS users' guide, statistics. SAS Institute, Cary, North Carolina

Smith H, Nilsson J-A (1987) Intraspecific variation in migratory pattern of a partial migrant, the Blue Tit (*Parus caeruleus*): An evaluation of different hypotheses. Auk 104: 109-115

Smith SM (1967) Seasonal changes in the survival of the black-capped chickadee. Condor 69: 344-359

Stacey PB, Ligon JD (1987) Territory quality and dispersal options in the Acorn Woodpecker, and a challenge to the habitat-saturation model of cooperative breeding. Am Nat 130: 654-676

Woolfenden GA, Fitzpatrick JW (1984) The Florida scrub jay: Demography of a cooperative-breeding bird. Princeton Univ Press, Princeton

Wynne-Edwards VC (1962) Animal dispersion in relation to social behaviour. Oliver and Boyd, Edinburgh

THE TIMING OF BREEDING AND CLUTCH SIZE OF BLUE TITS (*PARUS CAERULEUS*) IN TWO MONTANE HABITATS IN ALGERIA

Aïssa MOALI* and Paul ISENMANN**

* Institut de Biologie, Université de Tizi-Ouzou, DZ- 15000 Tizi-Ouzou, Algeria
** Centre d'Ecologie Fonctionnelle et Evolutive (CNRS), BP 5051, F- 34033 Montpellier-cédex, France

Northern Africa constitutes the southern breeding limit of the Blue Tit (*Parus caeruleus*), where a particularly small subspecies *ultramarinus* (Vaurie 1959, Martin 1988) occurs in various forest habitats (Heim de Balsac and Mayaud 1962). At the edge of their breeding limits species often show reduced fecundity, usually interpreted as the result of harsher environmental constraints (Slagsvold 1981, Järvinen 1986). To check this hypothesis, we collected data in Algeria on the Blue Tit for which other studies in the Mediterranean area have already shown a marked variation in breeding performances and a reduced fecundity (Blondel et al. 1987, Isenmann 1987).

MATERIAL AND METHODS

The study site is located near Tikjda, Northern Algeria (36°26'N, 03°55'E), in a mature Cedar forest (*Cedrus atlantica*) with 25 nestboxes, and in a nearby Holm Oak wood (*Quercus ilex*), with 15 nestboxes, at an elevation of 1300-1500 m a.s.l. The climate is sub-humid mediterranean with cool winters and warm summers, with snowfall between November and February/March. Data on time of breeding, size of first clutch, and breeding success were collected during four breeding seasons (1983-1986). With the Coal Tit (*Parus ater ledouci*), which does not use nestboxes, the Blue Tit was the most abundant breeding species. Thévenot (1982) found a natural density of 3 breeding pairs/10 ha in Cedar forests in Morocco. In Tikdja the species even reached 9 to 10 breeding pairs/10 ha due to the availability of nestboxes.

RESULTS

Laying dates

First clutches were initiated from April 30 to May 20 (Table I), with 77% of the laying dates of the first egg within the period 6-16 May. During 1983-

NATO ASI Series, Vol. G 24
Population Biology of Passerine Birds
Edited by J. Blondel et al.

1986 the laying period lasted 11 to 21 days. The annual mean laying dates did not differ significantly. The laying dates recorded in the Cedar forest were also not significantly different from those recorded in the Holm Oak forest (Table I).

Table I. Laying dates of first clutches of the Blue Tit in two habitats near Tikjda, Algeria.

	Laying date (first egg)	Limits (4 = April, 5 = May)	N
1983	8 May	30/4 - 20/5	18
1984	11 May	6/5 - 20/5	20
1985	12 May	5/5 - 15/5	20
1986	12 May	30/4 - 20/5	19
Mean value in Holm Oaks	9 May	30/4 - 15/5	35
Mean value in Cedars	10 May	2/5 - 20/5	42
Overall mean	10 May	30/4 - 20/5	77

Clutch size

The clutch size was small (6.4 eggs, Table II), 88% of clutches having 5 to 7 eggs. There were no significant differences between the clutch sizes in either the two habitats or the 4 years studied.

Breeding success

The values are particularly high (83-94 %) (Table II). Again there are no significant differences between the two habitats or between the years.

Table II. Clutch size and breeding success (young fledged from eggs laid) of the Blue Tit in two habitats near Tikjda, Algeria.

	Clutch size				Breeding success	
	mean	S.D.	Limits	N (clutches)	%	N (eggs)
1983	6.47	1.28	5 - 9	17	86	110
1984	6.10	0.96	5 - 8	20	84	122
1985	6.89	1.69	5 - 12	19	94	119
1986	5.84	0.89	4 - 8	19	83	111
Cedar	6.10	1.03	4 - 9	40	84	241
Holm Oak	6.66	1.43	5 - 12	35	90	221
Overall mean	6.36	1.25	4 - 12	75	87	462

DISCUSSION

The low mean clutch size of this Blue Tit population falls within the limits observed in the mediterranean area (Isenmann 1987). In Morocco, a mean clutch size of 6.8 eggs was found in a lowland Cork oak (*Quercus suber*) forest and of 6.7 eggs in a montane Cedar forest. Hence, there are no differences with our data. It is only on the Canary Islands that the clutch size drops to 3.5 eggs. Values recorded in the subspecies *caeruleus* from northern Mediterranean study sites, in Spain and continental France, are slightly higher, ranging mostly from 7 to 9 eggs (Isenmann 1987, Isenmann et al. 1990). The comparison with the population breeding at Mont Ventoux (Southern France) is particularly interesting, since Blue Tits lay there on the average 8.8 eggs, in a similar habitat at nearly the same elevation. The difference is significant (t-Test, $P<0.01$). It seems that the birds invest less energy in eggs in Northern Africa. These results agree with Ashmole's hypothesis which predicts that clutch size varies in direct proportion to the degree of seasonal fluctuation in the level of resources used by a population (Ricklefs 1980). This fluctuation declines toward the south, leading to a reduced spring flush of feeding resources for forest birds and making the environment harsher. The high breeding success suggests that the adjustment of breeding time as well as of clutch size to local conditions is good. Breeding performances in Cedar forest are as high as it is in a Holm Oak habitat, showing that a Cedar forest constitute a suitable habitat for this species (hence the subspecies *ultramarinus* differs strongly from "true" Blue Tits for which optimal habitats consist generally of oakwoods). A thorough discussion of the meaning of this reduced fecundity (no second clutches have ever been recorded) will only be possible when we have a better knowledge of the quality and seasonality of southern habitats, and of survival patterns of these populations.

ACKNOWLEDGEMENTS

Our thanks go to J. Thompson who checked our English.

REFERENCES

Blondel J, Clamens A, Cramm P, Gaubert H, Isenmann P (1987) Population studies on Tits in the Mediterranean region. Ardea 75: 21-34

Heim de Balsac H, Mayaud N (1962) Les Oiseaux du Nord-Ouest de l'Afrique. Lechevalier, Paris

Isenmann P (1987) Geographical variation in clutch size: the example of the Blue tit (*Parus caeruleus*) in the Mediterranean area. Vogelwarte 34: 93-99

Isenmann P, Alès E, Moreno O (1990) The timing of breeding and clutch size of Blue Tits (*Parus caeruleus*) in an evergreen Holm Oak habitat in Southern Spain. Rev Ecol (Terre Vie) 45, in press

Järvinen A (1986) Clutch size of passerines in harsh environments. Oikos 46: 365-371

Martin JL (1988) Structure de population et variations "périspécifiques" chez *Parus caeruleus* L. (*Aves*). Premiers résultats. Acta Oecologica (Oecol Gener) 9: 153-166

Ricklefs RE (1980) Geographical variation in clutch size among passerines birds: Ashmole's hypothesis. Auk 97: 38-49

Slagsvold T (1981) Clutch size and population variability in birds: a test of hypotheses. Oecologia 49: 213-217

Thévenot M (1982) Contribution à l'étude écologique des passereaux forestiers du Plateau Central et de la corniche du Moyen Atlas (Maroc). L'Oiseau et RFO 52: 97-152

Vaurie C (1959) The birds of the Palaearctic Fauna. I. Witherby, London

FACTORS AFFECTING CLUTCH-SIZE IN GREAT AND BLUE TITS

Christopher M. PERRINS

Department of Zoology, Edward Grey Institute, University of Oxford, South Parks Road, OX1 3PS Oxford, Great Britain

The study described here was started by Dr. D. Lack in 1947 in Wytham Wood near Oxford (Lack 1966). The original study was concentrated in a small area of the wood, Marley Wood; in the period 1958-64 it was extended to cover the rest of the wood.

The clutch-size of both Great and Blue Tits (*Parus major* and *P. caeruleus*) varies markedly. It varies between individuals, between birds of different age, between habitats and, within the same habitat, between years and between different times of the season (Kluijver 1951, Lack 1958, 1966, Perrins 1965). In our study area, in the period 1947-88, the mean of first clutches of Great Tits has varied, between years, from 7.7 to 12.1 and that of Blue Tits from 8.4 to 13.4.

Various explanations have been put forward for these differences. In particular, the differences between years have been thought to be related to the timing of laying (clutches being smaller in late years), to the density of breeding pairs (clutches being smaller in years of high breeding density) and perhaps to food supply. The last of these was supported in the early years of the study, but was then thought to be due to coincidence, the years of high and low caterpillar numbers being years of low and high densities respectively (Perrins 1965).

The factors affecting clutch-size of both Great and Blue Tit were re-examined by Perrins and McCleery (1989) who showed that the clutch-sizes of both species in Marley Wood were correlated with four factors: 1) the density of breeding Blue Tits, 2) the density of breeding Great Tits, 3) the mean laying date in the year and 4) the clutch-size in the previous year. We suggested that the correlation between clutch-size and laying date might result from there being more caterpillars present in early years; if clutch-size was related to caterpillar density, then a correlation between clutch-size and laying date could result, but would not be a causal relationship.

The aim of this paper is two-fold. First, I examine the variation in clutch-size in other areas of Wytham and see whether they vary in the same way as in Marley. Second, it has recently been possible to obtain further data on the caterpillar populations for Wytham, thanks especially to Dr. L. Cole, and the relationship between clutch-size and caterpillar densities is re-examined.

NATO ASI Series, Vol. G 24
Population Biology of Passerine Birds
Edited by J. Blondel et al.

METHODS

The data on tit laying-dates and clutch-size were obtained as part of the long-term study of the Wytham tit populations (e.g. Lack 1966, Perrins 1979). Different areas of the wood contain different densities of nest-boxes; in some, the density of nest-boxes is insufficient to enable all the pairs of Blue and Great Tits to obtain a nest-box. As a consequence of this, although we believe that all the Great Tits which want a nest-box can get one, some of the smaller Blue Tits are precluded from breeding in the nest-boxes in years of high density (Minot and Perrins 1986). Blue Tits seem to have little problem finding natural nest-sites in most areas of Wytham, so they are not prevented from breeding at all by a shortage of nest-boxes (East and Perrins 1988).

During the breeding season, nests are visited weekly and, since almost all birds lay at daily intervals and lay seven or more eggs, both laying date and clutch-size can be obtained from such visits. The following are the measures used in this paper:

1) Population size. This is the number of occupied boxes in each area each year. Repeat clutches (those laid by birds which had already lost a nest) and second clutches are discounted. This figure is normally fairly straightforward to obtain, but in years of high predation (Dunn 1976) with many repeat clutches, it can sometimes be more difficult and so open to some error. See also the earlier comment about the numbers of Blue Tits in nest-boxes in years of high density.

2) Laying-date. The mean of the date on which the first egg is laid by each pair. Again this is fairly straightforward to measure, though in years of many repeat clutches it is not always easy to distinguish between the repeat clutches of early laying pairs and the first clutches of late-laying ones, but errors in this are unlikely to lead to major errors in the mean laying date.

3) Clutch-size. The mean of the number of eggs laid in each nest. Replacement and second clutches and clutches of fewer than five eggs were excluded from such calculations.

4) The caterpillar data were collected by a variety of people, primarily by the late Prof. G. Varley and his team at the Hope Department of Entomology. The data used here are mainly for the Winter Moth *Operopthera brumata*. Although the tits take many species of caterpillars from the oaks, *Quercus* spp., in most years the Winter Moth is by far the commonest species of caterpillar in the trees in the wood. When fully grown, the caterpillar falls to the ground to pupate. It can be censussed by putting out trays containing water such that the caterpillars dropping into the trays are drowned and can be counted. The data are presented in terms of the numbers of caterpillars that fall per square metre of tray, i.e they represent the number of caterpillars present in a column of cross-section of a metre square projected vertically from the ground through the tree. These larvae are, of course, the ones which survive to reach the pupation stage, but Varley et al. (1973) showed that losses during the larval period were small, so they are likely to be representative of the populations available to the tits throughout the larval period. The timing of the larval period is measured by

the "half-fall" date, the date by which half the larvae which eventually fall into the trays have done so.

One of the other caterpillar species taken by the tits, the Green Tortrix *Tortrix viridana*, is sometimes quite numerous and may occasionally be more abundant than the Winter Moth; this was the case in the years 1951-1955 and again in 1980. Unfortunately, this species pupates by rolling up in a leaf and so cannot be censussed by the same method as the Winter Moth. It has been censussed by comparing its abundance in samples of oak leaves with that of the Winter Moth, but such estimates are open to some error and are more time-consuming to make; consequently, we have data on the abundance of this (and other) species for many fewer years than we have for the Winter Moth.

5) Weather data are collected for Oxford (c. 5 km east of Wytham) by the School of Geography.

RESULTS

Between area comparisons

The laying dates and clutch-sizes were analysed for two other areas of Wytham, Great Wood and Bean Wood (Fig. 1); records exist for these two areas from 1958 and 1961 respectively. Comparisons between these different areas of Wytham and Marley (Table I) show that, for both Great and Blue Tit, there are significant correlations between the population sizes, laying dates and clutch-size.

Table I. Between Wood Correlations for population size, laying date and clutch-size. Correlation coefficients are given for each species separately: Great Tit/Blue Tit. All samples are based on more than 25 years of comparisons, so that for >0.381 $p<0.05$, >0.445 $p<0.02$, $>.487$ $p<0.01$ and $>.597$ $p<0.001$.

	Marley	Great Wood
Population Size		
Great Wood	.572/.542	
Beanwood	.336/.580	.570/.542
Laying Date		
Great Wood	.873/.867	
Beanwood	.844/.910	.785/.853
Clutch-size		
Great Wood	.739/.640	
Beanwood	.788/.643	.709/.573

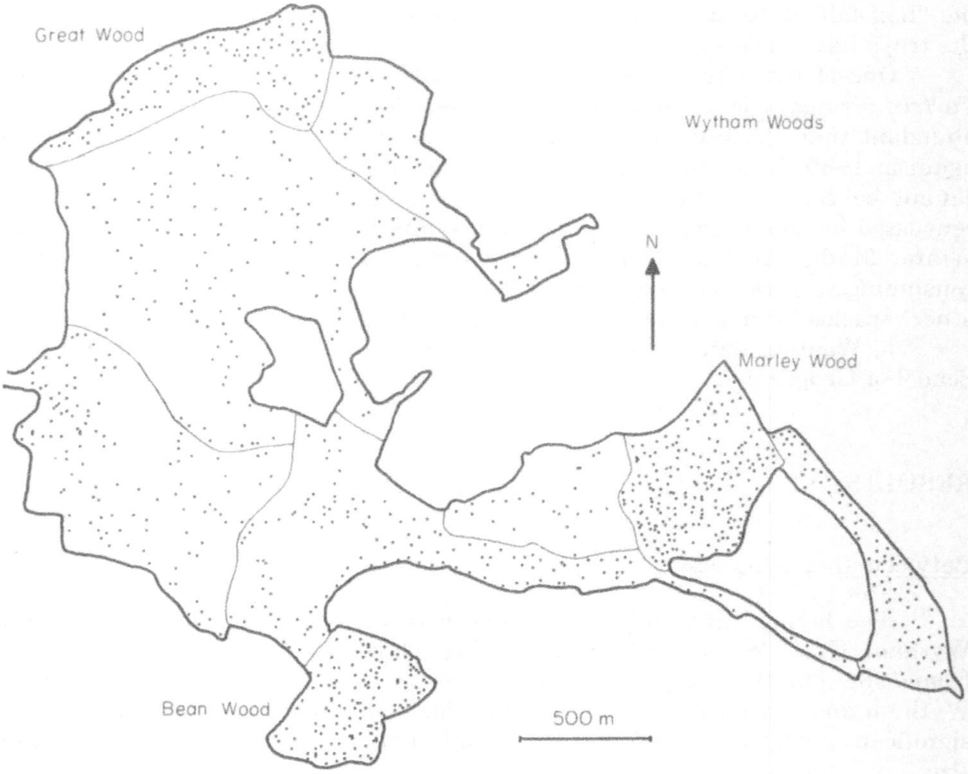

Fig. 1. Wytham Woods, Oxfordshire, showing the distribution of nest-boxes. The three study areas (Marley, Great Wood and Beanwood) are marked.

However, further analysis of the relationships of clutch-size with various factors does not consistently show the same relationships as were reported by Perrins and McCleery (1989) for Marley, or at least not so strongly (Table II).

There seem to be two factors which explain some of the differences. First, there are no correlations between clutch-size and density in Great Wood, though there are for Blue Tit in Bean Wood. This may be related to the densities of the nest-boxes in the two areas (4.1/ha in Great Wood, 7.9/ha in Bean Wood respectively, compared with 8.0/ha in Marley). As mentioned earlier, there are insufficient boxes in Great Wood to be able to accommodate all the birds and hence the numbers counted in boxes will not always be a true reflection of the total number of breeding pairs of tits present, though one might have expected Great Tit densities to be reasonably near to the total number of pairs of Great Tits nesting in the area. If this interpretation is correct, it emphasises that correlations between clutch-size and density should only be sought in study areas where there is a clear excess of nest-boxes over the maximum number of breeding pairs; whenever a significant proportion of the birds breed outside the nest-boxes, there is a risk that those that breed in the boxes will not be

Table II. Factors affecting clutch-size.

Wood	Species	GT Nos.	BT Nos.	Laydate	r^2
Marley	GT	-0.0292***	-0.0219***	-0.0685***	56%
	BT	-0.0390***	-0.0210 **	-0.0690***	51%
Great Wood	GT	-0.0155 ns	-0.1622 ns	-0.0149 ns	2%
	BT	0.0055 ns	-0.1164 ns	-0.0345 ns	0%
Beanwood	GT	-0.0085 ns	-0.0276 *	0.0262 ns	9%
	BT	0.0067 ns	-0.0568***	0.0127 ns	27%

Note: numbers given for each factor affecting clutch-size are the slopes of the regressions. The "*" denote the significance of the regression (*=$p<0.05$, **=$p<0.01$, ***=$p<0.001$). The final column shows the adjusted r^2 for the three factors combined. The degrees of freedom are 41 for Marley, 28 for Great Wood and 25 for Bean Wood. GT= Great Tit, BT= Blue Tit.

representative of the whole population. They may, for example, be the dominant birds with a larger than average clutch.

The second feature relates to the time-scale of the study. Perrins and McCleery (1989) showed that, for Marley, where we have forty years of data, the strength of the correlations between clutch-size and density varied markedly with time, being much stronger in the first and fourth decades of the study than in the middle two. The same pattern probably held true for Great Wood and Bean Wood; although we do not have data from the first period, the correlations are much stronger in the last ten years (Table III). We have no explanation for this apparent change in the relationship.

Table III. Factors affecting clutch-size: 1979-1989.

Wood	Species	GT Nos.	BT Nos.	Laydate	r^2
Marley	GT	-0.0514 *	-0.0306 ns	-0.0812 *	71%
	BT	-0.0475 *	-0.0605 *	-0.1039 **	78%
Great Wood	GT	-0.0567 *	-0.0406 ns	-0.0650 ns	30%
	BT	-0.0433 *	-0.0324 ns	-0.0387 ns	34%
Beanwood	GT	-0.0128 ns	-0.0942 **	-0.0082 ns	53%
	BT	-0.0001 ns	-0.0887 **	-0.0288 ns	67%

Note: numbers given are as for Table II. The degrees of freedom are 10 in all cases.

Relationship of clutch-size to caterpillar densities

The relationship between caterpillar densities and clutch-size of the two tit species is shown in Figure 2. In both species the clutch-size is strongly correlated with the densities of the Winter Moth, the relationship explaining about 38% of the variations in Great Tit clutch-size and about 40% of that in Blue Tits.

In all probability the relationship between clutch-size and densities of all caterpillar species would be much stronger than that shown here for Winter Moth alone. In the five years 1951-1955 densities of Green Tortrix exceeded 100 per square metre and in four of these years this species was markedly more abundant than Winter Moth. The points for these years are marked in Figure 2 from which it can be seen that, in these years, the clutch-size of the tits tended to be higher than would have been expected on the basis of Winter Moth numbers alone. Further, the highest mean clutch-size for both species during the years 1956-88 occurred in 1980. Although we do not have any accurate measurements of the caterpillar numbers in that year, they were clearly very high since many of the oaks were completely defoliated; again, from casual examination, it seemed that Green Tortrix was the most abundant species in that year. A point is included in Figure 2 for that year (though not included in the statistical analyses) against an arbitrary density of 1,000 caterpillars per square metre.

Further examination of the extended run of caterpillar numbers shows that the numbers in one year tend to be positively correlated with those in the previous one. A relationship between clutch-size and caterpillar density would therefore lead to clutch-size in one year being correlated with that in the previous year, an effect shown by Perrins and McCleery (1989) which they were not able to explain.

DISCUSSION

Overall, the three factors, Great Tit and Blue Tit numbers and caterpillar density, explain 61.5% of the variation in Blue Tit clutch-size and 70.8% of that of Great Tit clutch-size. Are these adaptive responses? Are the tits varying their clutch-size in ways which will help them to maximise their reproductive output? We cannot answer this question in a quantitative way, but we can say a little in qualitative terms.

First, it makes sense to have smaller clutches when conditions are not good for rearing young; we know that it is important for the young to be a reasonable weight at fledging if they are to have a good chance of surviving to breed (Perrins 1965). In qualitative terms, having small clutches in circumstances when the birds have difficulty raising young (such as in poor habitats or late in the season), is more likely to enable the parents to raise heavy young than is having large clutches.

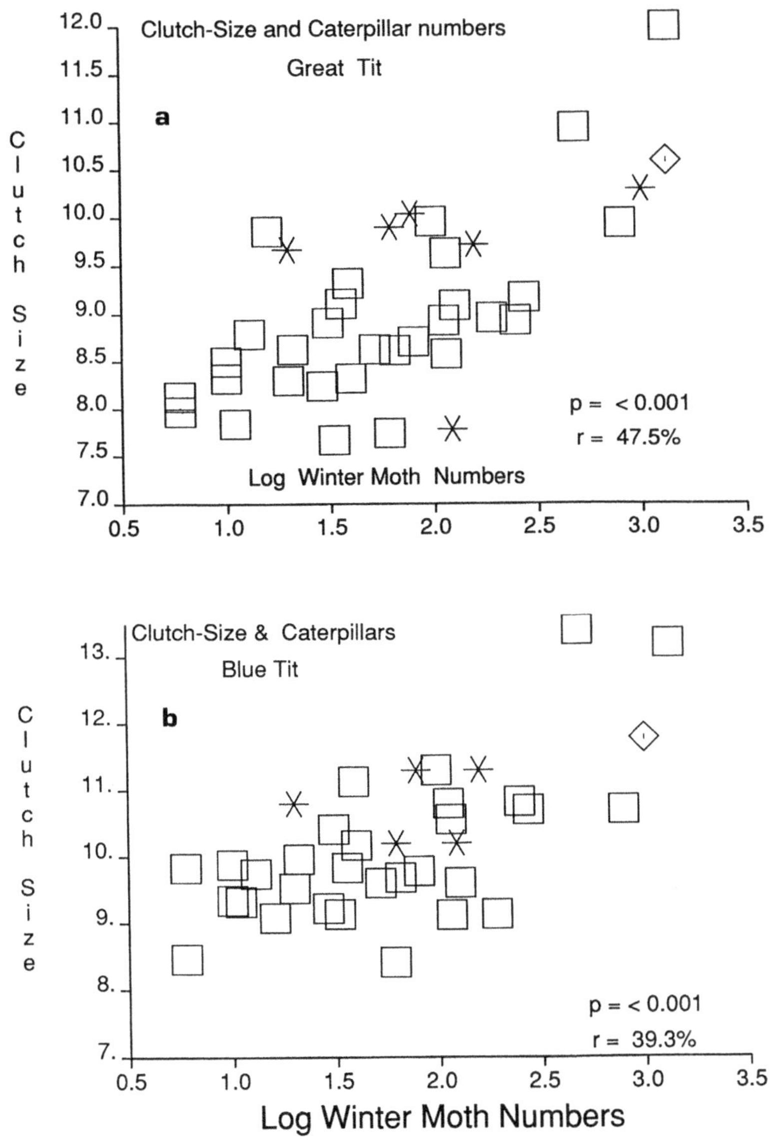

Fig. 2. The relationship between the clutch-sizes of the Great Tit (Fig 2a) and the Blue Tit (Fig 2b) and the number of Winter Moth caterpillars. The points marked (✳) show the years 1950-1955 when Green Tortrix was more abundant than the Winter Moth (points are still plotted against Winter Moth numbers) and the point marked (◇) denotes 1980 when the oak-trees were defoliated, primarily at least, by large numbers of Green Tortrix, but for which there are no actual counts of caterpillars.

If the tits can vary their clutch adaptively, are they doing so in relation to their own density? There is growing evidence (Dhondt 1981, Minot 1981) that the presence of the other species may affect the ease with which the two species can collect food for their young. Hence the presence of more pairs of their own species may also reduce the area within which they can collect food, either by reducing the area available for each pair to forage or by increasing the number of competitors over the area in which the individuals are foraging. It may therefore only be safe to raise large numbers of young in years when the density of breeding pairs is low. Reducing the clutch in years of high density would thus be an adaptive strategy. In this study however, the two tit species appear to be responding not only to the density of their own species, but also to the density of the other species. I know of no other case where the clutch size of one species is affected by the density of another.

When looking at the relationship between the tits and the caterpillar supply we are faced with a problem. Tits vary their clutch-size and their laying date. Provision of artificial food tends to lead to the birds advancing their laying date, but not laying larger clutches (Källander 1974, von Bromssen and Jansson 1980). Yet higher densities of caterpillars are associated with the opposite, namely larger clutches, but no change in laying date. Can we make sense of this?

Presumably the tits could respond to the numbers of caterpillars which are present when they are gathering the food to form their eggs. As already mentioned, losses of caterpillars between first instar, when the Winter Moth settle in the unfolding oak buds, and pupation are small. Hence a large number of first instar larvae is likely to mean a good food supply for the tits in the nestling period. Laying a larger clutch under these conditions could be an adaptive strategy.

The latest time at which the female can vary her clutch in respect to the numbers of caterpillars is when she "decides" to form the last egg which she does about 5 days before laying it. From then to having 10-day old young in the nest takes about 30 days (4 days from first egg, c14 to incubate, plus a nestling period of 10 days). The interval between hatching of the Winter Moth and the date of half-fall is about 40 days. Hence, in almost all years, it would be possible for the tits to adjust their clutch-size by laying more eggs when there were many small caterpillars around at laying time and have the young from those eggs in the nest when the same caterpillars are fully grown.

Why then do the tits not lay earlier in a season with a good caterpillar supply? There may be two reasons for this. First, the Winter Moth caterpillars are at this time exceedingly small and may not be much use as food themselves, the birds may have to rely on other food - as is suggested by the fact that they may lay earlier when food is artificially provided. Second, even if they are getting a lot of nutrient from the tiny caterpillars, they cannot lay early in relation to them, but rather have to wait until they are available.

Hence it would be perfectly possible for the tits to respond to high densities of caterpillars by laying larger clutches, but not by laying earlier. One important qualification is necessary. It is important to the parent tits to raise their young when the food supply is still plentiful, later broods are raised at a time when caterpillar numbers are declining and they fare badly (Perrins 1970).

However, although, on average, the Winter Moth take about 40 days to develop to pupation, this period varies markedly (from about 23-55 days, Perrins 1973). The main reason for this is variation in the temperature. Hot weather can greatly speed up the caterpillar development and result in them pupating much earlier than in a cooler summer. Plainly once the tits have completed their clutch there is no way in which they can shorten the incubation and nestling periods to match the variation in caterpillar development. The speed with which the caterpillars develop affects the survival of the young birds (van Noordwijk et al., in prep.).

ACKNOWLEDGEMENTS

I am particularly grateful to Dr. L. Cole for providing six more years of data for the Winter Moth populations in Wytham. I am also very grateful to the School of Geography for providing the weather data.

REFERENCES

Bromssen A von and Jansson C (1980) Effects of food addition to Willow Tit *Parus montanus* and Crested Tit *P. cristatus* at the time of breeding. Ornis Scand 11: 173-178

Dhondt AA and Eyckerman (1981) Competition and the regulation of numbers in Great and Blue Tit. In: Klomp H and Woldendorp JW (Eds) The Integrated Study of Bird Populations. Wageningen, Netherlands

Dunn EK (1976) Predation by weasels (*Mustela nivalis*) on breeding tits (*Parus* spp.) in relation to the density of tits and rodents. J Anim Ecol 46: 633-652

East M and Perrins CM (1988) The effect of nest-boxes on breeding populations of birds in broad-leaved temperate woodland. Ibis 130: 393-401

Källander H (1974) Advancement of laying of Great Tits by the provision of food. Ibis 116: 365-367

Kluijver HN (1951) The population ecology of the Great Tit, *Parus m. major* L. Ardea 39: 1-135

Lack D (1958) A quantitative breeding study of British tits. Ardea 46: 91-124

Lack D (1966) Population studies of birds. Clarendon Press, Oxford

Minot EO (1981) Effects of interspecific competition for food in breeding Blue and Great Tits. J Anim Ecol 50: 375-385

Minot EO and Perrins CM (1986) Interspecific interference competition - nest sites of Blue and Great Tits. J Anim Ecol 55: 331-350

Noordwijk A van, McCleery RH and Perrins CM (in prep.) Selection for Great Tits *Parus major* to synchronise their breeding with caterpillar abundance.

Perrins CM (1965) Population fluctuations and Clutch-size in the Great Tit *Parus major* L. J Anim Ecol 34: 601-647

Perrins CM (1970) The timing of birds' breeding seasons. Ibis 112: 242-255

Perrins CM (1973) Some effects of temperature on the breeding of the Great tit and the Manx shearwater. J Reprod Fert Suppl 19: 163-173

Perrins CM (1979) British Tits. Collins, London

Perrins CM and McCleery RH (1989) Laying dates and clutch-size in the Great Tit. Wilson Bull 101: 236-253

Varley GC, Hassell M and Gradwell GR (1973) Insect Population Ecology. Blackwell, Oxford

IS ADULT SURVIVAL OF THE BLUE TIT HIGHER IN A LOW FECUNDITY INSULAR POPULATION THAN IN A HIGH FECUNDITY MAINLAND ONE ?

Jacques BLONDEL and Roger PRADEL

Centre Louis Emberger, CNRS, B.P. 5051, 34033 Montpellier cédex, France

Population studies on tits have been in progress for 15 years in Corsica and in a mainland forest in Southern France. The island site is an old mature forest of the sclerophyllous Holm Oak *Quercus ilex* located near the sea. The mainland site is a mixed forest of conifers and broadleaved trees located in the southern slope of Mont-Ventoux, Provence (see characteristics of the two habitats in Table I and in Blondel 1985).

Because of its lower latitude, lower altitude and insular status, the Corsican climate is warmer, drier and more predictable (lower coefficients of variation) than the mainland one (Table I). Tits in this species-poor island forest exhibit several characteristics of a so-called insular syndrome which has been demonstrated at the community-wide level, especially density inflation and habitat-niche expansion (see Blondel et al. 1988). According to these insular features, several life history traits of the birds were expected to differ between this island and the nearby mainland. Theoretical predictions on the evolution of life history traits in benign insular bioclimates (MacArthur and Wilson 1967,

Table I. Characteristics of the habitats where survival of Blue Tit has been estimated. Climatic data for the years 1976-1988 combined, T= temperatures in °C, R= rainfall in mm, CV= coefficient of variation.

	Mainland (Ventoux)			**Island** (Corsica)		
Coordinates:	44°08'N - 5°11'E			42°24'N - 8°44'E		
Altitude:	800m - 1000m			100m - 200m		
Tree species:	*Cedrus, Pinus, Quercus pubescens*			*Quercus ilex*		
Climate:	T	(CV)	R	T	(CV)	R
March	5.0	(52)	65	10.0	(21)	69
April	8.2	(33)	79	12.1	(19)	40
May	11.4	(25)	88	15.8	(16)	79
June	15.5	(19)	69	19.5	(13)	34
July	19.0	(14)	69	23.5	(9)	13

NATO ASI Series, Vol. G 24
Population Biology of Passerine Birds
Edited by J. Blondel et al.

Pianka 1970, Cody 1966, 1971) propose that organisms in stable and predictable environments such as islands are selected for individual efficiency in crowded populations because of environmental stability, whereas organisms in unstable environments are selected for productivity in uncrowded populations because of environmental instability. The basic idea is that under island stable climatic conditions, the best reproductive strategy would be to lower the number of eggs laid and allocate the energy thereby saved into other components of fitness such as adult survival in crowded populations where density is supposed to be close to the carrying capacity K (Cody 1966, 1971). Given these premises, predictions on changes of life history traits on the island by comparison to the nearby mainland should include among others:

1) a delayed breeding season,
2) a shortened breeding season (absence of second broods),
3) a decrease of clutch-size,
4) an increase in adult survival rates.

The three first predictions are fulfilled (Fig. 1): Corsican tits lay on average three weeks later than their mainland counterparts, they have a shorter breeding season, and their clutch size is 27% lower. These results were first interpreted in the scope of the hypotheses stated above (Blondel and Isenmann 1979), and we predicted a higher survival rate of tits in Corsica than on the mainland (Fig. 1).

However, neither the adult/yearling ratio of breeders, considered to be an indication of adult survival rate (Hilden 1982), nor the recapture rate of previously ringed birds were higher on the island. These observations seriously weakened our hypothesis. Since the prediction of a higher adult survival in Corsica than on the mainland is a keystone of the theory on changes in life history traits on islands, we carried out a more careful analysis of survival probabilities.

MATERIAL AND METHODS

One hundred and thirty nestboxes were evenly distributed according to the same experimental design in each habitat in order to have comparable data sets. At each breeding season during 10 years (1980-1989) as many as possible of the birds breeding in the nestboxes were caught while feeding their young. The percentages of adults caught among those breeding in the nestboxes were 79% in Corsica and 76% on the mainland (all years combined). Capture sessions lasted about one month (May in Ventoux, June in Corsica). Each bird was banded, sexed, weighed and aged and all the nestlings were banded. The only possible determination of age, from plumage criteria, was between yearlings (one-year old birds), and older birds, hereafter called adults. A very small proportion of the nestlings were recaptured the following year(s): 91 out of 1836 for the two study sites combined. These birds have been considered as if they have been ringed as breeders at their first recapture. Therefore they have been put in the yearling group if they were observed as breeders for the first time

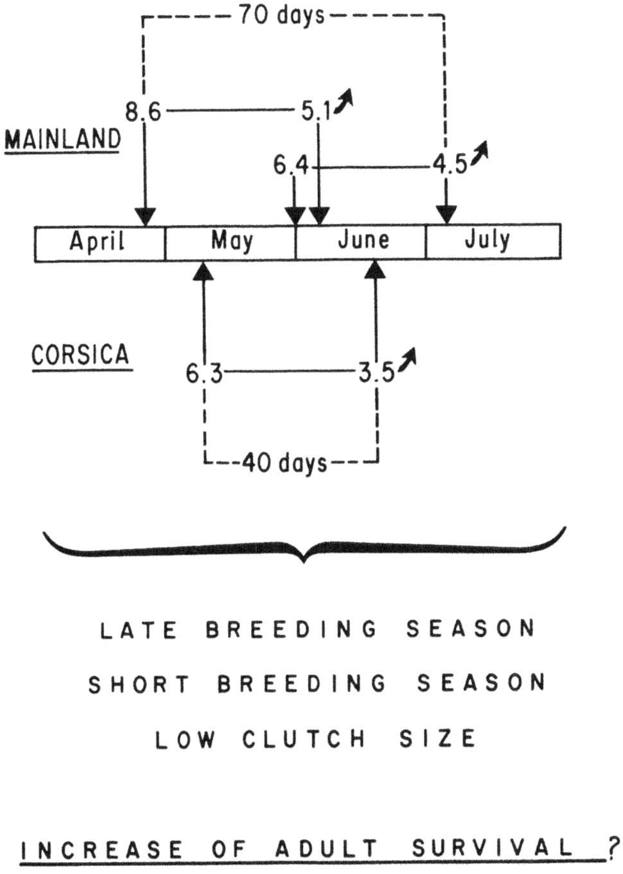

Fig. 1. Breeding patterns of the Blue Tit on the mainland and on Corsica. Figures from left to right are clutch size and numbers of fledglings (a small proportion of mainland tits attempt a second brood).

the year after they were born, and in the adult group if their first breeding event was recorded later. Without the birds marked the last year (1989), the data set includes 421 birds split in eight data subsets (Table II) according to age at first capture (yearlings or adults), site (Ventoux or Corsica), and sex (male or female).

There are now good methods to estimate survival rates from capture-recapture data in an efficient and flexible way (Pollock 1981, Burnham et al. 1987, Clobert et al. 1987, 1988, Lebreton 1989) generalized from that of Cormack, Jolly and Seber, hereafter denoted the CJS model (Cormack 1964, Jolly 1965, Seber 1965). The estimation of survival probabilities from recapture data relies on models which include both survival and recapture probabilities. Survival probability is the probability that a bird alive at sampling period $\underline{t}$ is

Table II. Numbers of birds ringed as yearlings (or first recaptured when marked as nestlings) and as adults in Mont Ventoux and in Corsica; 1979-1988.

		Yearlings	Adults	Total
Ventoux	Males	19	38	57
	Females	28	54	82
Corsica	Males	60	80	140
	Females	91	51	142
Total		198	223	421

still alive and in the area subjected to sampling effort at time t+1. Capture probability is the conditional probability that an individual which is alive and present in the area is caught in period t. Theses models explicitly include the possibility that an individual is alive but not caught. Therefore it avoids the biases of *ad hoc* estimates of survival probabilities based on ratios of the number of recaptured birds over the number of released birds. More precisely, such models avoid the biases resulting from the confounding of survival and recapture when estimating survival rates as the proportion of birds released in year i that are seen again in subsequent years ($\geq i+1$).

All estimates in this study were obtained using program SURGE (Lebreton and Clobert 1986), in a new version for multiple data sets (Pradel 1988). A detailed description of the procedure applied to a subset of the data set utilized in this paper is given by Pradel et al. (1990). The goodness-of-fit tests were performed using program RELEASE (Burnham et al. 1987). See Clobert et al. (1987) and Pradel et al. (1990) for more details on the statistical procedure.

RESULTS

Differences in survival rates and recapture probabilities were explored between sites, age classes, sexes and time.

Age effect on survival (Table III, Fig. 2)

Following the notation by Wilkinson and Rogers (1973) the model utilized is S_{a*s},P_s in Corsica and S_{a*s},P_{t+s} in Ventoux, where S= survival, a= age, s= sex, t= time, P= recapture probability. This means that survival depends on

Table III. Survival rates of Blue Tits on the mainland (Ventoux) and in Corsica according to age and sex; standard deviation in brackets (estimation from the model S_{a*s},P_s in Corsica and from the model S_{a*s},P_{t+s} in Ventoux, see text).

		Adults	Yearlings
Ventoux	Males	.68 (.049)	.73 (.128)
	Females	.49 (.050)	.68 (.097)
Corsica	Males	.54 (.046)	.68 (.101)
	Females	.57 (.041)	.63 (.060)

age, sex and their interaction in the two regions and that recapture probabilities depend on sex only in Corsica and on sex and time in Ventoux. In the four groups, survival is higher for yearlings than for adults. The overall difference between age classes across the four sex/site groups combined is significant (one-tailed test, standardized normal variable=1.958, P=0.025, Table III). However, since the age effect is not very strong, the CJS model (which does not account for age effect) is still acceptable for each sex in each site and we went on using the CJS model.

Sex and site effects on survival (Table III, Fig. 2)

There was no significant sex effect on adult survival in Corsica ($X^2_{(1)}$= 0.33, P=0.57) where survival of both sexes combined is 0.581±0.026 (mean±SD). On the other hand, males survived significantly better (0.691±0.045) than females (0.526±0.042) on the mainland ($X^2_{(1)}$=7.99, P=0.05). If one neglects the difference between sexes, the combined survival in Ventoux is 0.584, a figure strikingly similar to that in Corsica. The difference between regions is significant for males (standardized normal variable= 2.101, P=0.036) while it is not for females (standardized normal variable= 1.11, p=0.267). The most important result in the scope of this study is that whatever sex and age, survival rates of Blue Tits are not higher on the island than on the mainland.

Time effect on survival (Fig. 2)

There was no significant variation in survival over the years in Corsica. On the other hand, in Ventoux, a time effect is associated with the sex effect so

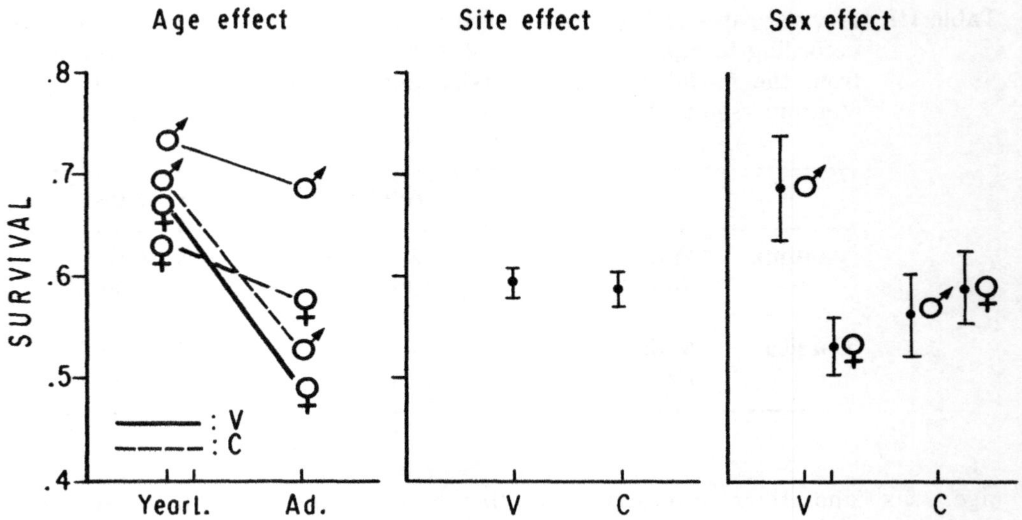

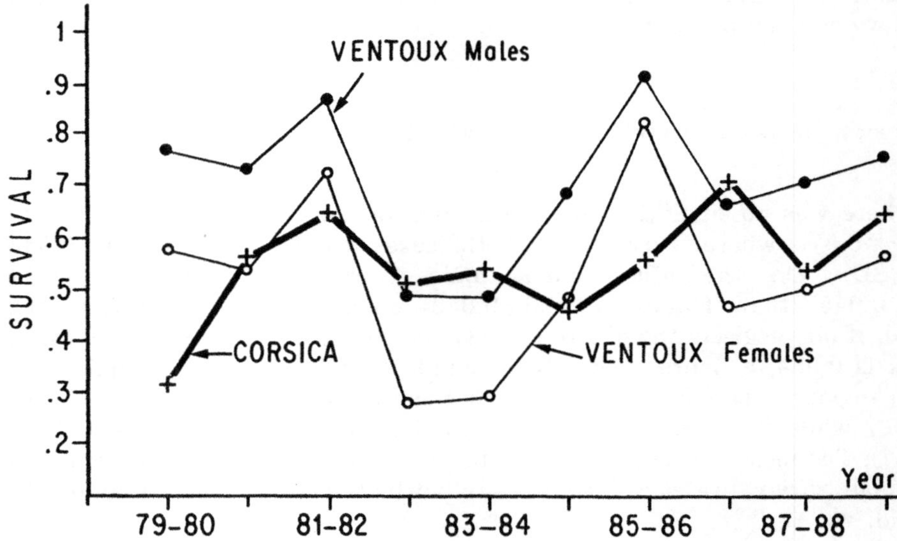

Fig. 2. Survival estimates of Blue Tits on the mainland (V) and in Corsica (C) according to age, site, sex and time (sexes combined for Corsica). Vertical bars = ±1 SD.

that the model with both factors (S_{t+s}, P_{t+s}) is better than a model with sex only ($X^2_{(9)} = 17.59$, $P = 0.04$). However the improvement is probably due to the high number of parameters in regard to the small sample size, especially during the first years where few birds were marked. An application of the Akaike's (1973) information criterion which penalizes models with many parameters (Lebreton et al. submitted) leads to the conclusion that the two models are nearly equivalent. Thus one may conclude that survival is more variable over time in Ventoux than in Corsica. On the whole, males survived worse on the mainland than on the island for two years (1982/83, 1983/84) and females survived better also for two years (1981/82, 1985/86) (Fig. 2).

Recapture probabilities

Recapture probabilities were sex dependent in both sites and females have significantly higher probabilities of being captured than males. Average figures are respectively 0.832 against 0.554 in Corsica and 0.854 against 0.612 in Ventoux ($X^2_{(1)} = 0.17$, $P = 0.68$). Moreover, in Ventoux there was a time component added on the sex effect (P_{t+s}).

DISCUSSION

At the scale of the Mediterranean region, with an average survival of 0.58 for both sexes and sites combined, survival rates of the Blue Tit are much higher than those reported from central and northern Europe, for example 0.41 (both sexes combined) in Finland (Hildén 1982), ca 0.40 (several figures pooled) in Belgium (Dhondt 1987, 1989), and 0.35 in Northern Europe (Snow 1956). These figures agree with the trend described by Snow (1956) of an increasing survival rate from north to south: tits survive better and produce fewer young in the Mediterranean region than in central and northern Europe.

However, at the between-habitats scale within the Mediterranean region, our data show that the problem is more complicated than expected from a simple clinal trend. In contrast to what was expected from theoretical predictions, Blue Tits do not survive better on Corsica than on the mainland, in spite of a much lower production of young and a much more predictable climate. Moreover, there is in Ventoux a large inter-annual variation in survival probabilities. Such a variation is consistent with the results of long term studies at higher latitudes, both in Blue Tits (Nur 1984, 1987, 1988) and Great Tits (Perrins 1965, Boyce and Perrins 1987, Clobert et al. 1987). The absence of variation in Corsica was expected in this stable insular environment.

These results question the long-held dogma of high survival rates to be found in the stable and predictable environments of islands. The problem is to know why survival rates are similar in the two habitats and how the demographical balance is met in two populations with similar survival rates but different breeding commitments.

Because most models of the evolution on reproductive rates assume a cost of reproduction, and because many studies have shown a direct relationship between reproductive effort and future survival of the parents (Drent and Daan 1980, Martin 1987, Nur 1988b, see Linden and Møller 1989 and Nur this volume for reviews), we assume that survival rates reflect a cost of reproduction. Therefore, adult survival must be discussed in relation to breeding effort (Stearns 1976, Tinbergen 1987, Nur 1988). Hence we address the following questions:

1) why is clutch size so low on Corsica?

2) why is adult survival similar in the two habitats,

3) what are the links and trade-offs between these demographic parameters?

Clutch size in Corsica is the lowest so far recorded in Europe and North Africa except the Canary islands. Clutch size is influenced by a number of factors such as age of the female, breeding density, date of laying, territory quality and food resources. Blondel (1985) has shown that clutch size in Corsica is much lower than expected if its reduction was only due to the first four factors. Clearly the most important factor affecting clutch size on Corsica is food. Accordingly clutch size must be interpreted as a component of the energetic parental investment in breeding (Klomp 1970, van Balen 1973, Martin 1987).

The low breeding commitment of Corsican tits suggests that for a similar cost of reproduction in the two habitats (reflected by similar survival rates), fewer offspring can be produced on the island. If feeding conditions ultimately operate to evolve optimal clutch size (Martin 1987), then clutch size reduction on this island should be an evolutionary response to poor trophic conditions. The problem is to investigate to what extent food conditions are poorer on the island. Several studies have shown that 1) The spring development of the vegetation and of its associated insect fauna occurs much later in sclerophyllous Mediterranean trees than in deciduous ones (Isenmann et al. 1987); 2) The food supply is very low in the sclerophyllous Mediterranean forests, especially on Corsica (Zandt et al. this volume, Blondel et al. in prep.).

A study of the diet of nestlings has shown that there are many fewer caterpillars in the diet of the Corsican nestlings and a greater diversity of prey than on the mainland (Blondel et al. in prep.). Assuming that caterpillars are the preferred and the most profitable prey for the tits (Perrins 1970, Royama 1980, van Balen 1973), the low abundance of caterpillars forces the birds to shift on other prey items such as spiders that are usually smaller and poorer in water content (Blondel et al. in prep.). A high diversity of water-poor prey items probably constitutes a severe constraint in Corsica, especially when temperatures are very high in the second half of the breeding cycle. Hence, presumably because of these climatic and trophic constraints, the birds start to breed later, stop earlier, have fewer eggs and produce fewer young in Corsica than on the mainland.

Adult survival rates must be interpreted in relation to breeding effort. The critical season for adult survival in Corsica is probably not the winter, which is mild and moist and probably not food limiting, but the hot and dry summer when evapotranspiration largely exceeds the water supply (Table I).

Although no survival data are available between fledging and the early autumn, it is hypothesized that the bottleneck for adult survival is the unfavourable Mediterranean summertime. During the recovery period after reproduction, the parent birds have to forage intensively to restore their nutritional level (Ricklefs 1977). Resource depletion within the territory while collecting food for nestlings might account for increased difficulties for the sedentary parents to find food after breeding once the young have dispersed (Greig-Smith 1982, Ekman and Askenmo 1986). Moreover the birds have also to complete their moult which is energy demanding (Ricklefs 1974). Some studies have shown a negative correlation between adult survival and the size of the juvenile population (Kluyver 1970, Ekman and Askenmo 1986, Arcese and Smith 1988), which is an indication of parent-offspring conflict for survival.

Assuming that the two populations are in equilibrium, the crucial point to explain the puzzle presented by similar survival rates and different fecundities in the two areas is that recruitment of fledged young differs between the two study sites. This is consistent with much less harsh weather conditions in the first winter of life on Corsica. Thus, other components of fecundity are involved in the trade-offs between clutch size and adult survival. Such components as the proportion of breeders, age at first breeding, and especially postfledging survival must be of particular importance (Bennett and Harvey 1988, Karr et al. 1990). Unfortunately our knowledge of these components is still very scanty. The main message of this study is that we need to study not only clutch size as the main component of fecundity, but also these other components which are probably subject to high variation between species, populations and habitats. This conclusion supports a model of Ricklefs (1977) testing the relationship between optimization of fecundity and components of mortality: "Fecundity is determined primarily by density-dependent response to adult mortality mediated through resources available for reproduction" (Ricklefs 1977).

CONCLUDING REMARKS

Because it is much easier to collect data on fecundity than data on survival, most estimations of survival rates on islands have so far been inferred from fecundity rates and environmental correlates (e.g. Higuchi 1976, Crowell and Rothstein 1981). In the general framework of the evolution of life history strategies on islands this study shows that 1) there is no simple relationship between fecundity and survival, and 2) survival rates are not necessarily higher in predictable island habitats than in more unpredictable continental ones.

The main message of our study is that survival estimates are valid only for that population and those environmental constraints for which they have been calculated. No generalisation on survival can be drawn from other traits such as fecundity or from environmental attributes until one has carefully studied pay-offs between reproductive cost and subsequent survival at the within-population level. Therefore the next steps will be 1) to analyze at the within population level the effects of reproductive cost on survival through brood size manipulations, and 2) to check by ecophysiological studies the importance of climatic and trophic constraints, especially the availability of water for the nestlings (Mertens in progress), in the shaping of life history strategies on Corsica.

ACKNOWLEDGEMENTS

Ph. Perret and M. Maistre provided much assistance in field work. J.D. Lebreton helped us in statistical calculations. The CRBPO agreed our project and supplied us with rings. R.H. McCleery and T. Slagsvold made very useful comments on a earlier draft of the manuscript. We are grateful to all of them.

REFERENCES

Akaike H (1973) Information theory and an extension of the maximum likelihood principle. In B.N. Petran and F. Csaki (eds) International symposium on information theory. (2nd ed.), p 267-281 , Akademiai Kiadi, Budapest

Arcese P, Smith JNM (1988) Effects of population density and supplemental food on reproduction in song sparrows. J Anim Ecol 57: 119-136

Balen JH van (1973) A comparative study of the breeding ecology of the Great Tit *Parus major* in different habitats. Ardea 61: 1-93

Bennett PM, Harvey PH (1988) How fecundity balances mortality in birds. Nature 333: 216

Blondel J (1985) Breeding strategies of the Blue Tit and the Coal Tit (*Parus*) in mainland and island Mediterranean habitats: a comparison. J Anim Ecol 54: 531-556

Blondel J, Chessel D, Frochot B (1988) Bird species impoverishment, niche expansion and density inflation in Mediterranean island habitats. Ecology 69: 1899-1917

Blondel J, Dervieux A, Maistre M, Perret P (in prep.) Feeding ecology of the Blue Tit *Parus caeruleus* in Mediterranean mainland and island habitats.

Blondel J, Isenmann P (1979) Insularité et démographie des Mésanges du genre *Parus* (Aves). C R Acad Sci Paris, Sér D, t 289: 161-164

Boyce M, Perrins CM (1987) Optimizing great tit clutch size in a fluctuating environment. Ecology 68: 142-153

Burnham KP, Anderson DR, White GC, Brownie C, Pollock KH (1987) Design analysis methods for fish survival experiments based on release-recapture. American Fisheries Society, Monograph 5

Clobert J, Lebreton JD, Allainé D (1987) A general approach to survival rate estimation by recaptures or resightings of marked birds. Ardea 75: 133-142

Clobert J, Perrins CM, McCleery RH, Gosler AG (1988) Survival rate in the Great Tit *Parus major* in relation to sex, age, and immigration status. J Anim Ecol 57: 287-306

Cody ML (1966) A general theory of clutch size. Evolution 20: 174-184

Cody ML (1971) Ecological aspects of reproduction. In Farner DS, King JR (eds) Avian Biology. Vol 1, Academic Press, New York, p 462-503

Cormack RM (1964) Estimates of survival from the sighting of marked animals. Biometrika 51: 429-438

Crowell KL, Rothstein SI (1981) Clutch-sizes and breeding strategies among Bermudan and North American passerines. Ibis 123: 42-50

Dhondt AA (1987) Reproduction and survival of polygynous and monogamous Blue Tit *Parus caeruleus*. Ibis 129: 327-334

Dhondt AA (1989) Blue Tit. In I Newton (ed) Lifetime Reproduction in Birds. Academic Press, New York, p 15-33

Drent R, Daan S (1980) The prudent parent: energetic adjustments in avian breeding. Ardea 68: 225-252

Ekman J, Askenmo C (1986) Reproductive cost, age-specific survival and a comparison of the reproductive strategy in two european tits (Genus *Parus*). Evolution 40: 159-168

Greig-Smith PW (1982) Dispersal between nest-sites by stonechats *Saxicola torquata* in relation to previous breeding success. Ornis Scand 13: 232-238

Higuchi H (1976) Comparative study on the breeding of Mainland and Island subspecies of the Varied Tit, *Parus varius*. Tori 25: 11-20

Hildén O (1982) Proportion of juveniles as a measure of adult mortality in a breeding population of Great tits *Parus major* and Blue tits *P. caeruleus*. Ornis Fenn 59: 149-151

Isenmann P, Cramm P, Clamens A (1987) Etude comparée de l'adaptation des mésanges du genre *Parus* aux différentes essences forestières du bassin méditerranéen occidental. Rev Ecol (Terre et Vie) Suppl 4: 17-25

Jolly GM (1965) Explicit estimates for capture-recapture data with both death and immigration-stochastic model. Biometrika 52: 225-247

Karr JR, Nichols JD, Klimkiewicz K, Brawn JD (1990) Survival rates of tropical and temperate forest birds: Will the dogma survive? Am Nat, in press

Klomp HW (1970) The determination of clutch size in birds. A review. Ardea 58: 1-124

Kluyver HN (1970) Regulation of numbers in populations of great tits (*Parus m. major*). In den Boer PJ, Gradwell GR (eds) Dynamics of numbers in populations. Oosterbeek, Centre for Agricultural Publishing and Documentation, Wageningen, p 507-523

Lebreton JD (1989) Statistical methodology for the study of animal populations. Bull Int Stat Inst 53(3): 267-282

Lebreton JD, Clobert J (1986) User's Manual for Program SURGE. CEPE/CNRS, Montpellier, France

Lebreton JD, Clobert J, Anderson DA, Burnham KP (in prep.) Modelling survival and testing biological hypotheses using marked animals.

Linden M, Møller AP (1989) Cost of reproduction and covariation of life history traits in birds. Tree 4: 367-371

MacArthur RH, Wilson EO (1967) The theory of Island Biogeography. Princeton Univ Press, Princeton, NJ

Martin TE (1987) Food as a limit on breeding birds: a life-history perspective. Annu Rev Ecol Syst 18: 453-487

Nur N (1984) The consequences of brood size for breeding blue tits. I. Adult survival, weight change and the cost of reproduction. J Anim Ecol 53: 479-496

Nur N (1987) Alternative reproductive tactics in birds: Individual variation in clutch size. In Bateson PPG, Klopfer PH (eds) Perspectives in Ethology. Vol 7, Plenum, NY, p 49-77

Nur N (1988a) The consequences of brood size for breeding Blue tits. III. Measuring the cost of reproduction: survival, future fecundity, and differential dispersal. Evolution 42: 351-362

Nur N (1988b) The cost of reproduction in birds: An examination of the evidence. Ardea 76: 155-168

Perrins CM (1965) Population fluctuations and clutch size in the great tit *Parus major* L.. J Anim Ecol 34: 601-647

Perrins CM (1970) The timing of birds'breeding seasons. Ibis 112: 242-255

Pianka ER (1970) On r and K selection. Am Nat 100: 33-46

Pollock KH (1981) Capture-recapture models allowing for age-dependent survival and capture rates. Biometrics 37: 521-529.

Pradel R (1988) Nouvelles approches statistiques en capture-recapture, un exemple: le test de parallélisme. DEA, Univ Montpellier, 63 p

Pradel R, Clobert J, Lebreton JD (1990) Recent developments for the analysis of multiple capture-recapture data sets: an example concerning two Blue Tit populations. The Ring, in press

Ricklefs RE (1974) Energetics of reproduction in birds. In RA Paynter (ed) Avian Energetics. Pub Nuttall Ornith Club, n° 15, p 152-292, Cambridge, Massachusetts

Ricklefs RE (1977) On the evolution of reproductive strategies in birds: Reproductive effort. Am Nat 111: 453-478

Royama T (1970) Factors governing the hunting behaviour and selection of food by the Great tit (*Parus major* L.). J Anim Ecol 39: 619-668

Seber GAL (1965) A note on the multiple-recapture census. Biometrika 52: 249-259

Snow DW (1956) The annual mortality of the Blue Tit in different parts of its range. Brit Birds 49: 174-177

Stearns SC (1976) Life-history tactics: a review of the ideas. Q Rev Biol 51: 3-47

Tinbergen JM (1987) Costs of reproduction in the Great Tit: intraseasonal costs associated with brood size. Ardea 75: 111-122

Wilkinson W, Rogers CE (1973) Symbolic description of factorial models for analysis of variance. Appl Statist 22: 392-399

Winkel W, Winkel D (1976) Über die brutzeitliche Gewichtsentwicklung beim Trauerschnäpper (*Ficedula hypoleuca*). J Orn 117: 419-447

Zandt H, Strykstra A, Blondel J, van Balen JH (1990) Food in two Mediterranean Blue Tit populations: Do differences in caterpillar availability explain differences in breeding behaviour? This volume

TWO MEDITERRANEAN BLUE TIT POPULATIONS: ARE DIFFERENCES IN THE TIMING OF BREEDING ASSOCIATED WITH CATERPILLAR AVAILABILITY?

Hagen S. ZANDT[1], Arjen M. STRIJKSTRA[2], Jacques BLONDEL[3] and Johan H. van BALEN[4]

[1] Zoologisches Institut, Rheinsprung 9, CH 4051, Basel, Switzerland
[2] Zoological Laboratory, University of Groningen, Haren, The Netherlands
[3] Centre Louis Emberger, CNRS, B.P. 5051, 34033 Montpellier, France
[4] Institute for Ecological Research, Boterhoeksestraat 22, 6666 GA Heteren, The Netherlands

A long term study on the population biology of tit species in Mediterranean habitats has documented a striking difference in breeding patterns between populations on the island of Corsica and the mainland (Blondel 1985, 1987, Blondel et al. 1987, 1989, Isenmann 1983, Isenmann et al. 1987). Blue Tit (*Parus caeruleus*) and Coal Tit (*P. ater*) populations on Corsica start laying three weeks later, have ca. 30% smaller clutches and suffer lower juvenile mortality than mainland tits. Blondel (1987) stated that it is not sufficient to focus on the community structure of these populations to understand the observed differences in breeding biology. Research should also concentrate on local environmental conditions: the availability of food, climatic factors and/or parasitism on nestlings by e.g. dipteran larvae (Blondel 1985, 1987).

The effect of the availability of food, an important ultimate and proximate factor in the timing of breeding in birds (Lack 1966, Perrins 1970, Drent and Daan 1980, Arcese and Smith 1988, Daan et al. 1989), on the start of egg laying has been intensively studied in tit species (*Parus* spp.), especially the Great Tit (*P. major*) and the Blue Tit. Caterpillars are the most important prey fed to nestling tits (Betts 1955, Gibb and Betts 1963, Royama 1970, van Balen 1973, Cowie and Hinsley 1988, Blondel et al. 1989). In two Great Tit populations the timing of the breeding season seems to have evolved so that young are raised when caterpillar availability reaches its maximum level (Perrins 1965, van Balen 1973). Whilst ambient temperature is an important proximate factor affecting laying date in tits (Kluyver 1951, 1952, Perrins 1965, Lack 1966, van Balen 1973, Slagsvold 1976, Schmidt 1984, Blondel 1985, 1987, Nager, this volume) part of the remaining variation might be attributed to variation in food supply in the pre-laying period, as indicated by supplemental feeding experiments (Källander 1974, Clamens 1987a). Blondel (1985) and Blondel et al. (1989) drew special attention to the potential role of food availability in the timing of the breeding season in Mediterranean tit populations, especially on Corsica where no correlation has been found between ambient temperature and laying date. This paper presents results of a study on

NATO ASI Series, Vol. G 24
Population Biology of Passerine Birds
Edited by J. Blondel et al.

the availability of food in two Mediterranean study sites, one on Corsica and one on the mainland in relation to the timing of the Blue Tit's breeding season.

METHODS

The availability of food and the breeding biology of Corsican Blue Tits were studied in Pirio (Corsica, 42°24'N, 08°44'E) and those of the mainland Blue Tits in Puéchabon (near Montpellier, France, 43°54'N, 04°01'E). The dominant tree species is the same in both sites (Holm oak *Quercus ilex*) but shrubs and herbs differ because the substrate is siliceous on Corsica and calcareous on the mainland. The altitude of both sites lies between 100 and 300 m a.s.l..

In the Mediterranean area the proportion (on the basis of biomass) of caterpillars in the diet of Blue Tit nestlings ranges from at least 50% on Corsica to 85% on the mainland (Blondel and Zandt, unpubl., Blondel et al. in prep.). Blondel et al. (1989) documented that other prey types are fed to nestlings in substantial numbers (spiders, grasshoppers), but on a biomass basis these species were unimportant (Blondel and Zandt, unpubl.). Hence we estimated the availability of food by estimating caterpillar abundance.

The abundance of caterpillars was estimated by measuring frass fall under oaks and converting it into caterpillar biomass (Schwerdtfeger 1931, Gibb 1950, Betts 1955, Tinbergen 1960, van Balen 1973, Liebhold and Elkinton 1988a). Ten oaks on the mainland and 28 oaks on Corsica were sampled, using one frass collector ($0.25 m^2$) per tree. Frass fall was collected once (mainland, 1988) or twice a week, over an episode spanning most of the tits' breeding season (Corsica 1988: days 66 to 122, 1989: 64 to 116; mainland 1988: 48 to 141, 1989: 45 to 123; day numbers: 1 = 1 March). Samples were air dried, stored and dried at 50°C during two days before analysis. Samples were sorted into two categories: frass produced by caterpillars of the genus *Lymantria*, and frass of all other caterpillars. *Lymantria* caterpillars (*L. dispar*, *L. monacha*) are hardly ever fed to nestling tits (photographic registrations, Blondel et al. in prep.), so these caterpillars are not relevant when estimating food availability for tits. Frass of third instar and older *Lymantria* caterpillars can be easily distinguished from other frass (Liebhold and Elkinton 1988a). Frass dry weights were measured and used to calculate frass production ($mg/24h/0.25m^2$).

The relation between frass production and caterpillar biomass was studied on Corsica in 1988 and 1989. Caterpillars, collected from Holm Oaks, were kept in small-mesh wire cages and jam jars, in an outside laboratory protected against wind and excessive sun. Every 1 to 5 days the caterpillars were weighed, and the frass produced was collected. Frass was handled as stated above and frass production calculated (mg/24h). The relation between frass production (FP) and biomass (B) of caterpillars was analysed separately for days with and without rain, using multiple linear regressions forced through the origin. The general model tested was $FP=a*B + b*B*T + c*B*R + d*B*T*R$, where a, b, c and d are constant, T is the average temperature during two frass collections from both collectors and cages, and R is the mean precipitation (mm) per day. Data on temperature and precipitation were available from meteorological stations at both study sites (Service Technique d'Etudes des Facteurs Climatiques de l'Environnement, Avignon) and were multiplied with biomass to avoid getting a nonsense value for frass production at zero biomass. The relationships were not significantly different between years, so the data were pooled: $FP=0.174B$ ($n=177$, $r^2=0.80$, $p<0.001$) for days without rain and $FP=0.177B - 0.015BR$ ($n=81$, $r^2=0.77$, $p<0.001$) for days with rain (Fig. 1). Ambient temperature was expected to affect frass production (e.g. Liebhold and Elkinton 1988b, Reichenbach and Stairs 1985) but its inclusion in the model did not significantly improve the relation between biomass and frass production. Both relations were used to calculate caterpillar biomass from frass fall data in Holm Oak at both sites assuming that larval development rates do not differ between Corsican and mainland caterpillar populations.

Data on the breeding biology of the tits were collected as described in Blondel (1985).

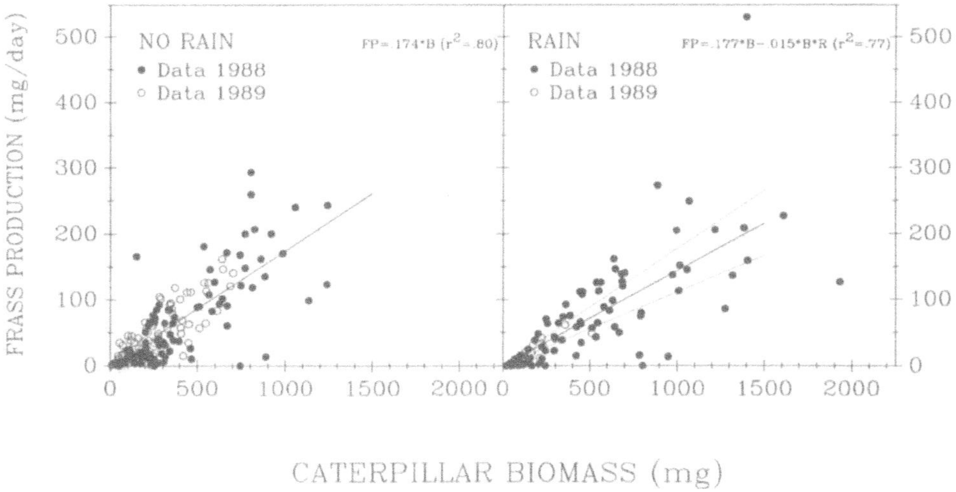

Fig. 1. Relation between frass production (mg/day) and caterpillar biomass (mg). Data from both years are shown separately, but were pooled for the analysis. Data were analysed for periods without rain (No Rain) and with rain (Rain) and least square regression lines are plotted, for RAIN for average precipitation (2.2 mm/day) and one deviation (resp. 0.0 and 4.4 mm/day). Both relations are significant ($p < 0.001$).

RESULTS

The mean abundance of caterpillars in the Holm Oak forests of Corsica and of the mainland is presented in Fig. 2 (older instars of caterpillars of the genus *Lymantria* are excluded). Peak dates in 1988 and 1989 were estimated from polynome fits (4-10 levels) through these data (Table I). There was little difference in peak date between years at either site. Within years caterpillar biomass tended to peak at a later date on the mainland than on Corsica.

The presence of caterpillars of the genus *Lymantria*, which were never observed to be fed to nestling tits, made the estimate of caterpillars exploitable by Blue Tits less accurate, especially on the mainland where they reached densities of up to 4000 mg/0.25m^2. Our estimate of the caterpillar peak date on the mainland might thus be affected by the presence of frass from the first two instars of *Lymantria*. However, the magnitude of this effect is at most a 7-10 days advance. Correcting for *Lymantria* frass on the mainland will result in a peak date rather similar to the Corsican peak date.

Average laying date, clutch size and date of maximum food demand by nestlings (estimated to be at the age of ca. 10 days (van Balen 1973, Blondel 1985, Blondel et al. 1987) are presented in Table I. Blue Tits on the mainland started laying ca. 3 weeks earlier than Corsican Blue Tits and raised their young before caterpillars became abundant.

Table I. Mean date of laying of first egg, mean clutch size and date of maximum food demand[1] by nestlings in two Blue Tit populations on Corsica and Puéchabon, and peak date of caterpillar availability. (Date: 1 = 1 March)

	Corsica			Puéchabon	
	1987	1988	1989	1988	1989
laying date	68	73	69	47	53
clutch size	6.7	6.4	6.6	8.2	8.0
max. demand[1]	98	103	99	78	84
peak date	102[2]	98	94	108	103

[1] Date of maximum food demand is estimated to occur at a nestling age of 10 days, and is calculated as (laying date-1)+ clutch size + 14 days incubation + 10.

[2] From Progress Report 1987, Institute for Ecological Research, p. 15.

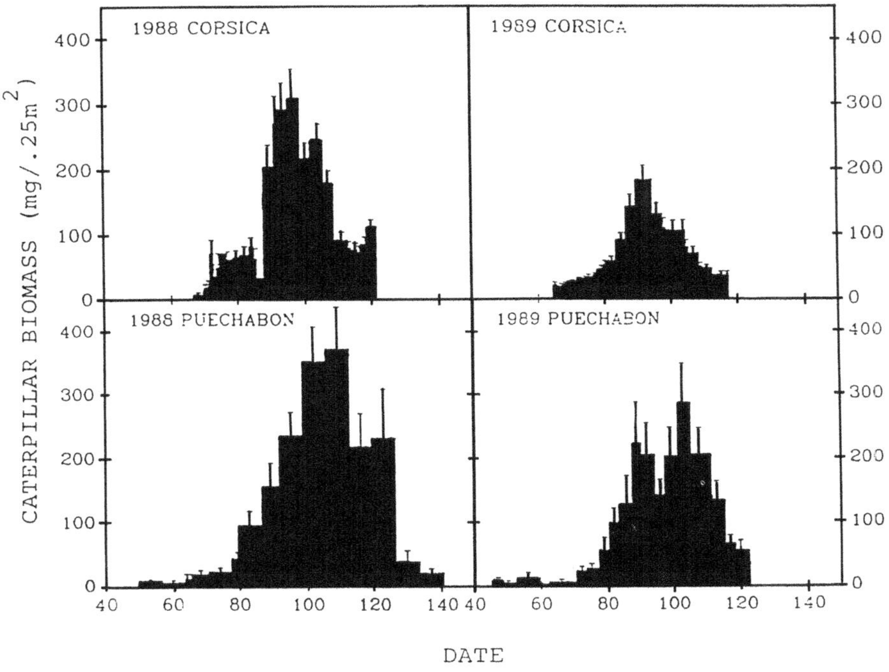

Fig. 2. Seasonal variation in caterpillar biomass (in mg/0.25 m^2) in time (date 1 = 1 March) during the 1988 and 1989 breeding seasons of Blue Tits. Biomass was calculated from frass fall data (frass produced by late-instar *Lymantria* was excluded) and the relations stated in Fig. 1.

DISCUSSION

Several studies on the breeding biology of tit species in the Mediterranean area have documented a rather similar start of egg laying among mainland populations and a significantly later start at Corsican sites despite the more southerly latitude (e.g. Isenmann 1983, Blondel 1985, 1987; Blondel et al. 1987, 1989; Clamens 1987b; Isenmann et al. 1987). Thus, on the mainland only slight differences between populations were observed despite the fact that the study sites differred in altitude and vegetation composition. For our study we chose two habitats, similar in altitude (between 100 and 300m) and vegetation (Holm Oak forests).

In Great Tit populations the timing of the breeding season is considered to have evolved in such a way that young are raised when caterpillars are most abundant (Perrins 1965, 1979, van Balen 1973). Central Trend Lines (Ricker 1981) of laying date on caterpillar peak date in two Blue Tit populations in oakwood indicate the same to be true for this species (Table II and Fig. 3 (dashed lines)). Observed peak dates of caterpillar abundance differred only

Table II. Peak date of caterpillar abundance (P) and mean laying date of first eggs (L) in seven Blue Tit populations (date: 1 = 1 March).

	SITE	YEAR	P	L	Reference
a.	Pirio,	1987	102	68	this study
	Corsica	1988	98	73	
	(*Quercus ilex*)	1989	94	69	
b.	Puéchabon,	1988	108	47	this study
	mainland	1989	103	53	
	(*Q. ilex*)				
c.	Oosterhout,	1957	81	42	van Balen 1973,
	Holland	1958	87	62	J.M. Tinbergen, unpubl.
	(*Q. robur*)	1959	75	46	
		1960	77	51	
		1961	74	43	
		1962	101	57	
		1963	91	56	
		1964	81	52	
		1965	85	54	
		1966	80	58	
		1986	86	59	
		1987	87	52	
		1988	79	52	
d.	Wytham/Marley	1948	78	41	Lack 1955, 1958,
	England	1949	85	53	Perrins 1965
	(*Q. robur*)	1950	92	53	
		1951	101	67	
		1952	82	55	
		1953	90	59	
		1954	88	53	
		1955	95	60	
		1956	93	63	
		1957	82	46	
		1958	95	65	
		1959	85	53	
		1960	87	51	
		1961	75	50	
e.	Hoge Veluwe,	1985	88	57	van Balen and
	Holland	1986	94	63	J.M. Tinbergen, unpubl.
	(*Q. robur*)	1987	90	55	
		1988	82	55	
f.	Liesbos,	1961	75	46	van Balen 1973
	Holland	1962	102	57	
	(*Q. robur*)				
g.	Hard (Basel)	1988	73	48	Henrich 1989 (p.130)
	Switzerland				van Noordwijk, unpubl.
	(*Q. robur*)				
h.	Hoge Veluwe,	1957	125	57	van Balen 1973
	Holland	1959	106	49	
	(*Pinus sylv.*)	1961	116	48	

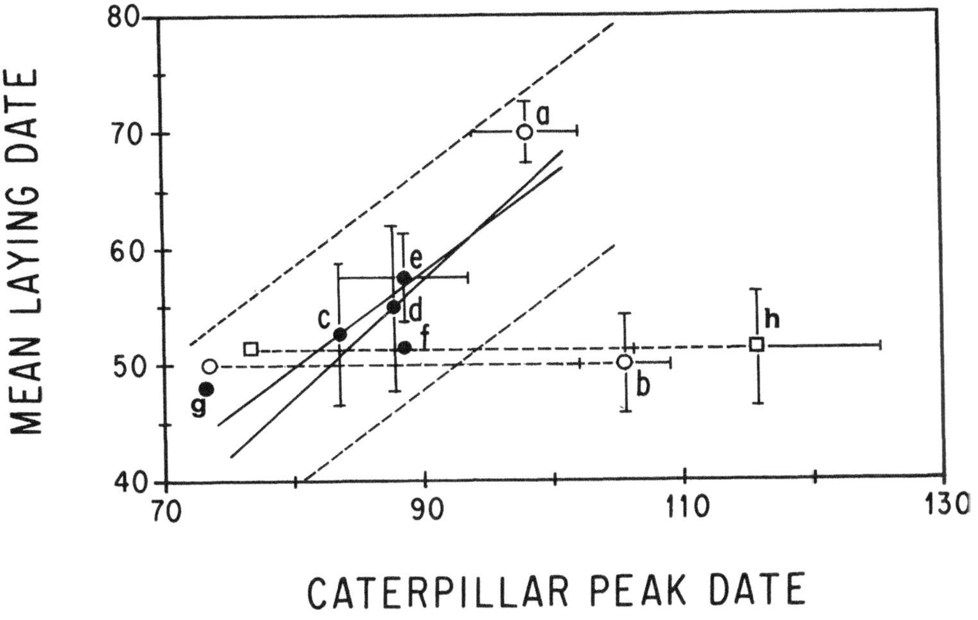

Fig. 3. Correlation between caterpillar peak date and mean laying date in 8 different Blue Tit populations (a-Pirio, n=3; b-Puéchabon, n=2; c-Oosterhout, n=13; d-Wytham, n=14; e-Hoge Veluwe/*Quercus robur*, n=4; f-Liesbos, n=2; g-Hard, n=1; h-Hoge Veluwe/*Pinus sylvestris*, n=3). Date 1 = 1 March. A Central Trend Line was calculated on data from populations c-g (n=34). A 95% interval of this relation is indicated by the solid lines. The dashed lines indicate Central Trend Lines based on Oosterhout (n=13) and Wytham (n=14) data. At both localities the correlation between caterpillar peak date and mean laying date was statistically significant (Spearman: $r_s=0.63$, $p<0.05$ and $r_s=0.85$, $p<0.002$, resp.). Mean laying dates for populations 2 and 7 are shown in relation to caterpillar peak date and laying dates in neighbouring oak habitats (see text).

slightly between our two sites, so we expected the timing of egg laying to be similar and the date of maximum food demand by nestlings to correspond with the peak date. However, mainland Blue Tits started breeding three weeks earlier than Corsican Blue Tits (Table I), and so raised young before the peak of caterpillar availability.

We tried to solve this puzzle by comparing the timing of breeding in Puéchabon and Corsica with that in other populations, with caterpillar availability as the likely ultimate factor. A search in the literature, combined with own unpublished data, resulted in 42 sets of data on mean laying date and caterpillar peak date. These are presented in the Table II and Figure 3. In the latter the 95% confidence intervals together with the mean value for each of 8 populations are plotted. The earlier observed relationship between caterpillar peak date and laying date in oakwoods (dashed lines) is not only valid within populations, but seems to be true also between populations (black dots). The present data from Corsica (dot a) fit nicely into this relationship whereas data from our mainland study site (dot b) and a *Pinus sylvestris* dominated forest at Hoge Veluwe (dot h) fall outside the relationship.

A comparison of the habitats of populations that fit the relationship with the neighbouring habitats of the "aberrant" populations shows the former to have earlier caterpillar peak dates and higher abundance of caterpillars (Perrins 1965, van Balen 1973, Blondel et al. 1987). Their density of breeding pairs is higher, fewer or no second broods are started and more recruits (nestlings that reproduce in a next breeding season), are produced (Perrins 1965, van Balen 1973, Blondel 1985, Blondel et al. 1987, Cowie and Hinsley 1987, Tinbergen and van Balen 1989). From the literature two explanations emerge which might explain why Blue Tits in various habitats time their breeding season so differently with respect to the caterpillar peak. These explanations are based on the occurrence of second broods and on the assumption of a genetic component in the timing of breeding.

Multiple breeding could be advantageous if individual birds thereby increase their fitness. In the Great Tit a higher proportion of second broods was found in habitats with a late caterpillar peak (van Balen 1973, Tinbergen and van Balen 1989). In addition, an early start of laying and a high abundance of caterpillars during the first brood favours the onset of second broods (Tinbergen and van Balen 1989). Blue Tits in pinewood and in the mainland Holm Oak site might start early to be able to produce second broods. However, no second broods are started in Corsica, despite a late caterpillar peak. This might indicate unfavourable (environmental) conditions early in spring, making early broods unproductive and/or of low value (few recruits).

Van Balen (1973) showed that Great Tit laying dates in pinewoods were too early for this habitat, but well timed for oakwood. This implies either absence of genetic variation and adaptation to oaks, or substantial gene flow. Van Noordwijk et al. (1981) calculated the genetical component of the timing of the breeding season to be 20-40% and the first possibility can thus be omitted. Asymmetric gene flow could arise from the various habitat types being present in different amounts and/or from differences in productivity between them. For example, if 90% of immigrants in pinewoods and 95% of immigrants in oakwoods come from oak habitats, an adaptation to oak would be expected.

Assuming a genetical component for the timing of the breeding season in the Blue Tit, it is possible to explain the outlaying populations in Figure 3: both are in less productive habitats. In the Dutch study sites both pine and oak habitats are equally available, but productivity is much lower in the former (van Balen and Potting, this volume). As a consequence, Blue Tits in pinewood time their breeding to the caterpillar peak in oakwood (Fig. 3). In southern France the common forests are Holm Oak and Downy Oak (*Quercus pubescens*) with a mixture of both in only a few cases. However, productivity is higher in Downy Oak forests, and an adaptation to this habitat is expected. In one year the caterpillar peak in Downy Oak was shown to be 32 days earlier than in Holm Oak (Clamens and Cramm in Blondel et al. 1987) and Blue Tits in the mainland site of this study seem to time well to this peak (Fig. 3). The Corsican population is different in that there is no other more productive habitat within its normal range of dispersal. Therefore gene flow must be small. Too few data exist on dispersal patterns between habitats to support our gene-flow-model directly with data. Once again it shows the importance of a study on short distance migration in tits.

Tits use proximate factors, related to caterpillar peak abundance and/or to other ultimate factors, to time their breeding season. The most important of these factors is probably temperature (Nager, this volume), because it correlates with bud burst in trees and with caterpillar development before and after hatching (Leffef 1988). In the Great Tit, mean date of laying is correlated with temperature (Perrins 1965, van Balen 1973, Schmidt 1984). Blondel (1985, unpubl.) found a similar correlation between the mean of temperatures from 1 April to laying date and the laying date in several Mediterranean Blue Tit populations; data from our mainland study area fit these correlations. The same correlation was found in a pinewood in The Netherlands, Blue Tits fitting the correlation of the nearby oakwood Oosterhout (van Balen, unpubl., Fig. 3). These correlations support both the model of multiple breeding, because of correlations between temperature and late-hatching caterpillars (van Balen 1973), and the model based on gene flow. On Corsica no correlation was found between temperature and the start of the breeding season. Blondel (1985) argued that this was due to a better predictability of the climate. Corsican tits, kept in aviaries on the mainland, started breeding at the same time as free-living birds, and three weeks later than mainland Blue Tits in neighbouring aviaries. This experiment (Blondel et al. 1990) indicates that a genetically determined component is prominently present in the timing of breeding in the Corsican population.

ACKNOWLEDGEMENTS

We are grateful to Philippe Perret, Marie Maistre and Hubert Kowalski for assistance in the field. Philippe Perret did some marvellous improvisation on the technical level. We thank Magda Kalk for her work on Corsica and M.A. Reglade for his work in Puéchabon in 1989. A.J. van Noordwijk and J.M. Tinbergen kindly provided unpublished data on caterpillar peakdates in their study areas. The manuscript benefited from discussions with, and useful comments by S. Daan, S.G. Henrich, H. Källander, E. Matthysen, R. Nager, A.J. van Noordwijk, J.M. Tinbergen and S. Verhulst.

REFERENCES

Arcese P, Smith JNM (1988) Effects of population density and supplemental food on reproduction in Song Sparrows. J Anim Ecol 57: 119-136

Balen JH van (1973) A comparative study of the breeding ecology of the Great Tit *Parus major* in different habitats. Ardea 61: 1-93

Betts M (1955) The food of titmice in oak woodland. J Anim Ecol 24: 282-323

Blondel J (1985) Breeding strategies of the Blue Tit and Coal Tit (*Parus*) in mainland and island Mediterranean habitats: a comparison. J Anim Ecol 54: 531-556

Blondel J (1987) From biogeography to life history theory: a multithematic approach illustrated by the biogeography of vertebrates. J Biogeogr 14: 405-422

Blondel J, Clamens A, Cramm P, Gaubert H, Isenmann P (1987) Population studies on tits in the Mediterranean region. Ardea 75: 21-34

Blondel J, Dervieux A, Kowalski H, Maistre M, Perret P, Zandt H (1989) Peuplements d'oiseaux et populations de Mésanges dans la forêt domaniale du Fango (Réserve MAB). Trav Sci Parc Nat Reg Rés Nat Corse 23: 17-55

Blondel J, Perret P, Maistre M (1990) On the genetical basis of the laying date in an island population of Blue Tit. J Evol Biol in press

Clamens A 1987a Rôle de la nourriture dans la détermination des paramètres démographiques des mésanges (Paridae) en chênaies vertes: résultats préliminaires. Alauda 55: 254- 266

Clamens A 1987b Importance de la territorialité dans la reproduction de la mésange bleue *Parus caeruleus* en chênaie verte. Rev Ecol (Terre vie) 42: 311-320

Cowie RA, Hinsley SA (1987) Breeding success of Blue Tits and Great Tits in suburban gardens. Ardea 75: 81-90

Cowie RA, Hinsley SA (1988) Feeding ecology of Great Tits (*Parus major*) and Blue Tits (*Parus caeruleus*) breeding in suburban gardens. J Anim Ecol 57: 611-626

Daan S, Dijkstra C, Drent R, Meyer T (1989) Food supply and the annual timing of avian reproduction. Proceedings XIXth Intern Ornith Congr, Ottawa, p 392-407

Drent RH, Daan S (1980) The prudent parent: energetic adjustments in avian breeding. Ardea 68: 225-252

Gibb J (1950) The breeding biology of the Great and Blue Titmice. Ibis 92: 507-539

Gibb JA, Betts MM (1963) Food and food supply of nestling tits (Paridae) in Breckland pine. J Anim Ecol 32: 489-533

Henrich SG (1989) The genetical ecology of nestling growth in the Great Tit (*Parus major* L.). DPh thesis, University of Basel

Isenmann P (1983) Zur Brutbiologie einer Blaumeisen-Population (*Parus caeruleus*) in Sued-Frankreich. Die Vogelwelt 104: 142-148

Isenmann P, Cramm P, Clamens A (1987) Etude comparée de l'adaptation des mésanges du genre *Parus* aux différentes essences forestières du bassin méditerranéen occidental. Rev Ecol (Terre Vie), Suppl 4: 17-25

Källander H (1974) Advancement of laying of Great Tit by the provision of food. Ibis 116: 365-367

Kluyver HN (1951). The population ecology of the Great Tit, *Parus m. major* L. Ardea 39: 1-135

Kluyver HN (1952) Notes on body weight and time of breeding in the Great Tit, *Parus m. major* L. Ardea 40: 123-141

Lack D (1955) British Tits (*Parus* spp.) in nesting boxes. Ardea 43: 50-84

Lack D (1958) A quantitative breeding study of British Tits. Ardea 46: 91-124

Lack D (1966) Population studies of birds. Clarendon Press, Oxford

Leffef F (1988) De gezondheid van de zomereik in relatie tot aantasting door insekten. Nederlands Bosbouw Tijdschrift 60: 414-420

Liebhold AM, Elkinton JS (1988a) Techniques for estimating the density of late-instar Gypsy Moth, *Lymantria dispar* (Lepidoptera: Lymantriidae), populations using frass drop and frass production measurements. Environ Entomol 17: 381-384

Liebhold AM, Elkinton JS (1988b) Estimating the density of larval Gypsy Moth, *Lymantria dispar* (Lepidoptera: Lymantriidae), using frass drop and frass production measurements: sources of variation and sample size. Environ Entomol 17: 385-390

Noordwijk AJ, van Balen JH, van Scharloo W (1981) Genetic variation in the timing of reproduction in the Great Tit. Oecologia (Berl.) 49: 158-166

Perrins CM (1965) Population fluctuations and clutch-size in the Great Tit, *Parus major* L. J Anim Ecol 34: 601-647

Perrins CM (1970) The timing of birds' breeding seasons. Ibis 112: 242-255

Perrins CM (1979) British Tits. Collins, London

Progress Report 1987/Institute for Ecological Research. Amsterdam, Kon Ned Akad Wetenschap., 1988

Reichenbach NG, Stairs GR (1985) Bioenergetics of the western spruce budworm (Lepidoptera: Tortricidae) with comments on endotherm and ectotherm population energetics. Can J Zool 63: 1330-1338

Ricker WE (1981) Computation and uses of central trend lines. Can J Zool 62: 1897-1905

Royama T (1970) Factors governing the hunting behaviour and selection of food by the Great Tit (*Parus major* L.). J Anim Ecol 39: 619-668

Schmidt KH (1984) Früehjahrstemperaturen und Legebeginn bei Meisen *Parus*. J Orn 125: 321-331

Schwerdtfeger F (1931) Die Ermittlung der Mortalität von Raupen während einer Insektenepidemie. Anz Schädlingsk 12: 85-90

Slagsvold T (1976) Annual and geographical variation in the time of breeding of the Great Tit and the Pied Flycatcher in relation to environmental phenology and spring temperature. Ornis scand 7: 127-145

Tinbergen JM, Balen JH van (1989) Food and multiple breeding. Proceedings XIXth Inter Ornith Congr p 380-391

Tinbergen L (1960) The natural control of insects in pinewood. I. Factors influencing the intensity of predation by songbirds. Arch néerl Zool 13: 265-343

VARIATION IN MORPHOLOGY, LAYING DATE AND CLUTCH SIZE BETWEEN NON-MEDITERRANEAN AND MEDITERRANEAN BLUE TITS

Jean-Louis MARTIN and Marie-Dominique BELLOT

CEPE/CNRS, B.P. 5051, 34033 Montpellier Cedex, France

The occurrence of differences among spatially segregated populations of a species is, since Darwin's time, essential in any reflection on species evolution and adaptation. With the development of multivariate analysis methods the study of variation in external characters has been greatly enhanced. Indeed, "geographic variation is not likely to be due to a single environmental variable, but is doubtless a multidimensional process involving the adaptation of many characters to a variety of independant environmental factors whose gradients and ranges probably overlap in rather complex fashion" (Sokal and Rinkel 1963). Progress has also been favoured by advances in ecomorphology, which allows a better understanding of the relations existing between geographic variation in morphology and variation in ecology over the distributional range of a species or a group of species (see Hespenheide 1973 or the review by Leisler and Winkler 1985).

While relations between variation in external characters and variation in the environmental conditions have been the subject of numerous studies (Snow 1954b, 1955, James 1970, 1982, 1983, Leisler 1982 for example), few attempts have been made to compare variation in external characters to variation in life history traits over entire distributional ranges. This is mainly because of the lack of suitable data on life history traits at the continental scale.

The present work intends to make such a comparison for a western Palearctic species, the Blue Tit (*Parus caeruleus*), for which several decades of life history studies over most of the species' range are now available. These studies provide a basis for comparing variation in the species laying date and clutch size with variation in external morphology between non-Mediterranean and Mediterranean populations. Interpretation will look for consistency between both kinds of variation and their biological meaning.

THE BIOLOGICAL MODEL

The Blue Tit is a Western Palearctic species which ranges from the Urals to the British Isles and from southern Scandinavia to North Africa and the Canary islands (Vaurie 1957). The species occupies mainly broad-leaved woodlands over a broad range of eco-climatic conditions (see Snow 1954a, Lack 1971, Perrins 1979). However, whereas it breeds mainly in deciduous forests in most of its breeding range, the habitat used in the Mediterranean part of that range is markedly different. There the species lives in broad leaved deciduous

NATO ASI Series, Vol. G 24
Population Biology of Passerine Birds
Edited by J. Blondel et al.

and evergreen habitats distributed in a mosaïc pattern. Finally, in the five western Canary islands the Blue Tits breed, mainly or exclusively, in long leaved Pine *Pinus canariensis* forests. In addition, the Mediterranean region shows important changes in seasonal constraints. It is probably the hot and dry summer rather than the winter that is the harshest period for passerines because of low levels in food and water resources (Blondel et al. 1987).

Vaurie (1957) distinguishes, on the basis of plumage colour, two different population groups. The first includes all populations from the Eurasian part of the species' distributional range, and the second includes all populations found in North Africa and the Canary islands (for a detailed description of these groups see Vaurie 1957, Etchecopar and Hüe 1964).

GEOGRAPHIC VARIATION IN MORPHOLOGY

In the Eurasian population group individual size varies from large sized birds in the north-northeast to small sized birds in the Mediterranean (Martin 1988a, 1988b). This variation in size has been explained (Martin 1988b) by the hypothesis of Geist (1987). This hypothesis states that latitudinal size increase in homeotherms is a consequence of higher seasonal resource peak at higher latitudes in synchrony with the growing period of the offspring. This seasonal peak increases from 0 to 60° latitude and decreases further North. In North Africa the smallest birds are found in the most arid habitats (resource poor habitats) (Martin 1988a, 1988b).

Variation in the shape of the birds has been shown in both the Eurasiatic and North African population groups. In north-northeastern Eurasia, birds have stout bills and short tarsus relative to body size (Martin 1988b). There the birds breed in deciduous broad leaved woodlands (smooth foraging substrate). In Mediterranean Eurasia Blue Tits have thinner bills and increased relative tarsus length. The pattern is the strongest for the large isolated Mediterranean islands (Martin and Mazurek 1986, Martin 1988a, Martin 1988b). This result has been related to the progressive increase of evergreen broad leaved trees in the habitats used by these populations (harsh foraging substrate). In the North African population group, bill shape, tarsus length and wing shape dramatically change with changes in habitat structure (Martin 1988b). The longest bills, longest tarsi and most rounded wings are found in the Canary island populations restricted to long leaved pine woods (Grant 1979, Martin 1988a, 1988b).

Analysis of morphological variation suggested, therefore, a close relation between changes in environmental conditions and changes in morphology, a point already discussed by Partridge (1976), Partridge and Pring-Mill (1977) and by Grant (1979) for the Canary island Blue Tits. Variation in size has been related to variation in resource level. Variation in bill length and depth, in tarsus length and in wing shape, independently from variation in the bird's overall size, modify the bird's shape and are related to changes in habitat structure.

Another point apparent from these results is the relation between sudden variation in body measurements (marked by gaps in cline of variation) and geographic isolation. The importance of isolation in the emergence of geographic variation in Parids has already been pointed out by Snow (1954a) (see also Slatkin 1981, 1987 on the role of gene flow in natural populations).

GEOGRAPHIC VARIATION IN LAYING DATE

On the basis of climatic data from Cour and Richard (1986), Bellot (1989) (see Fig. 1) has shown that, on a continental scale, there is a trend in southern Blue Tit populations to lay towards the end of the thermal spring, while northern populations lay at the beginning of the thermal spring (see also Slagsvold 1975, 1976). This can be interpreted as the result of latitudinal differences in the spring increase in trophic resources available for the female. These resources increase at the beginning of the thermal spring in the north and develop later in the thermal spring in the Mediterranean (see Cramm 1982, Blondel et al. 1987 for leafing phenology and carterpillar frass fall phenology in Mediterranean evergreen and deciduous broad leaved woodlands). Furthermore, studies on a local scale in the Mediterranean region have shown that this delay in laying date relative to the thermal spring is more pronounced in evergreen habitats than in deciduous habitats. Indeed, Isenmann et al. (this volume) show that over a period of seven years Blue Tits laid significantly earlier in deciduous woods (Downy oak) than in nearby evergreen woods (Holm oak), with an average difference of seven days (Blondel et al. 1987). These differences have been experimentally related to a delay in the spring development of food resources for the laying female in evergreen habitats (Clamens and Isenmann 1989). This trend of delayed laying dates in Mediterranean conditions is stronger in the geographically isolated populations of Corsica (Blondel 1985, Blondel et al. 1987) and Morocco (Baouab et al. 1986) where the species mainly breeds in evergreen habitats. Perret et al. (1989) have shown that delayed breeding in Corsica is genetically determined.

GEOGRAPHIC VARIATION IN CLUTCH SIZE

On a continental scale, Isenmann (1987), Isenmann et al. (1987) and Blondel (unpublished), comparing clutch sizes published for the Blue Tit, stress 1) that average clutch sizes are significantly higher in non Mediterranean Europe (m=10.9, 12 samples) than in the Mediterranean region (m=7.7, 11 samples) but 2) that between-sample heterogeneity is higher in the Mediterranean region (s.d.=1.9) than outside the Mediterranean (s.d.=0.5). The lower average clutch size in the Mediterranean is thought to reflect lower clutch sizes in evergreen habitats due to lower levels of spring resources in these habitats. Deciduous broad leaved habitats in the northern Mediterranean area

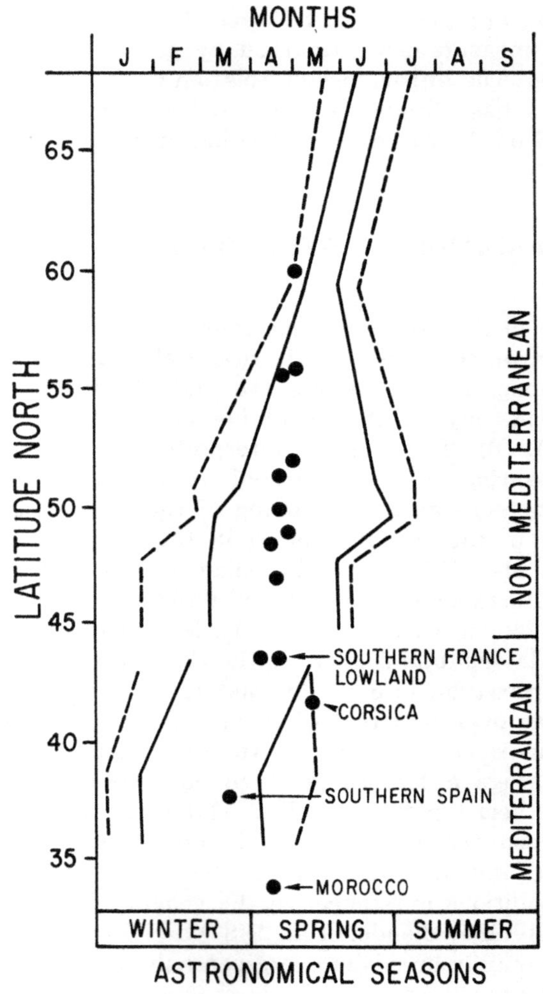

Fig. 1. Relationship between thermal spring and laying dates of the Blue Tit according to latitude (after Bellot 1989). Thermal spring (solid line = mean dates of beginning and end; broken lines = extreme dates of beginning and end) is the period of time between January and July for which weekly average temperatures range between 5 and 15°C (thermal diagrams after Cour and Richard 1986). Black dots = laying dates (from Baouab et al. 1986, Blondel 1985, Dhont et al. 1984, Hilden this volume, Hribek 1984, Isenmann unpubl., Frederiksen et al. 1972, Källander 1983, Lack 1966, Leclercq 1975, Neub 1977, Schmidt 1984).

show clutch sizes comparable to those from non-Mediterranean deciduous habitats (Isenmann 1987). Higher heterogeneity in Mediterranean values may therefore reflect the higher heterogeneity in habitats (resource poor evergreen habitats versus resource rich deciduous habitats).

On a local scale these conclusions are illustrated by the results of Isenmann et al. (this volume, see also Blondel et al. 1987) which show a significant difference of about 2.3 eggs in average clutch size between a continental Mediterranean deciduous forest (Downy oak) and a nearby evergreen forest (Holm oak). This reduction in clutch size in evergreen habitats is paralleled by lower spring peaks in caterpillar frass fall (Cramm 1982, Blondel et al. 1987, Clamens et al. 1987). In the isolated Corsican population which mainly breeds in habitats dominated by the evergreen Holm oak, clutch size is lower, on average, by about two eggs when compared to mainland reference Holm oak study sites (2.3 when compared to Holm oaks on limestone, Blondel 1985, Blondel et al. 1987; 1.7 when compared to Holm oaks on acid soils, Ph. Orsini in litt.).

CONCLUSIONS

The comparison of geographic variation observed in external morphology with the geographic variation in laying date and clutch size leads to the following conclusions: 1) in both morphological and life history traits marked differences appear when going from non-Mediterranean to Mediterranean populations. Mediterranean populations have smaller bodies, thinner bills, longer tarsi, lay later, and produce fewer eggs; 2) these differences are more pronounced in Mediterranean geographical isolates with little or no gene flow from non-Mediterranean populations. These results suggest that later and lower levels of resources available to the birds before or during laying, or lower resource levels during the individual's growing period in the Mediterranean region explain variation in laying date, clutch size and body size respectively. Comparative studies, on a local geographic scale, between evergreen and deciduous oak woodlands in the Mediterranean region (Cramm 1982, Blondel et al. 1987, Isenmann et al. 1987, Clamens and Isenmann 1989) suggest that the impact of the general latitudinal trend for delayed spring resource increase and lower spring resource peaks in the Mediterranean is amplified in evergreen habitats. The latter show lower and later spring increase in resources than nearby deciduous broad leaved woodlands.

ACKNOWLEDGEMENTS

The study on geographic variation in morphology was made possible by the help and assistance of numerous Natural History Museums. It also benefited from a F.M. Chapman grant (AMNH, New York). Understanding of life history traits largely benefited from discussions with the members from the group working on bird biology in Montpellier.

REFERENCES

Baouab RE, Thevenot M, Aguesse P (1986) Dynamique des populations de la Mésange bleue en chênaies de Mamora et du Moyen Atlas. Bulletin Institut Scientifique (Rabat) 10: 165-183

Bellot MD (1989) Relation entre la température et la date de ponte chez la Mésange bleue (*Parus caeruleus*) en chênaies méditerranéennes. Unpub report, University of Montpellier

Blondel J (1985) Breeding strategies of the Blue and Coal Tit (*Parus*) in mainland and island Mediterranean habitats: a comparison. J Anim Ecol 54: 531-556

Blondel J, Clamens A, Cramm P, Gaubert H, Isenmann P (1987) Population studies on tits in the Mediterranean region. Ardea 75: 21-34

Clamens A, Isenmann P (1989) Effect of supplemental food on the breeding of Blue and Great tits in Mediterranean habitats. Orn Scan 20: 36-42

Clamens A, Cramm P, Isenmann P (1987) Modalités de reproduction et ressources alimentaires: le cas des mésanges (*Aves, paridae*) dans les chênaies du Languedoc. Actes Coll Nat CNRS "Biologie des Populations": 397-401

Cour P, Richard P (1986) Latitudinal and interannual variations in pollen shedding periods: impact of thermal seasons along a transect from Abisko (68,4°N) to Oran (35,6°N). Third International Conference on Aerobiology, 1986, Basel

Cramm P (1982) La reproduction des mésanges dans une chênaie verte du Languedoc. L'Oiseau RFO 52: 347-360

Dhondt AA, Eyckerman R, Moermans R, Huble J (1984) Habitats and laying date of the Great and Blue Tit, *Parus major* and *P. caeruleus*. Ibis 126: 388-397

Etchecopar RD, Hüe F (1964) Les Oiseaux du Nord de l'Afrique. Boubée, Paris

Frederiksen KS, Jensen M, Larsen EH, Larsen VH (1972) Some data illustrating time of breeding and number of eggs in Tits (*Paridae*). Dansk Orn For Tidsskr 66: 73-85

Geist V (1987) Bergmann's rule is invalid. Can J Zool 65: 1035-1038

Grant PR (1979) Ecological and morphological variation of Canary Island Blue Tit, *Parus caeruleus* (*Aves:Paridae*). Biol J Linnean Society 11: 103-129

Hespenheide HA (1973) Ecological inferences from morphological data. Ann Rev Ecol Syst 4: 213-229

Hribek M (1984) Some notice from breeding biology of the Blue Tit, *Parus caeruleus* L. Zpravy Moravskeho, Ornitologickemo Sdruzeni 42: 93-106

Isenmann P (1987) Geographical variation in clutch size: the example of the Blue Tit (*Parus caeruleus*) in the Mediterranean area. Vogelwarte 34: 93-99

Isenmann P, Cramm P, Clamens A (1987) Etude comparée de l'adaptation des mésanges du genre *Parus* aux différentes essences forestières du Bassin Méditerranéen occidental. Rev Ecol (Terre et Vie) Suppl 4: 17-25

James FC (1970) Geographic variation in birds and its relation to climate. Ecology 51: 365-390

James FC (1982) The ecological morphology of birds: a review. Ann Zool Fennici 19: 265-275

James FC (1983) Environmental component of morphological differentiation in birds. Science 221: 184-186

Källander H (1983) Aspects of the breeding biology, migratory movements, winter survival, and population fluctuations in the Great Tit (*Parus major*) and the Blue Tit (*P.caeruleus*). Thesis, University of Lund, Sweden

Lack D (1966) Population studies of birds. Oxford University Press, Oxford

Leclercq B (1975) Contribution à l'étude expérimentale de l'écologie des mésanges en fûtaie de Chênes. Thèse, Univ Dijon

Leisler B (1982) Morphological aspects of ecological specialization in bird genera. Ökol Vögel 2: 199-220

Leisler B, Winkler H.(1985) Ecomorphology. In: Current Ornithology, 2: 155-186

Martin JL, Mazurek H (1986). Etude des variations géographiques de la morphologie d'un passereau, *Parus caeruleus*. Mappemonde 3: 22-25

Martin, JL (1988a) Structure de population et variations "périspécifiques" chez *Parus caeruleus* L. (*Aves*): premiers résultats. Acta Oecologica, Oecol Gener 9: 153-166

Martin JL (1988b) Variation géographique, adaptation et spéciation: l'exemple de *Parus caeruleus* (*Aves*). Thèse es Science, Montpellier

Neub M (1977) Evolutionsökologische Aspekte zur Brutbiologie von Kohlmeise (*Parus major*) und Blaumeise (*P. caeruleus*). Thesis, University of Freiburg, FRG

Partridge L (1976) Some aspects of the morphology of Blue tits (*Parus caeruleus*) and Coal Tits (*Parus ater*) in relation to their behaviour. J Zool 179: 125-133

Partridge L, Pring-Mill F (1977) Canary island Blue tits and English Coal tits: convergent evolution ? Evolution 31: 657-665

Perret P, Blondel J, Dervieux A, Maistre M, Colomb B (1989) Composante de la date de ponte chez la Mésange bleue (*Parus caeruleus*) (*Aves*). C R Acad Sc Paris 308 (III): 527-530

Perrins C (1979) British tits. Collins, London

Schmidt KH (1984) Frühjahrstemperaturen und Legebeginn bei Meisen (*Parus*). J Orn 125: 321-331

Slagsvold T (1975) Breeding time of birds in relation to latitude. Norw J Zool 23: 213-218

Slagsvold T (1976) Arrival of birds from spring migration in relation to vegetational development. Norw J Zool 24: 161-173

Slatkin MK (1981) Estimating levels of gene flow in natural populations. Genetics 99: 323-335

Slatkin MK (1987) Gene flow and the geographic structure of natural populations. Science 236: 787-792

Snow DW (1954a) The habitats of Eurasian tits (*Parus sp.*). Ibis 96: 565-585

Snow DW (1954b) Trends in geographical variation in Palearctic members of the genus *Parus*. Evolution 8: 19-28

Snow DW (1955) Geographical variation in the Coal tit *Parus ater* L. Ardea 43: 195-226

Sokal RR, Rinkel RC (1963) Geographical variation of alate *Pemphilus populi-transversus* in Eastern North America. Univ Kansas Sci Bull 44: 467-507

Vaurie C (1957) Systematic notes on the Palearctic birds, number 26. *Paridae*. The *Parus caeruleus* complex. Am Museum Novitates 1833: 1-15

;TWEEN-INDIVIDUAL VARIATION

WITHIN POPULATIONS

THE VARIABLE NICHE HYPOTHESIS REVISITED AN ANALYSIS OF INTRA- AND INTER-SPECIFIC DIFFERENCES IN BILL VARIATION IN *PARUS*

Andrew G. GOSLER

Edward Grey Institute, Department of Zoology, South Parks Rd., Oxford, OX1 3PS, UK

"Often a theory gains acceptance almost for no other reason than its inherent plausibility, the case in point being what has been called the 'niche-variation model'"

Soulé and Stewart 1970

"The niche variation hypothesis has a superficial plausibility, but superficially plausible processes (such as natural selection) sometimes are true."

van Valen and Grant 1970

The "Variable-Niche" or "Niche-Variation" hypothesis (van Valen 1965) was propounded as one of several explanations for the maintenance of genetic variation in wild populations. It suggests that different individuals in a population may be differently suited to different parts of the species' niche-space. The model was formulated to consider the maintenance of continuous variation in populations and had three requirements. First that "individuals of some set *a* survive or reproduce better than those of some set *b* in some environment *A*, while the reverse is true in some environment *B*". Secondly that "the above difference between *a* and *b* is in part genetic", and thirdly that "some appropriate mechanism of distributional or mating preference exists". Van Valen suggested that this last requirement could be met either by appropriate differences between groups in habitat preferences, or by non-random mating with respect to these groups. Despite an insistence that the model could explain continuous variation, it is easiest to see how it might work if the groups represent discrete classes within the population, especially sex (Willson 1969).

Central to the hypothesis is the argument that competition for resources between individuals of the same or different species results in the ecological separation of species or intra-specific groups. The principal prediction of the hypothesis is that in populations which occupy broader niches because of the differential sub-niche occupancy of its members, we should expect to see greater variation in those characters which are relevant to the exploitation of that niche. In particular this is likely to mean trophic characters. In other words, there should be a correlation between ecological variation and morphological variation across populations, and we expect differences in ecology between populations to arise partly from differences in the competitive environments of those populations.

NATO ASI Series, Vol. G 24
Population Biology of Passerine Birds
Edited by J. Blondel et al.

Van Valen tested the hypothesis by comparing the coefficients of variation of bill measurements from island and mainland populations of six passerine species (7 population comparisons) for which information on relative niche breadth was available. He reasoned that in the absence of competing species, individuals in island populations should show greater morphological divergence due to the relatively greater importance of intra- rather than inter-specific competition. Indeed in all cases he found greater variation in the population with the broader niche. However, he carried out separate comparisons for each sex (14 comparisons in all) stating that these were independent tests. This, together with the statement that no significant difference in the degree of sexual dimorphism was found between island and mainland populations (and a general suspicion of the niche concept) quickly earned the hypothesis rather more criticism than was justified (Willson 1969, Soulé and Stewart 1970).

Soulé and Stewart (1970) sought to test the hypothesis by a comparison of relative bill variation in euryphagous (generalist foragers) and stenophagous (specialist foragers) species. Van Valen and Grant (1970) quickly pointed out that the results of this analysis were irrelevant as it failed to recognise that the relatively expanded niche of euryphagous species will only be reflected in greater morphological variation if different members of the population/species exploited a different part of the resource spectrum, otherwise one should expect only that all individuals will possess a "general-purpose" bill which need not vary more between birds than would that of the specialist forager.

Rothstein (1973a) tested the hypothesis more rigorously than any of the aforementioned authors. He made two predictions from the model. First that within a community, abundant species should experience relatively more intra-specific competition than scarcer species which should suffer relatively more inter-specific competition. Hence abundant species should be more variable than scarce species. Secondly that niche breadth should be inversely proportional to the degree of sympatry which should reflect the intensity of inter-specific competition. Hence species which frequently occur in sympatry should be less variable than those which do not. Sympatry was assessed as the percentage of random locations where a species occurred with a congener. In a series of tests he showed both predictions to be supported, and further that morphological variation was most readily increased by the divergence of discrete classes such as age or sex. In a discussion paper on the subject, Rothstein (1973b) points out that bill characters are consistently found to be more variable than body-size characters (such as tarsus, toe or wing-length) and that this is difficult to explain without recourse to the variable niche hypothesis. These studies probably remain the most convincing support for the hypothesis (Schoener 1986).

Beever (1979) ignores both of Rothstein's papers in a review of van Valen's analysis in which he suggests that in view of the shortcomings of the original presentation, the hypothesis should be dropped from general usage. I suggest that such an act would amount to "throwing out the baby with the bath water", since the fact that a particular test of a hypothesis is ill-conceived, does not necessarily mean that the hypothesis was.

In the rest of this paper I shall test the variable niche hypothesis by both inter- and intra-specific comparisons of bill character data from four syntopic species of tit *Parus* collected from the same area. The species studied were the Coal Tit *Parus ater*, the Marsh Tit *P. palustris*, the Blue Tit *P. caeruleus* and the Great Tit *P. major*. Although the Willow Tit *P. montanus* is also found in the study area, it occurs at very low densities and no data were available for this analysis. The ecological background to and reasoning behind the analyses will be presented below after outlining the general methods used.

METHODS

The birds were mist-netted between October and March in each winter from 1981/82 to 1988/89 as part of the routine winter trapping programme of the Edward Grey Institute's tit population study at Wytham Woods near Oxford (Perrins 1979, Gosler 1987a). Tit populations at Wytham are markedly influenced by the abundance of tree seeds in winter, especially that of Beech *Fagus sylvatica* (Perrins 1966, 1979). When possible, birds were trapped whilst feeding on beechmast; when this was not possible, birds were trapped at feeders baited with fat or peanuts.

The birds were ringed, aged by the criteria described in Svensson (1984) and sexed by the criteria described in Svensson (1984) for Great and Blue Tits respectively, and in Gosler and King (1989) for Marsh and Coal Tits. Wing-length was taken as a measure of body-size and was measured to 0.5 mm as the maximum chord (Svensson 1984) on a stopped rule. Bill-length and depth were taken to 0.1 mm with a Vernier caliper as described in Gosler (1987a) and bill index calculated as bill-depth / bill-length (Gosler 1987a). The bill index, which is a measure of bill shape rather than size, has been used throughout in this paper as its relevance to feeding efficiency has been demonstrated (Gosler 1987a).

Despite the long period of the study, bill measurements were not taken from Coal, Marsh or Blue Tits until the winter 1986/87 so that the data used come from different periods depending on the species and this has greatly affected the sample sizes available. The reason for this was the inability to sex Blue Tits reliably before 1984 (see Svensson 1984) or Coal or Marsh Tits before 1987 (see Gosler and King 1989). The Great Tit data come from the whole study period. Since bill measurements of *Parus* may vary significantly between seasons (Gosler 1987a), the data were first standardised by ANOVA to remove between-winter variation. In fact, this procedure did not change the results of any analysis significantly. Statistical procedures follow Sokal and Rholf (1981).

INTER-SPECIFIC COMPARISONS

In winter the four *Parus* species form mixed flocks with other species such as Treecreeper *Certhia familiaris*, Goldcrest *Regulus regulus*, Long-tailed

Tit *Aegithalos caudatus* and Nuthatch *Sitta europaea* (Morse 1978). In general the relative abundance of *Parus* species is Coal Tit, Marsh Tit, Blue or Great Tit (in order of increasing density) and this is reflected in the sample sizes obtained for analysis. When beechmast is available, all four species aggregate together to feed on it and the importance of mast even to the most insectivorous - the Coal Tit - has been shown by its influence on movements of this species in winter (Glue 1982). A clear dominance hierarchy based on relative body size develops between species when flocking together. In order of declining dominance this is Great Tit, Blue Tit, Marsh Tit and Coal Tit. The non-*Parus* flock members are either sub-dominant to or do not interact with these (Morse 1978). Agonistic interactions over food generally take the form of supplanting attacks and this is the most obvious form of competition between individuals at this time of year (Morse 1978). Table I shows the relative number of attacks on members of each species made by members of the same or another species. The figures were calculated from data presented by Morse (1978). Note that intra-specific interactions increase across species with increasing body size. If we can assume that these figures reflect the relative importance of intra- and inter-specific competition to the competitive environment of these species in winter, then we should predict that the relative variation in bill characters should be greatest in the Great Tit and least in the Coal Tit. Further that, since intra-specific interactions are partly mediated by sex (Drent 1983, Gosler 1987a), the greater variation of the Great Tit should be due in part to sexual dimorphism. Note that these differences do not reflect the total niche breadth of the species which is notoriously difficult to define and thus compare (Soulé and Stewart 1970), but the degree to which the sexes have diverged ecologically within that niche space. Gosler (1987a, 1987b) has shown that bill size and feeding ecology are inextricably linked in the Great Tit so that a close relationship between bill variation and ecology can be assumed.

Table I. The percentage of attacks on members of the same (intra-specific) or another (inter-specific) *Parus* species in winter flocks in Wytham Woods. The figures are calculated from data in Morse (1978).

	Species attacked			
Species attacking	Coal	Marsh	Blue	Great
Same species	42%	78%	96%	100%
Different species	58%	22%	4%	0%

To test the prediction that sub-dominant species of *Parus* should be less variable in bill characters than dominant species, I have calculated the coefficient of variation for bill index for each species. To remove age effects only first-winter birds have been used in these analyses (although the results are the same if all birds are included). A similar analysis was carried out on the wing-length data to establish whether differences observed in bill index between species were due simply to differences in the variation in body-size as a whole. Table II shows the relative variation in bill index and wing-length for each species and the difference between sexes as a % of the mean male value together with values of "F" to test the significance of sexual dimorphism in each character in each species. Note that for wing-length there is little difference between species in the total variation and all show highly significant sex differences in wing-length. However, for bill index, the species differ significantly (between Great Tit and Coal Tit, $F_{889,29}=2.78$ $P<0.001$) in the total variation observed, and differ by more than seven times in the % difference between sexes. These measures are independent of sample size.

Table II. Relative variation and sex differences in bill index and wing-length in first-winter birds of four *Parus* species. Data were first standardised across winters. Only sexed birds were used throughout this analysis.

		Wing Length				Bill Index			
Species	n	% c.v.	% sex diff.	F	P	% c.v.	% sex diff.	F	P
Coal Tit	30	2.61	2.83	15.0	0.000	2.78	0.49	0.19	n.s.
Marsh Tit	23	2.95	4.81	63.1	0.000	2.96	1.74	4.32	0.047
Blue Tit	112	2.55	3.93	132.6	0.000	4.19	1.96	8.26	0.005
Great Tit	890	2.68	4.23	1061.0	0.000	4.64	3.76	267.30	0.000

INTRA-SPECIFIC COMPARISONS

The analyses above showed that the Great Tit is highly variable in bill index and that a substantial part of that variation was due to sex. This species has a well documented dominance hierarchy based on age, sex and size, whereby males are dominant to females and adults to first-years (Drent 1983). I have shown (Gosler 1987a) that in winter, niche differences may occur between sexes depending on the abundance of beechmast and the availability of alternative feeding sites. I showed that when mast was abundant, members of all age and sex classes made use of it whilst when mast became scarcer, sub-dominant classes were forced to expand their niche to exploit other resources as predicted by Morse (1974). Hence dominant classes forage in a narrower range of sites than sub-dominant classes when resources are limited. Given the ability of bill size in the Great Tit to respond rapidly to a changing feeding environment (Gosler 1987a), the variable niche hypothesis predicts that when flocking on mast, all age and sex classes should show a similar amount of variation in bill index and that this should be relatively low. However, when mast is not available to all classes, variation in bill index should increase in sub-dominant classes. To test this I divided each winter into two periods - October to December and January to March as the use of mast made by the birds differred greatly between these times within a winter (Gosler 1987a). This gave 15 "seasons" in which bill measurements were taken between 1982 and 1989, and each season could readily be scored for whether the birds were "on" or "off" mast. This gave five seasons when the birds were on mast and ten off. Table III shows the relative variation in bill index in each class when the birds were trapped on mast or off mast. I have made the same standardisation for winters that was made between species, however, since part of that variation is due itself to the beech crop, I present the un-standardised results also.

Table III. Differences in the relative variation (% c.v.) in bill index in four age and sex classes of Great Tits trapped on or off beechmast. F ratio tests are given within classes comparing data for "on" or "off" seasons, and between adult males and first-year females in each season type. 1Y= first year.
Probability levels are given as * <0.05, ** <0.01, *** <0.001.

Class	n		Un-standardised data				Winter standardised			
	off	on	off	on	F	P	off	on	F	P
1Y female	150	295	6.26%	3.68%	2.89	***	5.17%	3.65%	2.01	***
Ad female	127	122	5.41%	3.65%	2.19	***	4.78%	3.54%	1.82	***
1Y male	159	292	5.01%	4.09%	1.50	**	4.48%	4.17%	1.15	n.s
Ad male	199	167	4.09%	3.91%	1.09	n.s.	3.85%	3.88%	1.02	n.s
F between 1Y fem.	F		2.34	1.12			1.80	1.13		
and Ad males	P		***	n.s.			***	n.s.		

DISCUSSION

The results presented above provide strong support for the variable niche hypothesis. In the Great Tit, the sub-dominant classes expanded their niche when food was limiting and showed a corresponding increase in bill variation. However, for the variable niche hypothesis to be a valid explanation of this we must assume that the expanded niche of sub-dominants was brought about by increased intra-class competition and that whereas the class as a whole appeared to generalise, each individual actually specialised within that broader resource spectrum. This is difficult to test without feeding observations of known individuals. However, there is a tendency towards a bimodal distribution in bill size in sub-dominant classes when mast is absent, which suggests the presence of another resource which is exploited at this time. This will be explored further elsewhere.

In the inter-specific case, the degree of sexual dimorphism and therefore of total variation in bill index increased with body-size across species while that of wing-length itself did not. This agrees with the variable niche hypothesis if we consider the effect of competition on morphological variance directly rather than through the intermediate of niche volume. This might seem an unreasonable test of a hypothesis which predicts explicitly a correlation between the niche and morphology. But it is in fact a direct test of the underlying assumptions of the mechanism by which such a relationship might come about. After all, it is more the comparison of niche volumes among populations or species than the assumed mechanism which has brought the hypothesis into disrepute. In this regard, it is interesting that previous tests which addressed the hypothesis from the question of niche space either failed to find support for it (Willson 1969, Soulé and Stewart 1970) or have been discredited (van Valen 1965, Beever 1979), whilst the single attempt which considered the competitive mechanisms directly has provided convincing evidence in favour of the hypothesis (Rothstein 1973a, Schoener 1986).

Whether or not the Coal Tit occupies a greater or lesser realised niche than the Great Tit is perhaps irrelevant since even if we were able to measure niche volume in a meaningful way, it would still be impossible to standardise between species. For example, we might measure niche volume in terms of niche dimensions such as feeding height which are shared by all species involved. However, when the niche volume of a species is "compressed" by competition with other species, it may be forced to exploit new dimensions which were unforeseen. In the present case, it is only the sub-dominant species which hoard food and this may be regarded as a niche expansion along a new dimension to relieve competition (Haftorn 1960), but how should this be incorporated into the comparison of total niche volume between species?

In conclusion, many of the analyses which claimed to discredit the variable niche hypothesis were ill-conceived and the time to drop it from our theoretical armoury is not yet ripe.

ACKNOWLEDGEMENTS

I thank Doug Morse for allowing me to rearrange his data in the form of Table I and Christopher Perrins, Caroline Jackson-Houlston, Robin McCleery, Hans Källander and Paul Isenmann for commenting on an earlier draft.

REFERENCES

Beever JW (1979) The niche variation hypothesis: an examination of assumptions and organisms. Evol Theory 4: 181-191

Drent PJ (1983) The functional ethology of territoriality in the Great Tit (*Parus major* L.). Unpublished D Ph thesis, University of Groningen

Glue D (1982) The Garden Bird Book. Macmillan London

Gosler AG (1987a) Pattern and process in the bill morphology of the Great Tit *Parus major*. Ibis 129: 451-476

Gosler AG (1987b) Sexual dimorphism in the summer bill-length of the Great Tit *Parus major*. Ardea 75: 91-98

Gosler AG and King JR (1989) A sexually dimorphic plumage character in the Coal Tit *Parus ater* with notes on the Marsh Tit *Parus palustris*. Ring and Migr 10: 53-57

Haftorn S (1960) The proportion of Spruce seeds removed by the tits in a Norwegian Spruce forest in 1954-1955. K Norske Vidensk Selsk Forh 32: 121-125

Morse (1974) Niche breadth as a function of social dominance. Am Nat 108: 818-827

Morse DH (1978) Structure and foraging patterns of flocks of tits and associated species in an English woodland during winter. Ibis 120: 298-312

Perrins CM (1966) The effect of Beech crops on Great Tit populations and movements. Br Birds 59: 419-432

Perrins CM (1979) British Tits. Collins London

Rothstein SI (1973a) The niche variation model - is it valid? Am Nat 107: 598-620

Rothstein SI (1973b) Relative variation of avian morphological characters: relation to the niche. Am Nat 107: 796-799

Schoener TW (1986) Resource partitioning. In: Kikkawa J and Anderson DJ (eds) Community Ecology Pattern and Process. Blackwell London

Sokal RR and Rohlf FJ (1981) Biometry, 2nd edn. WH Freeman and Co, San Francisco

Soulé M and Stewart BR (1970) The "niche-variation" hypothesis: a test and alternatives. Am Nat 104: 85-97

Svensson L (1984) Identification Guide to European Passerines, 3rd edn. Naturhistoriska Riksmuseet, Stockholm

van Valen L (1965) Morphological variation and width of ecological niche. Am Nat 99: 377-390

van Valen L and Grant PR (1970) Variation and niche width re- examined. Am Nat 104: 589-590

Willson MF (1969) Avian niche size and morphological variation. Am Nat 103: 531-542

TEMPORAL AND SPATIAL VARIATION IN FOOD AVAILABILITY AND ITS EFFECTS ON FLEDGLING SIZE IN THE GREAT TIT

Sabine G. HENRICH-GEBHARDT

Zoologisches Institut, Rheinsprung 9, CH-4051, Basel, Switzerland *

Long-term studies on Great Tits in the Netherlands and England have revealed the importance of various environmental influences on the breeding biology of this species (see e.g. Perrins, 1965 and van Balen, 1973). Using Great Tits as a model I discuss the influences of temporal and spatial variation in food availability on fledgling size and weight. Temporal variation consisted of variation between three consecutive years as well as variation in hatching dates within years. Additionally, experimental brood size manipulations were performed to demonstrate the interaction between the main effects and the additional environmental conditions.

In this paper I discuss the influences of the hatching date, brood size manipulation, and spatial variation in food availability on fledgling size and weight during three consecutive years.

Hatching date is assumed either to be synchronized with the emergence of caterpillars as the main food source (Perrins, 1979) or to be determined by the food supply at the time of egg-laying (Perrins, 1970). In a broad-leaf forest in England, early broods consistently produced more and heavier fledglings even when the weather seemed unfavorable for early broods (Perrins, 1979, although this is not supported by data from 1960, 1962, and 1963 in Fig. 7 in Perrins, 1965). Selection differentials of laying date correlate with the degree of synchronisation between Great Tit breeding and the abundance of caterpillars (van Noordwijk et al., ms). The degree of synchronisation is affected by the temperatures after egg-laying and may be determined by the limited ability of the birds to adjust their breeding schedule to the population dynamics of the caterpillars after the onset of incubation (van Noordwijk et al., ms). The development and growth of caterpillars are temperature dependent. High temperatures result in rapid growth of caterpillars and late breeding birds may miss the peak of their abundance. Interestingly, no net selection differential for laying date was found in a different habitat (van Noordwijk et al., 1981).

Over the last few years forest damage, presumably through pollutants, especially "acid rain", has received much attention. Whatever the extent and irreversibility of tree damage, possible effects on other organisms have received little attention. A study on the distribution of forest birds shows that Great Tits tend to be more common in more heavily damaged forest sections and Blue Tits tend to be more common in less damaged sections (Naef-Daenzer and Blattner,

* present address:
107 Livestock-Poultry Bldg, University of Georgia, Athens, GA 30602, USA

NATO ASI Series, Vol. G 24
Population Biology of Passerine Birds
Edited by J. Blondel et al.

1989). Investigations of the effects of tree damage on caterpillars and hence on Great Tits are currently under way near Basel (see van Noordwijk, this volume). Here, I present preliminary data on the correlation of tree condition with fledgling weights.

MATERIALS AND METHODS

This study was performed in the Birsfelder Hard, a deciduous forest consisting of mainly oak, hornbeam and beech trees near Basel (NW Switzerland), during the breeding seasons in 1986-88. About 300 nestboxes in an area of 80 ha were checked at least once a week during the egg-laying period, to get the exact day of the first egg, onset of incubation, and day of hatching. Nestlings were individually marked, weighed with a Pesola spring balance, and their tarsus length was measured with calipers until they were 15 days old. Great Tits reach their approximate fledging weights about 15 days after hatching.

Parents were caught while feeding their one to two week old young and their weight, wing length, and tarsus length were measured. During nestbox inspections in Nov./Dec. and Feb./March after the breeding season weights were obtained on roosting birds, here denoted as "winter weights".

Brood size experiments (some broods were enlarged by 1/3 and some reduced by 1/3) were carried out when the oldest nestling of a brood was two days old.

In 1988, densities of caterpillars were estimated by collecting and counting pieces of frass fall. Two wooden plates (50 cm x 50 cm) covered with a sheet of paper with insect glue were placed under the nearest two oak trees for each brood. The sheets of paper were collected every day when the nestlings were measured. For the analyses all pieces of caterpillar frass in five squares of 10 cm x 10 cm on the paper were counted.

Data on the condition of trees within small plots of the study area, based on aerial infra-red photographs from the summer of 1984, were kindly provided by the Kantonsforstamt Basel-Landschaft. Damage intensities of each homogeneous plot were expressed in five classes of damage, from 0 = healthy to 4 = dead (Naef-Daenzer and Blattner, 1989). Data from each brood were correlated with the variables of the plot in which the nestbox was located. When more than one brood were raised in the same plot, the mean value of all broods in this plot was used in the analyses, i.e. each plot was only counted once. In multiple regressions all broods were used as independent points.

Only data from first broods were used. Second broods were very rare in this study area.

All sums of squares given in the analyses of variance are of Type III of SAS Statistical Package (SAS Institute, 1985). They are valid for an unbalanced design.

RESULTS

Temporal variation

Nestlings reached significantly different fledging weights in the three years of this study (Table I). The highest weights were found in 1987. Besides variation between years there was variation within years. The regression of fledging weight on hatching date was highly significantly negative and explained more than 30% of the variation in fledging weight for 1988 (slope $b=-0.39$ g/day, $r^2=0.31$, $p=0.0001$). Tarsus length on day 15 was also affected by hatching date, although less strongly ($b = -0.076$ mm/day, $r^2=0.18$, $p=0.0001$). As shown elsewhere (Henrich and van Noordwijk, ms), brood size manipulation also affected fledging weights and tarsus length. Both factors (hatching date and brood size manipulation) as well as their interaction showed significant effects on fledging weight (Table II). To visualize this interaction broods were classified as early or late depending on whether they had hatched before or after May 10. This dichotomy is justified by a difference in mean fledging weight (early broods = 17.38 g ± 1.23, late broods = 15.49 g ± 2.21).

Table I. Mean weight and tarsus length of 15 day-old nestlings in the years 1986-1988. Only unmanipulated broods are reported. SE = standard error. N = number of broods.

	weight [g]	SE	tarsus [mm]	SE	N
1986	17.64	0.14	19.69	0.09	37
1987	18.06	0.13	19.95	0.07	40
1988	16.19	0.31	19.62	0.10	48

Table II. Analysis of variance for the effects of brood size manipulation and hatching date on fledging weight (N=106 broods). Sums of squares are of Type III of SAS (allowing for an unbalanced design). Data are from 1988.

Source	df	SS	F-value	p
manipulation	2	208.07	4.41	0.0146
hatching date	1	1357.80	57.58	0.0001
interaction	2	254.42	5.39	0.0060
residual	100	2358.24		

Broods that had been reduced by 1/3 two days after hatching were hardly affected by hatching date, whereas broods that had been enlarged were the most affected (Fig. 1). Mean tarsus length was not affected by brood size manipulation. The reduction in fledging weight and tarsus length over time coincided with a decline in caterpillar numbers as measured from their frass production (Fig. 2). Although the amount of caterpillar frass widely varies, there was a clear reduction from May 15 onwards and the amount was about 1/8 of earlier values after May 20.

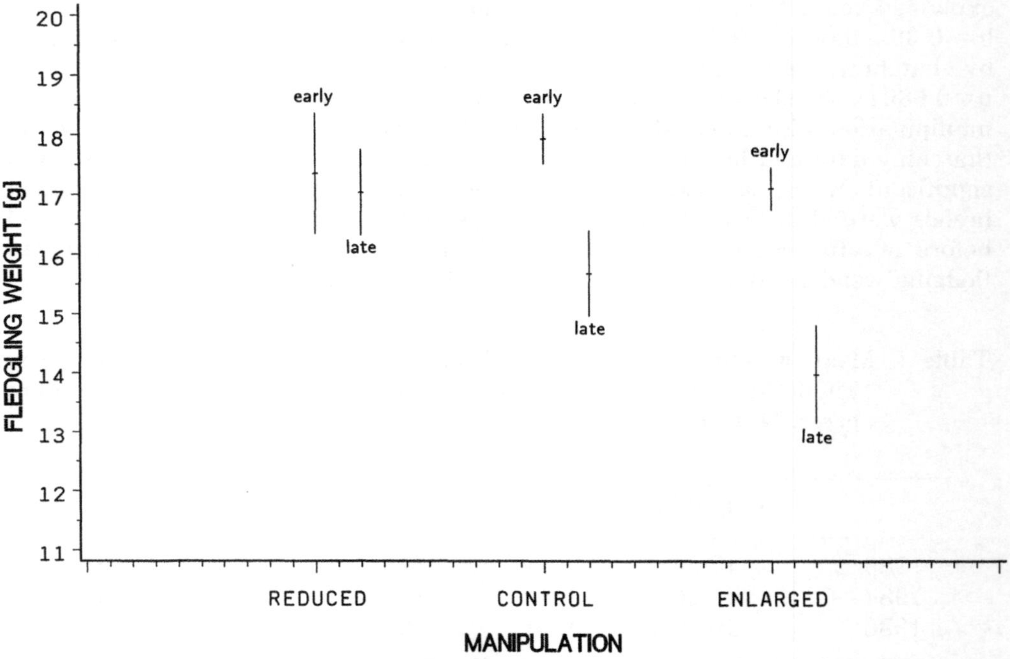

Fig. 1. Fledging weights in 1988 of control, experimentally reduced, and enlarged broods. Broods were either classified as early or late (see text). The vertical bars represent standard errors of the mean.

In 1986, fledging weights were also negatively correlated with hatching dates indicating that nestlings from earlier broods were on average heavier than nestlings that had hatched later ($b=-0.09$ g/day, $r^2=0.08$, $p=0.01$). In contrast to 1988, tarsus length was not significantly correlated with hatching date. An optimum hatching date for weight was present in 1985: early and late broods had lower fledging weights than broods with medium hatching dates (Fig. 3). The quadratic term of the regression of weight on hatching date was highly significant (linear and quadratic terms $p<0.002$). Again, there was no correlation between tarsus length and hatching date. In 1987, there was no significant relation between hatching date and fledging weight or tarsus length (Tab. III).

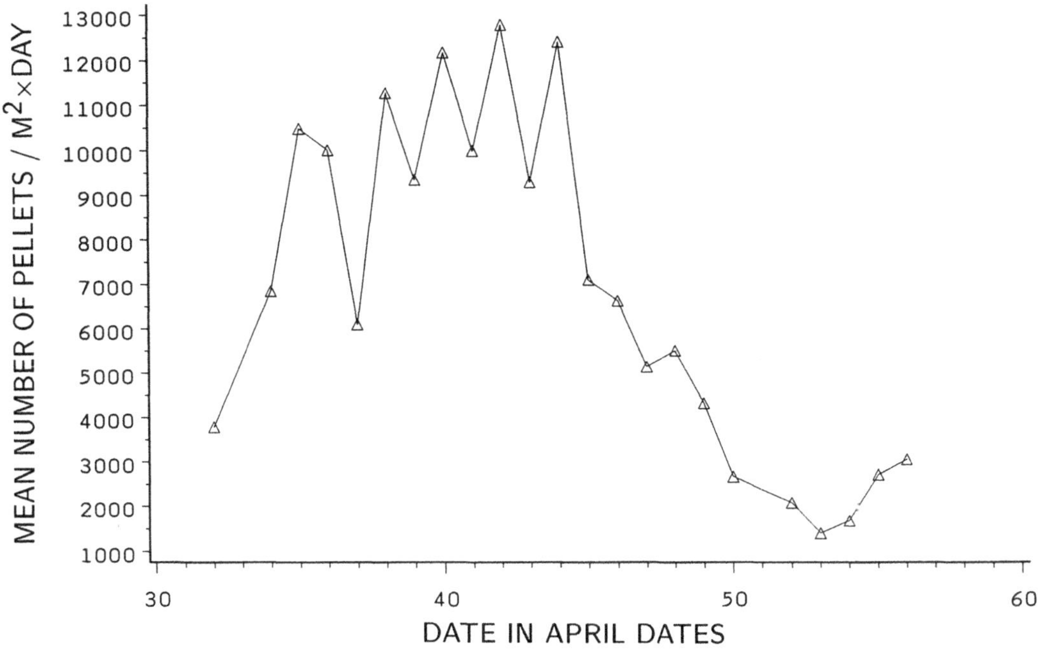

Fig. 2. Number of caterpillar frass pellets as estimates of caterpillar abundance during the month of May 1988.

Table III. Regressions of fledging weights on hatching date in different years show the varying effect of hatching date. The relationship between hatching date and weights was significant in three out of four years, in one year the quadratic term of the regression equation was significant. * = $p<0.05$.

year	linear effect	quadratic effect
1985	*	*
1986	*	NS
1987	NS	NS
1988	*	NS

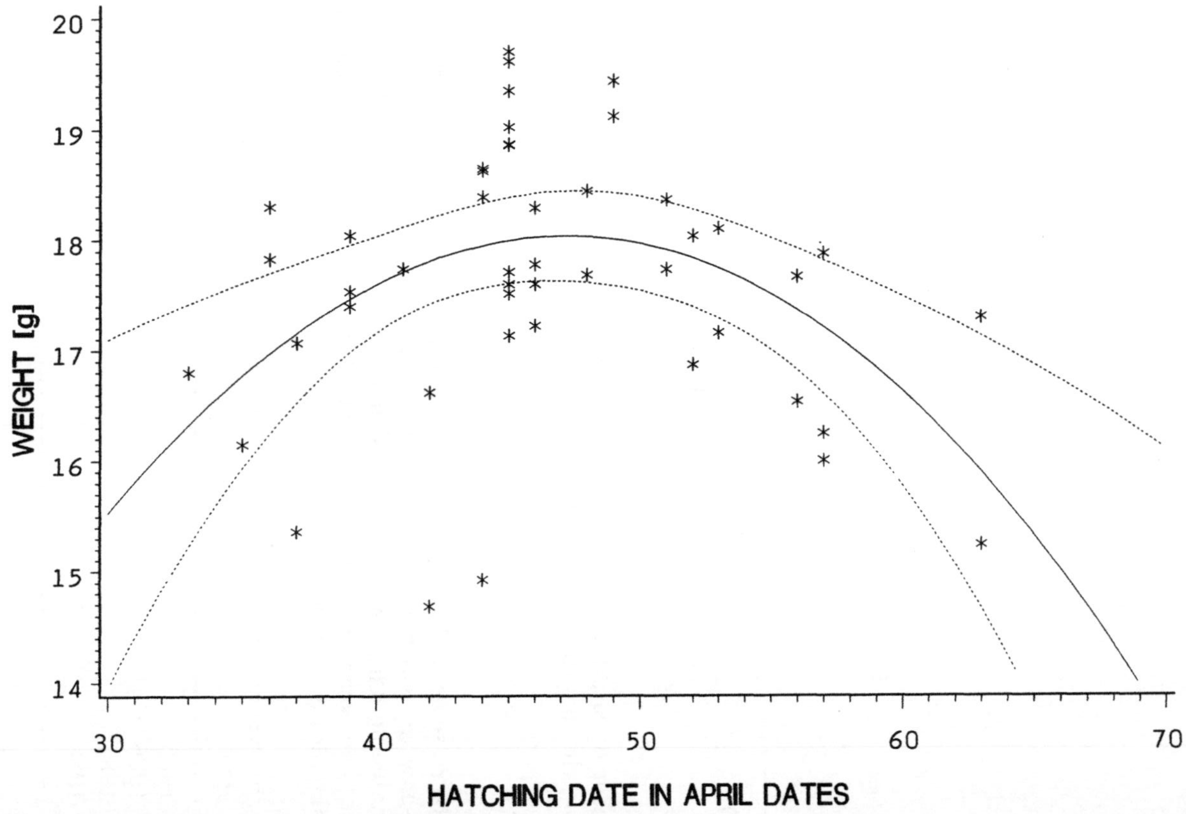

Fig. 3. Fledging weight in relation to hatching date in 1985. The solid line is the quadratic regression, the dotted lines show the 95% confidence interval of the regression line.

The availability of caterpillars as measured by their frass production correlated with fledging weight (Tab. IV). The amount of frass when nestlings of a brood were 8-10, 10-12, and 12-15 days old were positively correlated. The relationship between number of frass pellets and nestling weight was non-linear (Fig 4). A quadratic curve could be fitted well and explained most of the variation in fledging weight when the amount of caterpillar frass between day 8 and 10 and day 10 and 12 were used (Table IV). The same trend, but not as pronounced and not statistically significant, was found for tarsus length.

Table IV. Quadratic regression of mean fledging weight on the number of caterpillar pellets when the broods were between 8 and 10, 10 and 12, and 12 and 15 days old.

time	r^2	F-value	p	N
8 - 10	0.74	9.782	0.009	9
10 - 12	0.69	9.894	0.005	11
12 - 15	0.29	2.206	0.157	13

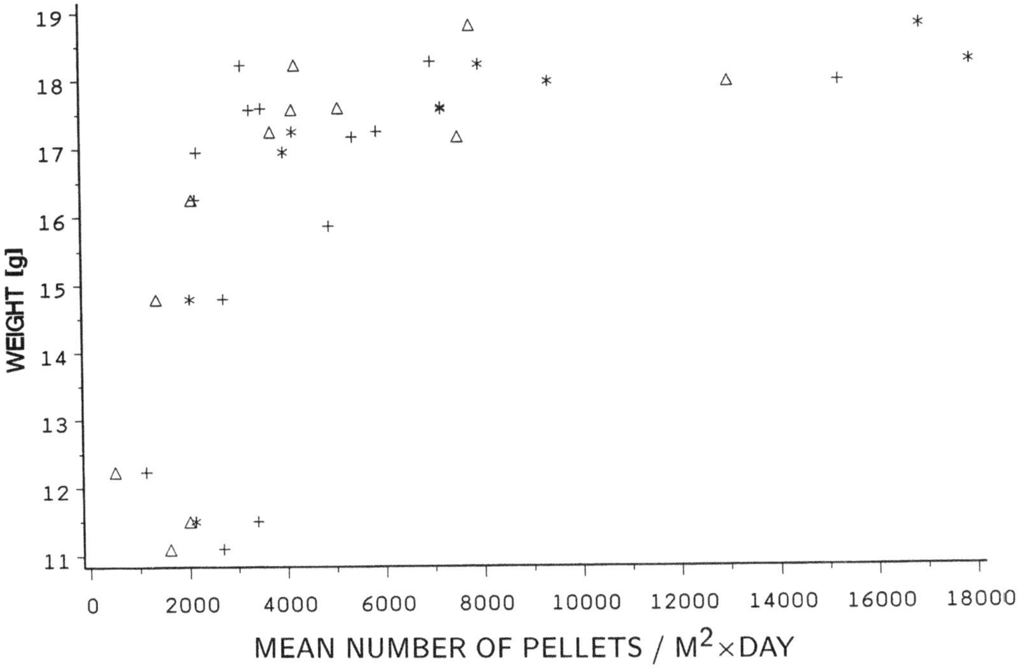

Fig. 4. Number of caterpillar frass pellets collected when the nestlings were between 8 and 10 days old (*), 10 and 12 (Δ), and between 12 and 15 (+) and fledging weight.

Food availability as represented by caterpillar abundance from day 10 onwards was strongly related to hatching dates (frass 10-12: $b=-8.2$ pieces/m^2 day, $p=0.001$; frass 12-15: $b=-0.68$, $p=0.008$, $N = 13$). The correlation was negative, the earlier the hatching date, the higher the caterpillar abundance.

Spatial distribution

Tree condition has been linked to caterpillar abundance. Branch samples from oaks that were differently damaged revealed higher caterpillar numbers on more damaged trees (R. Meury, unpublished). Therefore variation in tree condition was used as an indicator for variation in caterpillar abundance.

In 1986 both fledging weight and tarsus length were positively correlated with the intensity of tree damage (weight: $r^2=0.16$, $p=0.004$, tarsus: $r^2=0.12$, $p=0.02$). There were no interactions between damage intensity and brood size treatment for either weight or tarsus length ($p>0.85$). The use of the discrete variable "damage class" gave the same results. Weights and tarsus lengths were on average greater for broods from plots of higher damage classes (weight: $p=0.004$, tarsus: $p=0.04$). Again, this was true for all brood size treatments. There was no interaction between class of damage and brood size treatment ($p>0.9$).

Combined effects of timing and damage intensities

This section will only deal with data from 1986 since fledging weights were not correlated with tree damage intensities in 1987. Possible reasons for that will be discussed later. In 1986 hatching date and damage intensity were weakly negatively correlated ($r^2=0.05$, $F=4.28$, $p=0.04$). That means that broods in more damaged plots hatched on average earlier than broods in less damaged plots. When both factors, hatching date and damage intensity, were used as predictors for fledging weight in a multiple regression, both were significant (hatch: $p=0.05$, damage intensity: $p=0.001$). However, the two factors did not act simultaneously as is shown by the separate analyses of the different treatments (Table V). Tree damage intensity had a significant effect using control and reduced broods, whereas hatching date but not damage intensity had a significant effect in enlarged broods.

Parental influences

It is well established that fledging weight also has a heritable component meaning that the weight of the parents is a significant predictor of offspring weight (van Noordwijk et al., 1988, Henrich and van Noordwijk, ms). When

Table V. Multiple regression of fledging weight on hatching date and tree damage intensity for broods of different treatments. Data are from 1986 only.

	Predictor	b	F	p	N
reduced broods	hatch	0.1	1.88	0.18	22
	damage	1.06	7.99	0.01	
control broods	hatch	-0.03	0.32	0.53	34
	damage	1.68	7.07	0.01	
enlarged broods	hatch	-0.13	4.95	0.03	24
	damage	0.63	1.39	0.25	

the three factors (parents, hatching date, and tree damage intensity) were analyzed in a multiple regression all three factors were significant in 1986 (parents: $b=0.59$, $F=6.1$, $p=0.02$; hatching date: $b=-0.099$ g/day, $F=5.21$, $p=0.03$; damage intensity: $b=0.92$ g/unit damage intensity, $F=5.06$, $p=0.03$, $N=62$).

DISCUSSION

The most striking result is that the effect of various environmental factors on body size depended very much on other environmental conditions. All the significant factors affected food availability. General statements and predictions about the importance of these factors are difficult to make.

Hatching date was one of the most important factors correlated with fledging weight in 1988, somewhat important in 1985 and 1986, and not important at all in 1987. Whereas the regression of weight on hatching date showed an optimum in 1985, it was linear in 1986 and 1988. Experimental brood size manipulation interacted with hatching date in 1988, and the effects of manipulation were small in early broods but severe in late broods. When food is abundant, as it was for early broods, parent birds were still able to feed an enlarged brood adequately. When the timing and duration of caterpillar abundance differ between years, the influence of hatching date on fledging weight will differ as well.

The same applies to tree damage. Although correlations between tree damage and fledging weight were strong in 1986, they were non-existent in 1987. There are several possible explanations. First, data on tree condition were collected in the summer of 1984. The condition of a tree can change from year to year and the data from 1984 may not have been applicable in 1987. The condition of a forest depends strongly on the amount of rain. 1986 was rather wet and the condition of the forest had markedly improved in 1987 and 1988

(Orsini, pers. com.). Second, as with other environmental factors, the influence of tree damage may depend on the general situation. 1987 was a favorable year for Great Tits and during those years some factors may not be important although they could be under worse conditions. Without further data, any influence of tree condition on fledging weight must be regarded only as a hypothesis.

The importance of different environmental factors related to food differed between years and affected breeding success of Great Tits to different degrees. More detail is needed to detect causal relationships between environmental variables, the expression of genetic variation and reproductive success (see van Noordwijk, this volume). Only then may it be possible to predict long term influences of single factors, like e.g. tree damage, on the reproductive biology of Great Tits.

ACKNOWLEDGEMENTS

This study was part of my Ph.D. thesis (University of Basel, Switzerland). I thank A.J. van Noordwijk for his inspiration and help thoughout the study. Numerous comments of A.J. van Noordwijk and S.C. Stearns greatly improved the manuscript. I also thank the numerous field assistants for their effort.

The data on forest damage were kindly provided by the Kantonsforstamt Basel-Landschaft in Liestal. The financial support of the Swiss National Fonds (3.131-085 to A.J. van Noordwijk) is gratefully acknowledged. Thanks also to an anonymous reviewer for helpful comments.

REFERENCES

Balen JH van (1973) A comparative study of the breeding ecology of the Great Tit *Parus major* in different habitats. Ardea, 61: 1-93

Henrich SG and Noordwijk AJ van The genetical ecology of nestling growth in the Great Tit. I. Heritability estimates under different environmental conditions. Manuscript

Naef-Daenzer B und Blattner M (1989) Die raeumliche Verteilung waldbewohnender Vogelarten in Abhaengigkeit von Waldstruktur und Schaedigung. I. Eichenreiche Laubmischwaelder der Region Basel. Der Ornithologische Beobachter 86: 307-327

Noordwijk AJ van, Balen JH van, and Scharloo W (1981) Genetic variation in the timing of reproduction in the Great Tit. Oecologia (Berl) 49: 158-166

Noordwijk AJ van, Balen JH van, and Scharloo W (1988) Heritability of body size in a natural population of the Great Tit (*Parus major*) and its relation to age and environmental conditions during growth. Genet Res Camb 51: 149-162

Noordwijk AJ van, McCleery RH and Perrins CM The synchronization of Great Tit (*Parus major*) breeding with caterpillar growth, using temperature as a predictor. Manuscript

Perrins CM (1965) Population fluctuations and clutch size in the Great Tit, *Parus major* L.. J Anim Ecol, 34: 601-647

Perrins CM (1970) The timing of birds' breeding seasons. Ibis 112: 242-255

Perrins CM (1979) British Tits. Collins, Glasgow

SAS Institute Inc SAS User's Guide: Statistics, Version 5 Edition. Cary, NC: SAS Institute Inc, 1985

ON THE EFFECTS OF SMALL SCALE VARIATION IN TEMPERATURE AND FOOD AVAILABILITY ON LAYING DATE AND EGG SIZE IN GREAT TITS (*PARUS MAJOR*)

Ruedi NAGER

Zoologisches Institut, Rheinsprung 9, CH-4051 Basel, Switzerland

Extensive long term studies on Great Tits (*Parus major*) identified several factors which affect their timing of reproduction and their reproductive output (see Perrins 1979 for a review). Variation in breeding parameters such as date of first egg laying or clutch size could be successfully correlated with variation in temperature, breeding density, food availability and other proximate factors (e. g. Kluyver 1952, Perrins 1965, van Balen 1973, Dhondt et al 1984, Schmidt 1984, Perrins and McCleery 1989). Although mechanisms which act on the individuals are envisaged, the relationships are based on mean values of large groups. Breeding parameters vary within a breeding population by about the same magnitude as between different years. Variation among individuals could be partly attributed to such factors as, for example, age (Perrins 1965 and Dhondt 1989 for laying date and clutch size), body size (Jones 1973 for laying date) or genetic differences among them (Perrins and Jones 1975 and van Noordwijk et al. 1981a for laying date, van Noordwijk et al. 1981b for clutch size, van Noordwijk et al. 1981c and Ojanen et al. 1979 for egg size). But a considerable part of the variation among individuals in a breeding population remains unexplained. Differences in local environmental conditions could help us to explain more of this remaining variation, as environmental conditions can explain some of the differences between years.

I will here present relationships between within-year variation in laying date and egg size and variation in local environmental conditions such as temperature and food supply from two breeding seasons. Kluyver (1952), Perrins (1965), van Balen (1973), Schmidt (1984), Perrins and McCleery (1989) and others showed a strong relationship between mean laying date of a population and the temperature sum in March and April. Food supply can affect the onset of egg laying, too, which is demonstrated by experiments where some pairs received supplemental food in the period prior to laying. These pairs started their clutches earlier than unfed control pairs (Källander 1974, Brömssen and Jansson 1980, Clamens 1987, but see also Krebs 1971 and Jones 1973). The available energy may not only influence the decision of a bird to start egg laying but it could also affect size of the eggs. However much less information is available on factors affecting egg size. Van Noordwijk (1984) demonstrated an influence of temperature averaged over the three days preceeding laying (that corresponds approximately to the period of egg formation) on variation in egg size within as well as among clutches. Järvinen and Pryl (1989) found that female body weight and warm weather during the egg laying period had positive effects on egg size.

NATO ASI Series, Vol. G 24
Population Biology of Passerine Birds
Edited by J. Blondel et al.

For the understanding of how an individual phenotype is formed in a particular environment the relative importance of the different factors contributing to the phenotypic variation should be known (van Noordwijk 1989). First indications on the relative effects of temperature and food supply on laying date and egg size can be drawn from the data presented here.

METHODS

The study was conducted at the Blauen, a northern slope of the Jura about 10 km south of Basel (Northwest-Switzerland). Beech (*Fagus sylvatica*) is the most common tree in the study area. There are about 430 nestboxes at altitudes from 380 to 770 m. Data were collected in the two breeding seasons 1988 and 1989.

Environmental conditions were measured at 6 sites in 1988 and 8 sites in 1989. The sites were about 150 to 200 m in diameter and contained about 10 nestboxes. The sites differred in altitude, exposure and forest type. In the centre of each site three neighbouring nestboxes received a second nestbox with a fine grid covering the entrance hole. These second nestboxes contained minimum-maximum thermometers. These thermometers were checked daily from early March onwards. A mean daily temperature was calculated as average of minimum and maximum temperatures and averaged over the three boxes per site.

At each site branch samples were taken every week from three to five trees (mainly beech, but at some sites also oak, *Quercus robur*). In 1988 the samples were taken with a ladder of about 6m length and in 1989 with a small skylift of 16 m maximal height.

The samples were subsequently handled in the laboratory. All insects (mainly caterpillars) and spiders, but excluding small items such as mites, were counted and weighed within a day of the date of collection. The length of the branches was measured and the food availability was expressed as freshweight per 10 m branch length. The weekly branch sampling was started at around the time of bud burst in both years and continued as long as there were nestlings of first broods.

The breeding parameters I will use in this paper are only from first clutches of the nestboxes at the sites where the local environmental conditions were measured. Date of first egg and the laying sequence were determined by daily visits of the nests. For each site laying dates were averaged over the pairs breeding there. The length (L) and breadth (B) of the eggs were measured to the nearest 0.1 mm using callipers a few days after the clutches were completed. The egg volumes were calculated by the following formula: $\text{volume} = 0.495LB^2 - 0.017$ (van Noordwijk et al. 1981c). The egg volume is closely correlated with fresh egg weight (van Noordwijk et al. 1981c).

RESULTS

Variation in laying date

In both years the spread of laying dates in the study area was about 30 days. This is similar to the observed spread of annual mean laying dates (e. g. Perrins and McCleery 1989). The best prediction of annual means in laying dates is given by the temperature sums over the second half of March and whole April (Kluyver 1952). The spring temperatures (from March 17 to April 28) of 1988 as well as 1989 at the different sites in our study area were significantly correlated with the laying dates at the corresponding sites (1988: $r=-0.913$, $n=6$, $p<0.05$ and 1989: $r=-0.757$, $n=8$, $p<0.05$).

In order to compare the relationship between temperature and laying date of the two years, temperatures and laying dates were expressed as deviation of the mean over all sites of the corresponding year (Fig. 1). In both years Great Tits at warmer sites laid earlier than Great Tits at cooler sites. The regression lines of 1988 ($y=-4.931x-0.105$, $r^2=0.843$, $n=6$, $p<0.01$) and 1989 ($y=-4.647x+1.361$, $r^2=0.522$, $n=8$, $p<0.05$) indicated in Figure 1 do not differ ($F[1,10]=0.017$, $p>0.75$).

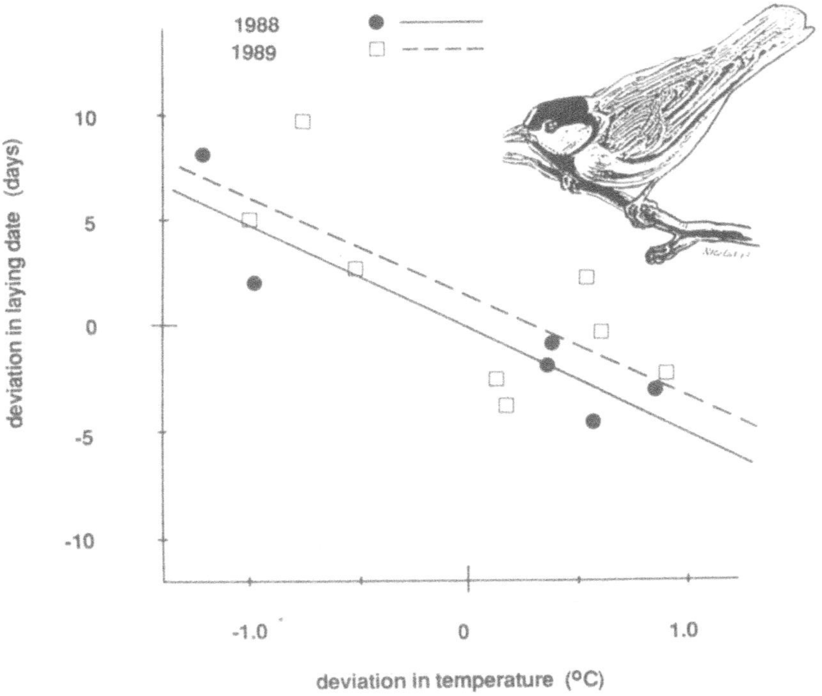

Fig. 1. There is a negative relationship between laying date and spring temperature within years. Temperature and laying date are expressed as deviations from the yearly means. Spring temperature is the mean daily temperature averaged over the period from March 17 to April 28.

In Figure 2 the same relationship between temperature and laying date is shown in a different way. The phenotypic variation at the different sites is expressed by the scattering of the black dots. The variation among the three temperature values per site and the variation in laying dates at the corresponding site are plotted as axes of an ellipse. The ellipses form a congruent pattern around the regression lines in both years (1988: $y=-4.833x+66.630$, $n=6$, $r^2=0.833$, $p=0.011$ and 1989: $y=-4.455x+56.302$, $n=8$, $r^2=0.618$, $p=0.021$). The regression lines of the two years differ only in their intercept, but in 1989 early March was exceptionally warm. Including also this relatively high temperature in early March in 1989 this lead to a shift of the regression line of 1989 by 1.5°C to the right. With this correction there is a close correspondence between the two years 1988 and 1989.

No relationship was detectable between the amount of food available at the sampling sites when the females were starting egg laying and the laying date. This was the case for the fresh weight of all potential prey items available per 10 m branch (1988: $r_s=0.714$, $n=6$, $p=0.111$ and 1989: $r_s=-0.048$, $n=8$, $p=0.911$) as well as for the biomass of small caterpillars (less than 10 mg of weight) in 1989 which will constitute the main food resource three to four weeks later when the large amount of food must be fed to the nestlings (1989: $rs=0.000$, $n=8$, $p>0.99$).

Variation in egg size

In 1988 and 1989 the laying sequences of 7 and 11 clutches, respectively, were determined. For each of these clutches there was also information available about local temperatures and the amount of food in the period of egg formation. Females did not lay eggs smaller than the one of the preceeding day when there was a decline in temperature between the two periods of egg formation (three days preceeding the laying) (Tab I, Spearman rank correlation, 1988: $r_s=-0.066$, $n=36$, $p=0.704$ and 1989: $r_s=0.060$, $n=70$, $p=0.624$). To check if there was a relationship between temperatures and absolute egg size as the one van Noordwijk (1984) found in a Dutch population of Great Tits, I first plotted egg size against temperatures during the period of egg formation and then, to compensate for the small sample sizes, I took the mean temperature and weighed mean egg size for every three points adjacent in temperature (but not necessarily adjacent in time). In most cases one data point is the average of eggs laid by about 6 to 8 females. Although there was a cold spell during several days during the laying period of 1989, the mean egg volume in 1988 (1.582 ± 0.197 cm^3) and in 1989 (1.545 ± 0.080 cm^3) were not different from each other (t-Test, $t=-0.102$, $p=0.906$). There was no relationship between mean temperature during the period of egg laying and egg volume (1988: $r=0.237$, $n=7$, $p=0.610$ and 1989: $r=-0.242$, $n=9$, $p=0.531$).

In Figure 3, egg size is plotted against the amount of available food (expressed as fresh weight per 10 m branch) at the time the female was starting the clutch. The mean egg volumes did not differ between 1988 and 1989, but in 1989 some females experienced an obviously richer food supply and

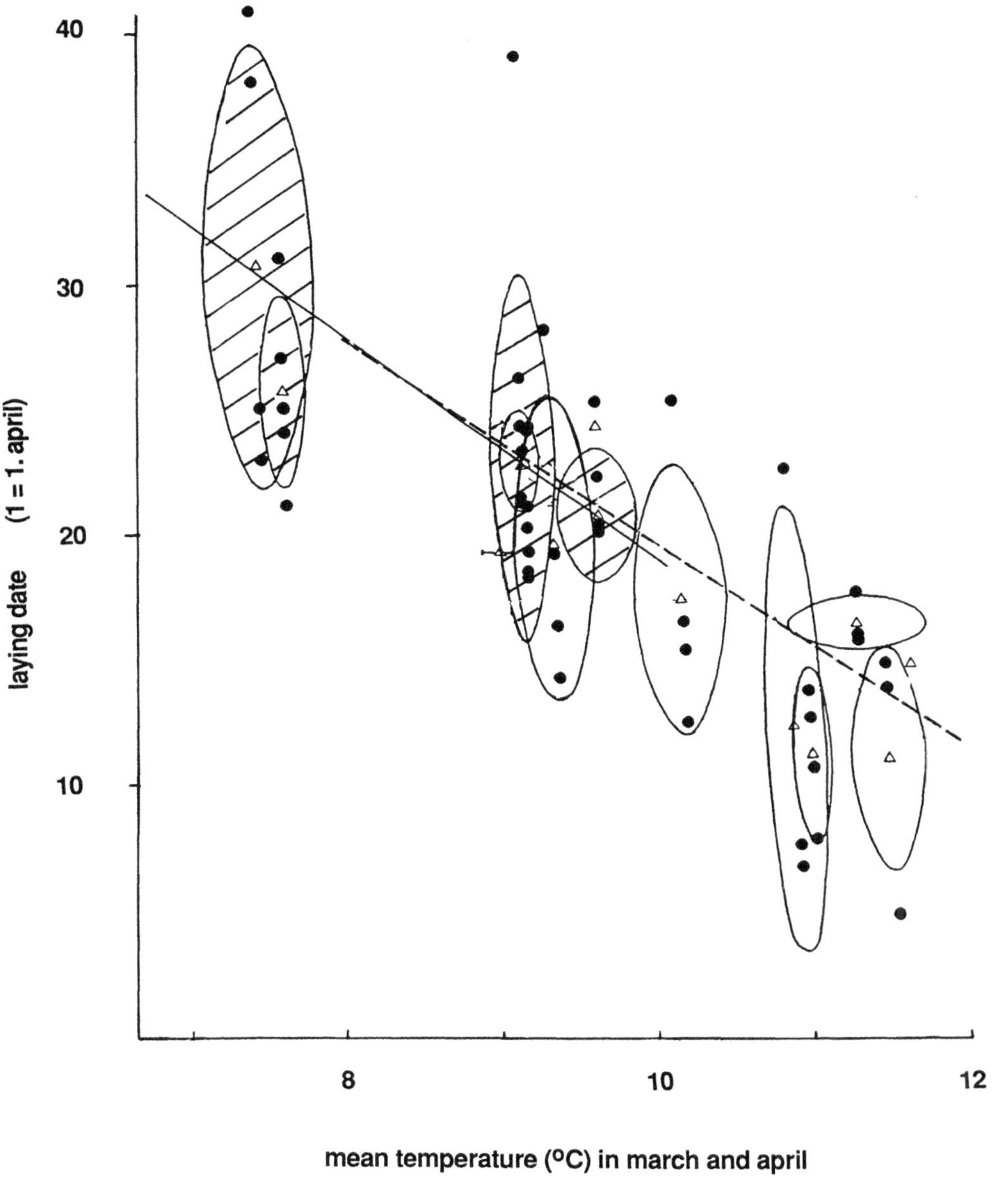

Fig. 2. The relationship between laying date and spring temperature. The standard deviation in temperature among the three measurements per site and the standard deviation in laying dates are plotted as the axes of an ellipse (dots = individual laying events, triangles = mean value of a site). Ellipses of 1988 are shaded. The ellipses of 1989 are shifted by 1.5°C to the right along the temperature axes (see text). The regression lines are based on the mean values of the sites.

Table I. There are no consistent changes in egg size of individual females compared to the size of the preceeding egg in response to changes in ambient temperatures, averaged over the three days preceeding the laying of the egg. Data of 1988 from 7 clutches, data of 1989 from 11 clutches.

	Egg Size	Temperature Increase	No change	Decrease
1988	Increase	1	3	8
	No change	6	2	7
	Decrease	2	2	5
1989	Increase	10	9	5
	No change	10	10	8
	Decrease	5	4	9

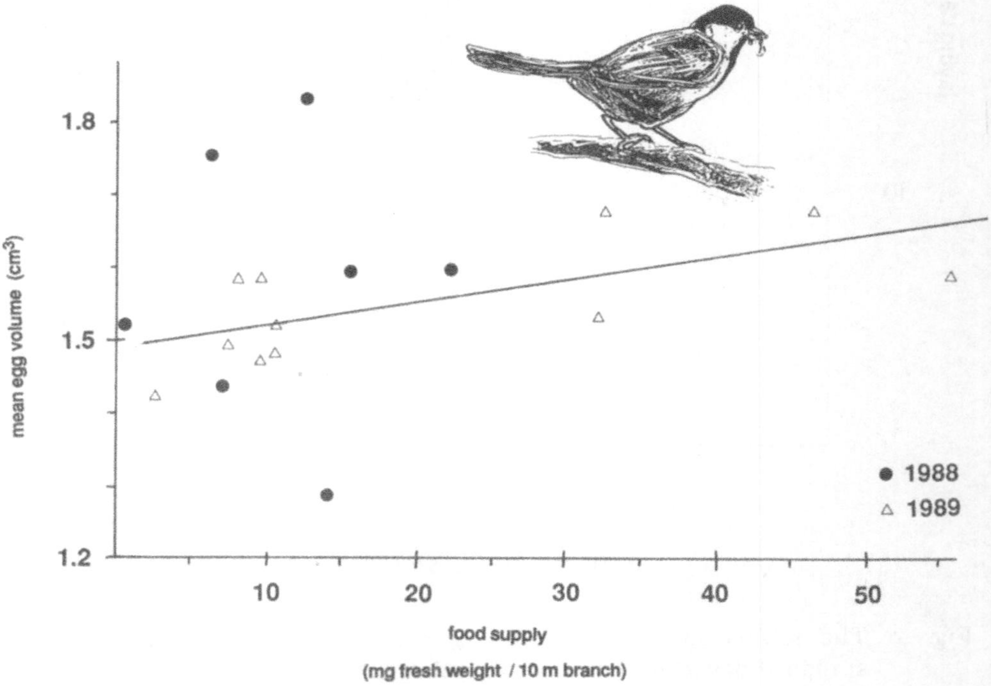

Fig. 3. There is a positive relationship between food supply and average egg volume of a clutch in 1989, but not in 1988 (regression line for the data of 1989). The total amount of food, expressed as fresh weight per 10 m branch length, was taken from the sampling date which was closest to the date where the first egg was laid.

they laid slightly larger eggs ($y=0.003x+1.491$, $r^2=0.48$, $n=11$, $p=0.02$, Fig. 3). In 1988 there is no detectable effect of food availability on egg size ($r^2=0.042$, $n=7$, $p>0.05$, Fig. 3).

DISCUSSION

Several long term studies could explain a high proportion of between-year variation in laying dates. Perrins (1973) predicted an advancement in mean laying date of 4 days with an increase in mean spring temperature of 1°C. The present study showed that small scale variation in temperature could also explain a high proportion of the within-year variation in laying date, confirmed in two years. The slope of the regression line for the within-year relation is approximately 4.5 to 5 days / 1°C.

Temperature can act in two different ways on laying date. O'Connor (1978) proposed a direct effect of temperature on the rapid growth of follicles. At lower temperatures the energy expenditure for basal metabolism will be increased and less energy is available for the process of growth of gonads and egg formation. O'Connor (1978) calculated that the increased energy expenditure which is associated with a decrease of 0.5°C in temperature would prolong the phase of rapid follicle growth by about 0.6 days. This could account for 25-30% of the variation in laying date among individuals. The effect of temperature can also be indirect. The development of caterpillars, the main food resource during the nestling period, depends on the spring temperatures. Caterpillars reach their highest densities earlier in years with warmer spring temperatures (van Balen 1973). For a small bird of the size of a Great Tit egg formation might be a costly process. It needs about an additional 40% to the daily basal metabolic rate to produce an egg (Walsberg 1983, Rahn et al. 1988) and the bird probably has to cover all of this energy demand by its daily energy income. Unfortunately there is hardly no information in the literature on food availability during early spring when Great Tits lay their eggs and on their diet during these days. In most years early laying females have a higher reproductive success in terms of numbers of recruits than late laying females (Perrins 1965, 1970; but see van Noordwijk et al. 1981a). But the absence of a consistent selection for laying earlier suggests that there are factors which prevent females to do so. Perrins (1970) suggested that females are energy limited in the production of their eggs. This seems to be supported by the finding of Jones (1973) that smaller females start laying earlier and by experimental manipulation of the food supply, in which pairs which received supplemental food laid earlier than unfed control pairs (Källander 1974, Brömssen and Jansson 1980, Clamens 1987). Our estimates of available food by the time of the first egg did not correlate with laying date. However there are also supplemental feeding experiments which showed no effect on laying date (Krebs 1971, Jones 1973). They used sunflower seeds as food whilst Källander (1974), von Brömssen and Jansson (1980) and Clamens (1987) offered mealworms to their experimental pairs. Perhaps not only the amount of energy is crucial, but also the quality of the food might be important for the ability of the females to form eggs (Arcese and Smith 1988). There are indications that some Great Tits are limited in their ability to form eggs by low levels of available calcium (Drent and Woldendorp 1989).

However we do not know whether the energy needed to produce eggs prevents laying earlier or whether the timing is determined by synchronising the peak energy demand of nestlings with the maximum caterpillar availability. This last point is the widely accepted ultimate explanation of the timing of birds' breeding season (Lack 1966). It might be that Great Tits react to the appearance of the first caterpillars hatching in spring as a clue that the favourable time to rear young is approaching. We recalculated the correlation between laying date and food availability for 1989 excluding caterpillars heavier than 10 mg. There was no detectable effect of small caterpillars on the timing. Generally there is a close relationship between the mean laying date of tits and the peak date of caterpillar availability (Perrins 1965, van Balen 1973, Zandt et al., this volume). The observation that the synchronisation of tits to the caterpillar peak results in strong selection pressures on laying date and that the selection differential depends strongly on temperature (van Noordwijk, McCleery and Perrins, in prep.) suggests a crucial importance of the ultimate explanation in the proximate determination of laying date.

A way to learn more about how a trait is expressed in a varying environment is the representation of the trait as a reaction norm (van Noordwijk 1989, van Noordwijk, this volume). Genes can be seen as programs which process information from the environment and a reaction norm describes all the possible phenotypes that a specific genotype can form under different environmental conditions (see for example Stearns 1989). If genotypes have different reaction norms, a natural population would be represented by a bundle of reaction norms (van Noordwijk 1989). Figure 2 can be interpreted as a translation of variation in environmental conditions into variation in laying date. The band of ellipses would represent the bundle of reaction norms of one population. Assuming that variation in some environmental factors which explains an important part in laying date is plotted on the horizontal axes, the width of the band of ellipses reflects the importance of genetic variation and the variation due to other environmental factors. Van Noordwijk et al. (1981a) estimated a heritability of laying date of 40% in two Dutch populations of Great Tits. In fact the observed laying dates in Figure 2 varied approximately in the range 40% around the mean values, but it should be kept in mind that heritability estimates are only valid for the population and the environmental conditions under which they were measured (Falconer 1981). Under these circumstances temperature seems to explain a considerable part of the remaining variation in laying date (60-80%, Fig. 2). This point further strengthens the view that temperature is a very important factor in explaining variation in laying dates within years. The shape of the bundle of reaction norms gives information about physiological mechanisms. From Figure 2 we can conclude that, over the observed range of temperatures, laying date and temperature are linearly related.

Less information is available on factors which have a proximate effect on intra-clutch variation in egg size. Van Noordwijk (1984) demonstrated an effect of ambient temperature during the three days preceeding the laying of the egg on its fresh weight. But only temperatures below 10°C affected the egg size. Smaller Great Tits (but not necessarily in poorer conditions) laid smaller eggs (Järvinen and Pryl 1989). Slagsvold and Lifjeld (1989) found that females of the Pied Flycatcher laid smaller eggs in poorer habitats. In none of the various experiments with supplemental feeding of Great Tits in the pre-laying period were any effects of extra food on egg size reported. In at least one year we were able to demonstrate that food supply was more important as an environmental factor for egg size than temperature. However the branch sampling in 1988 was more primitive and samples are likely to be less accurate. It is clear that we need more data on egg size variation.

We may conclude that the relationship between temperature and laying date within years is very similar to that based on the analysis of yearly means (see e.g. Perrins and McCleery 1989). Using the concept of reaction norms, both between year and within year variation can be described by the same relationship. This suggests that there is a short causal link from temperature to laying, which should further be investigated by field experiments manipulating temperatures.

ACKNOWLEDGEMENTS

This study was made possible by the combined efforts of many people collecting the data. I would like to thank C. Eichenberger, L. Keller, J. Klahre, A. van Noordwijk and C. Ruegger for their help in the field and laboratory. I thank D. Ebert, S. Henrich, U. Leugger, A. van Noordwijk, H. Zandt and two anonymous referees for their comments on an earlier version of the manuscript. Financial support from the Swiss Nationalfonds (3.131.88 to A. van Noordwijk) is gratefully acknowledged.

REFERENCES

Arcese P and Smith JNM (1988) Effects of population density and supplemental food on reproduction in Song Sparrows. J Anim Ecol 57: 119-136

Balen JH van (1973) A comparative study of the breeding ecology of the Great Tit (*Parus major*) in different habitats. Ardea 61: 1-93

Brömssen A von, Jansson C (1980) Effects of food addition to Willow Tit (*Parus montanus*) and Crested Tit (*Parus cristatus*) at the time of breeding. Ornis Scand 11: 173-178

Clamens A (1987) Rôle de la nourriture dans la détermination des paramètres démographiques des mésanges (*Paridae*) en chênaie verte: résultats préliminaires. Alauda 55: 254-266

Dhondt AA (1989) The effect of old age on the reproduction of Great Tits (*Parus major*) and Blue Tits (*Parus caeruleus*): Ibis 131: 268-280

Dhondt AA, Eyckerman R, Moermans R, Huble J (1984) Habitat and laying date of Great and Blue Tit (*Parus major* and *P. caeruleus*). Ibis 126: 388-399

Drent PJ, Woldendorp JW (1989) Acid rain and eggshells. Nature 339: 431

Falconer DS (1981) Introduction to Quantitative Genetics. 2nd edn. Longman, London

Järvinen A, Pryl M (1989) Egg dimensions of the Great Tit (*Parus major*) in southern Finland. Ornis Fennica 66: 69-75

Jones PJ (1973) Some aspects of the feeding ecology of the Great Tit (*Parus major*). Unpublished PhD Thesis, Oxford University, Oxford, England

Källander H (1974) Advancement of laying of the Great Tits by the provision of food. Ibis 116: 365-367

Kluyver HN (1952) Notes on body weight and time of breeding in the Great Tit (*Parus major*). Ardea 40: 123-141

Krebs JR (1971) Territory and breeding density in the Great Tit (*Parus major*). Ecology 52: 2-22

Lack D (1966) Population studies of birds. Clarendon Press, Oxford

Noordwijk AJ van (1984) Quantitative genetics in natural populations of birds illustrated with examples from the Great Tit (*Parus major*). In Woehrman K, Loeschke V (eds) Population biology and evolution. Springer Verlag, Berlin. p 67-79

Noordwijk AJ van (1989) Reaction norms in genetical ecology. BioScience 39: 453-459

Noordwijk AJ van, Balen JH van, Scharloo W (1981a) Genetic variation in the timing of reproduction in the Great Tit. Oecologia 49: 158-166

Noordwijk AJ van, Balen JH van, Scharloo W (1981b) Genetic and environmental variation in clutch size of the Great Tit. Neth J Zool 31: 342-372

Noordwijk AJ van, Keizer LPC, Balen JH van, Scharloo W (1981c) Genetic variation in egg dimensions in a natural population of Great Tit. Genetica 55: 221-232

Ojanen M, Orell M, Vaisanen RA (1979) Role of heredity in egg size variation in the Great Tit (*Parus major*) and the Pied Flycatcher (*Ficedula hypoleuca*). Ornis Scand 10: 22-28

Perrins CM (1965) Population fluctuations and clutch size in the Great Tit (*Parus major*). J Anim Ecol 34: 601-647

Perrins CM (1970) The timing of birds' breeding seasons. Ibis 112: 242-255

Perrins CM (1973) Some effects of temperature on breeding in the Great Tit and Manx Shearwater. J Reprod Fert, suppl 19: 163-173

Perrins CM (1979) British Tits. Collins, London

Perrins CM and Jones PJ (1975) The inheritance of clutch size in the Great Tit. Condor 76: 225-229

Perrins CM, McCleery RH (1989) Laying dates and clutch size in the Great Tit. The Wilson Bulletin 101: 236-253

Rahn H, Sotherland PR, Paganelli CV (1987) Interrelationships between egg mass and adult body mass and metabolism among passerine birds. J Ornithol 126: 263-271

Schmidt KH (1984) Frühjahrstemperaturen und Legebeginn bei den Meisen. J Ornithol 125: 321-331

Slagsvold T, Lifjeld JT (1989) Constraints on hatching asynchrony and egg size in Pied Flycatchers. J Anim Ecol 58: 837-849

Stearns SC (1989) The evolutionary significance of phenotypic plasticity. BioScience 39: 436-446

Walsberg GE (1983) Avian ecological energetics. In: Farner DS, King JR (eds) Avian biology, Vol VII, Academic Press, New York

THE ESTIMATION OF LOCAL IMMATURE SURVIVAL RATE AND OF AGE-SPECIFIC PROPORTIONS OF BREEDERS IN BIRD POPULATIONS

Jean CLOBERT[1], Jean Dominique LEBRETON[2] and Gilbert MARZOLIN[3]

[1] Laboratoire d'Ecologie, Ecole Normale Supérieure, 46 rue d'Ulm, 75231 Paris-cédex 05, France
[2] Centre d'Ecologie Fonctionnelle et Evolutive, C.N.R.S., route de Mende, B.P.5051, 34033 Montpellier-cédex, France
[3] 14 Chemin de la Latte, 57157 Marly, France

Local recruitment rate is an important component of lifetime reproductive success (review by Clutton-Brock 1988), particularly in bird populations (van Balen et al. 1987; McCleery and Perrins 1988). This is not surprising, since the recruitment rate combines several very distinct components: survival during the first year of life, fidelity of the survivors to their natal area, breeding activity of one year old individuals, survival rate of one year old non breeders, etc... Since these components are influenced by different biotic and abiotic variables (with, as a result, different selection pressures), the resulting overall recruitment rate is likely to show important variations, unless strong negative trade-offs exist between its components.

Unfortunately most available data do not allow easily estimation of the components of the recruitment rate. For instance, while it is possible in general to estimate the *local* immature survival rate, splitting it into the proportion of surviving birds which are faithful to their natal area and the immature survival rate is almost impossible, except perhaps with large scale experimental designs. Similarly, reobservations as breeders of birds of known age are needed to separate the local immature survival rate from the age-specific proportion of breeders. In this case, even when appropriate data are available, few statistical methods exist. Simple ratio estimates have been proposed by Barrat et al. (1976), Lebreton (1978, 1981) and Clobert (1981). Lebreton et al. (in press) have generalized some of the above approaches in a general statistical model for the estimation of age-specific proportions of breeders, for the particular case when individual recapture histories are not known, i.e. for cohort-characteristic marking.

The purpose of this paper is to present a model for estimating the local immature survival rate and age-specific proportions of breeders, from reobservations as breeders of birds marked individually at birth. We restrict our attention to the case of full breeding at age 2, the general case being presented elsewhere (Clobert et al., in prep.). An application to the dynamics of a population of Dipper *Cinclus cinclus* is given.

NATO ASI Series, Vol. G 24
Population Biology of Passerine Birds
Edited by J. Blondel et al.

PARAMETER DEFINITION AND ASSUMPTIONS

In the general case, we consider a population in which the first breeding attempts take place at age m, and in which the proportion of breeders equals 1 from age n onwards ($m \leq n$). For most Passerines one may assume, at least as a working hypothesis $m=1$ and $n=2$. This is the case treated here.

We consider k-1 cohorts of birds born in years $i=1,...,k-1$, and marked at fledging. For a bird marked in year i, several steps separate the time of fledging from its first breeding attempt:

1) It has first to reach the age of maturity $m=1$. Let s_{i1} be the survival rate from marking to age 1 for a bird born in year i (Tab. I). It may survive one or more further years. A key hypothesis is that the survival probability Φ_i from year i to year i+1 of animals older than $m=1$ does not depend on their age and of the number of previous breeding attempts.

2) It has to come back to its natal area. Let π_i be the probability, for a bird born in year i and alive at age 1, of being present in the natal area. "Present" includes here, and in what follows, individuals which would have been present had they "decided" to breed. We assume that the choice of the definitive breeding area is done at age $m=1$, even if the first breeding attempt of a particular individual takes place later. Most of the time, in the absence of relevant data, only the local immature survival rate $\Phi_{i1}=s_{i1}\pi_i$ can be estimated.

3) It may breed at age 1. Let a_{i1} be the proportion of breeders in year i+1 among individuals born in year i, present and alive at the beginning of year i+1. Since we assume that all birds born in year i breed in year i+2 ($n=2$), $a_{i2}=1$. For the sake of simplicity, the proportions of breeders among one year old birds of cohorts 1,... k-1, $a_{11},..a_{k-1,1}$, concerning events which take place in years 2,... k, are noted $a_2,... a_k$ in what follows. When having bred once, each individual is supposed to breed (if still alive) every year in the study area.

4) It has to be caught or sighted when breeding. Let $p_2,... p_k$ be the capture rate of breeders in year 2,... k. The capture rate is supposed to be independent of the previous history of the individual. In particular, it is independent of its number of previous breedings and of its age.

In this scheme, a one year old individual may breed, and be captured with the same probability as any breeder, or not breed, and be captured with probability 0. The overall recapture probability in year i is thus lower for one year old individuals than for older ones. This will be the case in general until age n. Let p'_{i1} be the probability of capture of an individual born in year i, at age 1, i.e. in year i+1. p'_{i1} is therefore the capture rate of one year old individuals which survived and are present (in the above meaning).

Since one year old individuals in year i+1 have a probability a_{i+1} of breeding and then p_{i+1} of being observed, one has $p'_{i1}=a_{i+1}p_{i+1}$, a key relationship in our model (Tab. I). We have assumed that all the two year old individuals breed. Following the above notation, this hypothesis reduces to $p'_{i2}=p_{i+2}$ (p'_{i2} is the capture rate of inviduals born in year i, present and alive at the beginning of year i+2, and not yet observed breeding). More general models follow by relaxing this hypothesis. A statistical comparison with the simplest model makes it possible to test the hypothesis $n=2$. When $n=2$, the notation can be simplified by denoting $p'_{1\ 1}, \ldots p'_{k-1\ 1}$, which concerns reobservations taking place in years $2, \ldots k$, as $p'_2, \ldots p'_k$. Estimates of $a_2, \ldots a_k$ can be obtained from estimates of $p'_2, \ldots p'_k$ and $p_2, \ldots p_k$ using the relationship $a_i = p'_i / p_i$.

DATA STRUCTURE AND PARAMETER ESTIMATION

In our model, the survival and recapture parameters applied to a particular individual differ before and after its first capture (Table I). Until the first reobservation, both survival and recapture probabilities are time-dependent, and partly age-dependent, since they take particular values for the first year (Table IA). The first reobservation can be thought of as the marking of an adult of unknown age. From this first reobservation onwards, survival and recapture parameters are time-dependent as in the classical Cormack-Jolly-Seber model (Table IB).

Table I. Survival (Φ) and recapture (p) parameters for birds marked as young, starting reproduction at age $m=1$, and reaching full reproduction at age $n=2$.

A: Until first reobservation as breeders (=first fictitious group of individuals)

year of marking (cohort) = i (≥1)	time of reobservation i+1	i+2	i+3 ...
surv.param.	<--- $\Phi_{i1} = s_{i1}\pi_i$ --->	<--- Φ_{i+1} --->	<--- Φ_{i+2} ---> <---...
recapt.param.	$p'_{i+1} = a_{i+1}p_{i+1}$	p_{i+2}	p_{i+3} ...

B: After first reobservation as breeders(=second fictitious group of individuals, equivalent to newly marked adults)

year of 1st reobs. j (≥2)	time of reobservation j+1	j+2	j+3 ...
surv.param.	<--- Φ_j --->	<--- Φ_{j+1} --->	<--- Φ_{j+2} ---> <---...
recapt.param.	p_{j+1}	p_{j+2}	p_{j+3} ...

Recent advances in capture-recapture models for several data sets offer the opportunity to frame our problem into an existing context. In these models, one can take into account equality constraints between parameters pertaining to different groups of individuals (Pradel et al. in press; Lebreton et al. in prep., Lebreton 1989, for a review). This approach can be used to estimate parameters in our model after having split the original data into the two fictitious groups described in table I. The individuals in group 1 must be considered as removed at the time of their first recapture since they are present thereafter in group 2. In turn a_i, the probability of breeding at age 1 at date i, is estimated as $\hat{p}'_i/\hat{p}_i$. Sampling variability may make such an estimate greater than one. The first reobservation of marked individuals is, in the most favourable case, in year 2 (Table IB), and there will be no estimate of p_2 : a_2 cannot thus be estimated. One can estimate a_2 if the time dependence in capture probabilities can be removed ($p_3 = ... = p_k = p$), assuming then $p_2 = p$. Unfortunately, this assumption, reasonable in many cases, cannot be tested. Estimating a_2 in the time-dependent case becomes possible only if data on individuals first caught as breeders (="adults") are available from the beginning of the study. They constitute a third group of data, with the same time-dependent parameters as the re-observed breeders of Table IB. One may then use the estimate of p_2 obtained from this third group to estimate a_2. One has however to assume that the capture rate is independent of age and of the number of previous capture or breeding attempts.

In the framework of capture-recapture models for several data sets, our model (and models for general m and n) can be fitted, using the maximum likelihood method, using computer programs SURVIV (White, 1983), from which a direct estimate of a_i can be obtained, and, with more user-friendliness, SURGE 4.0 (Pradel et al., in press, Lebreton et al., in prep.). From estimates of the asymptotic variances and covariances of the p_i and p'_i, an estimate of the variance of a_i can be obtained by standard approximation formulas (see e.g. Mood et al., 1974) :

$$\hat{\text{var}}\left(\frac{\hat{p}'_i}{\hat{p}_i}\right) = \frac{\hat{p}'^{\,2}_i}{\hat{p}_i}\left\{\frac{\text{var}(\hat{p}'_i)}{(\hat{p}'_i)^2} + \frac{\text{var}(\hat{p}_i)}{(\hat{p}_i)^2} - \frac{2\,\text{cov}(\hat{p}'_i,\hat{p}_i)}{(\hat{p}'_i)(\hat{p}_i)}\right\}$$

HYPOTHESIS TESTING AND GOODNESS-OF-FIT TESTS

It is necessary to select an appropriate model for group 2, and for group 3 if present. In particular, in group 2 the age at first "capture" is known, since "capture" is in fact the first reobservation as a breeder. One can thus test for age-dependence in survival and/or capture probability. In the same way, the independence of present fate upon past capture history can be assessed, in a specific way by Sandland's and Kirkwood's (1981) approach to trap-dependence, or more generally by the goodness-of-fit tests of Burnham et al. (1987) available in program RELEASE (Burnham et al., op. cit.).

At last, we assumed that full breeding was reached at age two (n=2, or $p'_{i2}=p_{i+2}$). A model in which full breeding is reached at age three can be built by moving the constraint between p' and p one year later (n=3 or $p'_{i3}=p_{i+3}$, with $p'_{i2}<p_{i+2}$). A one-sided Likelihood Ratio Test between these two models tests the hypothesis n=2. The procedure to derive estimates of the age-specific probabilities of breeding a_{ij} from the estimates of capture probabilities when n>2 is then more involved than in the simple case n=2, and is described elsewhere (Clobert et al., in prep.).

LOCAL JUVENILE SURVIVAL RATE AND PROPORTION OF BREEDERS AMONG ONE YEAR OLD INDIVIDUALS IN THE DIPPER

In a study of the European Dipper (*Cinclus cinclus*), one of us (GM) collected capture-recapture data for 6 years (1981-87) in the Lorraine region, eastern France (see Marzolin 1988, for description of the study site). Breeding males and females were caught every year by mist-netting near their nests, from the end of February to the beginning of June. Young were marked a few days before fledging. 207 adults and 892 young were marked during the study period. The data for the three groups (group 1: first marked as young, group 2: first recaptured as breeders considered as initial release, and group 3: first marked as breeders) are presented in table II.

In a previous analysis of the capture and survival rate of breeding adults, Lebreton et al. (in prep.) showed that the apparent survival rate was significantly lower in flood years (noted f) than in non flood years (respectively 0.469 ± 0.043 in 1982-83 and 83-84 and 0.607 ± 0.031 in 1981-82 and from 84-85 to 86-87) without any difference between sexes. The capture rate was independent of sex, cohort and time. We noted this model (Φ_f, p) (notation by Sandland and Kirkwood, 1981, expanded by Clobert et al., 1987 and Lebreton et al., in prep.). This model fitted the data correctly (Lebreton et al., in prep.).

Comparison of group 2 and group 3

The first step was to check whether survival and capture rates of birds in group 2 (first marked as young and recaptured as breeders) and group 3 (first marked as breeders) were similar or not. No time-dependence in capture and survival rates was detected in group 2, even if reduced to a flood effect (X^2_1=1.53, P=0.216, table III), although the estimates were in the order expected (normal years: 0.623 ± 0.063, flood years: 0.438 ± 0.135). The small number of birds in this group (48, among which 18 were never seen again) prevented detecting a flood effect. Under the hypothesis that the model (Φ_f, p) holds for both groups, survival and capture rates did not differ significantly between groups (model (6) vs model (5) + (3), X^2_3=1.19, p=0.757). The model (Φ_f, p) was kept to carry on the analysis.

Table II. Capture-recapture data of Dipper (Marzolin, 1988). B_i: number of individuals caught in year i. A_{ij}: number of individuals caught in year i which are recaptured in year j. C_{ij}: number of individuals caught in year i and never seen again after year j. D_{ij}: number of individuals caught in year i, recaptured in year j and which are not released (see Cormack, 1964 and Clobert et al., 1987 for further details).

B_i	A_{ij}						C_{ij}						D_{ij}					
Group 1: first marked as young, until first reobservation as breeder																		
62	4	1	1	0	0	0	56	4	1	1	0	0	0	4	1	1	0	0
140		7	4	0	0	0		129	7	4	0	0		0	7	4	0	0
99			9	0	0	0			90	9	0	0			0	9	0	0
181				11	1	0				169	11	1				0	11	1
219					10	4					205	10					0	10
191						10						181						0
Group 2: first marked as young, after first reobservation as breeder																		
0	0	0	0	0	0	0	0	0	0	0	0	0						
4		2	1	1	0	0		2	1	0	1	0						
8			3	2	0	0			5	1	2	0						
14				8	5	3				5	4	2						
11					9	5					2	4						
11						6						5						
Group 3: first marked as breeder																		
22	11	6	5	3	2	0	9	6	2	2	1	2						
45		18	7	6	4	1		27	10	2	2	3						
44			19	9	5	2			24	11	4	3						
31				17	10	7				12	7	5						
30					16	11					14	5						
35						17						18						

Modelling the proportion of breeders among one year old individuals

Birds first marked as nestlings (group 1) were now part of the analysis. The starting model kept time variation both in the local juvenile survival rate (Φ_{t1}) and in the capture rate of one year old individuals (p'_t). The Likelihood Ratio Test of model 1 (n=2, with parameters (Φ_{t1}, p'_t, Φ_f, p)) vs model 2 (n=3, parameters (Φ_{t1}, p'_{t1}, p'_{t2}, Φ_f, p)) was not significant ($X^2_4=3.95$, $p=0.206$, one sided test, Table IV). In the latter model, p'_{t1} denotes time-dependent capture probabilities of one year old individuals and p'_{t2} time-dependent capture probabilities of two year old individuals, observed for the first time as breeders. While the hypothesis n=2 (full breeding at age two) was accepted, the

Table III. Model comparisons for group 2 and 3. See text and Sandland and Kirkwood (1981), Clobert et al. (1987) and Lebreton et al. (in prep.) for model notation. lnL: logarithm of the Likelihood function. Np: number of distinct parameters estimated.

Model	Np	-2lnL	Hypothesis	Test description
Group 2				
$\Phi_t\ p_t$ (1)	9	112.87		
$\Phi_t\ p$ (2)	6	114.81	capture rate constant	(2) vs (1) $X^2_3=1.94$ p=0.585
$\Phi_f\ p$ (3)	3	115.31	survival rate constant within flood and non flood years	(3) vs (2) $X^2_3=0.50$ p=0.918
$\Phi\ p$ (4)	2	116.84	survival rate equal between flood and non flood years	(4) vs (3) $X^2_1=1.53$ p=0.216
Group 3				
$\Phi_f\ p$ (5)	3	541.97		
Group 2 + 3				
$\Phi_f\ p$ (6)	3	658.47	capture rates equal between groups, survival rates within flood and non flood years equal between groups	(6) vs (5)+(3) $X^2_3=1.19$ p=0.757

hypothesis n=1 (model 3 with parameters (Φ_{t1}, Φ_f, p), hence with $p'_t=p$) was strongly rejected (model 3 vs model 1, $X^2_5=18.87$, p=0.001, one sided test, Table IV).

The flood occurred in the middle of the trapping period in 1983. Consequently, both the 1982-83 and the 1983-84 survival rates were affected (Lebreton et al., in prep.). If there is competition for breeding sites, one may expect a higher rate of recruitment of one year old birds in 1984, as more sites were vacant than average. This effect may have persisted in 1985. The low number of young fledged in 1983 (99 in 1983 against 140 and 181 resp. in 1982 and 84, because of nest destruction by the flood) may have been insufficient to fill up all nest sites made available. The estimates of p'_i were indeed significantly higher in 1984 and 85 than in any other years (model 5, parameters Φ_{t1} p'_f Φ_f p, vs model 4, parameters, Φ_{t1} p' Φ_f p, $X^2_1=7.69$, p=0.006, Table IV). This particular flood effect correctly explained the time-dependence in p'_t (model 5 vs model 1, $X^2_3=1.55$, p=0.671, Table IV).

Table IV. Modelling age specific proportions of breeder and local immature survival rates. *: some parameters are extrinsically aliased. g: age effect; f: flood effect; g+f: additive effect of g and f; g*f: g+f+g.f where g.f is the interaction between g and f; p'_t and p'_{t2}: capture rate of one year old individuals; p'_{t3}: capture rate of two years old individuals; [1]: one sided test. See table IV and text for the explanation of other symbols.

Model with parameters		Np	-2lnL	Hypothesis	Test description
Age specific proportion of breeders					
$\Phi_{t1}, p'_t, \Phi_f, p$	(1)	14	1158.69		
$\Phi_{t1}, p'_{t2}, p'_{t3}, \Phi_f, p$	(2)	18*	1154.74	full breeding age at two years ($p'_{t3}=p$)	(2) vs (1) $X^2_4=3.95$ $p=0.206$[1]
Φ_{t1}, Φ_f, p	(3)	9	1177.56	full breeding age at one year ($p'_t=p$)	(3) vs (1) $X^2_5=18.87$ $p=0.001$[1]
Φ_{t1}, p', Φ_f, p	(4)	10	1167.93	capture rates of one year old individuals (p'_t) are constant	(4) vs (1) $X^2_4=9.24$ $p=0.055$
$\Phi_{t1}, p'_f, \Phi_f, p$	(5)	11	1160.24	p'_t are different between flood and non flood years	(5) vs (4) $X^2_1=7.69$ $p=0.006$
				P't are equal within flood and within non flood years	(5) vs (1) $X^2_3=1.55$ $p=0.671$
Local immature survival rate					
Φ_1, p'_f, Φ_f, p	(6)	6	1162.63	local immature survival rates are constant	(6) vs (5) $X^2_5=2.39$ $p=0.793$
Φ_{g*f}, p'_f, p	(7)	7	1162.24		
Φ_{g+f}, p'_f, p	(8)	6	1165.31	breeders and local immature survival rates fluctuate in parallel across years	(8) vs (7) $X^2_1=3.07$ $p=0.080$

Modelling the local immature survival rate

There was no evidence of time-dependence in the local immature survival rate (model 6, parameters Φ_1 $p'_f \Phi_f$ p, vs model 5, $X^2_5=2.39$, p=0.793). It was useless to investigate intermediate models since this X^2 was below significance even for one df, and could not be partitioned to yield a significant result.

Table V. Parameter estimates under the hypothesis of the model, with parameters Φ_1, p'_f, Φ_f and p, accepted. Sd: standard deviation.

	Estimates	Sd	Asymmetric confidence Interval
Local immature survival rate			
$\hat{\Phi}_1$	0.089	0.0117	0.068 - 0.115
Survival rate of breeders in flood (Φ_f) and non flood (Φ) years			
$\hat{\Phi}_f$	0.483	0.0424	0.399 - 0.568
$\hat{\Phi}$	0.597	0.0315	0.535 - 0.658
Capture rate of one year old birds in flood (p'_f) and non flood (p') years			
$\hat{p}'_f$	0.540	0.0856	0.367 - 0.704
$\hat{p}'$	0.910	0.0822	0.578 - 0.987
Capture rate of breeders			
$\hat{p}$	0.892	0.0288	0.818 - 0.938

Estimates of the proportion of breeders among one year old individuals in flood (a_f) and non flood (a) years

$$\hat{a}_f = \hat{p}'_f/\hat{p} = 0.540/0.892 = 0.605 \pm 0.09701$$
$$\hat{a} = \hat{p}'/\hat{p} = 0.910/0.892 = 1.020 \pm 0.09695$$

Further testing

The local immature survival rate and the survival rate of breeders did not seem to have the same response to the flood. This could have been looked at by testing beforehand the interaction term f.g (f, for flood effect; g, for age effect, accounting for a possible difference in year i between local immature survival rates Φ_{i1} and the breeders' survival rate Φ_i) like in an analysis of variance (Lebreton et al., in prep.). Although it addresses a general alternative hypothesis (all possible changes in the differences Φ_{i1}-Φ_i between flood and non flood years), the interaction test (model 7 vs model 8, Table IV) is almost significant ($X^2_1=3.07$, $p=0.080$), further suggesting that immatures and breeding adults seem not to have responded identically to the flood.

As a more parsimonious model was at last available (fewer parameters and consequently better power in testing), we checked again the hypothesis that all adults bred at age two. This hypothesis was still accepted (model 6, Table IV, vs a similar model with $n=3$, $X^2_1=1.63$, $p=0.202$).

Model 6 (Table IV), with parameters Φ_1 p'_f Φ_f p, appears as the most relevant. Parameter estimates are given in Table V. The estimated proportions of breeders among one year old birds, deduced from the estimates of recapture probabilities, were respectively 60.5% ± 19.40 for non-flood years and 102.02% ± 19.39 for flood years (Table V). Recruitment was thus at its maximum potential value after the flood. The estimate of immature survival rate was low (0.0890), indicating that an important proportion of the young do not return to their natal area to breed, despite the regional scale of the study.

DISCUSSION

The various demographic parameters inherent in recruitment reflect different steps in the life cycle, with, most likely, very different selective pressures. Most long term population studies of birds provide data on individuals marked as young and recaptured later as breeders. These data are often used to estimate the local immature survival rate by dividing the number of young recaptured as breeders by the total number of young marked (Dhondt and Eyckerman 1980; Garnett 1981; Perrins 1981; Gustafsson 1986; Harvey et al. 1988; Nur 1988; see Clutton-Brock 1988, for more examples). This ratio is called usually the local recruitment rate. This way of estimating recruitment does not separate the local immature survival rate from the age-specific proportions of breeders and depends on the value of the adult survival rate. For instance, even if the capture rate of breeders equals 1, a local recruitement rate of 0.25 can result from very different local immature survival rates, as soon as less than 100% of first year individuals breed (Table VI). Moreover, it does not take into account age and time variation in capture and survival rates, and the fact that capture rates are generally below 1. The bias in the local recruitment rate tends then to be even greater. In our example, the local recruitment rate (0.0695) is 22% lower than the estimated local immature survival rate (0.0890).

The model presented here attempts to separate local immature survival from age at first reproduction and adult survival. The reduction of this problem to a particular survival and recapture model for several data sets has several advantages :

(i) data on birds first marked as young and first marked as breeders can be analysed simultaneously (Lebreton et al., in prep.; Pradel et al., in press);

(ii) existing goodness-of-fit tests (Burnham et al., 1987) can be used. Age and/or time-dependence in capture and/or survival rate can be addressed by appropriate modelling and testing procedures (Clobert et al., 1987; Lebreton et al., in prep.);

Table VI. Differences between local immature survival and expected return rate, under various hypotheses on the proportion of breeders among one year old individuals. The capture rate of breeders is supposed equal to 1 and the adult survival rate equal to 0.5.

local immature survival rate	proportion of first years breeding	apparent return rate
.250	1.0	.250
.333	0.5	.250
.500	0.0	.250

(iii) estimation can be done using existing computer programs such as SURVIV (White, 1983) or SURGE (version 4.0, Pradel et al., in press);

(iv) local immature survival rate can be related to external variables like in multiple regression (Clobert and Lebreton, 1985). More generally, main effects and interactions of various factors can be investigated (Lebreton et al., in prep.);

(v) the hypothesis that full breeding is reached at age two (or more generally n) can be checked by straightforward tests.

Three major assumptions of our model require discussion : the definition of π_i, a component of local immature survival, the stabilization of the breeding probability, and the equality of adult survival between breeders and nonbreeders.

The inclusion of π_i into Φ_{i1} assumes that the probability of coming back into the study area is the same for one year old breeders and for birds which settle only later. This is the case, in particular, if the breeding area is selected before or at age 1, even if the first breeding attempt is postponed. This assumption is difficult to assess in the absence of relevant data (Clobert and Lebreton 1990). Nevertheless, some observations seem to give credit to this hypothesis. The post-fledging wandering (particularly in long-distance migrants) could help to select a future breeding area (Lörhl 1959; Berndt and Winkel 1981; Baker 1984). Coquillart (1981) and Vansteenwegen (1987) demonstrated that juvenile Swallow (*Hirundo rustica*) marked when wandering around breeding sites were more likely to be caught again as breeders on these sites than other birds. In species with a short generation time, it could be advantageous to save time at the beginning of the breeding season by coming back directly to a breeding site selected in the previous summer.

The assumption of full breeding at age 2, or more generally at age n, is necessary to assume $p'_i = p_i$ from this age onwards. This assumption should be understood as an hypothesis of stabilization of the breeding rate at a value denoted a, valid for birds having already bred. When a is lower than 1, our model provides estimates of a_i/a instead of a_i. Unfortunately a is unknown most of the time. One can at least test for the temporary absence of breeders as an indication of $a<1$. This problem is also discussed by Lebreton et al.(1990).

The assumption of equality of survival between breeders and non-breeders seems rarely testable, because estimates of survival of non-breeders, when available, will frequently be affected by permanent emigration. The validity of our estimates a_i depend strongly on this assumption. Nevertheless, moderate differences in survival will not affect any major conclusion on progressive accession to reproduction (see Lebreton et al. 1990).

In the case of the Dipper, full breeding seems to be reached at age two. The proportion of first-years breeding varies over time, with higher values in years following a flood which reduced adult survival rate. This suggests that competition for nest sites was relaxed in these years, and that most first-years are in general able to breed. Local juvenile survival and adult survival rates do not seem to fluctuate in the same way, the former having not been affected by the flood. This can result from :

(i) a limited relative precision of the estimate of local immature survival, based on fewer recaptures than the estimate of adult survival;

(ii) an increased fidelity of the young to their natal area (π_i) to compensate for an increased mortality. This would require some kind of anticipation which is hardly conceivable;

(iii) a higher survival for non-breeders in the years affected by the flood; since such survivors will be recruited when two years old, variations in the local immature survival Φ_{i1} for their cohort might have been masked;

(iv) more generally any possible effect may have been buffered or hidden by the role of other variables.

The potential for this methodology is considerable: for instance, the reproductive success of various groups of females can be compared by splitting the recapture data of young according to their mother's group. This should in particular help investigating trade-offs between demographic parameters. More generally, the estimation and comparison between components of the lifetime reproductive success (excluding the period at nest) from recapture data is based on informal statistical approaches. Like for survival studies (Clobert et al. 1987, Lebreton et al. in prep.), we advocate for such estimations and comparisons of recruitment parameters the use of statistical models of the kind developed here. Data concerning the nestling period (as well as other types of information such as winter recaptures...) could also be incorporated using recapture models for several data sets with an appropriate parametrization. Future developments should consider dispersal distances and interconnected populations.

ACKNOWLEDGEMENTS

We thank J.D. Nichols, N. Nur and J. Tinbergen for valuable comments on an earlier version.

REFERENCES

Baker RR (1984) Bird migration: the solution of a mystery. Hodder and Stoughton, London

Barrat RJ, Barre H and Mougin JL (1976) Données écologiques sur les grands Albatros *Diomedea exulans* de l'île de la Possession (archipel Crozet). L'oiseau et la Revue Française d'Ornithologie 46: 143-155

Berndt R and Winkel W (1981) Field experiments on problems of inprinting to the birth place in Pied Flycatcher. Proceedings of the XVII International Ornithological Congress, Berlin, p 156-171

Brownie C, Hines C and Nichols JD (1986) Constant parameter capture-recapture models. Biometrics 42: 561-574

Burnham KP, Anderson DR, White GC, Brownie C and Pollock, KH (1987) Design and analysis methods for fish survival experiments based on release-recapture. American Fisheries Society Monograph n°5

Clobert J (1981) Etude de la dynamique d'une population d'Etourneaux sansonnets (*Sturnus vulgaris* L.). Unpublished Thesis, Université Catholique de Louvain, Louvain-la-Neuve

Clobert J and Lebreton JD (1985) Dépendance de facteurs de milieu dans les estimations de taux de survie par capture-recapture. Biometrics 41: 1031-1037

Clobert J and Lebreton JD (1990) Estimation of demographic parameters in bird population dynamics. In: Perrins CM, Lebreton JD, Hirons GJM (Eds) Bird population studies: their relevance to conservation and management. Oxford University Press, Oxford (in press)

Clobert J, Lebreton JD and Allainé D (1987) A general approach to survival rate estimation by recaptures or resightings of marked birds. Ardea 75: 133-142

Clobert J, Lebreton JD, Allainé D and Gaillard JM (in prep.) The estimation of age-specific breeding probabilities from recaptures or resightings in vertebrate populations II.Longitudinal models.

Clutton-Brock TH (Ed) (1988) Reproductive success. Studies of individual variation in contrasting breeding system. Chicago University Press, Chicago

Coquillart H (1981) Modélisation de la dynamique des populations d'hirondelles de cheminée (*Hirondo rustica*) à l'échelle régionale. Unpublished Thesis, Université Lyon I, Villeurbanne

Dhondt AA and Eyckerman R (1980) Competition and the regulation of numbers in Great and Blue Tit. Ardea 68: 121-132

Garnett MG (1981) Body size, its heritability and influence on juvenile survival rate among Great Tits, *Parus major*. Ibis 123: 31-41

Gustafsson L (1986) Lifetime reproductive success and heritability: empirical support for Fisher's fundamental theorem. American Naturalist 128: 761-764

Harvey PH, Stenning MJ and Campbell B (1988) Factors influencing reproductive success in the Pied Flycatcher. In: Clutton-Brock TH (Ed)

Reproductive success. Studies of individual variation in contrasting breeding system. Chicago University Press, Chicago, p 189-200

Lebreton JD (1978) Un modèle probabiliste de la dynamique des populations de Cigogne Blanche (*Ciconia ciconia* L.) en Europe occidentale. In: Legay JM and Tomassone R (Eds) Biométrie et Ecologie. Société française de Biométrie, Paris, p 277-343

Lebreton JD (1981) Contribution à la dynamique des populations d'oiseaux. Modèles mathématiques en temps discret. Unpublished Thesis, Université Lyon I, Villeurbanne

Lebreton JD (1989) Statistical methodology for the study of animal populations. Bulletin of the International Statistical Institute 53(1): 267-282

Lebreton JD, Burnham KH, Clobert J and Anderson DR (in prep.) Modeling survival and testing biological hypotheses using marked animals: case studies and recent advances.

Lebreton JD, Hemery G, Clobert J and Coquillart H (1990) The estimation of age-specific breeding probabilities from recaptures or resightings in vertebrate populations I.Transversal models. Biometrics (in press)

Lörhl H (1959) Zur frage des zeitpunks einer pragung auf die heimatregion beim Halsbandschnapper *Ficedula albicollis*. Journal für Ornithologie 100: 132-140

Marzolin G (1988) Polygynie du Cincle plongeur (*Cinclus cinclus*) dans les côtes de Lorraine. L'Oiseau et la Revue Française d'Ornithologie 58: 277-286

McCleery RH and Perrins CM (1988) Lifetime reproductive success of the Great Tit, *Parus major*. In: Clutton-Brock TH (Eds) Reproductive success. Studies of individual variation in contrasting breeding system. Chicago University Press, Chicago, p 136-153

Mood AM, Graybill FA and Boes DC (1974) Introduction to the theory of statistics. McGraw-Hill, New-York, p 181

Nur N (1988) The consequences of brood size for breeding Blue Tits III Measuring the cost of reproduction: survival, future fecundity, and differential dispersal. Evolution 42: 351-362

Perrins CM (1981) The survival rate of young Great Tits, *Parus major*. Proceedings of the XVII International Ornithological Congress, Berlin, p 159-174

Pollock KH (1981) Capture-recapture models allowing for age-dependent survival and capture rates. Biometrics 35: 521-529

Pradel R, Clobert J and Lebreton JD (1990) Recent developments for the analysis of multiple capture-recapture data sets. An example concerning two Blue Tits populations. The Ring (in press)

Sandland RL and Kirkwood P (1981) Estimation of survival in marked populations with possibly dependent sighting probabilities. Biometrika 68: 531-541

van Balen JH, van Noordwijk AJ and Visser J (1987) Lifetime reproductive success and recruitment in two Great Tit populations. Ardea 75: 1-11

Vansteenwegen C (1987) Dynamique d'une population d'hirondelles de cheminée (*Hirondo rustica* L.). Unpublished Thesis, Université Catholique de Louvain, Louvain-la-Neuve

White GC (1983) Numerical estimation of survival rates from band-recovery and biotelemetry data. Journal of Wildlife Management 47: 716-728

THE EFFECTS OF FOREST DAMAGE ON CATERPILLARS AND THEIR EFFECT ON THE BREEDING BIOLOGY OF THE GREAT TIT, AN OVERVIEW

Arie J. van NOORDWIJK

Zoologisches Institut, Rheinsprung 9, CH 4051 Basel, Switzerland

One of the main questions in micro-evolution is how genes interact with the environment in forming a phenotype, and to what extent the environmentally induced phenotypic variations are adaptive. One can think of the genes as coding for a program that processes information from the environment. This program is subjected to natural selection. Especially under field conditions it is often difficult to determine to what extent phenotypic variation is caused by different genotypes and to what extent by different environmental input. The main practical difficulty here is the manpower needed to measure the environmental conditions in sufficient detail for a large enough sample.

One reason for the study of micro-evolution is that human activity causes changes in the environment of almost every species. This leads either to evolutionary changes or to extinction. At present our understanding of the processes involved is not sufficient to predict which species have how much potential for change and thus for survival. The poor condition of both coniferous and deciduous trees in much of Europe, which is often referred to as forest decline or forest damage, is an example of a major man-made change in the habitat of many species. The opportunities to study the micro-evolutionary changes in the Great Tit (*Parus major*), a common forest bird, are good, because detailed population ecological data are available over about 40 years from several populations. Moreover, the genetics of several ecologically important traits has already been studied (Jones 1973, van Noordwijk 1980). This gives us the opportunity to study micro-evolutionary processes on two scales in the same system: by analysing both small scale spatial variation and the effects over decades.

In addition to the importance of such a study for obtaining a better general understanding of micro-evolutionary processes, there is also a pressing need for descriptive work on the effects of forest damage on forest ecosystems. In particular, that trees in poor condition are probably better food for their herbivorous insects increases the risk of herbivore outbreaks that could lead to defoliation and thereby greatly amplify the direct effects of air pollution. This possibility has hardly been investigated as yet. The food chain is illustrated in Figure 1.

I will first describe some of the data that suggest measurable effects of so called tree damage on birds and their main prey species. I then give an overview of what we expect in the plant-insect interface, in the proximate

NATO ASI Series, Vol. G 24
Population Biology of Passerine Birds
Edited by J. Blondel et al.

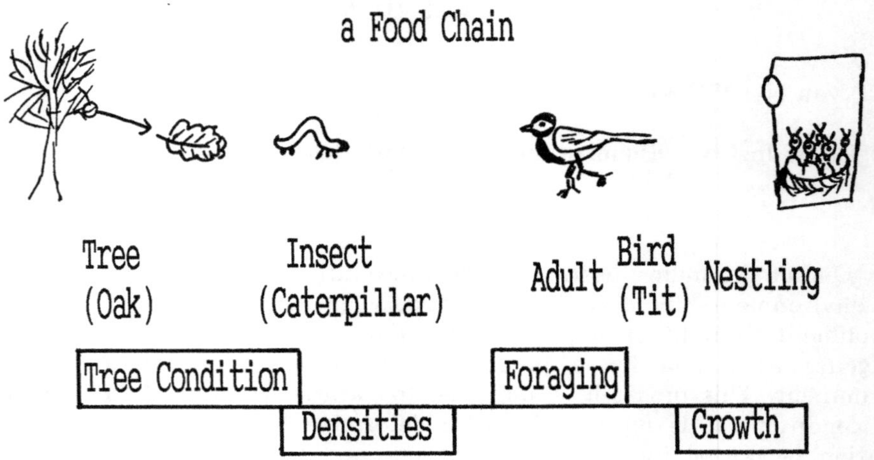

Fig 1. A schematic representation of the study system

determination of clutch size and laying date and then sketch a framework for the study of proximate mechanisms in a genetic context.

DOCUMENTED EFFECTS OF TREE CONDITION ON FOREST BIRDS

From 1985 to 1987 inventories of bird densities were made with standardized methods in 44 selected woods in Switzerland (Mosimann et al. 1987). Significant correlations were found between densities of Great and Blue Tits (*Parus caeruleus*) and the average degree of assessed damage in broadleaved woods, but it is, of course difficult to interpret such correlations, especially since Great Tits were positively and Blue Tits negatively associated with more damage. In 1986 and 1987 foraging behaviour of the Nuthatch (*Sitta europea*) and tits was observed in winter. A correlation between the parts of the trees used and tree damage was found (Blattner et al. in prep.). In 1987 two quantitative assessments of arthropod density on 11 Oaks (*Quercus robur* and *Q. petraea*) were made in the Hard study area, near Basel. The tree damage was assessed by two independent observers later in the summer. Significant positive correlations between degree of damage of oaks and caterpillar densities on them were found in 1987 and 1988 (Meury et al. unpubl.). In 1989 the same correlation was dependent on taking the tree phenology into account, which may have been due to an unusually long period of bud-break. Attempts were also made in 1987 to record whether tits and nuthatches foraged more on damaged than on undamaged trees in the breeding season. This was not successful because of low visibility of the birds in the foliage but led to plans for monitoring foraging behaviour with radiotelemetry. In 1988 a pilot study gave indeed some indication that more heavily damaged trees were more frequented by tits (Naeff et al. in prep.).

The degree of tree damage was also assessed from aerial photographs in the Sana-silva project in 1984. A first analysis of potential relations between tree damage and prey density effects on the tits was made by coding the degree of tree damage of the sub-areas in which the nestboxes are hanging. A significant positive correlation was found between nestling weights and degree of damage for the 1986 data (see Table in Henrich, this volume) after correction for the effects of hatching date (see Henrich, this volume).

All of these observations are consistent with the idea that more heavily damaged trees have higher caterpillar densities, and that these higher prey densities affect the breeding biology of the tits. In the next two sections I will inspect the supposed mechanisms more closely.

EFFECTS OF AIR POLLUTION ON TREES AND INDIRECTLY ON HERBIVOROUS INSECTS

The effects of air pollutants on trees (e.g. Flückiger et al. 1986, Landolt and Keller 1985), plant defense mechanisms against herbivorous insects (e.g. Haukioja 1982), and the effects of plant condition on insect growth and on population dynamics (e.g. Crawley 1983, White 1984) are important links in a chain from air pollutants to insect biomass. Very briefly, herbivorous insects are generally nitrogen limited (Mattson 1980, White 1984). Stress in plants, caused by air pollutants or lack of water, tends to increase the soluble nitrogen content of the leaves, and especially the amino-acid bound nitrogen that is most critical for the insects (e.g. Schroeder and Malmer 1980). At the same time, the secondary metabolites that are poisonous to insects may be present in lower concentrations in stressed plants (e.g. Strong et al. 1984). In some Lepidopterans a capacity to sense either nitrogen content or a combination of available sugars and nitrogen has been shown to affect the egg-laying behaviour of females; far more eggs were laid on those plants that were a better medium for the larvae (e.g. Myers 1985).

Extensive measurements were made on the concentrations of nutrients in the leaves of beech (*Fagus sylvatica*) in northwestern Switzerland (Flückiger et al. 1986). A high N/K ratio in damaged trees was accompanied by an increased infestation by parasites. It could be shown that this nutrient imbalance is a consequence of an increased impact of N deposition and an increased leaching of potassium induced by air pollution (Flückiger et al. 1986).

Hard data on the effect of acid rain on Lepidopteran larvae, the prey items for birds, are still almost non-existent. There are however experimental studies on aphids, whose densities have been observed to have increased dramatically in areas that are especially affected by air pollutants. Aphids grow faster and produce larger larvae (the aphids are viviparous) when feeding on plants that have been treated with acidified water (Neuvonen and Lindgren 1988) or with polluted air (Braun and Flückiger 1984, 1985, Bolsinger and Flückiger 1987, 1988), even though the direct effect of acid mist on the animals was largely negative (Braun and Flückiger 1988). There are thus several factors that would predict better growth of herbivorous insects on damaged trees.

PROXIMATE DETERMINATION OF CLUTCH SIZE AND LAYING DATE IN THE GREAT TIT

In many textbooks, a negative correlation between mean annual clutch size and population density in the Great Tit is cited as a mechanism of population regulation (e.g. Krebs 1972 p. 317). Although this negative correlation has been found both by Kluyver (1951) for several Dutch populations and for the population in Wytham Wood (Lack 1966, Perrins 1965), it was not found in other populations (van Balen 1973, van Noordwijk et al. 1981). Moreover, the correlation found for Wytham Wood 1947-1962 was not observed in subsequent years (McCleery and Perrins 1985).

One explanation of the original significance of the correlation lies in the fact that both mean clutch size and densities are strongly correlated to those of the next year (van Noordwijk 1987). One should probably shed new light on the correlation found (but dismissed as an explanation) by Perrins (1965) between annual mean clutch size and the density of the winter moth, which is one of the important prey species during the breeding season. At the moment, it seems likely that much more of the annual variation in clutch size can be explained by caterpillar densities than had been thought (Perrins 1965, van Noordwijk 1989).

These analyses are, however, based on annual means. If the mechanisms invoked are true, then similar patterns should also be visible on the scale of individual territories. It would therefore be very useful to collect data on the variation in environmental parameters such as temperature or food availability at the level of the individual territory. In itself, the idea to search for patterns on the scale of territories is not new. An a posteriori small-scale analysis looking at the effect of local density on clutch size, using Theissen Polygons to characterize territory size, did not show any effects of bird density on clutch size (McCleery and Perrins 1985). The same data on food availability at a very local scale could be used to look at the variation in nestling growth, clutch size, egg size and laying dates.

The proximate effect of the timing of caterpillar growth on Great Tit laying dates is not clear. To begin with, we do not know whether the timing is fully determined by a synchronisation of the peak energy demand of nestlings with maximum caterpillar availability, or whether the energy needed to produce eggs prevents earlier laying. The latter explanation is widely held (e.g. Perrins 1970, Drent and Daan 1980), and is supported by consistent selection for laying early. In some experiments, additional food leads to advancement of laying (Källander 1974, Arcese and Smith 1988) and temperature has proximate effects on intra-clutch variation in egg-size. In other populations, no consistent selection for earlier laying was found and some experiments with additional feeding, were not successful - simply providing extra energy seems insufficient (Arcese and Smith 1988, Jones 1973).

The proximate explanation is also inconsistent with genetic variation for laying date. While the mean laying dates can be predicted from the temperature sums over various periods (see Perrins and McCleery 1989), there are no data explaining the relatively large spread of laying dates within a season, (but see Nager, this volume). A recent re-analysis of the relation between peak caterpillar abundance and the selection for laying dates shows that the synchronisation of the tits with the caterpillar peak is very important and results in selection differentials of up to one phenotypic standard deviation, which is equivalent to 50 % culling (van Noordwijk, McCleery and Perrins, in prep.). This suggests that the ultimate explanation of laying date may well be of overriding importance in the proximate determination because there is sufficient genetic variation for laying dates for this selection to act on. It is an open question however, how genetic mechanisms produce the observed phenotypic plasticity.

THE EXPRESSION OF GENETIC VARIANCE IN HETEROGENEOUS ENVIRONMENTS

A key question in the study of evolution is what limits its speed. Is evolutionary change limited by the absence of genetic variation or by the absence of directional selection? This dichotomy is too simple, because the genotypes are also to some extent plastic, i.e. leading to different phenotypes under different environmental conditions. The basic question should then be rephrased as how much of the variation among individuals is due to the same reaction to different environmental conditions, and how much is due to different reactions to the same environmental conditions. The consequences of the various quantitative answers have great implications for the ability of organisms to adapt to changing conditions brought about by man.

A very useful concept in the analysis of these problems is the reaction norm, defined as the set of phenotypes displayed by a single genotype under different environmental conditions. Reaction norms thus combine genetic variation among individuals with the phenotypic plasticity shown by each genotype. Interest in reaction norms is growing and a rapid development in this area is to be expected in the next decade (see e.g. Dodson 1989, van Noordwijk 1989, Scharloo 1989, Schlichting 1989, Stearns 1989).

There are several main lines of research using reaction norms. One approach is to predict optimal reaction norms from knowledge of physiology and selection regime (Stearns and Koella 1986). A second approach analyses the genetic basis of reaction norms, by measuring the genetic component of reaction norms in the laboratory (e.g. Gupta and Lewontin 1982), or more directly through selection on the reaction norm. A discussion of various experimental designs for analysing reaction norms is given elsewhere (van Noordwijk and Hillesheim, 1990).

In a third approach the environmental variation is reduced to two or at most three discrete environments, which allows the calculation of genetic correlations for the same traits between environments (e.g. Via 1987).

A fourth line takes more physiology into account. The aim here is to uncover the Genotype Environmental factor Mapping Function (see Scharloo 1987). One of the open questions in an ecological context is which elements in the natural environment trigger the development of a particular phenotype. Especially for phenotypes that are formed repeatedly by a single individual, such as laying date, egg size or clutch size in birds, it would be very useful to have more accurate information on which aspects of the environment are critical.

One aspect of reaction norms that has, so far, been largely neglected, is the timing of the various processes. In a controlled constant environment, this aspect can be neglected. In a variable environment it becomes important to know whether the environmental variation in the phenotype reflects the current environment, yesterdays environment or the environment during growth and development. Juvenile growth leading to adult body size is a good model system, because growth processes are relatively well studied, and different measures of body size (tarsus length, weight and wing length) are expected to reflect different timing of environmental influences.

CONCLUSIONS

We are still in a preliminary stage of collecting the data necessary to answer these questions. I have mainly outlined the general framework and the overall strategy. Findings such as the reported effects of increasing acidity on the ability to produce good eggs (Drent and Woldendorp 1989) show how urgently we need to know more. It is a standard practice in science to disturb the system under study in order to learn how the system functions. The large scale disturbances that occur everywhere are unwanted and ill-designed as experiments -for lack of suitable controls. Nevertheless I feel that we should try to make the most of the opportunities provided by these disturbances, to gain the basic knowledge that is among other things needed to avert or reduce further disturbances.

ACKNOWLEDGEMENTS

I gratefully acknowledge financial support by the Schweizerischer Nationalfonds.

REFERENCES

Arcese P, Smith JNM (1988) Effects of population density and supplemental food on reproduction in song sparrows. J Anim Ecol 57: 119-136

Balen JH van (1973) A comparative study of the breeding ecology of the Great Tit, *Parus major*, in different habitats. Ardea 61: 1-93

Bolsinger M, Flückiger W (1987) Enhanced aphid infestation at motorways: The role of ambient air pollution. Entomol exp appl 45: 237-243

Bolsinger M, Flückiger W (1988) Ambient air pollution induced changes in amino acid pattern of phloem sap in host plants - relevance to aphid infestation. ms

Braun S, Flückiger W (1984) Increased population of the Aphid *Aphis pomi* at a Motorway: Part 1 - Field Evaluation. Env Poll (Series A) 33: 107-120

Braun S, Flückiger W (1985) Increased population of the Aphid *Aphis pomi* at a Motorway: Part 3 - The effect of exhaust gases. Env Poll (Series A) 39: 183-192

Braun S, Flückiger W (1988) Effect of ambient ozone and acid mist on aphid development. ms

Crawley MJ (1983) Herbivory, the dynamics of animal-plant interactions. Blackwell, Oxford

Dodson S (1989) Predator-induced Reaction Norms. Bioscience 39: 447-452

Drent PJ, Woldendorp JW (1989) Acid rain and egg shells. Nature, 339: 431

Drent RH, Daan S (1980) The prudent parent, energetic adjustments in avian breeding. Ardea 68: 225-252

Flückiger W, Braun S, Flückiger-Keller H, Leonardi S, Asche N, Bühler U , Lier M (1986) Untersuchungen über Waldschäden in festen Buchenbeobachtungsflächen der Kantone Basel-Landschaft, Basel-Stadt, Aargau, Solothurn, Bern, Zürich und Zug. Schw Zeitschr Forstwesen 137: 917-1010

Gupta AP , Lewontin RC (1982) A study of reaction norms in natural populations of *Drosophila pseudobscura*. Evolution 36: 934-948

Haukioja E (1982) Inducible defences of white birch to a geometrid defoliator, *Epirrita autumnata*. Proc 5th Int Symp Insect-Plant Relationships, Wageningen, 1982, Pudoc, Wageningen, p 199-203

Jones PJ (1973) Some aspects of the feeding ecology of the Great Tit (*Parus major* L.). D Phil thesis, Oxford, England

Källander H (1974) Advancement of laying of Great Tits by the provision of food. Ibis 116: 365-376

Kluyver HN (1951) The population ecology of the Great Tit, *Parus m. major* L. Ardea 39: 1-135

Krebs CJ (1972) Ecology. 1st ed, Harper and Row, New York

Lack D (1966) Population studies of birds. Oxford University Press, Oxford

Landolt W, Keller Th (1986) Uptake and effects of air pollutants on woody plants. Experientia 41: 301-310

Mattson JM (1980) Herbivory in relation to plant nitrogen content. Ann Rev Ecol Syst 11: 119-161

McCleery RH, Perrins CM (1985) Territory size, reproductive success and population dynamics in the Great Tit, *Parus major*. In RM Sibly , RH Smith (eds) Behavioural Ecology, Blackwell, Oxford, p 353-373

Mosimann P, Naef-Daenzer B, Blattner M (1987) Die Zusammensetzung der Avifauna in typischen Waldgesellschaften der Schweiz. Orn Beob 84: 275-299

Myers JH (1985) Effect of physiological condition of the host plant on the ovipositional choice of the cabbage white butterfly, *Pieris rapae*. J An Ecol 54: 193-204

Neuvonen S, Lindgren M (1987) The effect of simulated acid rain on the performance of the aphid *Euceraphis betulae* (Koch) on Silver Birch. Oecologia (Berlin) 74: 77-80

Noordwijk AJ van (1980) On the genetical ecology of the Great Tit (*Parus major* L.). Thesis, University of Utrecht

Noordwijk AJ van (1987) On the implications of genetic variation for ecological research. Ardea 74: 13-19

Noordwijk AJ van (1989) Reaction norms in genetical ecology. BioScience 39: 453-459

Noordwijk AJ van, Balen JH van, Scharloo W (1981) Genetic and environmental variation in clutch size of the Great Tit (*Parus major*). Neth J Zool 31: 342-372

Noordwijk AJ van, Hillesheim E (1990) Reaktionsnormen: Wechselwirkungen zwischen Genotyp und Umwelt. in B Streit (ed) Evolutionsprozesse im Tierreich. Birkhäuser Verlag, in press

Perrins CM (1965) Population fluctuations and clutch size in the Great Tit, *Parus major* L. J An Ecol 34: 601-647

Perrins, CM (1970) The timing of birds' breeding seasons. Ibis 112: 242-255

Perrins CM, McCleery RH (1989) Laying dates and clutch size in the Great Tit. Wilson Bulletin, 101: 236-259

Scharloo W (1987) Constraints in selection response. In V Loeschcke (ed) Genetic constraints to adaptive evolution. Springer Verlag, p 123-149

Scharloo W (1989) Development and physiological aspects of reaction norms. BioScience 39: 465-471

Schlichting C (1989) Phenotypic integration and environmental change. BioScience 39: 460-464

Schroeder LA, Malmer M (1980) Dry matter, energy and nitrogen conversion by lepidoptera and hymenoptera larvae fed leaves of black cherry. Oecol (Berl) 45: 63-71

Stearns SC (1989) The evolutionary significance of phenotypic plasticity. BioScience 39: 436-446

Stearns SC, Koella JC (1986) The evolution of phenotypic plasticity in life-history traits: predictions of reaction norms for age and size at maturity. Evolution 40: 893-913

Strong DR, Lawton JH, Southwood TRE (1984) Insects on plants, community patterns and mechanisms. Harvard University Press, Cambridge Massachusetts

Via S (1987) Genetic constraints on the evolution of phenotypic plasticity. In V Loeschcke (ed) Genetic constraints to adaptive evolution. Springer Verlag, p 47-71

White TCR (1984) The abundance of invertebrate herbivores in relation to the availability of nitrogen in stressed plants. Oecol (Berl) 63: 90-105

GENETIC PARASITISM IN THE EUROPEAN STARLING

Harry W. POWER

Department of Biological Sciences,Rutgers University
P.O. Box 1059,Piscataway, NJ 08855-1059 USA

Genetic parasitism in birds has two components, cuckoldry and conspecific brood parasitism (CBP). Genetic parasitism is important because it negatively affects the reproductive success of hosts and implies the existence of more than one reproductive strategy, i.e., an individual may provide parental investment to some offspring while leaving others to be raised by conspecifics. Trivers (1972) called this a "mixed reproductive strategy."

My graduate students and I have been studying genetic parasitism for over 10 years, using European Starlings (*Sturnus vulgaris*) as a model system. This paper is a synopsis of our work. The students (now all graduated) who worked on the project were Elizabeth Litovich, Michael P. Lombardo, Ann S. Hoffenberg, Linda C. Romagnano, Philip C. Stouffer, and E. Dale Kennedy.

Starlings are an appropriate model system for studying genetic parasitism because they nest in nestboxes (permitting large scale, long-term studies), and they withstand many experimental manipulations without deserting.

A nest box trail was established on the Kilmer Campus of Rutgers University in New Jersey, USA, in 1975 by Crossner (1977) and enlarged by Litovich (1982), beginning in 1978. We rebuilt this trail with new nest boxes specially designed to trap birds for biopsy, banding and collecting, beginning in 1983. We studied genetic parasitism by a combination of behavioural observation, nest censusing, electrophoresis and post ovulatory follicle (POF) analysis. POF analysis involves comparing the number of post ovulatory follicles (the split, yellowish chambers left in the ovary after ovulation of each egg) with the number of eggs in a nest. When the number of POFs is less than the number of eggs, CBP is indicated. When the number of POFs exceeds the number of eggs, conspecific egg removal (CER) is indicated. All methods are described in detail elsewhere (Power et al. 1981, Hoffenberg et al. 1988, Kennedy et al. 1989, Lombardo et al. 1989, Power et al. 1989, Romagnano et al. 1989, Romagnano et al. 1990).

CUCKOLDRY STUDIES

Our behavioral study of cuckoldry focused on incubation behavior (Power et al. 1981). Prior to our study it was known that (i) males take turns incubating with females (Kluijver 1933, Kessel 1957), (ii) incubation begins

NATO ASI Series, Vol. G 24
Population Biology of Passerine Birds
Edited by J. Blondel et al.

before clutch completion (Crossner 1977), and (iii) copulation continues throughout laying (Kessel 1957). Viewing these facts in the context of cuckoldry risk, we predicted that (i) males would not incubate before clutch completion in order to guard their mates from insemination by other males, thereby leaving only females to incubate at that time, and (ii) males would monitor their mates more closely during laying than after clutch completion.

We observed 25 nests during the laying and incubation periods until two days after we had determined that both sexes incubated at each nest (Power et al. 1981). Both predictions were corroborated: (i) in all cases, only females incubated during the laying period, and (ii) males monitored their mates more closely before than after clutch completion.

Corroboration of our predictions presented a paradox, viz. that males reduce their chances of being cuckolded in such a way that they thereby increase their chances of suffering from CBP and CER (Power et al. 1981). This happens because males can guard only one thing at a time. When they guard their mates away from their nests, they cannot prevent conspecific intruders from laying or removing eggs. These are not trivial risks because both CBP and CER have high frequencies (see below). The apparent trade-off of risks by males suggested to us that either (i) the cost per egg, (ii) the number of eggs per incident, or (iii) the number of incidents per clutch is greater for males (in all three cases) with cuckoldry than with CBP or CER. While we do have reasons for believing that the cost per egg to the male is greater with cuckoldry, we have no direct evidence for this and do not have the space to develop our indirect evidence in this paper. Instead we shall consider the second and third possibilities. The second was directly tested in our electrophoretic study of parentage (Hoffenberg et al. 1988).

Our electrophoretic study was one of the largest of its kind. We biopsied 550 wild birds (83 males, 102 females, and 365 nestlings) from 118 families, of which 106 were complete for both putative parents and all young. The number of families exceeded the number of adults of either sex because some birds had more than one brood within or between years. (Hoffenberg et al. (1988) reported only 95 families because Hoffenberg herself did not participate in the first year of the study, 1983, when 11 complete families were biopsied). All birds were biopsied for both blood and pectoral muscle ($n=1100$ tissue samples).

We used polyacrylamide gel electrophoresis because we could not replicate our results using starch but we could using polyacrylamide, and polyacrylamide has higher resolving power (Coyne et al. 1979, Ramshaw et al. 1979). Thirty-three loci were resolved but only three were genetically polymorphic. Moreover, only one of these (plasma amylase) had the approximately equal ratio of alleles considered ideal for parentage studies. Thus our parentage results are conservative because otherwise genetically different birds could have been identical at the three polymorphic loci (Hoffenberg et al. 1988, Romagnano et al. 1990).

Only two of the 106 complete families (1.9%) had unambiguous cases of cuckoldry. Another six cases could have been either cuckoldry or CBP. If they are included, eight of 106 families (7.5%) had cases of cuckoldry. Both of the unambiguous cases involved two nestlings each.

The low frequency of cuckoldry (1.9-7.5%) is partly an artifact of the conservatism of electrophoresis, but it also suggests the effectiveness of mate guarding. The fact that both unambiguous cases of cuckoldry involved two nestlings each while only 12.1% of the cases of CBP and 8% of the cases of CER involved more than one egg each (see below), supports the second possibility that males preferentially guard against cuckoldry because it involves more than one egg more often than CBP or CER do. Of course our sample of only two unambiguous cases of cuckoldry is insufficient to demonstrate this possibility.

The third possibility (that the number of incidents per clutch is greater with cuckoldry than CBP and CER) is logistically reasonable. In the absence of mate guarding, an act of cuckoldry requires only that a would be cuckolder find a fertilizable female; hundreds could be found simultaneously foraging during the fertilizable period. But in the absence of nest guarding by males, an act of CBP or CER would still require that a would be parasite or egg remover find a nest unguarded by a female; determining whether a nest is guarded or not requires inspection of individual nests one at a time, and is therefore time consuming and possibly dangerous.

CBP STUDIES

We have published varying estimates of the frequency of CBP depending upon our criterion of effective parasitism. Thus Power et al. (1981) reported a CBP rate of 46% but this included parasitic eggs laid after host clutch completion as well as those laid before that time. As we shall see below, those laid after host clutch completion always died. Therefore we subsequently reported only the frequency of parasitic eggs laid before host clutch completion because only these had the potential to harm the host or help the parasite. All further estimates of the frequency of CBP in this paper will be based only on parasitic eggs laid during the host laying period.

We have also published varying estimates based upon the number of years of study, the techniques, and the category of nest. Thus nest censusing alone revealed that only 14.9% of all nests received at least one parasitic egg (Romagnano et al. 1990). When the results of electrophoresis were included, this value rose to 20.3% (Romagnano et al. 1990). When only early nests were considered (early nests are those begun in April when there is a highly pronounced peak of synchronous laying (Romagnano et al. 1990) and 92% of all brood parasitism occurs (Power et al. 1989)), the parasitism rate varied between 23.2% (Lombardo et al. 1989, Power et al. 1989) and 27.8% (Romagnano et al. 1990), depending upon the years of study included. POF analysis revealed a rate of 23% for all nests and 36.4% for early nests (Kennedy et al. 1989). Since almost all brood parasitism occurs in early nests and the fledging success of both hosts and parasites peaks at that time (Romagnano et al. 1990), early nests are the decisive ones for determining the efficacy of host and parasite reproductive strategies. Therefore we emphasize early nests in this paper.

Comparison of techniques suggests that POF analysis is the most accurate and cost effective (Kennedy et al. 1989, Power et al. 1989). Therefore we will use POF measurements as the basis for most discussion. Yet all techniques are conservative, meaning that the true rate of CBP is higher than any measured frequency (Romagnano et al. 1989, Kennedy et al. 1989), even that determined by POF analysis.

CBP was not a very effective reproductive strategy (Romagnano et al. 1990). Parasitic eggs laid during the host laying period had the same fledging rate as host eggs, but those laid before the host laying period were always thrown out by the host (Stouffer et al. 1987) while those laid after host clutch completion never fledged, either because they didn't hatch or failed to survive in competition with their older nestmates. Since only 49% of parasitic eggs were laid during the host laying period, parasitic eggs had only half the fledging success of nonparasitic eggs.

Parasitic females probably were not able to compensate for their low productivity per egg by laying more eggs. Kennedy et al. (1989) found no female that ovulated more than six eggs, suggesting that this is the upper limit to clutch size in our population. If so, then parasitic females could not have laid the double clutch sizes required to compensate for their eggs having only 50% of the productivity of host eggs. (Modal host clutch size in our population is five eggs (Power et al. 1989)).

The pattern of appearance of host and parasitic eggs was inconsistent with the hypothesis that they were normally laid by the same females (Romagnano et al. 1990). The peak of parasitic laying fell within the peak of host laying. This implies that parasites were not normally females that had nests of their own and either (i) laid parasitic eggs before laying in their own nests, (ii) laid in their own nests before laying parasitic eggs, or (iii) laid parasitic eggs midway through the period in which they laid in their own nests. If parasites normally followed tactic (i), the peak of parasitic laying would have preceded the peak of host laying. If they normally followed tactic (ii), the peak of parasitic laying would have followed the peak of host laying. And if they normally followed tactic (iii), there would have been frequent gaps in the host laying sequences because birds can lay only one egg per day. In fact, there were almost no gaps in host laying sequences and egg removal could not be ruled out as the cause of any of those gaps. Of course, this line of evidence does not rule out the possibility that some parasites were simultaneously also hosts, but it does rule out the possibility that most parasites were. Thus most parasites appeared to be females without nests of their own. Circumstantial evidence suggests that the bulk of these were yearlings.

Given (i) the low productivity or parasitic eggs, (ii) the probable inability of parasites to compensate for this low productivity by laying more eggs than hosts, (iii) the fact that parasitism was seldom if ever a supplementary strategy to nesting (i.e., part of a mixed reproductive strategy within years), and (iv) the probability that parasites were almost always females without nests of their own (and thus birds of low productivity), parasitism appears to persist in our population not because it is superior to raising one's own young but because it is superior to laying no eggs at all.If this analysis is correct, then prasites may be able to compensate for the low productivity of their parasitic years by having nests of their own in earlier or later years. In that way, they could engage in a mixed reproductive strategy between years.

CBP only moderately depressed host reproductive success. When all early nests were considered, there was no statistically significant difference in the fledging success of host young from parasitized and nonparasitized nests (Power et al. 1989). However, when only the electrophoresis sample was considered, significantly fewer host young fledged from parasitized than nonparasitized nests (Romagnano et al. 1990).

The reason that CBP had such an equivocal effect on host reproductive success was that both parasite and host act to mitigate the potential overcrowding effects of CBP. Parasites do this by removing host eggs (Lombardo et al. 1989). During the host laying period, CER occurred at 19% of all early nests and at 34% of parasitized early nests but at only 14.2% of nonparasitized early nests. Altogether, 4.2% of all eggs were removed from early nests during the host laying period. In 92% of the cases of CER during the host laying period, only one egg was taken but at the remaining 8% two eggs were taken. At 75% of the nests where both CER and CBP occurred, they happened either on the same day (41.7%) or CER happened one day before CBP (33.3%). This suggests that CER is a tactic in a parasitic strategy in which the parasite substitutes its own egg for that of the host, thereby preventing its own young from being reared in an overcrowded clutch. (CER also appears to be a tactic in a strategy of nest usurpation. Perhaps females without nests really seek to usurp nests from those who have them, but settle for CBP if the host doesn't yield to the parasite's depredations. All eggs were removed from 5 of 138 early clutches (3.6%) during the host's post-laying incubation period; these nests may have been the victims of successful usurpation attempts but alternative explanations could not be ruled out).

Hosts attempt to mitigate the potential overcrowding effects of CBP by leaving space for parasitic eggs as described by the "parasitism insurance hypothesis" (Power et al. 1989). Leaving space is not an altruistic act evolved to aid parasites. Rather it is a form of insurance in which the host protects herself from reproductive failure as a back-up to the tactic of preventing parasites from laying in the first place. For this tactic to be favored, the frequency of "net CBP" must exceed a threshold amount.

Net CBP is the number of eggs added to a nest by parasitism less the number subtracted by removal (Powers et al. 1989). Net CBP is critical because a host will not have to rear an overcrowded clutch insofar as egg removal results in a clutch size that is the same as or smaller than the one that the host laid. A host will have to rear an overcrowded clutch only when the net effect of CBP and CER is to produce a clutch size beyond the capacity of the host to rear without excess mortality or poor condition of young at the termination of parental care.

The threshold frequency of net CBP is the frequency where the alternative tactics of leaving space for parasitic eggs and not leaving space produce equal reproductive success (Power et al. 1989). Leaving space is favored when the threshold is exceeded but disfavored when it is not.

We tested the parasitism insurance hypothesis partly by determining the frequencies and productivities of different clutch sizes, and the frequency and effect of net CBP (Power et al. 1989). Then we used these results to generate values for two models that we developed to determine whether leaving space is favored. The first or "either-or" model generates a precise value for the threshold frequency of net CBP. It treats net CBP as an either-or event: either it happens or it doesn't. The second or "graded" model treats net CBP as an incremental event: net CBP can add 1 to 4 eggs per nest in our population. The graded model establishes whether the reproductive success of leaving space exceeds that of not leaving space no matter how many parasitic eggs are added or what the relative effects are of different net numbers of parasitic eggs.

The either-or model generated a threshold frequency of net CBP of 26.4%, well below the observed net CBP frequency of 33.3% in the POF sample, implying that leaving space was favored in our population. The graded model showed that leaving space for parasitic eggs was favored in our population even when the negative impact of multiple acts of net CBP were taken into account (21.2% of the females having early clutches in the POF sample had a net increase of only one egg each but 12.1% had a net increase of two to four eggs each).

RELATION TO OTHER STUDIES

Our results coincide with other studies in showing (i) the importance of genetic parasitism in birds, and (ii) ways that birds have evolved to avoid or mitigate its costs. Because of space restriction, we unfortunately cannot cite those studies here. Readers are referred to McKinney et al. (1984) and Hoffenberg et al. (1988) for papers on cuckoldry, to MacWhirter (1989) for papers on CBP, and to Lombardo et al. (1989) for papers on CER.

COMPARISON WITH EVANS' STUDY

I have delayed any reference to Evans' (1980, 1988) parallel study of starlings in Great Britain until now in order to make that comparison in detail. Evans performed a large scale electrophoretic study on two populations, one in England and the other in Scotland. He used starch gel electrophoresis whereas we used polyacrylamide gel electrophoresis (Hoffenberg et al. 1988). We each found three genetically variable loci. One of his loci, esterase-1, may be the same as our pectoral muscle esterase 1 or 2 (he biopsied only blood; we biopsied both blood and muscle). His other two loci (6-phosphogluconate dehydrogenase and glucosephosphate isomerase) were monomorphic in our study, and our other two loci (plasma amylase and either pectoral muscle esterase 1 or 2) were not reported in his. This failure to find the same variability is not surprising. This is because all of Evans' loci were nearly monomorphic (the most common allele at each of his loci had a frequency of 94% or greater), and two of our three loci were also nearly monomorphic (the most common allele at each had a frequency of at least 92%). Thus it is conceivable that all three of Evans' loci were completely monomorphic in our population because of the founder effect (our population descends from the English population). Moreover, starch gel electrophoresis has less resolving power than polyacrylamide gel electrophoresis (Coyne et al. 1979, Ramshaw et al. 1979), leaving open the possibility that Evans' technique was insufficient to detect at least two of our three loci.

Evans (1980) biopsied a large number of birds for electrophoretic comparison: 358 individuals from 60 complete families. We biopsied 550 individuals from 106 complete families (Romagnano et al. 1990). Thus there is a good quantitative basis for comparison between studies.

Evans (1988) found that 24.2% of his first nests were parasitized. We found that 27.8% of our early nests were parasitized. (I here use Romagnano et al.'s (1990) estimate for early broods because it is based on the methods most similar to Evans'). This is a remarkably close agreement given the differences between studies in electrophoretic medium, location and data partitioning.

Both studies also found (i) that the frequency of CBP increased from year to year as our respective nest box trails became increasingly saturated (Evans 1988, Romagnano et al. 1990); (ii) that host eggs fared better than parasitic eggs; and (iii) that parasitic females remove host eggs in an apparent attempt to reduce the effects of nestling competition on parasitic nestlings (Evans 1988, Lombardo et al. 1989).

However, there are some points of disagreement. We have a major disagreement on host removal of parasitic eggs. Evans (1988) believes that hosts can recognize parasitic eggs, particularly in small clutches, and then remove them. However, we experimentally showed that hosts never remove any eggs once they themselves begin to lay although they always remove eggs added to their nests before they themselves have begun to lay (Stouffer et al. 1987). Thus hosts are unable to discriminate host from parasitic eggs as such; instead, they discriminate on the basis of the phenology of their own laying cycle, i.e., all eggs laid before host clutch initiation are treated as parasitic eggs while all eggs laid after that time are treated as host eggs.

We also have a major disagreement about the importance of cuckoldry. Although he recognized that cuckoldry could not be ruled out as a possibility, Evans (1988) chose to interpret all genetic mismatches as CBP on the grounds that extra-pair copulations were not observed in an unpublished study in Britain, and that Power et al. (1981) had documented intense mate guarding (see above). That extra pair copulations have not been reported but mate guarding has does not mean that cuckoldry does not occur; it means that it is difficult for birds to achieve and scientists to observe. More importantly, we disagree with Evans about the importance of cuckoldry because we found it in our population (Hoffenberg et al. 1988).

We have a minor disagreement on the identity of parasites. Evans (1988:1284) provides a list of four possible categories of parasite. We are puzzled by his third and fourth categories: "(3) an unpaired (probably young) female who has copulated with a male that already has a mate and a nest site; and (4) a 'professional' nest parasite who distributes some of her eggs around other nests either before or after laying in a nest of her own, or, at its extreme, distributing all her eggs and not making a nest of her own at all." These categories are ambiguous partly because they overlap. Why stipulate that category 3 females copulated with paired males? Why couldn't they copulate with unpaired males? What is the definition of a professional parasite? How can a parasite be a "professional" if she can either have a nest of her own (making her primarily a potential host) or not have a nest (giving her only the reproductive potential of a parasite)?

We have simplified categorization of parasites by presenting evidence that most parasitism is done by females without nests of their own, possibly mostly yearlings (Romagnano et al. 1990). But we do not believe that this makes such parasites "professionals" because we do not believe that parasitism alone is sufficient to insure the evolutionary success of a genetic lineage. As stated above, we believe that parasitism is a strategy used by females having only the choice of parasitizing or not reproducing at all. When these females have the opportunity, we believe they care for their own young in their own nests. If we are correct, then parasites are always "amateurs", never "professionals". We stipulate nothing about which males sire parasitic eggs because we have no information on that subject.

Despite our differences with some of Evans' (1988) interpretations, we are impressed by the fact that his and our studies agree far more than they disagree, particularly on purely factual matters, e.g., the frequency of CBP. This lends credence to the idea that genetic parasitism can be objectively studied. It also shows that the comparative method works.

REFERENCES

Coyne JA, Eanes WA, Ramshaw JAM, and Koehn RK (1979) Electrophoretic heterogeneity of α-glycerophosphate dehydrogenase among many species of *Drosophila*. Syst Zool 28: 164-175

Crossner KA (1977) Natural selection and clutch size in the European Starling (*Sturnus vulgaris*). Ecology 58: 885-892

Evans PGH (1980) Population Genetics of the European Starling (*Sturnus vulgaris*), D. Phil. thesis. Univ Oxford, England

Evans PGH (1988) Intraspecific nest parasitism in the European Starling (*Sturnus vulgaris*). Anim Behav 36: 1282-1294

Hoffenberg AS, Power HW, Romagnano LC, Lombardo MP and McGuire TR (1988) The frequency of cuckoldry in the European Starling (*Sturnus vulgaris*). Wilson Bull 100: 60-69

Kennedy ED, Stouffer PC and Power HW (1989) Postovulatory follicles as a measure of clutch size and brood parasitism in European Starlings. Condor 91: 471-473

Kessel B (1957) A study of the breeding biology of the European Starling (*Sturnus vulgaris*) in North America. Am Midl Nat 58: 257-331

Kluijver IHN (1933) Bijdrage tot de biologie en de ecologie van den spreeuw (*Sturnus vulgaris* L.) gedurende zinj voortplantingstijd. Versl Meded Plantenziekt, Wageningen, 69: 1-145

Litovich E (1982). Experiments on the Basis of Parent-offspring Conflict in Starlings, Ph D thesis. Rutgers Univ, New Brunswick, NJ

Lombardo MP, Power HW Stouffer PC, Romagnano LC and Hoffenberg AS (1989) Egg removal and intraspecific brood parasitism in the European Starling (*Sturnus vulgaris*). Behav Ecol Sociobiol 24: 217-223

MacWhirter RB (1989) On the rarity of intraspecific brood parasitism. Condor 91: 485-492

McKinney F, Cheng KM and Bruggers DJ (1984). Sperm competition in apparently monogamous birds. In: Smith RL (ed) Sperm Competition and the Evolution of Animal Mating Systems. Academic Press, Orlando, Florida ,p: 523-545

Power HW, Kennedy ED, Romagnano LC, Lombardo MP, Hoffenberg AS, Stouffer PC and McGuire TR (1989) The parasitism insurance hypothesis: why starlings leave space for parasite eggs. Condor, in press

Power HW, Litovich E and Lombardo MP (1981) Male starlings delay incubation to avoid being cuckolded. Auk 98: 386-389

Ramshaw JAM, Coyne JA and Lewontin RC (1979) The sensitivity of gel electrophoresis as a detector of genetic variation. Genetics 93: 1019-1037

Romagnano LC, Hoffenberg AS and Power HW (1990) Intraspecific brood parasitism in the European Starling. Wilson Bull, in press

Romagnano LC, McGuire TR and Power HW (1989) Pitfalls and improved techniques in avian parentage studies. Auk 106: 129-136

Stouffer PC, Kennedy ED and Power HW (1987) Recognition and removal of parasite eggs by starlings. Anim Behav 35: 1583-1584

Trivers RL (1972) Parental investment and sexual selection. In: Campbell B (ed). Sexual Selection and the Descent of Man, 1871-1971. Aldine, Chicago, p. 136-179

COSTS AND BENEFITS OF REPRODUCTION

LIFE-HISTORY TRADE-OFFS AND OPTIMAL CLUTCH SIZE IN RELATION TO AGE IN THE COLLARED FLYCATCHER

Lars GUSTAFSSON

Department of Zoology, Uppsala University, Box 561, S-75122 Uppsala, Sweden

Reproduction is a demanding activity which has often been demonstrated to have negative effects on both survival and future fecundity. Both probability of survival and fecundity often show age-specific variation. There is therefore a trade-off between current and future reproduction, and consequently, fecundity at all ages cannot be simultaneously maximized. The fact that some empirical studies have failed to demonstrate any costs associated with reproduction has caused a debate over the importance of cost of reproduction (Partridge and Harvey 1988). These negative results usually occur in studies where individuals are allowed to reproduce at their own chosen rate, and subsequent survival and fecundity are recorded. Conversely, costs of reproduction have been recorded most often when reproductive rate has been experimentally manipulated (Reznick 1985, Partridge and Harvey 1988). One problem with non-experimental studies is that confounding variables are not controlled for. Individuals may differ in their ability both to reproduce and to survive, either because of their own physiological condition or their surrounding environment (van Noordwijk and de Jong 1986). Under such circumstances one would predict a positive association between survival and future fecundity. It is therefore necessary to allocate individuals randomly to experimental groups whose reproductive rates are then experimentally varied to measure any costs associated with reproductive rates.

There are two main reasons why reproductive rate may change with age. First, individual phenotype may change during life in a manner that alters the costs or benefits of a given level of reproduction. Social dominance or foraging efficiency could increase with age so that more offspring could be produced without additional costs. The second reason is senescence, which could be expressed as lower survival probabilities and fecundity with advancing age (Partridge 1989, Dhondt 1989). Therefore analyses of experimental results should preferably take into account age of manipulated individuals.

Ever since Lack (1954) originally suggested that the most frequent clutch size should be the one which yields the highest number of viable offspring, there has been considerable interest in tests of these ideas. There are, however, many examples showing that the clutch size that produces the most fledglings is actually larger than the most frequent size (Lessells 1986, Murphy and Haukioja 1986). This discrepancy has been accounted for by suggesting that high fecundity may result in a reduction in fledgling survival (Lack 1954), juvenile survival (Perrins and Moss 1975), adult survival (Williams 1966, Nur 1984, Reid 1987), subsequent fecundity (Williams 1966), or the fecundity of the

NATO ASI Series, Vol. G 24
Population Biology of Passerine Birds
Edited by J. Blondel et al.

offspring (Andersson 1978), i.e. some kind of trade-off. Alternatively there may be a relationship between parent quality or territory quality and brood size (Högstedt 1980), and female nutritional condition may affect both clutch size and clutch productivity (Drent and Daan 1980). The result would be that an apparent directional selection on clutch size in effect is stabilizing (Price and Liou in press) towards different individual optimum.

Most experimental studies of the determination of clutch size have made use of enlarged and reduced broods (Murphy and Haukioja 1986, Gustafsson and Sutherland 1988, Pettifor et al. 1988, Partridge 1989), and have thus disregarded the potential impact of clutch size on breeding success, namely, in terms of costs arising during incubation (Nur 1986). Assuming that variation in short term correlates of lifetime reproductive success (e.g. hatching success, time spent incubating or female condition loss) has predictable consequences on lifetime reproductive success itself, the optimal brood size as judged from clutch manipulations could be smaller than the optimal brood size as judged from brood manipulations. If the difference between incubation costs for small and large clutches is significant and consistent (e.g. Haftorn and Reinertsen 1985), it will be a target of natural selection and become a component in the evolution of different life-histories. Clutch size in that case would be geared to the interaction between the incubating capacity of the parents and their later capacity to take care of dependent young.

The Collared Flycatcher *Ficedula albicollis* on the island of Gotland forms an isolated population with a high site fidelity among both adults and young such that the lifetime reproductive success can be assessed with an accuracy rarely possible under natural conditions (Gustafsson 1989). This paper is a synthesis of various results gathered here from several papers previously published or in preparation. It uses data from 1980 to 1988 from this population to show which trade-offs between life-history parameters that might eventually be important in the evolution of the clutch size of this species. I will also demonstrate the importance of taking the age of the individual into account in this context.

THE STUDY

The study has been going on since 1980 (see Gustafsson 1985 for a detailed description). Parent birds have been caught, in their nest boxes while feeding their young and marked with numbered rings. Most birds were already ringed by us and their age was known. Unringed birds could be classified as yearlings or older on basis of their plumage. Laying date, clutch size, hatching date, number of hatched young and number of fledglings in each broods were recorded. Thirteen days after hatching, each young was ringed and their weight and tarsus measured. Return rates of adults and the number of offspring recruited into the breeding population were determined by catching all breeding pairs within the study area in subsequent years. This provides a measure of local recruitment and return rate which to some degree underestimates true survival. However, birds that survived but did not breed or remained unpaired in the following year went undetected, to a large degree, in that year. On the other hand these birds in fact often had zero fitness in that year, making them equivalent to non-survivors in an evolutionary context. Of surviving adults we probably miss very few that actually has a subsequent successful breeding attempt, since almost all suitable breeding habitat is covered by our nest box plots and the study area is large enough to catch up most of those birds that disperse between plots (Pärt and Gustafsson 1989, Pärt 1990).

As in many other non-manipulative studies (see Murphy and Haukioja 1986, Partridge 1989), our data on non-manipulated clutches suggest that clutch size in the Collared Flycatcher is not associated with any costs. Gustafsson and Sutherland (1988) examined the relationship between six different parameters and natural clutch size (summarized in Table I) and were unable to find any evidence that clutch size had any effect on adult survival, juvenile survival or future reproduction of either the parents or the young (but see Nur 1988b). As a consequence, larger clutches produce more recruits, and birds with high lifetime reproductive success on average have larger clutches (Fig. 1). This, however, does not disprove the occurrence of possible trade-offs.

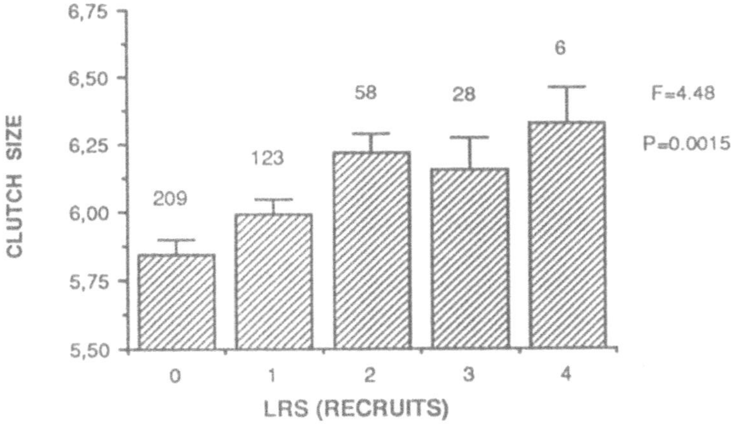

Fig. 1. Average clutch size in relation to total number of recruits produced during life (LRS) of Collared Flycatcher females.

CLUTCH SIZE EXPERIMENT

During three years (1983-85) the clutches of 320 Collared Flycatchers were manipulated (Gustafsson and Sutherland 1988). Clutches with the same hatching date were treated in pairs and either one or two young, at two days of age, were transferred from one nest to the other. In the control group two young were swapped between nests. Exchange of young took only a few minutes since distances were never more than a few hundred meters between suitable nest boxes. I included also unhatched eggs and dead young as a possible tranfer, among the live young. This was done by placing them in a row and randomly choosing the young to be transferred by using a table of random digits. I did this because I wanted to take into account any random or non-random occurrence of dead young or egg as a component of natural variance. This is an important feature of the experimental design since each egg has a certain probability of ending up as an unhatched egg or of producing a young which dies during the first two days. Small scale manipulations, as in the present case, have the disadvantage that large sample sizes are required to detect any effects, but at the same time the essential advantage that the clutch size variation achieved is close to the heritable natural variation. Obviously, this is crucial in a study of the evolution of clutch size.

Significant effects of manipulated clutch sizes on life history parameters were demonstrated by Gustafsson and Sutherland (1988) and are summarized in Table I. The return rate of both male and female was unaffected by the manipulation unlike female Blue Tit (Nur 1984). Consistent with our observations concerning natural clutches, artificially enlarged clutches produced more fledged young. However, these larger clutches resulted in lighter and smaller fledglings (Fig. 2), which were less likely to return as recruits. The result was that the natural clutch size produced more recruits than did enlarged or reduced clutch sizes (Gustafsson and Sutherland 1988). Both the reproductive success in terms of number of recruits in the second generation and the subsequent fecundity (clutch size and fledged young) of manipulated females was affected by the brood size manipulations (Gustafsson and Sutherland 1988). This experimental effect on adult female fecundity persisted for several years and was the largest for females with enlarged broods. Their clutch size were clearly reduced at older ages (5.33 ± 0.29, $n=9$, three years after the experiment) compared with control birds (6.28 ± 0.16, $n=18$, Gustafsson and Pärt, MS).

The fact that female clutch size in the subsequent year was affected by the experiment raises the question whether this was an effect of some

Table I. Relationship between different life history parameters and clutch size in the Collared Flycatcher. Only direction of relationship, from regression analyses, is given (-,0,+, intermediate optimum=<>). Data from Gustafsson and Sutherland (1988).

	Natural clutch size			Experimental clutch size		
	relation	n	p	relation	n	p
Male survival	0	1629	>.6	0	288	>.7
Female survival	0	1832	>.4	0	303	>.8
No. fleglings	+	2135	<.0001	+	308	<.001
No. recruits	+	1826	<.0001	<>	308	<.02
Female fecundity	0	344	>.1	-	36	<.001
Recruits fecundity	0	359	>.3	-	59	<.05

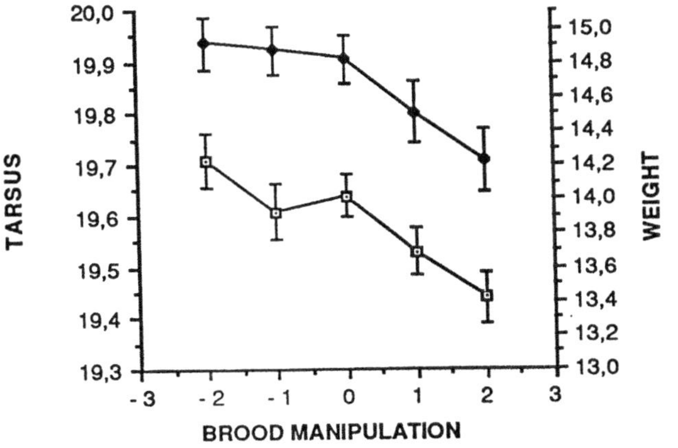

Fig. 2. Weight (filled) and tarsus (open) of fledged Collared Flycatcher young in relation to brood size manipulation. Regression for tarsus on brood size change, y=19.77-0.62x, n=283, F=15.1, $p<0.0001$. Regression for weight on brood size change y=15.19-0.17x, n=283 , F=11,3, $p<0.001$.

physiological costs in relation to brood size or a behavioural response to the relative success of that breeding attempt. The alternative to the physiological cost hypotheses is that some birds with reduced broods had an easy time raising all the nestlings in the brood up to healthy independent young and reacted to this experience through a mechanism causing them to lay one or two eggs more the next year. Likewise some females with enlarged broods may have realized that they had too many nestlings to be able to raise them to healthy surviving young and as a consequence reduced their clutch size in the subsequent breeding season. It may be fruitful in future research on clutch size variation to consider these possible mechanisms when designing experiments.

INCUBATION COSTS

In 1988 we performed an experiment designed to examine whether incubating clutches of different sizes entail any fitness costs (Moreno, Gustafsson, Carlson and Pärt, MS). The only possible way of determining the fitness cost of incubation per se is to perform a clutch size manipulation experiment and brood size manipulation experiment at the same time in the same population and to calculate the eventual differences in reproductive success between the two groups. In 25 cases when 3 or more clutches of the same size were completed on the same day, we transferred 2 eggs from one clutch to another and 2 other eggs from the second to a third clutch on the day after the last egg had been laid. These clutches constituted the reduced, control and the enlarged groups, respectively, in this clutch size manipulation experiment.

The result was that females with enlarged clutch size had a significantly higher daily energy expenditure (see also Haftorn and Reinertsen 1985). However, this had only a slight effect, if any, on reproductive success. The only significant difference we found was that hatching success tended to be lower in the enlarged group. We found no effect when we compared number and weight of fledglings between the clutch and the brood manipulations (Moreno, Gustafsson, Carlsson and Pärt MS).

AGE, REPRODUCTION AND OPTIMAL CLUTCH SIZE

In this part I will discuss age dependent reproduction and survival and optimal clutch size. For data about the age dependence of fitness components we have used cross-sectional data of all females of known age breeding between 1980 and 1987 standardized within year (Gustafsson and Pärt, MS). Ideally, longitudinal data should be used, but that would reduce sample sizes enormously since very few individuals survive to old age (adult yearly survival varies from 40 to 50 %). A problem may be that individuals surviving to old age could be better or worse parents at all ages. However, we checked for any survival selection in relation to components of fitness by calculating selection differentials, but found no significant differences between surviving and non-surviving individuals.

The relationship between six fitness components and age is shown in Table II. All reproductive parameters show a clear age dependence, increasing during early life, reaching a peak and decreasing at older ages. This pattern has also been observed in many other bird species (Dhondt 1989, Saether in press). Thus Collared Flycatcher females show clear signs of senescence with reproductive performance declining with age. However, adult return rate seems not to be related to age.

Table II. Relationship between life history parameters and age in the Collared Flycatcher. Indication as in Fig. 1 (from Gustafsson and Pärt MS).

	Relation	n	p
Laying date	<>	871	<.001
Clutch size	<>	869	<.001
% successful broods	<>	700	<.01
No. fledglings	<>	700	<.001
No. recruits	<>	484	<.1
Female survival	0	666	>.5

These results suggest that individuals of different ages may have different optimal clutch sizes. The question now arises: do individuals of all ages lay the optimal clutch size? We know that individuals cannot maximize their reproductive rate at all ages because of the trade-offs discussed earlier. But do they lay the number of eggs that is optimal at their current age? When trying to define the optimal clutch size we usually employ a random sample of the population. However, if different groups, say, different age classes, have different ***sensitivities*** in relation to the manipulation, we might interpret the results incorrectly. For instance, if young birds derive higher success from having a reduced clutch and older from having an enlarged clutch, the average optimum might be the natural clutch size, although no bird is laying the optimal clutch size. For the term optimal clutch size to have any biological meaning, we have to analyse if indeed different categories of birds lay their optimal clutch size.

The above mentioned problem to some degree exists in the Collared Flycatcher data since one year old birds in the brood size experiment described above do not seem to lay an optimal clutch size. They score about the same number of recruits regardless of the size of their manipulated clutch (Fig. 3a). On the other hand, older birds clearly lay the clutch size that maximizes their production of recruits (Fig. 3b). The reason why we obtain a statistically significant effect for the population sample is because the very strong effect from the older birds. However, we might had obtained a non-significant result for the population if the effect from the older individuals had been weaker, but anyhow of evolutionary significance.

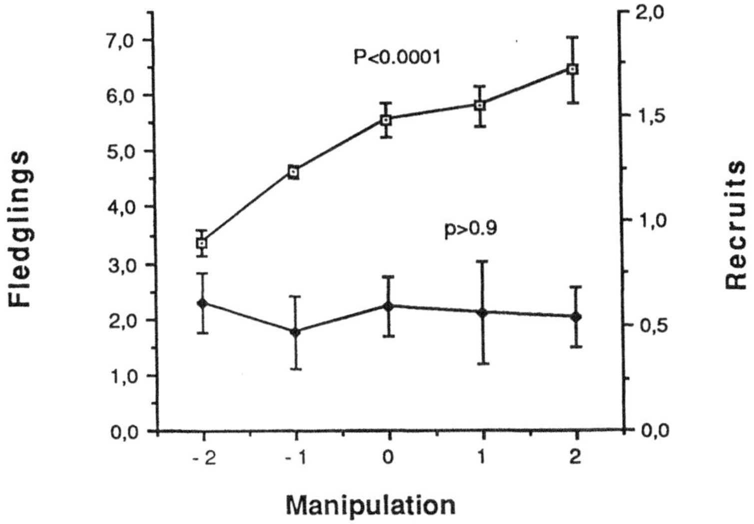

Fig. 3a. Number of fledlings (open) and recruits (filled) in relation to brood manipulation of one year old Collared Flycatcher females. ANOVA (fledglings) F=9.3, n=94, p<0.0001. ANOVA (recruits) F=0.1, n=95, p>.9.

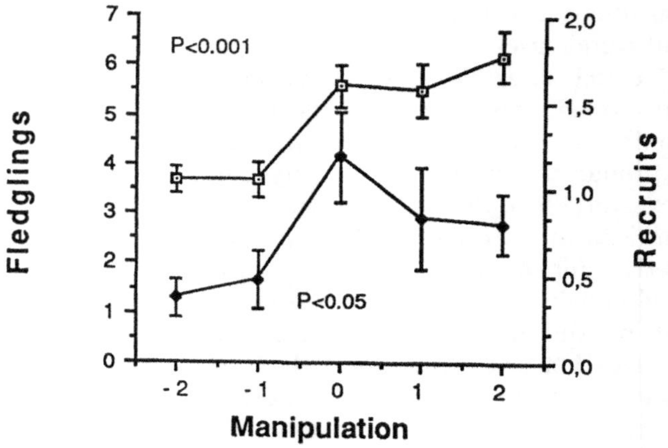

Fig. 3b. Number of fledglings (open) and recruits (filled) in relation to brood manipulation of old Collared Flycatcher females (>1 year). ANOVA (fledglings) $F=8.1$, $n=86$, $p<0.0001$. ANOVA (recruits) $F=2.7$, $n=86$, $p<0.05$.

This result raises further questions about the mechanism behing the determination of clutch size and adds strength to the notion that research on clutch size variations has much to gain from examining the mechanism behing the age-dependent and between-year differences in clutch size, as described above.

A further complicating factor is that the very old individuals(> 5 years) show a different pattern from both the yearlings and the 2 to 4 year old birds. They actually reproduce at a rate that have for them severe costs in that female survival is negatively correlated with reproductive success (Pärt, Gustafsson and Moreno in prep.). This may be an effect of senescence, i.e. of their lower future probability of successful reproductive events, because of which they possibly put more effort into their "terminal" reproduction.

CONCLUSIONS

The difficulties of determining the trade-offs of importance for life history evolution are clearly illustrated by the data from Collared Flycatchers. Within a population, correlations between estimates of life-history parameters may fail to reveal the nature or strength of the trade-offs, presumably because the specific form of the trade-offs differs between phenotypes and individuals adjust their clutch size accordingly. Experimental manipulations provide the only reliable means of determining the costs of reproduction. Such experiments have to be small scale to be relevant for the natural situation and to determine which costs are actually important for the evolution of clutch size. The possible trade-offs so far investigated and their relative importance are summarized in Table III. This

Table III. An overwiew of possible trade-offs against clutch size and their effect on fitness in the Collared Flycatcher, based on Gustafsson and Sutherland (1988), Moreno, Gustafsson, Carlsson and Pärt MS, Gustafsson and Pärt MS, and Pärt, Gustafsson and Moreno (in prep.).

	Effect	Relative contribution
Male survival	no	non
Female survival	no, except	non
	neg. for >5 year old	very minor*
Hatchlings surv. (fledglings)	neg.	minor
Fledgling surv. (recruits)	neg.	major
Female fecundity	neg.	minor
Female senescence	pos.	very minor*
Recruits fecundity	neg.	minor

* the effect is strong but its fitness consequences are small since very few individuals survive to such an old age.

experiment of actual clutch size showed that adult survival did not change with clutch size. The main costs were that fledglings from enlarged clutches were of lower quality and survived to breed at a much lower rate. Also, their future fecundity was affected by clutch size. Incubation costs seemed to be of minor importance. The experiment also affected the future fecundity of the adults There was a clear difference between young and old birds in their ability to predict their own optimal clutch size. These findings raise several questions about the causes of between year clutch size variation. In the future, experiments should be designed to allow an investigation of the relative effects of behavioural responses vs. true physiological cost responses. These mechanisms might differ considerably between individuals of different ages.

To conclude, the strongest force in the evolution of clutch size in the Collared Flycatcher is how a change in clutch size affects the number of recruits produced to the next generation. This relationship is actually a stabilizing selection towards each individual's optimal clutch size.

ACKNOWLEDGEMENTS

This paper is based on research financed by the Swedish Natural research Council who also contributed with a special travel grant to participate in the NATO Workshop on Corsica. I thank T. Pärt for discussions and S. Ulfstrand for comments on the MS.

REFERENCES

Andersson M (1978) Natural selection of offspring numbers. Am Nat 112: 762-766

Dhondt AA (1989) The effect of old age on the reproduction of great and blue tit. Ibis 131: 268-280

Drent RH, Daan S (1980) The prudent parent; energetic adjusment in avian breeding. Ardea 76: 225-252

Gustafsson L (1985) Fitness factors in the collared flycatcher *Ficedula albicollis* Temm. D Ph Thesis, Uppsala University, Uppsala

Gustafsson L (1989) The collared flycatcher. In: Newton I (Ed) Lifetime reproduction in birds. Oxford University Press, Oxford

Gustafsson L and Sutherland WJ (1988) Cost of reproduction in the collared flycatcher *Ficedula albicollis*. Nature 335: 813-817

Gustafsson L and Pärt T Reproductive costs accelerate senescence in the collared flycatcher. Manuscript

Haftorn S and Reinertsen RE (1985) The effect of temperature and clutch size on the energetic cost of incubation in a free-living blue tit (*Parus caeruleus*). The Auk 102: 470-478

Högstedt G (1980) Evolution of clutch size in birds: adaptative variation in relation to territory quality. Science 210: 1148-1150

Lack D (1954) The natural regulation of animal numbers. Clarendon, Oxford

Lessells CM (1986) Brood size in Canadian geese: a manipulation experiment. J Anim Ecol 55: 669-689

Moreno J, Gustafsson L, Carlsson A and Pärt T Optimal clutch size in the collared flycatcher: on the cost of incubation. Manuscript

Murphy EC and Haukioja E (1986) Clutch size in nidicolous birds. In: Johnston RF (Ed) Current Ornithology. Vol 4: 141-180, Plenum press, New York

Nur N (1988a) The consequences of brood size for breeding blue tits. III Measuring the cost of reproduction: survival, future fecundity, and different dispersal. Evolution 42: 351-362

Nur N (1988b) The cost of reproduction in birds: An examination of the evidence. Ardea 76: 155-168

Partridge L (1989) Lifetime reproductive success and life history evolution. In: Newton I (Ed): Lifetime reproduction in birds. Academic press, London

Partridge L and Harvey PH (1988) The ecological context of life history evolution. Science 241: 1449-1455

Perrins CM and Moss D (1975) Reproductive rates in the great tit. J Anim Ecol 44: 659-706

Pettifor RA, Perrins CM and McCleery RH (1988) Individual optimization of clutch size in great tits. Nature 336: 160-162

Price T and Liou L (1989) Selection of clutch size in birds. Am Nat (in press)

Pärt T and Gustafsson L (1989) Breeding dispersal in the collared flycatcher (*Ficedula albicollis*): possible causes and reproductive consequences. J Anim Ecol 58: 305-320

Pärt T (1990) Natal dispersal in the collared flycatcher (*Ficedula albicollis*): possible causes and reproductive consequences. Ornis Scand (in press)

Reid WV (1987) The cost of reproduction in the glaucous-winged gull. Oecologia 74: 458-467

Reznik D (1985) Cost of reproduction; an evaluation of the empirical evidence. Oikos 44: 257-267

Saether BE (1989) Age dependent reproduction in birds. In: Johnston RF (Ed): Current Ornithology, vol 4, Plenum press, New York

van Noordwijk AJ and de Jong G (1986) Acquisition and allocation of resources: their influence on variation in life history tatics. Am Nat 128: 137-142

William GC (1966) Natural selection, the cost of reproduction and a refinement of Lack's Principle. Am Nat 100: 687-690

EGG SIZE INVESTMENTS OF TITS; DOES NUMBER CONFLICT WITH SIZE?

Jan EKMAN* and Anita JOHANSSON-ALLENDE

Department of Zoology, University of Gothenburg, Box 250 59, S-400 31 Gothenburg, Sweden.
*Present address: Department of Zoology, University of Stockholm, S-106 09 Stockholm, Sweden.

When different components of reproduction cannot be maximized simultaneously, parents are forced to make a decision over the allocation of resources. For instance, it is not possible to maximize both present and future reproduction (Fisher 1930) and resources have to be allocated between reproduction and parental maintenance.

Another allocation problem for parents is that it may only be possible to invest more in one offspring by giving less to the rest of the brood. Parents then have to choose between few and large or many but small offspring (MacArthur and Wilson 1967). The optimal investment in each offspring depends on how their performance later in life relates to the initial investment (Smith and Fretwell 1974) and the total effort (Winkler and Wallin 1987).

A trade-off between the number and size of offspring implies that offspring size should decrease in larger clutches. For three *Parus* species, the Coal Tit *P. ater*, the Willow Tit *P. montanus* and the Crested Tit *P. cristatus*, of the temperate coniferous forest we compared investments in size and number of eggs between species as well as between females. These species are similar in morphology and share habitat. One feature of the Old World avifauna is the large number of *Parus* species of comparable size which coexist without habitat separation (Lack 1971). Due to morphological and ecological similarities the cost of offspring production may be comparable. This is supported by survival data for Willow Tits and Crested Tits (Ekman and Askenmo 1986).

METHODS

Eggs were weighed during laying in tit populations inhabiting coniferous forest habitats about 40 km E of Gothenburg, W Sweden. The tits used nest-boxes erected in the area and they were inspected daily during laying. Willow Tits and Crested Tits normally excavate their holes and they readily accepted nest-boxes when filled with coarse saw-dust. Eggs were weighed on the day they were laid. Hence, all weight data come from measurements taken before incubation started, and the sequence of laying was known. Eggs were individually marked with a pencil, and weighed to the nearest 0.05 g using a 5 g "Pesola" spring balance.

NATO ASI Series, Vol. G 24
Population Biology of Passerine Birds
Edited by J. Blondel et al.

Eggs were weighed during the 1981 breeding season. This year the population was only partially marked by the start of breeding, but all Crested and Willow Tit females could be captured, weighed and marked during the breeding season. Females were sexed by their brood patch. The predation on Coal Tit nests was substantial and only two nests out of the seven which we studied escaped predation before we had the opportunity to catch and mark the female. Hence, complete weight data of Coal Tit females are missing.

Estimates of clutch size were collected for the years 1974 to 1984. Inspections of the nest-box area provided information on the size of first and second clutches and the rate of second clutches. Estimates of the incidence of second clutches were taken in breeding populations which were completely banded after the first clutch, but the value should be conservative as some second clutches may have been overlooked.

RESULTS

Clutch size

Reproductive investments differ between species. Crested Tits lay the smallest clutch and Coal Tits the largest (Table I). Differences in the annual reproductive rate are augmented by a high rate of second clutches in Coal Tits, while second clutches are rare among Crested Tits in particular (Table I). Second clutches are smaller than first clutches (Table I).

Table I . Reproductive investments in number of eggs.

Species	Clutch size				Proportion second clutches	Annual egg production
	first clutch (N)	±SE	second clutch (N)	±SE (N)		
Crested tit	5.02 (97)	±0.085	5 (1)		0.01 (73)	5.07
Willow tit	8.24 (111)	±0.11	6.00 (7)	±0.40	0.07 (92)	8.66
Coal tit	9.00 (20)	±0.20	8.22 (9)	±0.21	0.90 (19)	16.40

Significant difference in clutch size between species
($F_{2,225}=283$, $P<0.0001$)

Species differences

Tit species which lay larger clutches seem to invest less in each egg. In absolute size, Coal Tits lay the smallest egg and Crested Tits the largest of the three (Table II). After taking body mass differences (Table II) into account there is substantial residual species variation in the relative egg investment (Fig. 1). In a comparison including also data for the Great Tit *P. major* and the Blue Tit *P. caeruleus* there is a significant negative slope of the relation between relative egg investment and clutch size among European tit species. Still, the decrease in relative egg size of species ($b=-0.318$) is appreciably and significantly lower than $b=-2$ which corresponds to the case where all resources invested in one egg had to be withdrawn from another. Of course, part of this difference could result from differences in the total investment.

Table II. Egg size investments (±S.E.).

Species	Egg mass (g) (N=clutches;eggs)	Female body mass (g) (N)
Crested tit	1.31±0.016 (9;51)	11.14±0.13 (9)
Willow tit	1.20±0.0092 (8;70)	10.86±0.41 (8)
Coal tit	0.96±0.011 (7;37)	9.14±0.083 (32)

Significant difference between species in egg mass ($F_{2,155}=167.8, P<0.0001$), and female body mass ($F_{2,46}=95$, $P<0.001$).

Variations within species

There is substantial variation in egg size within species which, to a large extent, results from differences between clutches (Table III). In contrast to interspecies differences there is no obvious relationship between egg size investment and female body mass within species (Fig. 2), but variation between clutches may rather reflect the quality of the habitat. Laying sequence and temperature have minor effects on egg mass (Table III; see also Ojanen et al. 1981). Females lay smaller eggs in periods of low ambient temperature during egg formation (Fig. 3), and the general trend during laying appears to be increasing egg mass. Laying sequence had a significant effect only in the Willow Tit where eggs laid late appear to be heavier (Fig. 4).

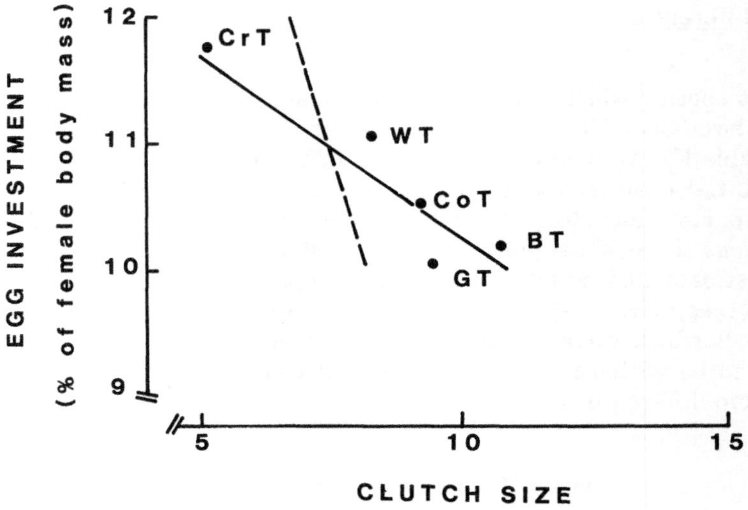

Fig. 1. The relative investment in the size of a single egg among some European tit species (the mass of average sized egg in percent of female body mass). Hatched line - expected slope of regression for a constant total investment (b=-2); solid line - regressions from data (b=-0.318, P<0.02). Significant difference (t=6.24, P<0.001) between slopes for observed relationship and expected from a constant total investment. BT-Blue Tit, CoT-Coal Tit, CrT-Crested Tit, GT-Great Tit, WT-Willow Tit. Data for Great Tit and Blue Tit from O'Connor (1977) and Lehikoinen (1986).

Table III. Proportions in egg size variation explained (r^2) by ambient temperature, the laying sequence and differences between clutches.

	Significant factor (r^2)		Total variation explained
Species	First	Second	Third
Crested tit	clutch	ambient	
	temperature	-	0.89
(r^2=0.76***)	(r^2=0.074**)		
Willow tit	clutch	laying	ambient
	sequence	temperature	0.80
(r^2=0.72***)	(r^2=0.049**)	(r^2=0.023*)	
Coal tit	clutch		
(r^2=0.52**)	-	-	0.53

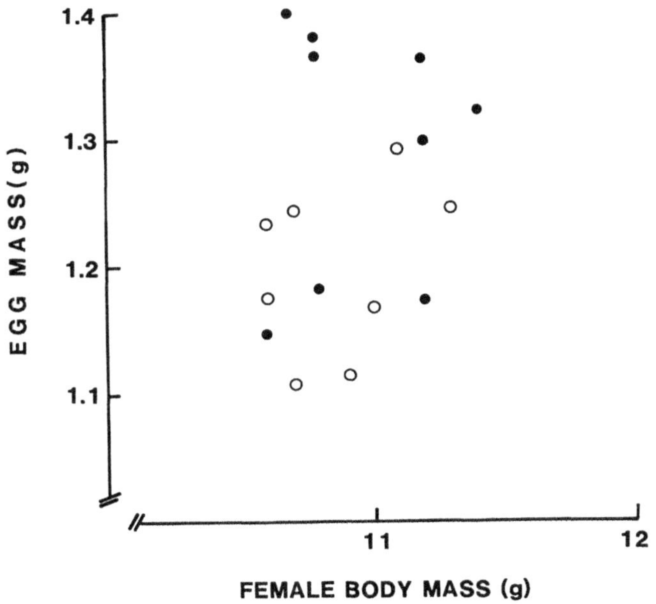

Fig. 2. Variation of egg mass (clutch mean) and female body mass in Crested Tit (filled dots) and Willow Tit (open dots).

Reductions in egg mass when ambient temperature fall suggest a conflict between energy going into the egg and energy required by the female. In 1978, laying female Crested Tits were exposed to an exceptional decrease in ambient temperature, which one night dropped by about 20°C, so that the morning temperature was -5°C against about +15°C by dusk. Within the next three days six out of 16 Crested Tit nests were deserted. It seems likely that all the females from the deserted nests died as they were never reobserved while their males were all resighted alone within a week. This loss of laying females is exceptional. In the six other years between 1974 and 1980 there were only two losses out of 50 females followed during laying ($X^2_{(2)}=9.81$, df=2, $P<0.01$), and both these females were taken by predators.

DISCUSSION

The decline in egg mass for tit species laying larger clutches does not appear to be simply the result of trading size against number for eggs produced from a resource "capital". First, the negative relationship is too weak to be explained simply by the fact that eggs are produced from a resource capital. Second, there is no evidence that females which lay a large egg early in the sequence should suffer a cost in smaller eggs afterwards. The size of single eggs may sometimes have to be reduced in adverse weather (Fig. 3, Ojanen et al.

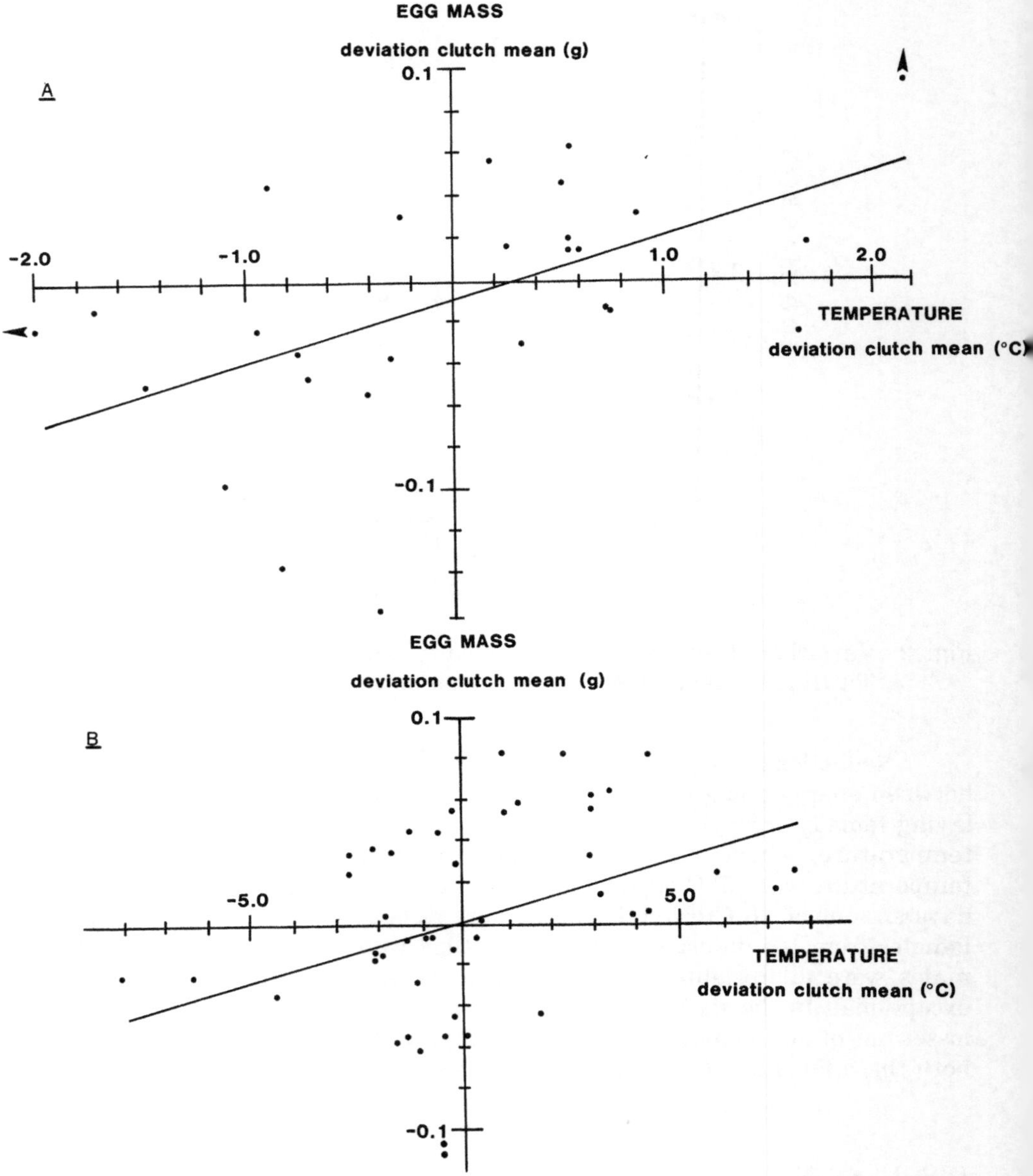

Fig. 3. The effect of ambient temperature on egg mass. Variables normalized for variations between clutches in egg mass and laying date (ambient temperature). Y-axis gives deviations from mean egg mass within clutches; relative deviations in cumulated mean daily temperatures during the four days of egg formation before laying on x-axis. A) Crested Tit ($r=0.53$, $df=25$, $p<0.05$), and B) Willow Tit ($r=0.37$, $df=43$, $p<0.05$).

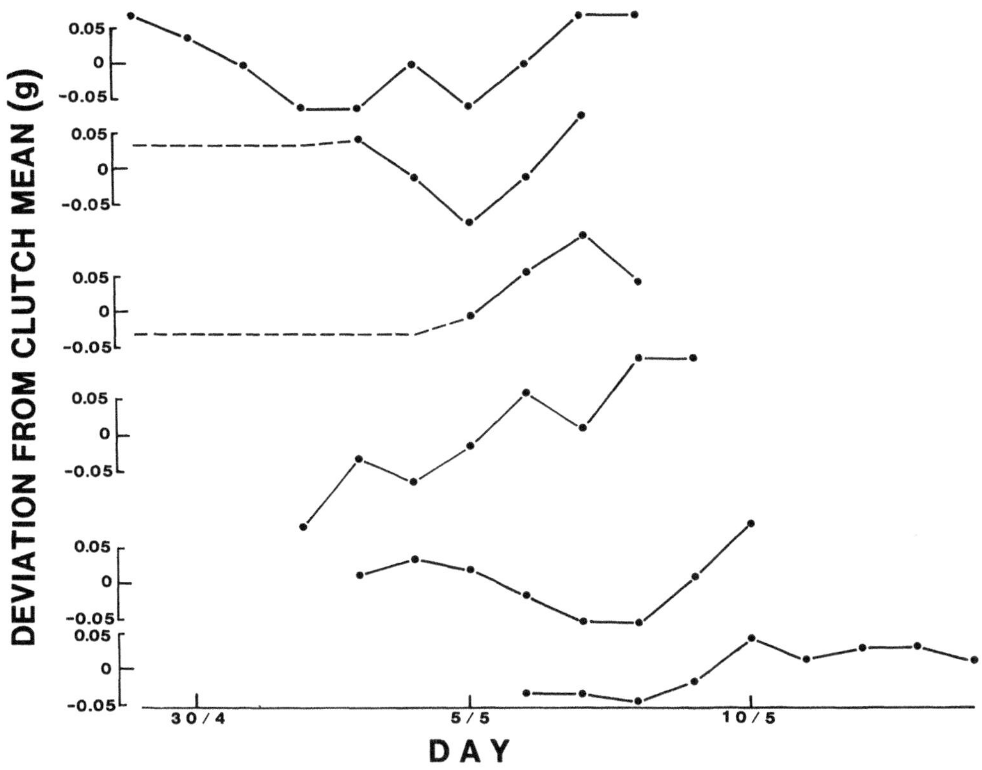

Fig. 4. Variation in egg mass during the laying sequence of Willow Tits. The size of single eggs for six clutches is given in deviations from clutch mean, and laying sequence is given in absolute time. Hatched lines - means for eggs where the laying sequence is unknown.

1981), but there are no apparent lasting effects. Despite the fact that the size of single eggs may have to be reduced in adverse conditions, egg size of tits appear to increase during the laying sequence (Fig. 4, Winkel 1970), which has been found also for other small hole-nesting passerines (Slagsvold et al. 1984, Slagsvold and Lifjeld 1989). As egg mass declines when the female's metabolic rate increases in cold weather, this increase should accordingly show that access to energy generally increases towards the end of laying.

Egg size does affect the performance of tits later in life (Schifferli 1973) and parents should benefit from making large eggs. Hence, reduced egg mass at low ambient temperature suggests a conflict between egg mass and the female's own requirements. The incidence of a large scale disappearance of Crested Tit females in the midst of laying after a cold spell in 1978 also suggests that energy drainage through egg formation could sometimes be fatal. Conceivably, species differences in the relative investments may depend on this trade-off between the size of each egg and the female's wellbeing rather than a simple mechanical relationship between size and numbers. Each egg could represent a larger reproductive value for species with small clutch size, and a larger value of the egg could make laying females more prone to accept personal risks on behalf of the egg.

One reason for the absence of a trade-off between the size of current and future eggs could be that tit females form eggs from a daily "income" (Drent and Daan 1980). Then single eggs do not compete for the same resource and they can vary independently. The quality of females or their habitat is then more free to affect egg mass, which is consistent with the observation that differences between clutches appear to be the main source of variation in egg size among tits (Table III, Ojanen et al. 1981). It is also logical that small passerines like these tits need a daily energy surplus to be able to form an egg. Sometimes they may, in little more than a week, produce an egg mass corresponding to their own body mass, and they are not known to build up energy reserves to an extent which could account for the production of this egg mass from an energy "capital" (Tiainen 1980). Rather, it has been suggested that tits start laying when it is possible to acquire a sufficiently large surplus in their daily acquisition of energy (Perrins 1970) and that intra-clutch variation in egg size should reflect proximate constraints on egg formation (Slagsvold and Lifjeld 1989). Experiments also confirm that they have to postpone laying for lack of energy (Källander 1974, Jansson and von Brömssen 1980). Egg size adjustments and the timing of laying are both compatible with an "income" model suggesting that laying starts when the daily surplus exceeds the threshold necessary to form an egg.

Egg size may be determined in a trade-off with the wellbeing of the female herself when she does not have to balance the size of eggs against their number. Still, not all bird species may be released from a trade-off between number and size of eggs. The size of eggs will not conflict with the total investment only when eggs are produced from a daily energy surplus. Trade-offs affecting egg size and the total number of eggs that can be laid may thus differ between bird species depending on whether they produce eggs from a "capital" or from an "income". Obviously some birds like geese breeding in the arctic do produce eggs from a capital (Drent and Daan 1980). For such species, reproductive investments should be constrained by past conditions to a larger extent than for species producing eggs from an "income".

REFERENCES

Drent RH, Daan S (1980) The prudent parent: energetic adjustments in avian breeding. Ardea 68: 225-252

Ekman J, Askenmo C (1986) Reproductive cost, age-specific survival and a comparison of the reproductive strategy in two European Tits (genus *Parus*). Evolution 40: 159-168

Fisher RA (1930) The genetical theory of natural selection. Clarendon Press, Oxford

Jansson C, von Brömssen A (1980) Effects of food addition to Willow Tit *Parus montanus* and Crested Tit *P. cristatus* at the time of breeding. Ornis Scandinavica 11: 173-178

Källander H (1974) Advancement of laying of Great Tit by the provision of food. Ibis 116: 365-367

Lack D (1971) Ecological isolation in birds. Blackwell, Oxford

Lehikoinen E (1986) Winter ecology of passerines: significance of weight and size. Ph.D. thesis, University of Turku

MacArthur R, Wilson EO (1967) Island Biogeography. Princeton Univ Press, Princeton

O'Connor RJ (1977) Growth strategies in nestling passerines. The living bird 16: 209-238

Ojanen M, Orell M, Väisänen RA (1981) Egg size variation within passerine clutches: effects of ambient temperature and laying sequence. Ornis fennica 58: 93-108

Perrins CM (1970) The timing of birds' breeding season. Ibis 112: 242-255

Schifferli L (1973) The effect of egg weight on the subsequent growth of nestling Great Tits (*Parus major*). Ibis 115: 549-558

Slagsvold T, Lifjeld JT (1989) Constraints on hatching asynchrony and egg size in pied flycatchers. Journal of Animal Ecology 58:837-850

Slagsvold T, Sandvik J, Rofstad G, Lorentsen Ö, Husby M (1984) On the adaptive value of intraclutch egg-size variation in birds. The Auk 101:685-697

Smith CC, Fretwell SD (1974) The optimal balance between size and number of offspring. The American Naturalist 108:499-506

Tiainen J (1980) The adaptedness of the Willow Tit (*Parus montanus*) to the migratory habit. Ornis Fennica 57: 77-81

Winkel W (1970) Experimentelle Untersuchungen zur Brutbiologie von Kohl- und Blaumeisen (*Parus major* und *P. caeruleus*). Journal für Ornithologie 111: 154-174

Winkler DW, Wallin K (1987) Offspring size and number: a life history model linking effort per offspring and total effort. The American Naturalist 129: 708-720

MANIPULATION OF THE BROOD SIZE OF PIED FLYCATCHERS

Hans KÄLLANDER and Henrik G. SMITH

Department of Ecology, Animal Ecology, Lund University, Ecology Building, S-223 62 Lund, Sweden

Most studies on passerines have shown a positive correlation between clutch size and the production of fledglings. They have also shown that clutches larger than the modal one fledge more young than those of modal size (Klomp 1970, Murphy and Haukioja 1986, Alatalo and Lundberg 1989). This observation apparently contradicts Lack's (1947, 1954) original hypothesis of clutch size determination in nidicolous birds, according to which clutch size is determined by the largest number of young that the parents can adequately nourish.

This discrepancy between theory and observation can be explained in any of at least four ways: (1) birds are constrained to various degree during either laying or incubation, or (2) there is a cost to reproduction, as originally stated by Williams (1966) in his elaboration of Lack's hypothesis (see also Charnov and Krebs 1974). This cost could be reduced survival or future reproductive output of the adults. Alternatively, (3) fledglings from large clutches may have lowered chances of being recruited to future breeding populations. One further explanation has been proposed; (4) between-year environmental variation may result in a mean clutch size lower than the most productive one observed in any particular year (Gillespie 1977, Boyce and Perrins 1987). Since there is little evidence that clutch size in most passerines is proximally constrained, e.g. by resource scarcity or depletion of female body reserves during laying or incubation (Murphy and Haukioja 1986, but see Mertens 1987) or incubation efficiency (Briskie and Sealy 1989), it seems logical to analyse the ultimate determinants of clutch size.

The concept of reproductive cost has stimulated much experimental work during the 1980s; brood sizes have been manipulated and the effects on adults and young studied (for references, see Smith et al. 1989a). Reznik (1985) criticized this approach on the grounds that only phenotypic responses are measured. While we recognize that there may be some problems in evaluating the results of brood size manipulation experiments, we agree with Gustafsson and Sutherland (1988) and Lindén (1988) that clutch or brood size manipulations represent the best way to look for reproductive costs and thus reveal potential selection pressures operative in clutch size evolution. This is because birds may adjust clutch size to their breeding situation (Perrins and Moss 1975, Högstedt 1980), which automatically would lead to a positive correlation between clutch size and fledging success and mask the presence of any reproductive costs. For the same reason also adult survival may be unaffected by brood size or even show a positive relationship to it (Smith 1981).

NATO ASI Series, Vol. G 24
Population Biology of Passerine Birds
Edited by J. Blondel et al.

In the present contribution we report the results of brood size manipulations in the Pied Flycatcher *Ficedula hypoleuca* (see also von Haartman 1954, Askenmo 1977, 1979, Alatalo and Lundberg 1989). The study supplements our previous studies on the Great Tit *Parus major* (Smith et al. 1987, 1988, 1989a, 1989b).

METHODS

The study was carried out in the Revinge area, 17 km east of Lund, southernmost Sweden, and in Vombs Fure, 7 km SE of Revinge, in 1987 and 1988. Some supplementary data on parental feeding frequencies were collected at Revinge in 1989. The study area at Revinge is heterogeneous, mainly with small copses and plantations of deciduous trees; one small area consists of planted pines. Oak, beech and birch dominate in different parts, but most woods contain a mixture of tree species. Nestboxes were present at a density of about 2/ha. In Vombs Fure, a pine plantation, nestboxes were placed 40 m apart on alternating sides along roads in middle-aged stands with usually sparse undergrowth of birch, rowan, etc. (overall nestbox density ca. 1.5 boxes/ha). We found no statistically significant differences in various parameters, such as laying date, clutch size, feeding frequency or adult and nestling mass between Revinge and Vombs Fure. Therefore data from the two areas have been combined in some of the following analyses.

Nestboxes were checked regularly to determine date of laying of the first egg and clutch size for each clutch; at the time of expected hatching, boxes were controlled every day until hatching had been established. Five days after hatching (nestling day 6), parents were captured and ringed, and their wing length, tarsus length and weight determined. On the same day, young were transferred from one nest to another to create broods larger or smaller than the original clutch. Some broods hatching on the same day as the experimental broods were left as controls. Which of the broods should become reduced, enlarged or control, was determined by chance. In 1987 brood size was changed by ca. 50%, in 1988 and 1989 it was reduced or enlarged by two young, in relation to clutch size. A few clutches with several unhatched eggs were excluded from the study. There were no differences in clutch size or in number of nestlings between the three categories prior to manipulation ($p>0.15$ in all cases, ANOVA and t-test, respectively; Table I). In all, 70 broods were manipulated.

On nestling day 13, parents were recaptured and weighed, while nestlings were ringed and weighed, and their wing and tarsus lengths measured. All weights were to the nearest 0.1 g, wing lengths to the nearest 0.5 mm and tarsus lengths to the nearest 0.1 mm.

Parental feeding frequencies were determined on days 11-12, at which time they have reached their maximum (von Haartman 1954). At Revinge we used automatic recorders triggered by an infrared sensor in the entrance hole, usually operated over two days in 1987 and one day in 1988. They were started at night after the birds had been allowed to accustom to the equipment for

Table I. Mean ($\pm$SD) clutch and brood sizes of the three categories of broods in 1987-1989. Day 6A gives the number of nestlings before manipulation and day 6B gives the number of nestlings after manipulation on day 6.

	1987			1988			1989		
	$\bar{x}$	SD	n	$\bar{x}$	SD	n	$\bar{x}$	SD	n
Clutch size									
Reduced	6.50	0.71	10	6.60	0.82	20	7.17	0.75	6
Control	6.42	1.24	12	6.58	0.52	12	-	-	-
Enlarged	6.33	0.50	9	6.20	0.70	20	6.60	0.55	5
Day 6A									
Reduced	6.00	1.05	10	6.40	0.88	20	6.00	1.55	6
Control	6.25	1.06	12	6.36	0.67	11	-	-	-
Enlarged	5.89	0.78	9	6.00	0.92	20	6.40	0.55	5
Day 6B									
Reduced	4.11	0.78	9	4.20	1.10	11	4.17	0.75	6
Enlarged	8.33	0.50	9	8.20	0.70	20	8.60	0.55	5

about a day. Feeding frequencies were also recorded visually (both study areas) at each box for 1-3 h in 1987 and 1 h in 1988 to determine the relative contribution of the two sexes. The proportion of feedings made by each parent was multiplied by the automatically recorded number of daily feedings to obtain the total feeding frequency of each sex. The correlation between visually and (daily) automatically recorded feeding frequencies was high but not perfect ($r=0.77$, $df=41$, $p<0.001$).

Despite the possibility of clutch size adjustment of the kind suggested by Högstedt (1980), in most of the following analyses we use brood size rather than experimental categories as the independent variable. Statistical analyses were carried out following Sokal and Rohlf (1981).

RESULTS

Parental feeding frequencies

The frequency with which parents brought food to the nestlings increased with increasing brood size, explaining 41% of the variation in feeding frequency.

There was, however, a tendency towards a levelling off at brood sizes above 6 nestlings (Fig. 1). Thus, there was no difference in mean feeding frequency between parents of broods of 6-7 on the one hand and those of 8-9 on the other ($p>0.10$, t-test). Date had no significant effect on feeding frequencies.

Since a proportion of males may allocate some of their effort to a secondary nest (Lifjeld and Slagsvold 1989), one might perhaps expect males to show less clear responses to altered brood sizes than females. In 1987, 42% of the males at Revinge were shown to be polygynously mated, whereas in 1988 no case of polygyny was detected at either Revinge or Vombs Fure. Both males and females increased their feeding frequencies significantly in relation to the number of young in the nest, but there was considerably more scatter than in the data for combined daily feeding frequencies as determined automatically.

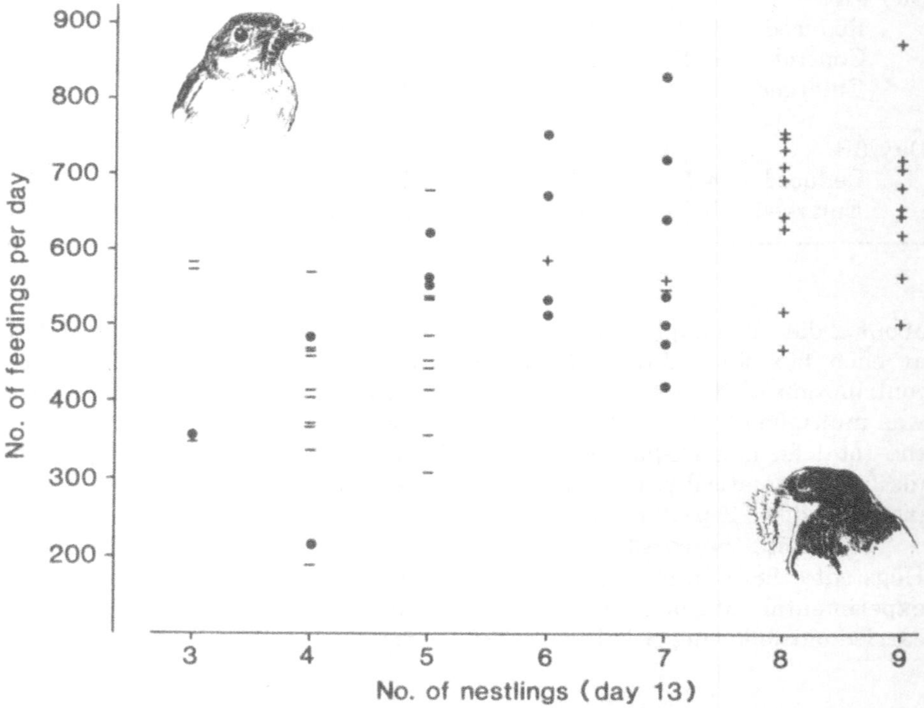

Fig. 1. Parental feeding frequency of nestlings in relation to brood size in the Pied Flycatcher. Combined data for Revinge in 1987-1989. - = reduced broods, • = control broods, + = enlarged broods.

Effects on parental weights

On nestling day 6, despite being larger (having longer wings), males weighed less than females ($\bar{x}$ 12.34±0.52(SD) g, n=82 vs 13.05±0.74 g, n=100), a difference that had decreased by day 13 (12.20±0.54 g, n=69 vs 12.43±0.66 g, n=69) (cf. Winkel and Winkel 1976). The absolute weight on nestling day 13 of each of the sexes was negatively related to brood size, being about 0.08 g lower for each additional nestling (males: $p<0.02$; females: $p<0.04$). However, the amount of mass lost between day 6 and day 13 showed a more inconsistent pattern. In males, mass loss was significantly correlated with the number of nestlings after manipulation on both day 6 and day 13, whereas it showed no statistically significant relationship with male feeding frequency. In females, by contrast, mass loss did not correlate with either the number of nestlings on day 6 or day 13, whereas it did with female feeding frequency ($p=0.04$).

In both sexes, mass loss was negatively related to the day 6 weight prior to manipulation (Fig. 2); heavy birds thus lost more mass than light birds (the slope of the regression line for day 6 mass on day 13 mass was significantly less than unity in both sexes, $p<0.001$). Females, which in general were heavier than males on day 6, lost more mass than males, and only 4 of 61 females increased in weight, whereas 16 of 62 males did so.

Effects on the young

Like in earlier studies (e.g. Nur 1984b, Tinbergen 1987, Lindén 1988, Smith et al. 1989, Alatalo and Lundberg 1989; but see Askenmo 1977), enlarged broods produced more fledged young than did controls ($p<0.001$ in both 1987 and 1988, Mann-Whitney U-test; Table II). Mean nestling losses

Table II. Mean (±SD) number of fledglings and mean (±SD) nestling mass (g) on day 13 in relation to manipulation category for 1987 and 1988. Sample sizes in parentheses.

	1987	1988
Mean no. of fledglings		
Reduced	3.70±1.50 (10)	4.20±1.11 (20)
Control	5.17±2.13 (12)	6.25±0.75 (12)
Enlarged	7.56±1.33 (9)	7.70±1.92 (20)
Mean nestling mass		
Reduced	14.12±0.83 (9)	14.36±0.47 (20)
Control	13.62±1.80 (12)	13.91±0.79 (12)
Enlarged	12.47±1.80 (8)	13.99±1.05 (19)

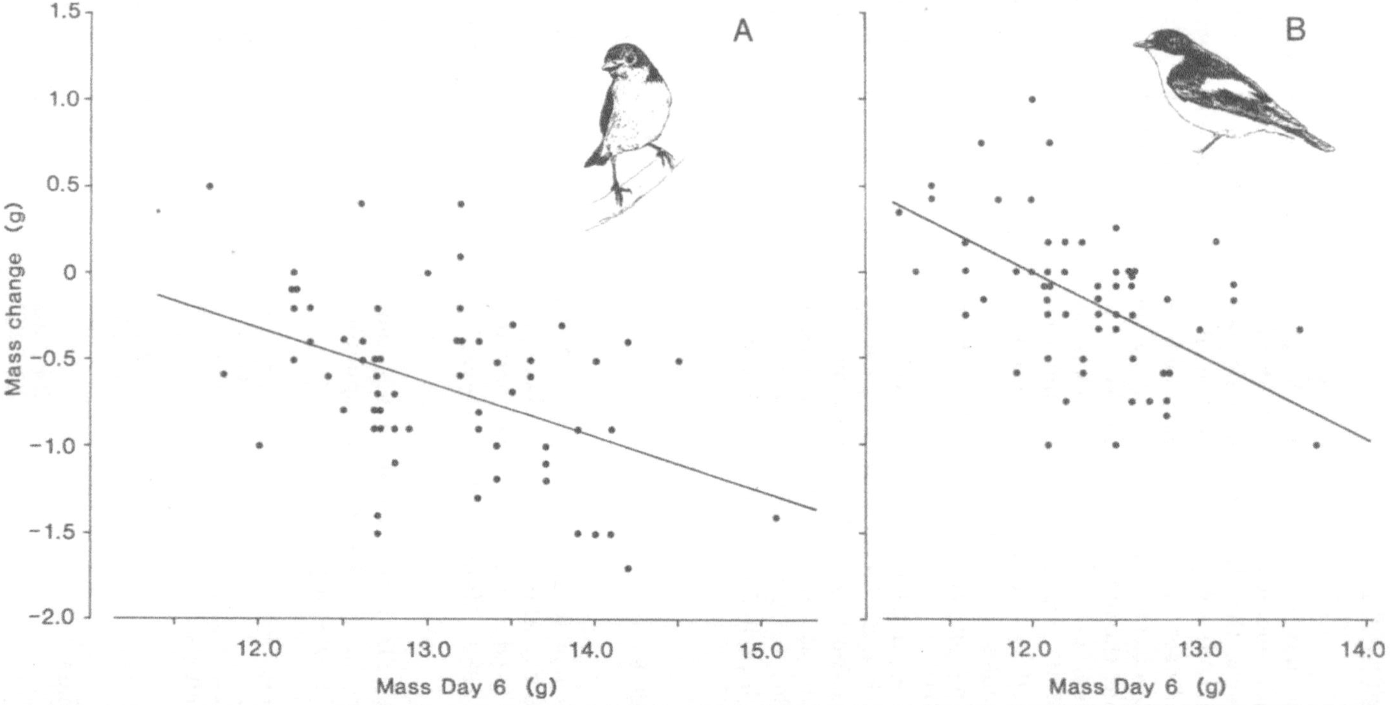

Fig. 2. Mass change of female (A) and male (B) Pied Flycatchers between nestling day 6 and day 13 in relation to their initial (day 6) weight. The regression equations are: females, $y = 3.33 - 0.31x$; males, $y = 5.76 - 0.48x$.

were similar in control and enlarged, but lower in reduced broods and were slightly larger in all three categories in the rainy summer of 1987 than in the sunny summer of 1988. Thus, in 1987 11% of the nestlings at Revinge died between day 6 and day 13, whereas the corresponding figure for 1988 was only 1%.

Contrary to earlier results of brood size manipulations in Great Tits (Lindén 1988, Smith et al. 1989), we found no effects of brood size on nestling wing length or tarsus length on day 13; the latter result differs from that of Alatalo and Lundberg (1989: Fig.2B), who reported shorter tarsi of young in enlarged Pied Flycatcher broods.

Nestling weights on day 13, however, decreased significantly with number of young in the nest after manipulation in both 1987 and 1988 ($p<0.02$ in both years; Fig.3); the decrease was significantly ($p<0.01$) steeper in 1987 than in 1988.

DISCUSSION

The relevant measure of reproductive costs, as indeed in all studies of adaptation, is lifetime reproductive success, which is dependent on both adult survival and offspring recruitment to future breeding populations. As in many other studies of reproductive costs (but see Nur 1984a, 1984b, Boyce and Perrins 1987, Gustafsson and Sutherland 1988), we lack data on these parameters and can only infer costs from parent and nestling weights and parental feeding frequencies of broods of different sizes (cf. e.g. Schifferli 1978, Hegner and Wingfield 1987, Alatalo and Lundberg 1989).

Parental feeding frequencies and weights

Even with a calculated linear increase in feeding frequency with increasing brood size (slope of regression = 0.45), each nestling would have received less food as brood size increased, as found in previous studies of both natural and manipulated broods (references in Nur 1984c, Grundel 1987), but the situation is somewhat complicated by the fact that the Pied Flycatcher is a multiple-prey loader. However, there are good reasons to believe that loads should decrease in size as feeding frequencies increase (von Haartman 1954, Lifjeld 1988), thus further enhancing this trend. Although more data are needed to test this, there was even an indication that parental feeding frequencies levelled off at the largest brood sizes, as one would expect if there is an energetic ceiling to parental work load (Drent and Daan 1980). This result is similar to that of Smith et al. (1988) but differs from that of Nur (1984c).

Most adult Pied Flycatchers lost mass between nestling days 6 and 13, and although mass loss was correlated with different variables in the two sexes, it seemed to be related to parental work load, as previously found in this species by Winkel and Winkel (1976) and Askenmo (1977); also see Lifjeld (1988) and,

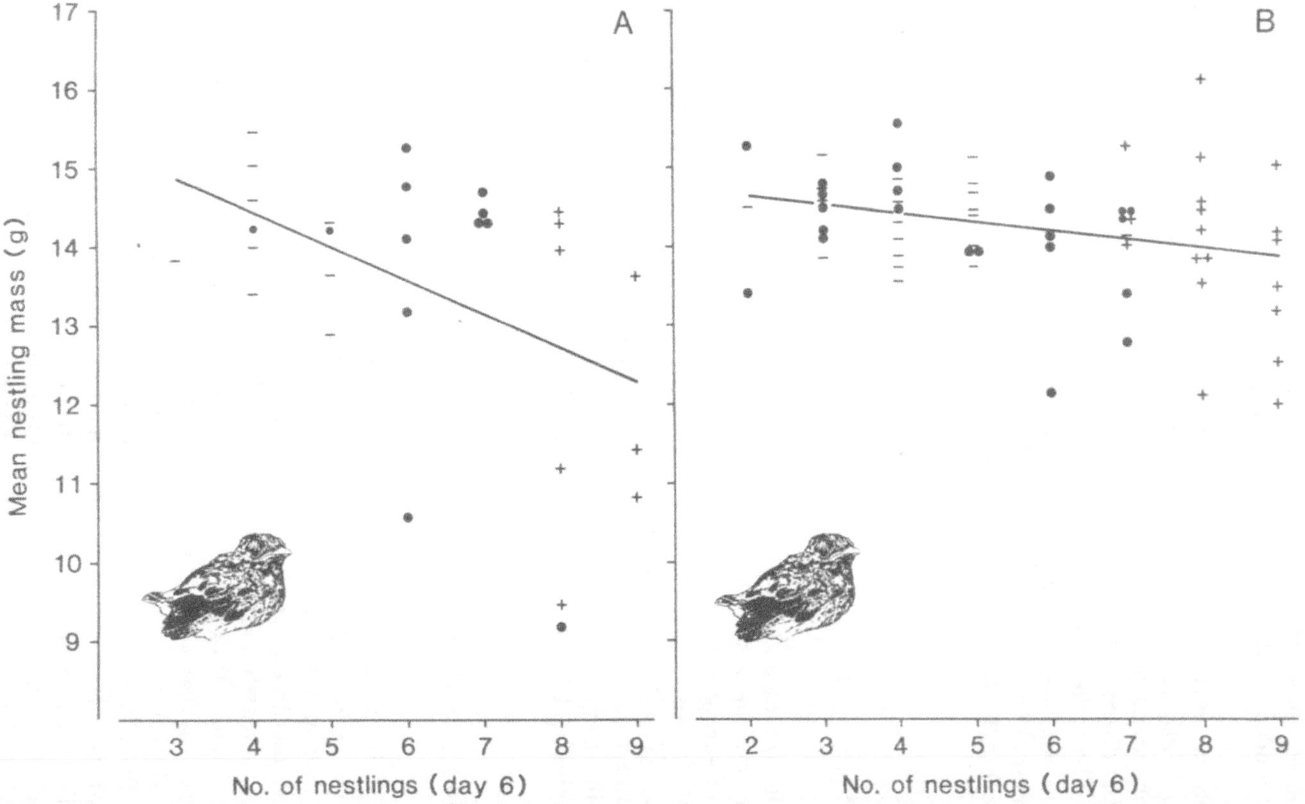

Fig. 3. Mean nestling weight on day 13 as a function of brood size after manipulation on day 6 in (A) 1987 and (B) 1988. The regression equations are: 1987, $y = 16.14 - 0.43\,x$, $P=0.010$; 1988, $y = 14.85 - 0.11\,x$, $P=0.015$. Symbols as in Figure 1.

for energy expenditure and brood size, Bryant and Tatner (1988). There was, however, also a negative correlation between the amount of mass lost and the initial (day 6) mass of the parent birds. Whether mass loss represents a cost or an adaptation (Norberg 1981, Freed 1981) is an open question. However, even if it were an adaptation, a cost must also be involved (cf. Nur 1984a), otherwise birds with a low work load would also lower their mass to save energy; both in this study and that of Smith et al. (1988) some birds with relaxed work loads increased in weight. Moreover, in this study, mass loss was documented during nestling feeding, not prior to it as in e.g. Hegner and Wingfield's (1987) study.

Males were on average 0.7 g lighter than females on day 6 and no male lost more than 1 g, whereas several females did so. In both sexes, the lowest weight recorded on day 13 was 11.0 g. This might indicate that there exists a lower mass level below which adult survival is jeopardized and which parents do not trespass even if nestlings remain unsatiated. In accordance with this, heavy Pied Flycatcher parents showed a tendency to feed hungry young more often than did light parents (Lifjeld 1988). Furthermore, it may be significant that the female with the lowest day 6 mass in our study, 10.8 g, was never seen again, the brood being raised by the male alone.

Effects on the nestlings

The effects of brood size on the nestlings differed between the two years, but in both more young fledged from enlarged than control broods. The 1987 breeding season was rainy, 1988 warm and sunny. This was reflected in the nestling weights: in 1988, the mean mass on day 13 was identical in control and enlarged broods; in 1987, the corresponding difference was 1.2 g. However, differences between categories were not statistically significant in either year. Nevertheless, linear regressions of nestling mass on number of young after manipulation were negative and statistically significant, the slope being less steep in 1988.

Large differences in fledgling production in relation to summer weather have previously been found in the Pied Flycatcher (e.g. Borgström 1983, Järvinen and Väisänen 1984, Virolainen 1984, Källander et al. 1987, Alatalo and Lundberg 1989). As there is little evidence that Pied Flycatchers adjust their clutch size to the breeding situation, e.g. habitat (Gezelius et al. 1984), with the exception for calendar date (von Haartman 1967), such brood-size related variations in productivity might act to maintain genetic variability for clutch size and, in particular, account for the fact that clutch sizes in the Pied Flycatcher seem to be smaller than the most productive one (Källander et al. 1987).

ACKNOWLEDGEMENTS

We acknowledge the financial support of this study by the Swedish National Research Board in Science and the Längman Cultural Foundation.

REFERENCES

Alatalo RV, Lundberg A (1989) Clutch size of the Pied Flycatcher *Ficedula hypoleuca* - an experiment. Ornis Fennica 66:15-23

Askenmo C (1977) Effects of addition and removal of nestlings on nestling weight, nestling survival, and female weight loss in the Pied Flycatcher *Ficedula hypoleuca* (Pallas). Ornis Scand 8: 1-8

Askenmo C (1979) Reproductive effort and return rate of male Pied Flycatchers. Am Nat 114: 748-752

Borgström E (1983) The effects of extreme weather conditions in summer 1981 on the breeding success of some nestbox-breeding passerines. Vår Fågelvärld 42: 153-160 (In Swedish with summary in English)

Boyce MS, Perrins CM (1987) Optimizing Great Tit clutch size in a fluctuating environment. Ecology 68: 142-153

Briskie JV, Sealy SP (1989) Determination of clutch size in the Least Flycatcher. Auk 106: 269-278

Bryant DM, Tatner P (1988) The costs of brood provisioning: Effects of brood size and food supply. Acta XIX Congr Int Ornithol: 364-379

Charnov EL, Krebs JR (1974) On clutch-size and fitness. Ibis 116: 217-219

Drent RH, Daan S (1980) The prudent parent: Energetic adjustments in avian breeding. Ardea 68: 225-252

Freed LA (1981) Loss of mass in breeding wrens: Stress or adaptation? Ecology 62: 1179-1186

Gezelius L, Grahn M, Källander H, Karlsson J (1984) Habitat-related differences in clutch size of the Pied Flycatcher *Ficedula hypoleuca* Ann Zool Fennici 21: 209-212

Gillespie JH (1977) Natural selection for variances in offspring numbers: a new evolutionary principle. Am Nat 111: 1010-1014

Grundel R (1987) Determinants of nestling feeding rates and parental investment in the Mountain Chickadee. Condor 89: 319-328

Gustafsson L, Sutherland WJ (1988) The costs of reproduction in the collared flycatcher *Ficedula albicollis*. Nature 335: 813-815

von Haartman L (1954) Der Trauerfliegenschnäpper. III. Die Nahrungsbiologie. Acta Zool Fenn 83: 1-96

von Haartman L (1967) Clutch-size in the Pied Flycatcher. Proc XIV Int Ornithol Congr: 155-164

Hegner RE, Wingfield JC (1987) Effects of brood-size manipulations on parental investment, breeding success, and reproductive endocrinology of House Sparrows. Auk 104: 470-480

Högstedt G (1980) Evolution of clutch size in birds: adaptive variation in relation to territory quality. Science 210: 1148-1150

Järvinen A, Väisänen RA (1984) Reproduction of Pied Flycatchers *Ficedula hypoleuca* in good and bad breeding seasons in a northern marginal area. Auk 101: 439-450

Källander H, Karlsson J, Rosenlund N, Svensson S (1987) Clutch-size and breeding success of the Pied Flycatcher *Ficedula hypoleuca* in two contrasting habitats in South Sweden. Acta Reg Soc Sci Litt Gothoburgensis. Zoologica 14: 76-83

Klomp H (1970) The determination of clutch-size in birds. A review. Ardea 58: 1-124

Lack D (1947) The significance of clutch-size. Ibis 89: 302-352

Lack D (1954) The Natural Regulation of Animal Numbers. Clarendon Press, Oxford.

Lifjeld JT (1988) Prey choice and nestling hunger: an experiment with pied flycatchers, *Ficedula hypoleuca*. Anim Behav 36: 134-139

Lifjeld JT, Slagsvold T (1989) Allocation of parental investment by polygynous Pied Flycatcher males. Ornis Fennica 66: 3-14

Lindén M (1988) Reproductive trade-off between first and second clutches in the great tit *Parus major*: an experimental study. Oikos 51: 285-290

Mertens JAL (1987) The influence of temperature on the energy reserves of female Great Tits during the breeding season. Ardea 75: 73-80

Murphy EC, Haukioja E (1986) Clutch size in nidicolous birds. In: Johnston RF (ed) Current Ornithology. Vol.4. Plenum Press, New York, p 141-180

Norberg RÅ (1981) Temporary weight decrease in breeding birds may result in more fledged young. Am Nat 118: 838-850

Nur N (1984a) The consequences of brood size for breeding Blue Tits I. Adult survival, weight change and the cost of reproduction. J Anim Ecol 53: 479-496

Nur N (1984b) The consequences of brood size for breeding Blue Tits II. Nestling weight, offspring survival and optimal brood size. J Anim Ecol 53: 497-517

Nur N (1984c) Feeding frequencies of nestling blue tits (*Parus caeruleus*): costs, benefits and a model of optimal feeding frequency. Oecologia (Berl.) 65: 125-137

Perrins CM, Moss D (1975) Reproductive rates in the great tit. J Anim Ecol 44: 695-706

Reznick D (1985) Costs of reproduction: an evaluation of the empirical evidence. Oikos 44: 257-267

Schifferli L (1978) Experimental modification of brood size among House Sparrows *Passer domesticus*. Ibis 120: 365-369

Smith HS, Källander H, Nilsson J-Å (1987) Effect of experimentally altered brood size on frequency and timing of second clutches in the Great Tit. Auk 104: 700-706

Smith HS, Källander H, Fontell K, Ljungström M (1988) Feeding frequency and parental division of labour in the double-brooded great tit *Parus major*. Behav Ecol Sociobiol 22: 447-453

Smith HS, Källander H, Nilsson J-Å (1989a) The trade-off between offspring number and quality in the great tit *Parus major*. J Anim Ecol 58: 383-401

Smith HS, Källander H, Nilsson J-Å (1989b) The significance of clutch overlap in great tits *Parus major*. Ibis 131: 589-600

Smith JNM (1981) Does high fecundity reduce survival in Song Sparrows? Evolution 35: 1142-1148

Sokal RR, Rohlf FJ (1981) Biometry. 2nd ed. Freeman, New York

Tinbergen JM (1987) Costs of reproduction in the Great Tit: Intraseasonal costs associated with brood size. Ardea 75: 111-122

Virolainen M (1984) Breeding biology of the Pied Flycatcher *Ficedula hypoleuca* in relation to population density. Ann Zool Fennici 21: 187-197

Williams GC (1966) Natural selection, the costs of reproduction, and a refinement of Lack's principle. Am Nat 100: 687-690

Winkel W, Winkel D (1976) Über die brutzeitliche Gewichtsentwicklung beim Trauerschnäpper (*Ficedula hypoleuca*). Ein Beitrag zur Frage "Belastung während der Fortpflanzungsperiode". J Ornithol 117: 419-437

FITNESS EFFECTS OF PARASITES ON PASSERINE BIRDS: A REVIEW

Anders P. MØLLER, Klas ALLANDER and Reija DUFVA

Department of Zoology, Uppsala University, Box 561, S-751 22 Uppsala, Sweden

Reproductive success has been the target of many field studies of birds, because it is supposed to relate closely to fitness. Nest-predation has been since a long time recognized as a major determinant of reproductive success in passerine birds (e.g. Nice 1957, Ricklefs 1969). However, until quite recently, infestation by parasites has generally been neglected as a selective agent affecting fitness components in birds. One major reason for this neglect may be the fact that a large fraction of generalizations on passerine reproduction originate from nest-box studies, where effects of ecto-parasites have been minimized efficiently by field workers removing old parasite-infested nests (Møller 1989). Some effects of parasites may also have gone undetected because nest losses due to parasitism were mis-classified as being due to nest-predation or starvation.

Many parasitologists have taken the view that parasites do not markedly reduce the fitness of hosts, because long-term coevolutionary interactions between parasites and hosts have reduced the virulence of parasites to a low level (e.g. Lincicome 1971; van Valen 1973; Davies et al. 1980; Wilson 1980; Alexander 1981). Here, we define parasites as species drawing their food from the living bodies of other species, most of the life of each individual parasite being spent in association with an individual host and some damage being done to the host (Price 1980). Extensive evidence for fitness loss among hosts due to parasitism exists and we will later quantify this fitness loss for passerine birds.

In this paper we start out by identifying the bird species most likely to suffer from parasitism. Secondly, we provide a review of the effects of parasites on the fitness of passerine birds. Thirdly, we provide a list of measures of defence taken by passerine birds against parasite infections, and, finally, we discuss the importance of parasitism as a selective agent in passerine birds and suggest that nest-site choice, life history traits and reaction norms of reproductive parameters have evolved in concert with parasite infestations.

WHICH BIRD SPECIES SUFFER PARTICULARLY FROM PARASITISM?

Parasites comprise a very large fraction of all living organisms and no other single method of resource acquisition is used by such a large number of

NATO ASI Series, Vol. G 24
Population Biology of Passerine Birds
Edited by J. Blondel et al.

species as parasitism (Price 1980). All other organisms are therefore probably affected by parasites in one way or another.

Two major factors promote the spread of parasites to hosts. First, most parasite species are severely restricted in their movements, and many rely on vectors for transportation or on "contact" transmission (Rothschild and Clay 1952; Baker 1967; Baker 1975; Marshall 1981). But even vectors have a restricted movement range. Close proximity of potential hosts therefore considerably increases the colonization rate of a large number of hosts, once the parasite has obtained access to a group of hosts. Frequent interactions between members of a group of hosts will soon spread parasites to all potential hosts. In fact Alexander (1974) suggested parasitism to be one of the major costs of social behaviour.

Second, since parasites generally have a limited ability to move from one site to another, re-use of a particular site by hosts will cause efficient spread of a parasite genotype to new generations of hosts (Rothschild and Clay 1952). Re-use of sites can be due to use of the same foraging habitats, roosts or breeding sites. Two major groups of birds are considered to be limited by the number of breeding sites, namely colonially nesting birds and hole-nesters (Lack 1968). The relatively high safety of breeding sites among hole-nesters and colonially-breeding birds has led to local population crowding of such birds and subsequently to intense intraspecific competition for sites. Frequently re-used breeding sites will also form the basis for a host population occurring predictably in time, and parasite infestations are therefore predicted to be particularly common among individual birds re-using nest-sites.

FITNESS EFFECTS OF PARASITES ON PASSERINE BIRDS

The study of parasitism as a determinant of reproductive success is in its infancy. Relatively few studies have been conducted, and even fewer have used an experimental approach allowing reproductive success of parasite- infested individuals to be compared with that of parasite-free individuals. In this section we present parasite effects on fitness components of their passerine hosts and list them in Table I.

Time of breeding

Only one study has addressed the question whether parasites retard the time of breeding of their hosts. Swallows (*Hirundo rustica*) build new nests and breed later, if old nests are infected with mites. Second clutches are also delayed if first clutch nests contain mites, because new nests have to be built for second clutches (Table I; Møller 1990). Collared Flycatchers (*Ficedula albicollis*) suffering from malaria take longer time to incubate their eggs (Widemo 1989).

Clutch-size

If parent birds can predict the degree of parasitism before the start of egg-laying, then they may be able to adjust clutch-size to this change in the nesting environment (Moss and Camin 1970). Female swallows reduce the size of second clutches if their first nests are infected with mites, but there is no indication of an adjustment of clutch-size to the extent of ecto-parasitism (Møller 1990).

Reproductive success

Most studies have shown a negative effect of parasites on reproductive success of hosts; in particular, nestling survival has been shown to be reduced because of the presence of parasites (Table I). In studies where the effects of parasitism have been directly quantified, the reduction in survival of host offspring range from 0 to 51% with a mean value of 25% (S.D. = 19, N = 10). Only one of the experimental studies did not find any effect of parasites on the reproductive success of the hosts.

Quality of offspring

The quality of host offspring in terms of body mass, body condition and body size was measured in most studies. The general conclusion is that body mass and body condition are reduced by parasitism, whereas feather and skeletal growth are not affected, although development of parasitized offspring is generally retarded. Body mass of the hosts' offspring is reduced by 5 to 15% in different studies (Table I).

Clutches per season

Only one study addressed the question of whether parasites reduced the frequency of second clutches laid by hosts (Møller 1990). An haematophagous mite reduced the frequency of second clutches in Swallow hosts by 16% relative to pairs having their nests sprayed with a pesticide. This effect is independent of laying date of first clutches.

Table I. Effects of parasites on fitness components of reproduction in passerine birds.

Host species	Parasite species	Study methods	Remarks	Reference
TIME OF BREEDING				
Swallow (*Hirundo rustica*)	Tropical fowl mite (*Ornithonyssus bursa*)	Experiment	Old nests re-used if not containing mites; if new nests are built, breeding is delayed	Møller (1990)
Collared Flycatcher (*Ficedula albicollis*)	Avian malaria (*Plasmodium* sp.)	Observation	Longer incubation periods	Widemo (1989)
CLUTCH-SIZE				
Purple Martin (*Progne subis*)	Martin mite (*Dermanyssus prognephilus*)	Experiment	No effect	Moss and Camin (1970)
Swallow (*Hirundo rustica*)	Tropical fowl mite (*Ornithonyssus bursa*)	Experiment	Reduction of second clutch by 8% when first nests are infected rather than sprayed	Møller (1990)
REPRODUCTIVE SUCCESS				
Great Tit (*Parus major*) Blue Tit (*Parus caeruleus*) Coal Tit (*Parus ater*)	Hen flea (*Ceratophyllus* sp.)	Observation	Heavy nestling mortality	Winkel (1975)
Tree Swallow (*Tachycineta bicolor*)	Blowfly (*Protocalliphora sialia*)	Experiment	No effect	Rogers et al. (1990)
Purple Martin (*Progne subis*)	Martin mite (*Dermanyssus prognephilus*)	Experiment	Reduction in fledging success by 14%	Moss and Camin (1970)

Host species	Parasite species	Study methods	Remarks	Reference
Swallow (*Hirundo rustica*)	Tropical fowl mite (*Ornithonyssus bursa*)	Experiment	Reduction in nesting success by 33% when nests are infected rather than sprayed	Møller (1990)
	Blowfly (*Protocalliphora hirundo*)	Observation	Reduction in nesting success by 17% in first and 57% in second clutches	Shields and Crook (1987)
Cliff Swallow (*Hirundo pyrrhonota*)	Swallow bug (*Oeciacus vicarius*) Swallow tick (*Argas cooleyi*) Swallow tick (*Ornithodoros concanensis*)	Experiment	Reduction in fledging success by 40% in infected nests	Chapman (1973)
	Swallow bug (*Oeciacus vicarius*) Swallow flea (*Ceratophyllus celsus*)	Experiment	Reduction in nesting success by up to 50% in infected nests	Brown and Brown (1986)
South African Cliff Swallow (*Hirundo spilodera*)	Tropical fowl mite (*Ornithonyssus bursa*)	Observation	Reduction in fledging success by 34%	Burgerjon (1964)
Pearly-eyed Thrasher (*Margarops fuscatus*)	Tropical fly (*Philornis deceptivus*)	Observation	Reduction in fledging success	Arendt (1985)
American Dipper (*Cinclus mexicanus*)	Blowfly (*Protocalliphora* sp.) Fowl mite (*Ornithonyssus sylviarum*)	Observation	Nestling mortality observed	Halstead (1988)

Table I (continued)

Host species	Parasite species	Study methods	Remarks	Reference
Hooded Crow (*Corvus cornix*)	Nematode (*Syngamus tracheae*)	Observation	Reduction in fledging success by 51% in infected nests	Loman (1980)
Starling (*Sturnus vulgaris*)	Red fowl mite (*Dermanyssus gallinae*)	Experiment	Reduction in nestling survival by 15%	Royall (1966)
	Tropical fowl mite (*Ornithonyssus bursa*)	Observation	No effect	Powlesland (1977)
Wagler's Oropendola (*Zarhynchus wagleri*) Yellow-rumped Cacique (*Cacicus cela*)	Botfly (*Philornis* sp.)	Observation	Reduced nesting success	Smith (1968)
QUALITY OF OFFSPRING				
Great Tit (*Parus major*)	Hen flea (*Ceratophyllus gallinae*)	Experiment	Reduced body mass	K. Allander and R. Dufva (pers.comm.)
Tree Swallow (*Tachycineta bicolor*)	Blowfly (*Protocalliphora sialia*)	Experiment	No effects	Rogers et al. (1990)
Purple Martin (*Progne subis*)	Martin mite (*Dermanyssus prognephilus*)	Experiment	Reduction in body mass of nestlings by 7%	Moss and Camin (1970)
Swallow (*Hirundo rustica*)	Tropical fowl mite (*Ornithonyssus bursa*)	Experiment	Reduction in body mass by 5% in first and by 6% in second broods	Møller (1990)

Host species	Parasite species	Study methods	Remarks	Reference
	Blowfly (*Protocalliphora hirundo*)	Observation	Reduction in growth rate	Shields and Crook (1987)
Cliff Swallow (*Hirundo pyrrhonota*)	Swallow bug (*Oeciacus vicarius*) Swallow tick (*Argas cooleyi*) Swallow tick (*Ornithodoros concanensis*)	Experiment	Reduction in body mass by 12%	Chapman (1973)
	Swallow bug (*Oeciacus vacarius*) Swallow flea (*Ceratophyllus celsus*)	Experiment	Reduction in body mass by 15%	Brown and Brown (1986)
Pearly-eyed Thrasher (*Margarops fuscatus*)	Tropical fly (*Philornis deceptivus*)	Observation	Reduction in body mass and retarded growth	Arendt (1985)
Beechey Jay (*Cyanocorax beecheii*)	Botfly (*Muscoidea*)	Observation	Retarded feather development	Winterstein and Raitt (1983)
Starling (*Sturnus vulgaris*)	Tropical fowl mite (*Ornithonyssus bursa*)	Observation	Reduced body mass	Powlesland (1977)
CLUTCHES PER SEASON				
Swallow (*Hirundo rustica*)	Tropical fowl mite (*Ornithonyssus bursa*)	Experiment	Reduction by 16% in frequency of second clutches	Møller (1990)

DISCUSSION

There are considerable fitness effects of parasites on reproductive parameters in a number of passerine bird species (Table I). Fitness of parasitized individuals is decreased because egg-laying is delayed, clutch-size is reduced, fewer offspring fledged, offspring are in a poor quality in terms of body mass and body condition and fewer clutches are initiated per season. Future survival prospects of nestlings are likely to be directly affected by their body mass and body condition (e.g. Perrins 1965). Survival of parasitized offspring is reduced after fledging (Arendt 1985; Gold and Dahlsten 1983; Møller 1990), partly because parasites often cause premature fledging of offspring in a poor body condition (Møller 1990).

The sample of species studied is very biased, since hirundines form a large proportion of the data sets. However, one can speculate about the relative importance of parasitism as a selective agent during reproduction in passerine birds. Mortality rates due to parasites are only slightly smaller than mortality rates due to nest predation in birds in general (e.g. Nice 1957; Ricklefs 1969). Since nest predation rates generally are small for hirundines and other species nesting in protected sites, parasitism accounts for a major fraction of the overall mortality (e.g. Shields and Crook 1987; Møller 1990). It is possible that overall nestling mortality is similar in species nesting in protected sites such as holes and/or colonies and in species nesting in less protected sites (open-nesting, solitarily-breeding species). However, different selection pressures may balance and account for the overall mortality; whereas nest predation is more important among open-nesting, solitary species, the effects of parasitism may be much more pronounced among hole nesters and colonially-breeding species (Nilsson 1986). In other words, the cost of parasitism is what hole nesters and colonial breeders pay for the safety from nest predators.

Since parasites frequently reduce the reproductive success of their hosts, anti-parasite behaviour should be well developed. Ectoparasite numbers could be reduced by the direct removal of parasites from the plumage and the nest. Particularly, several types of nest sanitation may have developed because of the associated reduction of parasite loads. Removal of fecal sacs, ejection of faeces over the rim of the nest and frequent renovation of the nest-lining material belong to this category (Bucher 1988). Choice of nest-lining material may also reduce parasite loads, since several plant species have repellent or pesticide properties (Winberger 1984, Clark and Mason 1985). Nesting associations with wasps and bees are frequent among tropical bird species and one major reason for such associations may be that hymenoptera greatly decrease the risk of being attacked by parasitic diptera (Smith 1968). Personal hygiene by parent birds may also reduce their own ectoparasite loads and, therefore, the subsequent loads of the nest and the offspring. Preening and allopreening are supposed to reduce ectoparasite loads of birds (e.g. Brooke 1985), and anting, whereby the plumage is treated with formic acid, may similarly reduce parasite infestations (Simmons 1966).

Parasites pose a strong selection pressure on their passerine hosts by reducing their reproductive success and survival prospects and selection for anatomical, physiological or behavioural means of resistance among hosts should lead to less severe effects of parasites (Wakelin and Blackwell 1988). Coevolution of parasites and their hosts should continuously generate new additive genetic variation in resistance to parasites and in the ability of parasites to exploit their hosts (Clark 1979). However, parasites should have a lead in this arms race because of their generally short life cycles relative to those of their hosts. Parasites and their hosts may therefore enter coevolutionary relationships and changes in host resistance to parasites are followed by similar changes in infectivity and virulence of parasites (Futuyma and Slatkin 1983; Wakelin and Blackwell 1988). Since parasites spread more efficiently among host individuals living in groups or re-using nest-sites, specialization of parasites to their hosts should occur particularly quickly during evolutionary time among such host species.

One way to partially circumvent any detrimental fitness effects of parasites is phenotypic plasticity among hosts in their response. Reaction norms describe all phenotypes developed by a single genotype exposed to a range of environments, and selection may cause changes in reaction norms. If traits are closely related to fitness such changes are particularly likely (Dobzhansky 1951). Reaction norms can be considered as an integral part of the life history of an organism (Caswell 1983) and they can be optimized in the same ways as other life history traits (Via and Lande 1985). A commonly encountered environmental gradient experienced by animals is infection rates with parasites. Individual birds should respond to variation in parasite infection by developing reaction norms that reduce the detrimental effects of parasites. Some of the studies, where parasite infestation levels have been manipulated experimentally, provide evidence for the presence of reaction norms for reproductive parameters to variable levels of parasitism (Moss and Camin 1970; Chapman 1973; Brown and Brown 1986; Møller 1990).

Nest-box studies of birds have been very successful because a large number of nests can easily be studied each year, and reproductive events of particular individuals can be followed throughout their lives. However, most nest-box studies have applied the practice of removing old nests in order to reduce infestations by ectoparasites (Møller 1989). This can have profound effects on the number and the quality of offspring produced and density-dependent effects of parasitism are simultaneously likely to be minimized (Møller 1989). Furthermore, nest-predation rates may sometimes be increased by the effects of parasites, if parasitism causes more frequent visits by parent birds to their nests, and if parasites cause nestlings to beg more and stay longer in the nest than nestlings not suffering from parasitism (Møller 1989). Most nest-box studies of birds have actually only provided data on reproduction in relatively parasite-free conditions, whereas hole-nesters are supposed to have evolved their reproductive strategies in heavily parasitized environments. One of the biologically most revealing part of the reproductive strategies of common hole-nesting bird species thus still awaits study.

The study of parasite effects on fitness components of birds is still in its infancy. We suggest that any progress in the study of passerine reproduction and its relation to parasitism relies on adoption of the following research programme: (1) the effects of parasites on reproduction and interactions between parasitism and nest-site choice, reproductive success and nestling growth should be examined in experimental studies, (2) the interaction between parasitism and nest-predation should be addressed, (3) reaction norms of life history traits of passerine birds to parasite infections should be studied by recording reproductive parameters at natural, but experimentally manipulated levels of parasitism, (4) effects of parasites on fitness components of hosts among frequently and less frequently infected species should be studied in experiments, and (5) defence mechanisms applied by hosts against their parasites should be studied both in frequently infected species (i.e. colonially-breeding and hole-nesting species) as well as in less frequently infected species of birds (i.e. solitarily breeding and open-nesting species).

ACKNOWLEDGEMENTS

The study was supported by a grant from the Swedish Natural Science Research Council.

REFERENCES

Alexander M (1981) Why microbial predators and parasites do not eliminate their prey and host. Ann Rev Microbiol 35: 113-133

Alexander RD (1974) The evolution of social behavior. Ann Rev Ecol Syst 5: 324-383

Arendt WJ (1985) *Philornis* ectoparasitism of pearly-eyed thrashers. I. Impact on growth and development of nestlings. The Auk 102: 270-280

Baker JR (1967) A view of the role played by the *Hippoboscidae* (*Diptera*) as vectors of endoparasites. J Parasitol 53: 412-418

Baker JR (1975) Epizootiology of some haematozoic protozoa of English birds. J Nat Hist 9: 601-609

Brooke M de L (1985) The effect of allopreening on tick burdens on moulting eudyptid penguins. The Auk 102: 893-895

Brown CR and Brown MB (1986) Ectoparasitism as a cost of coloniality in cliff swallows (*Hirundo pyrrhonota*). Ecology 67: 1206-1218

Bucher EH (1988) Do birds use biological control against nest parasites? Parasitol Today 4: 1-3

Burgerjon JJ (1964) Some census notes on a colony of South African cliff swallows *Petrochelidon spilodera* (Sundevall). Ostrich 35: 77-85

Caswell H 1983 Phenotypic plasticity in life history traits: demographic effects and evolutionary consequences. Amer Zool 23: 35-46

Chapman BR (1973) The effects of nest ectoparasitism on cliff swallow populations. Unpublished PhD thesis, Texas Tech University, Austin, Texas

Clark L and Mason JR (1985) Use of nest material as insecticidal and anti-pathogenic agents by the European starling. Oecologia 67: 169-176

Clarke BC (1979) The evolution of genetic diversity. Proc R Soc Lond B 205: 453-474

Davies JS, Hall GA, Target T and Murray M (1980) The biological significance of the immune response with special reference to parasites and cancer. J Parasitol 66: 705-721

Dobzhansky T (1951) Genetics and the origin of species. Columbia University Press, New York

Futuyma DJ and Slatkin M (eds)(1983) Coevolution. Sinauer, Sunderland

Halstead AJ (1988) American dipper nestlings parasitized by blowfly larvae and the northern fowl mite. Wilson Bull 100: 507-508

Hamilton WD and Zuk M (1982) Heritable true fitness and bright birds: a role for parasites? Science 218: 384-387

Lack D (1968) Ecological adaptations for breeding in birds. Chapman and Hall, London

Lincicome DR (1971) The goodness of parasitism: a new hypothesis. In: Cheng T (ed). Aspects of the biology of symbiosis, University Park Press, Baltimore, p 139-228

Loman J (1980) Reproduction in a population of the hooded crow *Corvus cornix*. Holarct Ecol 3: 26-35

Marshall AG (1981) The ecology of ectoparasitic insects. Academic Press, London, UK

Møller AP (1989) Parasites, predators and nest boxes: Facts and artefacts in nest box studies of birds. Oikos 56: 421-423

Møller AP (1990) Effects of parasitism by the haematophagous mite *Ornithonyssus bursa* on reproduction in the barn swallow *Hirundo rustica*. Ecology (in press)

Moss WW and Camin JH (1970) Nest parasitism, productivity, and clutch size in purple martins. Science 168: 1000-1003

Nice MM (1957) Nesting success in altricial birds. The Auk 74: 305-321

Nilsson SG (1986) Evolution of hole-nesting in birds: on balancing selection pressures. The Auk 103: 432-435

Perrins CM (1965) Population fluctuations and clutch size in the great tit, *Parus major* L.. J Anim Ecol 34: 601-647

Powlesland RG (1977) Effects of the haematophagous mite *Ornithonyssus bursa* on nestling starlings in New Zealand. N Z J Zool 4: 85-94

Price PW (1980) Evolutionary biology of parasites. Princeton University Press, Princeton

Ricklefs RE (1969) An analysis of nesting mortality in birds. Smithson Contrib Zool, 9: 1-48

Rogers CA, Robertson RJ and Stutchbury BJ (1990) Patterns and effects of parasitism by *Protocalliphora sialia* on Tree Swallow (*Tachycineta bicolor*) nestlings. In: Loye JE and Zuk M (eds). Ecology, behavior and evolution of bird-parasite interactions. Oxford University Press, Oxford, in press

Rothschild M and Clay T (1952) Fleas, flukes and cuckoos. A study of bird parasites. Collins, London, UK

Royall WC (1966) Breeding of the starling in central Arizona. The Condor 68: 196-205

Shields WM and Crook JR (1987) Barn swallow coloniality: a net cost for group breeding in the Adirondacks? Ecology 68:1373-1386
Simmons KEL (1966) Anting and the problem of self-stimulation. J Zool 149: 145-162
Smith NG (1968) The advantage of being parasitized. Nature 219: 690-694
van Valen L (1973) A new evolutionary law. Evol Theory 1: 1-30
Via S and Lande R (1985) Genotype-environment interaction and the evolution of phenotypic plasticity. Evolution 39: 505-522
Wakelin D and Blackwell JM (eds)(1988) Genetics of resistance to bacterial and parasitic infection. Taylor and Francis, London
Widemo F (1989) Effect of blood parasites on the Collared Flycatcher *Ficedula albicollis* . Honours Thesis, Dept of Zoology, Uppsala University
Wilson DS (1980) The natural selection of populations and communities. Benjamin-Cummings, Menlo Park, CA
Wimberger PH (1984) The use of green plant material in bird nests to avoid ectoparasites. The Auk 101: 615-618
Winkel W (1975) Vergleichend-brutbiologische Untersuchungen an fünf Meisen-Arten (*Parus* spp.) in einem niedersächsischen Aufforstungsgebiet mit Japanischer Lärche *Larix leptolepis* . Die Vogelwelt 96: 41-63, 104-114
Winterstein SR and Raitt RJ (1983) Nestling growth and development and the breeding ecology of the Beechey Jay. Wilson Bull 95: 256-268

THE COST OF REPRODUCTION IN BIRDS: EVALUATING THE EVIDENCE FROM MANIPULATIVE AND NON-MANIPULATIVE STUDIES

Nadav NUR

Point Reyes Bird Observatory, 4990 Shoreline Highway, Stinson Beach, CA 94970, USA

The cost of reproduction plays a pivotal role in theories of life history strategies (Cody 1966, Williams 1966, Charnov and Krebs 1974, Schaffer 1974, Stearns 1976, Ricklefs 1983). Four years ago, at the Conference of Causal and Evolutionary Aspects of the Determination of Bird Numbers, held in Wageningen, the Netherlands, I reviewed the evidence for the cost of reproduction in birds (Nur 1988a). Since then, empirical studies of the cost of reproduction have proliferated and debate concerning the significance of the cost of reproduction for the evolution of avian life history strategies has continued. For this reason, I think it worthwhile to review the current state of the empirical literature, emphasizing difficulties in evaluating the evidence for, or against, the cost of reproduction. In addition, I reanalyze data presented in Nur (1988b) concerning the cost of reproduction in the Blue Tit (*Parus caeruleus*).

There are three levels of questions one might ask regarding the cost of reproduction:

1. Is there a cost of nontrivial magnitude?
2. How is the cost expressed? For example, does the cost vary among years, does it differ between the sexes? Also, is the cost expressed as a reduction in subsequent survival or future fecundity, or both? Is it expressed within a breeding season or between breeding seasons?
3. What are the quantitative characteristics of the cost of reproduction? For example, what is the magnitude of the cost? Does an increase in reproductive effort result in an accelerating, decelerating or linearly increasing cost?

The life history theoretician desires answers to all three levels of questions (Stearns 1989). This review emphasizes the first level, regarding the presence/absence of the cost of reproduction, since here the debate has been the most heated. As in Nur (1988a) the review is concerned with costs bearing on Darwinian fitness, that is, costs expressed (ultimately) as a reduction in subsequent survival or future fecundity (i.e. reproductive success at subsequent breeding opportunities). Linden and Møller (1989) and Partridge (1989) have also recently reviewed the avian literature.

Reznick (1985) identifies several methods to identify costs of reproduction. Only two of these have been applied to birds, in the wild: 1) phenotypic correlational studies (i.e. naturalistic, nonmanipulative studies which correlate measures of present fecundity with survival and future fecundity), and 2) experimental phenotypic studies (in which an aspect of fecundity, such as brood

NATO ASI Series, Vol. G 24
Population Biology of Passerine Birds
Edited by J. Blondel et al.

size, has been experimentally manipulated). The primary disadvantage of the correlational approach is that if both clutch size and adult survival, say, are greater under good environmental conditions (or are greater if individual condition is good), this may lead to a spurious correlation between clutch size and survival or between present and future fecundity (Reznick 1985, Partridge and Harvey 1985, van Noordwijk and de Jong 1986, Reznick et al. 1986, Nur 1988a). Random assignment of individuals to experimental treatment (manipulation of clutch or brood size) can avoid the problem of confounding factors. An additional advantage of the experimental method is that it has greater statistical power to detect a cost of reproduction, compared to the nonmanipulative approach (all else being equal), because it is easy to create broods or clutches that are notably small or notably large, whereas such broods or clutches are relatively uncommon in nature. The virtues of the experimental approach are generally well-appreciated, but there are disadvantages to this approach as well (see below).

EXPERIMENTAL STUDIES: THE BLUE TIT

I begin my review of the empirical literature by summarizing my own experimental study, pointing out its deficiencies, and reanalyzing some of the key results. Nur (1984a, 1988b) reported that Blue Tit (*Parus caeruleus*) female recapture rate (proportion recaptured or resighted at least 3 months after fledging of the experimental brood) decreased with increasing experimental brood size in two out of three years. In the third year, the trend (non-significant) was for increasing recapture rate with increasing brood size. The overall relationship of brood size and recapture rate, averaged over the three years, was significant (linear regression analysis, weighted by sample size, $P<0.01$).

The analysis has been criticized on the grounds that the method of weighting was inappropriate and that logistic regression analysis is preferable (Cox 1970; reviewed in Kotz and Johnson 1985: 128-133). I therefore reanalyzed the data using two methods of logistic regression: 1) the minimum logit X^2 method of Berkson (Kotz and Johnson 1985: 518-522), also termed empirical logistic regression (Cox 1970), which is a form of weighted least squares regression (implemented with STATA, Computing Resource Center) and 2) an iterative maximum likelihood method (EGRET, Statistics and Epidemiology Research Corporation). Both methods can be implemented using the CATMOD procedure of SAS (Statistical Analysis System, 1985). They both yield approximate (large-sample) P-values only; asymptotically (with large enough sample sizes) the two methods yield the same results. Analysis using the minimum logit X^2 method (method 1) indicated a significant effect of brood size on female recapture rate, $P=.004$ (Table I); the maximum likelihood method (method 2) yielded a marginally significant effect of brood size, $P=.066$ (brood size log-transformed in the analyses). The estimate of the effect of log-brood size on female recapture rate was the same, however, using the two logistic regression methods, beta$=-0.53$ (Table I).

Table I. Logistic regression analyses of female and male Blue Tit survival in relation to brood size (log-transformed). Data from Nur (1988b).

	Method 1 Minimum Logit X^2	Method 2 Maximum Likelihood
A. Female survival (observed ≥3 mos post-fledging)		
Regression[a] Coefficient ± s.e.	-.526±.065	-.528±.286
Statistic[b]	F(1,3)=64.8	LRS(1)=3.38
P-value	P=.004	P=.066
B. Female survival (observed after 1 year manipulation)		
Regression[a] Coefficient ± s.e.	-.641±.144	-.647±.324
Statistic[b]	F(1,3)=19.8	LRS(1)=3.95
P-value	P=.021	P=.047
C. Male survival (observed ≥3 mos post-fledging)		
Regression[a] Coefficient ± s.e.	-.819±.514	-.970±.401
Statistic[b]	F(1,3)=2.55	LRS(1)=6.67
P-value	P=.209	P=.010

[a] Brood size log-transformed.
[b] F statistic for Method 1; Likelihood Ratio statistic for Method 2. Degrees of freedom in parentheses.

In the above analyses, females observed in the autumn or winter following the manipulation were counted as "survived" even though some may have perished before the next breeding season. If the criterion of survival is taken to be recapture or resighting in the next breeding season and the data reanalyzed, the effect of brood size on recapture rate is significant (P=.021, P=.047; Table I) using logistic regression methods 1 and 2, respectively. Thus using the more stringent criterion of survival, the results of the two methods converged, and both indicated a significant effect of experimental brood size on probability of female recapture.

Nur (1988b) reported that male recapture rate varied significantly with experimental brood size, and noted that males rearing broods of 9-15 were most likely to be recaptured. Reanalysis of the data using logistic regression yielded different conclusions depending on the method used: the minimum logit X^2 method indicated a non-significant trend of increasing male recapture probability with increasing brood size (P=.209; Table I) while the maximum likelihood method indicated that the trend is significant (P=.010). These results provide partial support for the notion of a benefit of reproduction. However, there are three potential biases which make it difficult to infer male survival on the basis of recapture data. First, some males were manipulated but never identified and thus their subsequent survival could not be assessed. At 22% of the nests, no male was identified. Moreover, males given broods of 3 and 6 were least likely, and males given broods of 9 were most likely to be captured and identified; overall, there were significant differences in capture probabilities with respect to the experimental brood size (Likelihood Ratio Statistic=9.58, df=4, P=0.048, maximum likelihood method of logistic regression). Note that females were caught and identified before the manipulation of brood size (and all females, save one, were caught) but males were only caught 8 to 12 days after the manipulation. The second potential bias concerns the probability that a male, alive the next year, was caught and identified. If the experimental manipulation influenced the probability a male was caught in the same year (as shown above), it is also possible that it influenced the probability he was caught in the next year. The second source of bias was of less concern for females, because nearly all breeding females were caught and identified, unlike males. The third bias is that dispersal of adults may have been influenced by the experimental manipulation (Nur and Clobert 1988, Breitwisch 1989). On the basis of known between-year dispersal distances within the study area, Nur (1988b) concluded that this bias was of minimal importance for females, but there were too few data regarding males.

Nur (1988b) also reported a significant future fecundity effect: the number of surviving offspring (i.e. observed at least 3 months after fledging) produced in the year following the experimental manipulation decreased as brood size increased. Proximally, the future fecundity effect was manifest as a reduction in the probability offspring (from next year's clutch) survived (shown in Figure 1). Clutch size, brood size (number of young hatched) and hatching date did not correlate significantly with experimental brood size in the previous year (Table 2 in Nur 1988b). Differential dispersal cannot be excluded as the explanation for differences in recapture probabilities among offspring. The future fecundity effect was similar among years and between the sexes (Nur 1988b). To summarize, two different types of costs are demonstrated in the Wytham Blue Tit population: a survival cost, evident only in females, and which varies from year to year, and a future fecundity cost, expressed in both sexes and which was consistent from year to year.

OTHER EXPERIMENTAL STUDIES

Table II summarizes the results of 18 experimental studies; all the published studies I am aware of, and an additional study in press. Each study

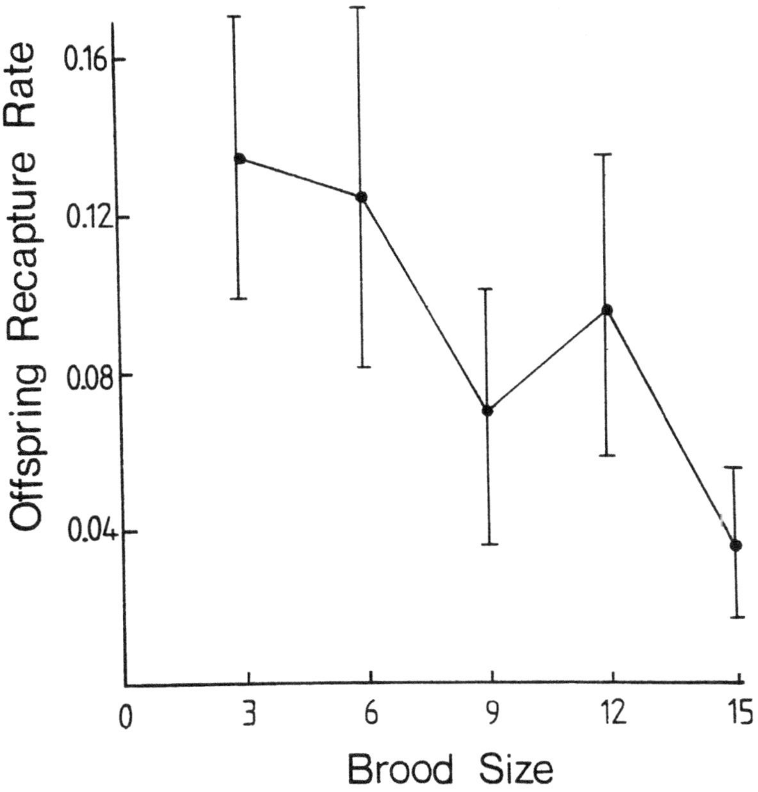

Fig. 1. Proportion of Blue Tit offspring surviving at least 3 months post-fledging in relation to parental experimental brood size the year before. Analysis of data reported in Nur (1988b).

manipulated brood size and examined one or more costs of reproduction. In Table II each experimental study is represented by one citation only. Thus Pettifor et al. (1988) is cited but not Boyce and Perrins (1987); Orell (in press) but not Orell and Koivula (1988); Nur (1988b) but not Nur (1984a).

None of the studies reported a positive impact of brood enlargement on subsequent survival and/or fecundity (but see reanalysis of Nur 1988b, above). Four studies found no compelling evidence of a cost of reproduction. Two of the four studies (De Steven 1980, Alatalo and Lundberg 1989) did not examine future fecundity; the other two examined fecundity in the year following manipulation, but only in terms of eggs laid or number of fledglings produced, not the number of surviving offspring or number of recruits produced. None of the four studies examined possible intra-year future fecundity effects. In two of the four studies (De Steven 1980, Korpimaki 1988), sample size was small and the manipulation was conducted in only one or two years. In Alatalo and Lundberg's (1989) study, few female Pied Flycatchers (*Ficedula hypoleuca*)

Table II. Summary of results from experimental manipulations of brood size in birds (effect of brood size manipulation on survival and future fecundity).

POSITIVE	NO EFFECT	NEGATIVE IMPACT ON:		
		future fecundity intra-year only	**future fecundity inter-year only**	**survival**
None	deSteven 1980* Tree Swallow	Slagsvold 1984 Great Tit	Røskaft 1985 Rook	Askenmo 1979 Pied Flycatcher
	Pettifor et al. 1988 Great Tit	Tinbergen 1987 Great Tit	Gustafsson & Sutherland 1988 Collared Flycatcher	Nur 1988b° Blue Tit
	Korpimäki 1988 Tengmalm's Owl	Smith et al. 1987 Great Tit	Orell in press Willow Tit	Reid 1987 Glaucous-winged gull
	Alatalo & Lundberg 1989* Pied Flycatcher	Lindén 1988 Great Tit		Dijkstra et al. 1990 Kestrel
		Hegner & Wingfield 1987 House Sparrow		
		(Finke et al. 1987)[i] (House Wren)	(Lessells 1986)[i] (Canada Goose)	
Number of Studies				
0	4	5 + (1)	3 + (1)	4

* No significant effect on adult survival. No analysis of future fecundity effect.
° Future fecundity as well.

()[i] Indirect effect only. See text.

returned (only 14 out of 193). This presents two problems: sample size was small and return rate greatly underestimates survival rate. Even among male Pied Flycatchers, return rate (17.9%) was much lower than the true survival rate. Only one of the four studies, that of Pettifor et al. (1988), equivalently Boyce and Perrins (1987), was conducted over several years, and obtained large sample sizes. Note that all four studies found evidence, albeit weak or inconsistent, of a survival cost, e.g. Korpimaki (1988) and Alatalo and Lundberg (1989) found that parents rearing enlarged broods were less likely to return than those rearing control broods (differences not significant) and De Steven (1980) found that females rearing reduced broods were more likely to return than those rearing control broods (not significant). The Wytham Great Tit study (Boyce and Perrins 1987, Pettifor et al. 1988,) yielded evidence for a survival cost, but it was inconsistent: females whose broods were enlarged by 4 nestlings survived significantly more poorly than those whose broods were diminished by 4 nestlings; there were no differences in recapture rate comparing the +3 and -3 treatments; no control broods were part of the experimental design (see Nur 1988a for discussion).

Five studies found evidence of a future fecundity cost within the same breeding season (Table II). In each case, the success (or presence) of second (or replacement) broods was adversely affected by parents' rearing a large first brood (Slagsvold 1984, Tinbergen 1987, Smith et al. 1987, Hegner and Wingfield 1987, Linden 1988). A sixth study Finke et al. (1987) demonstrated that a larger first brood delayed the start of the second brood. Because offspring survival is thought to decline strongly as the breeding season progresses (Daan et al. 1988, Pettifor et al. 1988), this suggests a future fecundity cost. Return rates of offspring were too low to analyze this cost directly. Because Finke et al. (1987) provides only indirect support of a cost, I have listed the study in parentheses. Only three of the six studies reported on parental survival in relation to brood manipulation: for example, in Smith et al.'s (1987) study on Great Tits, experimental brood size significantly affected female recapture rate (Smith 1988: 10-11). In Hegner and Wingfield's (1987) study, apparent survival rates were 0.58 (n=12), 0.44 (n=16), and 0.30 (n=10) for parents rearing reduced, control and enlarged broods, respectively. It would be hard to find better evidence of a survival cost, yet Hegner and Wingfield concluded, with reference to these data, that "adults that reared a greater number of young in 1984 were as likely to be seen again in 1985 as were those that reared fewer young" (p. 476). The difficulty is that small sample size did not allow Hegner and Wingfield to reject the null hypothesis, yet this is hardly reason to interpret their data as supporting the null hypothesis. Tinbergen (1987) found that recapture rate was not correlated with brood size.

Four studies found evidence of a future fecundity cost expressed between years (Table II: Røskaft 1985, Gustafsson and Sutherland 1988, Nur 1988b, Orell in press). A fifth study (Lessells 1986) demonstrated that Canada geese rearing larger broods bred later in the following year. If late breeding is associated with reduced offspring survival (see above) this implies a future fecundity cost. Again, demonstration of an indirect effect only is indicated in Table II by enclosing the citation in parentheses. Lessells (1986) also found that rearing a larger brood delayed molt, suggesting a second reproductive cost.

Conventional wisdom has been that future fecundity effects are unimportant, a point made by Murphy and Haukioja (1986: 162): "In general, it seems unlikely that reproductive attempt(s) in one breeding season could be affected by the reproductive output in the previous breeding season, particularly for small birds that have major daily cycles of accumulation and depletion of reserves...." The empirical record demonstrates that this widely-held view should be reconsidered.

In four studies, return rate was influenced by the experimental brood size (Table II). Only two of the studies (Reid 1987, Nur 1988b) also tested for a future fecundity effect. In Reid (1987) and Dijkstra et al. (1990) both sexes demonstrated similar survival cost; Askenmo (1979) could examine male return data only; as discussed above, Nur (1988b) found evidence of a survival cost in females but not males. In addition there is the study of Smith et al. (1987) discussed above, in which female recapture rate declined significantly with increasing brood size, but the decline in male recapture rate with increasing brood size was not significant (Smith 1988).

To summarize, most studies (12/16=75%, if we exclude the two studies which only demonstrated indirect effects) revealed at least one type of cost of reproduction. Relatively few studies failed to demonstrate a cost of reproduction. Those that did not demonstrate a cost were characterized by small sample size, were conducted in only one or two years, and either did not examine future fecundity cost at all, or only investigated a between-year effect. For that matter, none of the 18 studies comprehensively examined all costs, or even all types of costs, of reproduction. More studies revealed future fecundity costs than revealed survival costs. In fact, only 27% (4/15) of the studies examining survival costs reported consistent, significant differences in recapture rates. Nevertheless, studies demonstrating between-year effects (fecundity or survival) were more numerous than those demonstrating within-year effects (7 studies vs. 5).

CORRELATIONAL STUDIES

The difficulties in using correlational, i.e. nonmanipulative, studies to infer the cost of reproduction are well established (see above). As pointed out in Nur (1988a), phenotypic plasticity, e.g. if clutch size responds adaptively to a bird's condition or to environmental conditions (Perrins and Moss 1975, Hogstedt 1980, 1981) can mask the cost of reproduction. Individuals in poor condition (or breeding in poor environmental conditions) might also be constrained to lay small clutches (or rear small broods); such constraints could also mask the cost of reproduction. Thus, a positive or zero correlation between, say, clutch size and adult survival is ambiguous (see above); in such a case, we can draw no conclusions regarding the presence or absence of a cost of reproduction. However, a negative correlation between clutch size and adult survival does provide evidence of a cost of reproduction. Adaptive adjustment in clutch size and/or constraints on clutch size can turn a negative correlation (manifestation of a cost of reproduction) into a positive correlation, but cannot turn a positive correlation into a negative one (Nur 1988a).

Most nonmanipulative studies have correlated adult survival with a measure of fecundity; a small number have examined intra-year future fecundity costs (e.g. McGillivray 1983); fewer still have examined inter-year future fecundity costs. It seems unlikely that correlational studies would reveal inter-year future fecundity costs since individuals rearing a large clutch (or brood) in one year are likely to rear a large clutch (or brood) in the next year, if only because clutch size is, in part, genetically determined (van Noordwijk et al. 1981). Because few correlational studies have examined fecundity cost, in this review I consider only studies examining survival in relation to fecundity; Table III summarizes results from 22 such studies. The table is intended to be comprehensive with respect to studies published prior to October 1989, but

Table III. Summary of results from non-manipulative (correlative) studies of the cost of reproduction in birds. See text.

Correlation of survival and reproductive output

positive		
	Kluyver 1963	Great Tit
*	Högstedt 1981	Magpie
*	Smith 1981	Song Sparrow
	Safriel et al. 1984	Oystercatcher
*	den Boer-Hazewinkel 1987	Great Tit
	Rockwell et al. 1987	Lesser Snow Goose
*	Boyce & Perrins 1987	Great Tit
	Gustafsson & Sutherland 1988	Collared Flycatcher
	van Noordwijk & van Balen 1988	Great Tit
	Thomas & Coulson 1988	Kittiwake
	Korpimäki 1988	Tengmalm's Owl
*	Nur, Geupel, deSante 1989	Wrentit
near zero		
	Newton 1988	Sparrowhawk
	McCleery & Perrins 1988	Great Tit
negative		
*	Kluyver 1971	Great Tit
*	Bryant 1979	House Martin
*	Dow & Fredga 1984	Goldeneye Duck
*	Ekman & Askenmo 1986	Willow Tit
*	Pugesek 1987	California Gull
*	Clobert et al. 1987	Starling
*	Nur 1988a	Pied Flycatcher
*	Ollasson & Dunnett 1988	Fulmar

* statistically significant, $P<.05$

undoubtedly some studies have been inadvertently omitted (see also studies in Newton 1989). The table demonstrates that even though many studies obtained positive correlations between adult survival and fecundity, seven obtained significant negative correlations; these seven studies provide evidence of a cost of reproduction. As for studies which obtained positive correlations, I note that the absence of a cost of reproduction would not by itself lead to a positive correlation. Instead, a positive correlation supports the idea of adaptive adjustments in clutch size (or other aspect of fecundity) and/or constraints on fecundity. These possibilities can be tested experimentally (Hogstedt 1980, Nur 1988a).

DIFFICULTIES IN MEASURING THE COST OF REPRODUCTION

Table IV lists six difficulties with measuring the cost of reproduction that apply to experimental studies; points 1) - 4) potentially apply to nonmanipulative studies as well. Perhaps the dominant problem is 1) small sample size, a problem which characterizes the studies in Table II but not Table III. In many of the studies in Table II, the total number of enlarged (or diminished) broods was 20 or fewer. This makes it difficult or impossible to accurately estimate differences in survival and results in little statistical power to detect an effect of brood size manipulation on survival. Nur (1988a) makes

Table IV. Difficulties in measuring cost of reproduction: experimental manipulations.

1. Small sample size
   Little power to detect evolutionarily significant differences

2. Year to year variation in cost of reproduction

3. Need to look at future fecundity (intra-year and inter-year)
   not just survival of parents
   also survival/recruitment of young from future broods

4. Recapture/return may not reflect true survival

5. Manipulation might not alter reproductive effort
   i.e. experiment may not succeed

6. Some manipulations technically difficult/impossible
   cf. number of nestlings reared with number of eggs laid
   or number of broods reared

the point that a difference in survival rates of as little as 10%, e.g. adult survival of 0.44 vs. 0.40, may suffice to shift the optimal clutch size by 2 eggs. Even with sample sizes of a thousand individuals for each clutch-(or brood-) size group, the probability of detecting a significant difference between two groups whose survival rates differ by .04 is 44% (at the 0.05 significance level). A number of the studies in Table II found survival differences of small magnitude, consistent with the cost of reproduction, yet not statistically significant. We must not confuse evolutionary significance with statistical significance; we should consider the estimate of the effect (if any) of brood size on survival, and not just the P-value. The estimate of the effect of brood size on parental survival is greater in the House Sparrow (Hegner and Wingfield 1987), than it is in the Glaucous-winged Gull (Reid 1987), yet it is the latter which, because of large sample size, demonstrates statistical significance, not the former.

The problem of small sample size is compounded by problem 2) year-to-year variation in the cost of reproduction. Life history theory requires knowledge of how strong the cost of reproduction is in each year; this requires large sample sizes in every year. Whereas the experimental studies in Table II are characterized by short duration, nonmanipulative studies in Table III are not, and might better reveal year-to-year fluctuation in the cost of reproduction. Female Pied Flycatchers, for example, demonstrated significant year-to-year differences in the slope of recapture rate in relation to brood size (Nur 1988a; see also Tinbergen et al. 1985).

The need to assess the cost of reproduction with respect to future fecundity and not just parental survival (problem 3) has already been stressed. Even recent reviews of the cost of reproduction have overlooked this point (Alatalo and Lundberg 1989, Breitwitsch 1989).

Difficulties in using recapture (or return) rate to measure survival (problem 4) have been discussed above (see also Clobert et al. 1987b, Nur and Clobert 1988, Slagsvold in press). Some biases may be subtle, as exemplified by the nonrandom inclusion of males in the experimental sample in Nur (1988b) discussed above. A different but related problem is that nestbox studies may underestimate the cost of reproduction due to predation (Møller 1989, Alatalo and Lundberg this volume).

A problem specific to experimental studies is that the manipulation of brood or clutch size may fail to alter reproductive effort (problem 5). In that case, it is the experimental protocol that has failed; the results are inconclusive with respect to the cost of reproduction. Korpimaki (1988) suggested that male Tengmalm's owls did not alter their behaviour in relation to the addition or subtraction of one nestling and that perhaps this explained the lack of demonstrable costs of reproduction in his study.

A related point to problem 5 concerns the magnitude of the manipulation. Too large a manipulation may produce aberrant behavior, e.g. as a result of unnaturally large broods being created or if the parents respond to brood diminishment by abandoning the young. These possibilities can be assayed directly (Nur 1984b). However, too small a manipulation (e.g. changing brood size by one in a species with a natural brood of 10) may result in an effect of evolutionary importance yet too subtle to be detected without prohibitively large sample sizes (see problem 1, above).

The final problem is that not all costs of reproduction are accessible using the experimental method, e.g. the cost of laying and the cost of rearing two broods rather than one. The cost of incubation is open to experimental investigation; however, until recently few studies investigated ecological costs of incubation (but see Moreno and Carlson 1989, Smith 1989).

CONCLUSIONS

Most published experimental studies, as well as half a dozen nonmanipulative studies, have verified a cost of reproduction, either a future fecundity cost or a survival cost. These results demonstrate the generality, though not the universality, of a cost of reproduction. The empirical record answers the first-level question, "Is there a cost of reproduction?" in the affirmative, generally speaking, and brings us to the second-level question, "Which cost of reproduction?" A variety of costs of reproduction have been demonstrated, within and between years, involving both future fecundity and subsequent survival. Early treatment of the cost of reproduction (Kluyver 1963, Lack 1966, Williams 1966, Charnov and Krebs 1974) considered only survival costs, however future fecundity cost is an equally important component of the cost of reproduction, in practice and not just in theory. Whether or not there is a cost of reproduction is of fundamental importance to theories of life history strategies, but whether the cost is expressed as reduction in survival probability or reduction in future reproductive success is of secondary importance. How the cost of reproduction is expressed is a worthy topic in the study of life history strategies (Stearns 1989).

Investigators have generally underestimated the difficulty in measuring the cost of reproduction (Reznick et al. 1986). It is especially difficult to estimate survival costs (Table IV). The observation that few studies have demonstrated survival costs may reflect these difficulties or it may reflect the genuine lack of a tradeoff between survival and present fecundity in some populations. The overriding problem in measuring the cost of reproduction is that our estimates of the cost are incomplete, which makes it difficult, if not impossible, to adequately answer the third-level question: how strong is the cost of reproduction? We have enough information to know that we cannot ignore the cost of reproduction with regard to, for example, the evolution of clutch size (cf. Alerstam and Hogstedt 1984), but we have insufficient information to "solve" the questions of optimal clutch size or optimal parental investment.

ACKNOWLEDGEMENTS

I thank the participants of the NATO Advanced Research Workshop on Demographical, Physiological, Genetical and Behavioral Aspects of Population Biology of Passerine Birds for discussion, an anonymous reviewer for comments on the manuscript, and J. Blondel, the Point Reyes Bird Observatory and I. and

U. Nur for financial support. This is Point Reyes Bird Observatory Contribution No. 458.

REFERENCES

Alatalo RV and Lundberg A (1989) Clutch size of the Pied Flycatcher *Ficedula hypoleuca* - an experiment. Ornis Fennica 66: 15-23

Alatalo RV and Lundberg A (in press) Polygyny and nesting success of Pied Flycatchers breeding in natural cavities. In: J Blondel, A Gosler, JD Lebreton and R McCleery (Eds), this volume

Alerstam T and Hogstedt G (1984) How important is clutch size dependent adult mortality? Oikos 43: 253-254

Askenmo C (1979) Reproductive effort and return rate of male pied flycatchers. American Naturalist 114: 748-753

den Boer-Hazewinkel J (1987) On the costs of reproduction: parental survival and production of second clutches of the great tit *Parus major*. Ardea 75: 99-110

Boyce M and Perrins CM (1987) Optimizing great tit clutch size in a fluctuating environment. Ecology 68: 142-153

Breitwisch R (1989) Mortality patterns, sex ratios, and parental investment in monogamous birds. Current Ornithology 6: 1-50

Bryant DM (1979) Reproductive costs in the house martin (*Delichon urbica*). Journal of Animal Ecology 48: 655-675

Charnov EL and Krebs JR (1974) On clutch size and fitness. Ibis 116: 217-219

Clobert J, Bouchau V, Dhont A and Vansteenwegen C (1987a) Survival of breeding female Starlings in relation to brood size. Acta Oecologia, Oecologia generalis 8: 427-434

Clobert J, Lebreton JD and Allaine D (1987b) A general approach to survival rate estimation by recaptures or resightings of marked birds. Ardea 75: 133-142

Cody ML (1966) A general theory of clutch size. Evolution 20: 174-184

Cox DR (1970) The analysis of binary data. Methuen, London

Daan S, Dijkstra C, Drent R and Meijer T (1988) Food supply and the annual timing of avian reproduction. Acta XIX Congressus Internationalis Ornithologici, University of Ottawa Press, Ottawa, Canada, p 392-407

Dijkstra C, Bult A, Bijlsma S, Daan S, Meijer T, and Zijlstra M (1990) Brood size manipulations in the kestrel *Falco tinnunculus*: effects on offspring and parental survival. Journal of Animal Ecology 59: 269-285

Dow H and Fredga S (1984) Factors affecting reproductive output of the goldeneye duck *Bucephala clangula*. Journal of Animal Ecology 53: 679-692

Ekman J and Askenmo C (1986) Reproductive cost, age-specific survival and a comparison of the reproductive strategy in two European tits (Genus *Parus*). Evolution 40: 159-168

Gustafsson L and Sutherland WJ (1988) The costs of reproduction in the collared flycatcher *Ficedula albicollis*. Nature 335: 813-815

Hegner RE and Wingfield JC (1987) Effects of brood-size manipulations on parental investment, breeding success, and reproductive endocrinology of house sparrows. Auk 104: 470-480

Hogstedt G (1980) Evolution of clutch size in birds: adaptive variation in relation to territory quality. Science 210: 1148-1150

Hogstedt G (1981) Should there be a positive or negative correlation between survival of adults in a bird population and their clutch size? American Naturalist 118: 568-571

Kluyver HN (1963) The determination of reproductive rates in *Paridae*. Proceedings XIIIth International Ornithological Congress, p 706-716

Kluyver HN (1971) Regulation of numbers in populations of great tits (*Parus m. major*). p 507-523 in: PJ de Boer and CR Gradwell (Eds) Dynamics of Populations. Pudoc, Wageningen, Netherlands

Korpimaki E (1988) Costs of reproduction and success of manipulated broods under varying food conditions in Tengmalm's owl. Journal of Animal Ecology 57: 1027-1039

Kotz S and Johnson NL (1985) Encyclopedia of Statistical Sciences vol 5, Wiley and Sons, New York

Lack D (1966) Population Studies of Birds. Oxford University Press

Lessells CM (1986) Brood size in Canadian geese: A manipulation experiment. Journal of Animal Ecology 55: 669-689

Linden M (1988) Reproductive trade-off between first and second clutches in the great tit *Parus major*: an experimental study. Oikos 51: 285-290

Linden M and Møller AP (1989) Cost of reproduction and covariation of life history traits in birds. Trends in Ecology and Evolution 4: 367-371

McCleery RH and Perrins CM (1988) Lifetime reproductive success of the Great Tit, *Parus major*. p 136-153 in: TH Clutton-Brock (Ed) Reproductive Success. University of Chicago Press, Chicago

McGillivray WB (1983) Intraseasonal reproductive costs for the House Sparrow (*Passer domesticus*). Auk 100:25-32

Møller AP (1989) Parasites, predators and nest boxes: facts and artefacts in nest box studies of birds? Oikos 56: 421-423

Moreno J and Carlson A (1989) Clutch size and the costs of incubation in the Pied Flycatcher *Ficedula hypoleuca*. Ornis Scandinavica 20: 123-128

Murphy EC and Haukioja E (1986) Clutch size in nidicolous birds. p 141-180 in: Johnston RF (Ed): Current Ornithology, vol 4, Plenum Press, New York

Newton I (1988) Age and reproduction in the Sparrowhawk. p 201-219 in: TH Clutton-Brock (Ed) Reproductive Success. University of Chicago Press, Chicago

Newton I (1989) (Ed) Lifetime reproduction in birds. Academic Press, New York

Noordwijk AJ, van Balen JH and Scharloo W (1981) Genetic and environmental variation in clutch size of the great tit (*Parus major*). Netherlands Journal of Zoology 31: 342-372

Noordwijk AJ van and JH van Balen (1988) The great tit, *Parus major*. p 119-135 in: TH Clutton-Brock (Ed) Reproductive Success. University of Chicago Press, Chicago

Noordwijk AJ van and Jong G de (1986) Acquisition and allocation of resources: their influence on variation in life history tactics. American Naturalist 128: 137-142

Nur N (1984a) The consequences of brood size for breeding blue tits. I Adult survival, weight change and the cost of reproduction. Journal of Animal Ecology 53: 479-496

Nur N (1984b) Feeding frequencies of nestling blue tits (*Parus caeruleus*): Costs, benefits, and a model of optimal feeding frequency. Oecologia 65: 125-137

Nur N (1986) Is clutch size variation in the blue tit (*Parus caeruleus*) adaptive? An experimental study. Journal of Animal Ecology 55: 983-999

Nur N (1988a) The cost of reproduction in birds: An examination of the evidence. Ardea 76: 155-168

Nur N (1988b) The consequences of brood size for breeding blue tits. III. Measuring the cost of reproduction: Survival, future fecundity, and differential dispersal. Evolution 42: 351-362

Nur N and Clobert J (1988) Measuring Darwinin fitness in birds: A field guide. Proceedings XIX International Ornithological Congress, p 2121-2130

Nur N, Geupel GR and De Sante D (1989) Variability in the reproductive strategy of the wrentit, *Chamaea fasciata*: A model and its application to a coastal scrub population. Paper presented at 59th Annual Meeting, Cooper Ornithological Society, Moscow, Idaho

Ollason JC and Dunnet GM (1988) Variation in breeding success in Fulmars. p 263-278 in: TH Clutton-Brock (Ed) Reproductive Success. University of Chicago Press, Chicago

Orell M (in press) Brood size manipulations and reproductive success in the Willow Tit. In: J Blondel, A Gosler, JD Lebreton and R McCleery (Eds), this volume

Orell M and Koivula K (1988) Cost of reproduction: parental survival and production of recruits in the Willow Tit *Parus montanus*. Oecologia 77: 423-432

Partridge L (1989) Lifetime reproductive success and life-history evolution. p 421-440

Partridge L and Harvey PH (1985) Costs of reproduction. Nature 316: 20-21

Perrins CM and Moss D (1975) Reproductive rates in the great tit. Journal of Animal Ecology 44: 695-706

Pettifor RA, Perrins CM and McCleery RH (1988) Individual optimization of clutch size in great tits. Nature 336: 160-162

Pugesek B (1987) Age-specific survivorship in relation to clutch size and fledging success in California gulls. Behavioral Ecology and Sociobiology 21: 217-221

Reid WV (1987) The cost of reproduction in the glaucous-winged gull. Oecologia 74: 458-467

Reznick D (1985) Costs of reproduction: an evaluation of the empirical evidence. Oikos 44: 257-267

Reznick DN, Perry E and Travis J (1986) Measuring the cost of reproduction: a comment on papers by Bell. Evolution 40: 1338-1344

Ricklefs R (1983) Avian demography. Current Ornithology 1: 1-32

Rockwell RF, Findlay CS and Cooke F (1987) Is there an optimal clutch size in lesser snow geese. American Naturalist 130: 839-863

Røskaft E (1985) The effect of enlarged brood size on the future reproductive potential of the rook. Journal of Animal Ecology 54: 255-260

Safriel UN, Harris MP, Brooke M de L and Britton CK (1984) Survival of Oystercatchers *Haematopus ostralegus*. Journal of Animal Ecology 53: 867-877

Schaffer WM (1974) Selection for optimal life histories: the effects of age structure. Ecology 55: 291-303

Slagsvold T (1984) Clutch size variation in birds in relation to nest predation: on the costs of reproduction. Journal of Animal Ecology 53: 945-953

Slagsvold T (in press) Site fidelity in Pied Flycatchers: Do the males always return to their previous breeding area? in: J Blondel, A Gosler, JD Lebreton, R McCleery (Eds), this volume

Smith HG (1988) Reproductive costs and offspring quality: the evolution of clutch size in tits (*Parus*). PhD Dissertation Lund University, Sweden

Smith HG (1989) Larger clutches take longer to incubate. Ornis Scandinavica 20: 156-158

Smith HG, Källander H and Nilsson J-Å (1987) Effect of experimentally altered brood size on frequency and timing of second clutches in the Great Tit. Auk 104: 700-706

Smith JNM (1981) Does high fecundity reduce survival in song sparrows? Evolution 35: 1142-1148

Stearns SC (1976) Life history tactics: a review of the ideas. Quarterly Review of Biology 51: 3-47

Stearns SC (1989) Trade-offs in life-history evolution. Functional Ecology 3: 259-268

Steven D de (1980) Clutch size, breeding success, and parental survival in the tree swallow (*Iridoprocne bicolor*). Evolution 34: 278-291

Thomas CS and Coulson JC (1988) Reproductive success of Kittiwake Gulls, *Rissa tridactyla*. p 251-262 in: TH Clutton-Brock (Ed) Reproductive Success. University of Chicago Press, Chicago

Tinbergen JM (1987) Costs of reproduction in the great tit: intraseasonal costs associated with brood size. Ardea 75: 111-122

Tinbergen JM, Balen JH van and Eck HM van (1985) Density dependent survival in an isolated Great Tit population: Kluyver's data reanalyzed. Ardea 73: 38-48

Williams GC (1966) Natural selection, the costs of reproduction, and a refinement of Lack's principle. American Naturalist 100: 687-690

EFFECTS OF BROOD SIZE MANIPULATIONS ON ADULT AND JUVENILE SURVIVAL AND FUTURE FECUNDITY IN THE WILLOW TIT *PARUS MONTANUS*

Markku ORELL and Kari KOIVULA

Department of Zoology, University of Oulu, Linnanmaa A, SF-90570 Oulu, Finland

Lack's hypothesis (1954, 1966) that individual birds lay that number of eggs which will maximize the number of viable offspring produced has been the starting point in studies of the evolution of avian clutch size for many years. The hypothesis predicts that the modal clutch size in the population should be optimal. Many studies, however, have shown that larger clutches do not deviate from modal clutches with respect to nestling or postfledging survival (see e.g. the review of Murphy and Haukioja 1986). This discrepancy has been accounted for by explanations that are refinements of Lack's theory. The cost hypothesis predicts a trade-off between current and future reproductive success (e.g. Williams 1966, Charnov and Krebs 1974, Stearns 1976) i.e. investment in current reproduction will reduce the individual's chances of survival or reduce its productivity in the future reproductive effort. Hence if adult survivorship decreases with increasing clutch size the optimal clutch size will be smaller than the most productive clutch size when the lifetime reproductive success of the adult is concerned.

The results from correlative as well as from experimental studies have been contradictory in demonstrating the existence of a reproductive cost. That is why many researchers have been sceptical of its evolutionary significance (e.g. DeSteven 1980, Smith 1981, Alerstam and Högstedt 1984, Horn and Rubenstein 1984, Boyce and Perrins 1987, Gustafsson and Sutherland 1988, Pettifor et al. 1988). Others however have stressed its importance in shaping reproductive tactics (e.g. Reid 1987, Dijkstra 1988, Nur 1988). Validity of the reproductive cost hypothesis was tested by manipulating brood sizes of the Willow Tit *Parus montanus*. Subsequent survival of adults and their descendants was followed. Their fecundity was measured in the next breeding season.

MATERIAL AND METHODS

Willow Tit broods were manipulated during four seasons 1986-89 in the Oulu area, northern Finland. The research area (8 km^2 in 1986 and 12-13 km^2 thereafter) consists of a variety of habitats, mostly coniferous forests of different ages. For further information about the study area see Orell and Ojanen (1983) and Orell and Koivula (1988).

NATO ASI Series, Vol. G 24
Population Biology of Passerine Birds
Edited by J. Blondel et al.

All but three nests were located in natural decaying stumps or in birch stumps which were transported into the area. Each spring nest searches started in early April and continued during the whole breeding season until early July in order to find possible repeat nestings. Practically all the nests of the colour-ringed population (see Orell and Koivula 1988) were found each summer.

Start of laying, clutch size, time of hatching, number of hatched young and number of fledglings were recorded during routine visits to the research area. The nest checks were done with a small mirror and a torch through the entrances or making a break in the wall at the level of the nest cup. Altogether the size of 116 broods were altered. Two chicks were moved from one nest to the other at the age of less than three days (day 0 = hatching day of the oldest chicks). Most of the transfers took place when the young were newly hatched or one day old. The experimental broods produced in this way were within the normal range found in natural conditions (Orell and Ojanen 1983) and hence near the expected heritable variation (Gustafsson and Sutherland 1988). The experimental broods were randomly chosen provided that hatching was simultaneous and that they were located so close to each other that the transfer could be done without risking the chicks' condition. For this reason the broods which were not manipulated served as controls. First clutches and 5 repeat clutches commenced within the main laying season were included in the material. The sample of successfull nests consists of 102 experimental and 86 control broods.

When the oldest young of the brood were 13 days old the nestlings were marked. During the same visit the parents were captured. The young and the parents were weighed to an accuracy of 0.1 g with a Pesola spring balance. Their wing and tail were measured to the nearest 1 mm, following the maximum chord method (Svensson 1970, Busse and Kania 1970), respectively. The tarsus length was measured with a dial caliper (Svensson 1970).

Surviving adults and the number of young recruited into the breeding area were determined at the start of the next breeding season (early May). The birds could be individually identified by their colour rings which they got when being captured in winter. Material for survival analyses includes only successful breeding attempts. Willow Tits are faithful to their breeding territories; the birds overwinter and breed in the same area for their whole life (Ekman 1979, Koivula and Orell 1988). It is therefore highly probable that an individual was dead if it was not observed during the breeding season in the site it bred the year before. Areas bordering the study site were visited each spring to find out possible movements of older birds there.

RESULTS

Nesting success

In the ideal case, random assortment of broods into different manipulation categories would have resulted in the same initial number of eggs for different groups. This did not succeed every year when only the successful broods are considered. The control clutches in 1986 had significantly more eggs than the experimental clutches (Orell and Koivula 1988). In the three other years there were no significant differences in the initial clutch size between the manipulation categories.

Different mortality equalized the number of fledglings per brood in 1987 and 1988. This resulted in nonsignificant variation in the number of fledglings between the manipulation categories ($F(2, 39)=0.96$, $p=0.394$ and $F(2, 46)=2.56$, $p=0.088$, respectively). Logarithmic transformation was applied to the number of fledglings ($\log(x+1)$) to minimize correlation between the mean and the variance of the data. In the years 1986 and 1989 the enlarged broods fledged the most and the reduced broods the least number of young. Trends were significant ($F(2, 27)=32.97$, $p=0.000$ and $F(2, 72)=9.88$, $p=0.000$, respectively, log-transformated as above). The results show that Willow Tits had no difficulties in rearing extra young in 1986 and 1989, whereas some young even in reduced broods starved to death in 1987 and 1988.

Survival after breeding

According to the pooled material, adult survival was rather good (males 66.4 % and females 59.4 %). With respect to the manipulation categories the results were obscure in both sexes (Fig. 1). In 1987 there were virtually no differences between manipulation groups whereas in 1986 (females) and in 1988 (males) feeding extra young seemed to be costly. To test the cost hypothesis, logistic regression analyses were run to study survival in relation to manipulation. Fledgling number and year were included as the other independent factors. Neither manipulation ($b=-0.14$, $df=1$, $p=0.327$) nor brood size ($b=-0.08$, $df=1$, $p=0.486$) affected the survival of females ($n=111$). The year effect was almost significant ($b=-0.51$, $df=1$, $p=0.053$) reflecting the low survival after the season 1988 (Fig. 1). Similar results were obtained for male survival ($n=113$, manipulation effect: $b=-0.06$, $df=1$, $p=0.686$, brood size: $b=0.03$, $df=1$, $p=0.812$, and year: $b=-0.11$, $df=1$, $p=0.689$). The years 1986 and 1988 were analysed separately using female and male data, respectively. There was no significant relationship between survival and manipulation group (females: $b=-0.15$, $X^2=0.29$, $p=0.590$, males: $b=-0.28$, $X^2=1.75$, $p=0.186$).

Of the 743 fledglings altogether 51 young were later found breeding in the study area during 1987-89. The figure 2 implies that the year 1986 was a good one in producing surviving offspring. Part of the high recruitment rate can be due to enlargement of the study area the year after. There is no reason to suspect this having influenced the outcome of different manipulation groups. It could be so if the natal dispersal is not independent of manipulation. We could not find any such correlation (Koivula and Orell 1988). Natal dispersal of Willow Tits will be discussed in detail in another paper.

In 1986 there was a clear tendency for decreasing post-fledging survival from reduced through enlarged broods (Fig. 2). Logistic regression analysis did

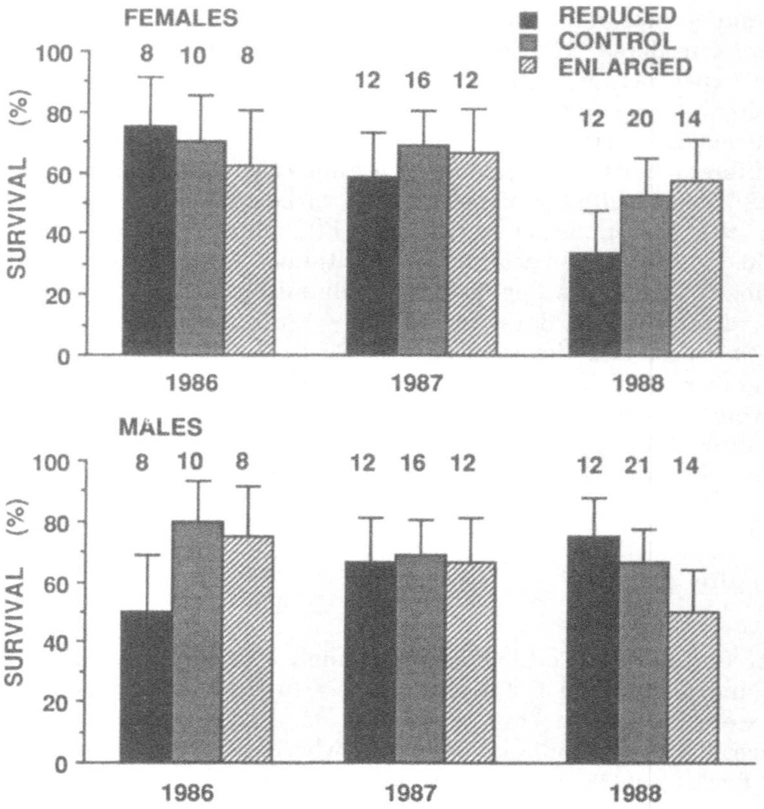

Fig. 1. Percentage (+ SE) of parents surviving till the next breeding season in relation to the manipulation category.

not result in a significant relationship, however (b=-0.23, df=1, p=0.124, n=204). In the two other years the results clearly deviated from that trend.

According to the pooled material, fledglings from reduced broods tended to survive better (8.5 % of fledglings were recovered) to the next May than those from control or reduced broods (6.4 and 6.2 %, respectively). Logistic regression, after the year effect was included in the model, resulted in a significant relationship between survival and year (b=-0.49, df=1, p=0.006, n=743). The effect of manipulation was not significant. The results point out that the manipulation was not enough to create quality differences between fledglings which could have contributed to selective mortality between the treatment groups. The year 1986 may be an exception in this respect. This result is somewhat surprising because post-fledging survival correlates positively with fledging weight and the weight decreased from reduced through enlarged broods each year (Orell and Koivula unpublished).

From the evolutionary point of view it is more important to analyse the contribution of each pair to the next generation i.e. the recruitment rate per brood. Again the results from different years vary (Fig. 2). Differences in the

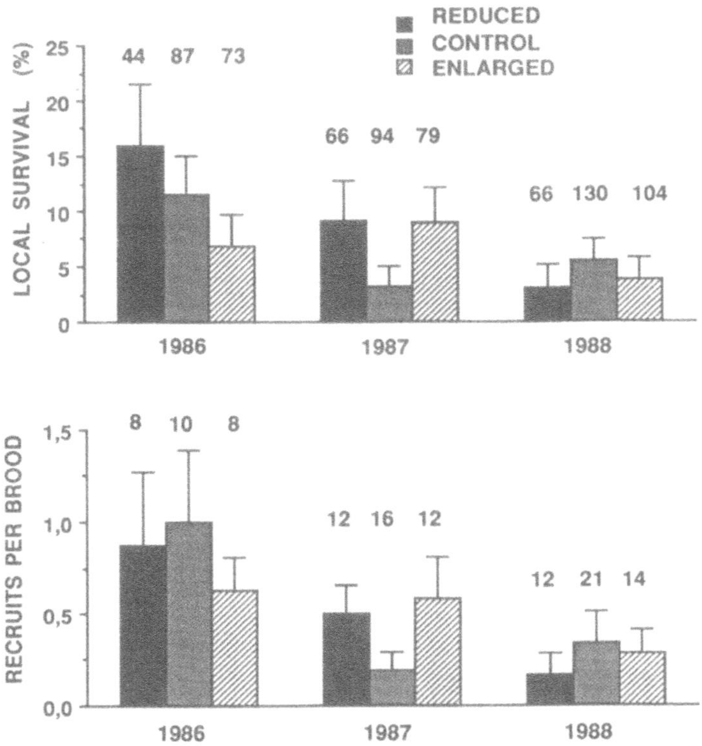

Fig. 2. Percentage (+ SE) of fledglings surviving to breed in the study area (upper graph) and average number of young recruited to the breeding population (lower graph) in relation to brood size manipulation. The numerals above the columns denote the number of fledglings (upper graph) and the number of broods (lower graph).

success of recruiting offspring into the subsequent population was analysed using log-linear models. The factors used were: 1) number young recruited (groups 0, 1, 2 or more young), 2) brood size at fledging (4-6, 7, 8, 9-11 young), and 3) manipulation category. The simplest fitting model explaining the observed distribution was (2*3), 1 (Likelihood ratio $X^2=30.52$, $df=22$, $p=0.106$). This means that there is a significant association between brood size and manipulation category, i.e. brood size increased from reduced to enlarged broods as was seen above. However, the analysis also shows that this dependence is conditional with respect to survival classes. This is the outcome of the fact that those broods producing recruits were larger than those which were not as successful in each manipulation category. These results mean that the correlation between recruitment and manipulation category, we are interested in here, does not improve the fit of the model significantly.

Fecundity in the next breeding season

The effect of brood size manipulation on subsequent fecundity was studied by comparing the clutch size of females from the different categories in the next spring. As the clutch sizes were quite different in the years 1987-89 the year effect was eliminated before pooling the data. This was done by standardizing the clutch sizes in such a way that the average for each year was zero and the standard deviation one.

Among birds older than one year subsequent clutch size (Fig. 3) was not significantly affected by brood size manipulation ($F(2, 8)=2.07$, $p=0.188$) nor was the female effect significant ($F(28, 8)=2.45$, $p=0.094$) in the two-way ANOVA. In the yearling birds the manipulation effect was significantly affected by the low fecundity among females with enlarged broods ($F(2, 18)=3.71$, $p=0.045$). So it seems that increased efforts can contribute to lowered fecundity of yearling females in their subsequent breeding attempt.

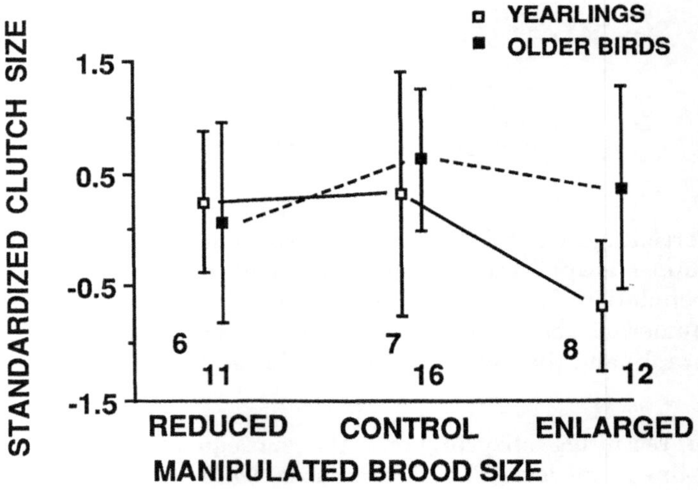

Fig. 3. Fecundity (mean ± SD) of experimental females in the subsequent breeding season expressed as standardized clutch sizes. Females are grouped according to their age when manipulated. The numerals above the x-axis denote the number of females; upper = yearlings, lower = older birds.

DISCUSSION

A prerequisite of success in this kind of experiment is that parents attending manipulated broods increase or decrease their efforts. To definitely prove this is difficult. Brood size at fledging can be regarded as one criterion.

This precondition was fulfilled well in 1986 and 1989 when enlarged broods fledged the largest number of young. In the two other years differences between manipulation categories were less pronounced. This fact causes some problems when drawing conclusions.

Despite mortality, especially in large broods, fledglings in the broods which were experimentally enlarged were still retarded compared with the other groups in 1987 and 1988 (Orell and Koivula unpublished). Body weight at fledging correlates positively with post-fledging survival in *Paridae* (e.g. Perrins 1965, Lack 1966, van Balen 1973, Garnett 1976, Smith et al. 1989, Orell and Koivula unpublished). So one could expect to find poorer juvenile survival among fledglings from enlarged than from the other broods in the Willow Tit. However the result did not accord with the hypothesis that greater quantity leads to lower quality. Local recoveries from a single year, 1986, suggested that it could be so (Orell and Koivula 1988) but inclusion of the material from the two other years cancelled this trend out.

Perhaps nestling starvation in control and enlarged broods especially in 1987 was likely to reduce the quality differences to the level where this factor lost its significance. It can be that the parents of those broods compensate for the slightly lower weight of their fledglings. This could be achieved by lengthening the period they cared for the young before their independence.

The results point out the importance of environmental variability in the evolution of clutch size. In some years clutches of about the average size are at an advantage whereas in other years small and/or large clutches are selected for. As a consequence no net selection for the trait results (van Noordwijk et al. 1981, Orell and Ojanen 1986, Boyce and Perrins 1987).

Parental survival was highly variable between the years with respect to the manipulation category and sex. Two of the six possible comparisons were in the direction predicted by the reproductive cost hypothesis. One can discuss the interpretation of the non-significant trends which seem to support the cost hypothesis. Although it is very difficult to have statistical support for small differences in survival probabilities the differences can be evolutionarily significant (DeSteven 1980, Nur 1988). It is also quite possible that a reproductive cost might only characterize some years (e.g. Nur 1988). In that case it is still more difficult to judge which trend is a result of reproductive cost and which one has been obtained by chance.

Tuomi et al. (1983) presented a threshold hypothesis which says that reproductive cost is not manifest if the stress due to raising extra young remains below a hypothetical threshold. It can be that Willow Tits have not exceeded that hypothetical level of stress while raising their broods especially in 1987 and 1988 when nestling starvation affected large broods.

Unless more data are available the results indicate that the cost of reproduction is unlikely to be a major determinant of clutch size of the Willow Tit (see also Boyce and Perrins 1987, Gustafsson and Sutherland 1988, Pettifor et al. 1988). There are other factors than those derived from breeding efforts that are more important in determining the survival chances of Willow Tits. On the other hand if the cost of reproduction is manifest only in certain years, a three year study like this is far too short for final conclusions. Relatively poor breeding success in 1987 and 1988 can confound the possible trade-off between current reproductive effort and future survival. That is why the survivorship results concerning the parents manipulated in 1989 will be of utmost interest.

Reproductive cost can work not only through a cost to survival but also through a cost to fecundity. Results available concerning the fecundity cost are contradictory (Slagsvold 1984, Røskaft 1985, Tinbergen 1987, Smith et al. 1987, Gustafsson and Sutherland 1988, Korpimäki 1988 and Pettifor et al. 1988). With the present material fecundity cost can be suspected among yearling Willow Tit females which attended enlarged broods. However, more data are needed before final conclusions can be drawn.

To conclude, the present results suggest that a hypothetical trade-off between clutch size and parental survival seems not to be important in shaping the life-history strategy of the Willow Tit, a small short-lived passerine. It is more likely that the cost works through recruitment success and perhaps also by lowering subsequent fecundity of the females. Making conclusions from 3-year data is risky. That is why experiments should be carried out for many years to test the hypotheses under varying environmental conditions.

ACKNOWLEDGEMENTS

We thank Mikko Mönkkönen and Seppo Rytkönen for valuable assistance in the field and discussions which improved the text. Arch McCallum made useful comments on the manuscript. The work was supported by Tauno Tönning Foundation, Finnish Cultural Foundation and Grants from the University of Oulu.

REFERENCES

Alerstam T, Högstedt G (1984) How important is clutch size dependent adult mortality? Oikos 43: 253-254

Balen JH van (1973) A comparative study of the breeding ecology of the Great Tit *Parus major* in different habitats. Ardea 61: 1-93

Boyce MS, Perrins CM (1987) Optimizing great tit clutch size in a fluctuating environment. Ecology, 68: 142-153

Busse P, Kania W (1970) Akcja Baltycka 1961-1967. Methody pracy (Operation Baltic 1961-1967. Working methods). Acta Orn 12: 231-267

Charnov EL, Krebs JR (1974) On clutch size and fitness. Ibis 116: 231-267

DeSteven D (1980) Clutch size, breeding success, and parental survival in the tree swallow (*Iridoprocne bicolor*). Evolution 34: 278-291

Dijkstra C (1988) Reproductive tactics in the kestrel *Falco tinnunculus*. A study in evolutionary biology. Ph D Thesis, Univ Groningen

Ekman J (1979) Seasonal changes in demographic parameters of two coniferous forest tit species during the non-breeding season, with aspects to their relation to the territorial system and population dynamics. Ph D Thesis, Dept Zool, Univ Gothenburg

Garnett MC (1976) Some aspects of body-size in the Great Tit. Ph D Thesis, Dept Zool, Univ Oxford
Gustafsson L, Sutherland WJ (1988) The cost of reproduction in the collared flycatcher *Ficedula albicollis*). Nature 335: 813-815
Horn HS, Rubenstein DI (1984) Behavioural adaptations and life history. In: Krebs JR, Davies NB (eds) Behavioural ecology. An evolutionary approach. Blackwell, Oxford
Koivula K, Orell M (1988) Social rank and winter survival in the Willow Tit *Parus montanus*. Ornis Fennica 65: 114-120
Korpimäki E (1988) Costs of reproduction and success of manipulated broods under varying food conditions in Tengmalm's Owl. J Anim Ecol 57: 1027-1039
Lack D (1954) The natural regulation of animal numbers. Oxford Univ Press, London, New York
Lack D (1966) Population studies of birds. Oxford Univ Press, Oxford
Murphy EC, Haukioja E (1986) Clutch size in nidicolous birds. In Johnston R F (ed.) Current Ornithology 4: 141-180
Noordwijk AJ van, Balen JH van, Scharloo W (1981) Genetic and environmental variation in clutch size of the great tit (*Parus major*). Neth J Zool 31: 342-372
Nur N (1988) The cost of reproduction in birds: an examination of the evidence. Ardea 76: 155-168
Orell M, Koivula K (1988) Cost of reproduction: parental survival and production of recruits in the Willow Tit *Parus montanus*. Oecologia (Berl.) 77: 423-432
Orell M, Ojanen M (1983) Breeding biology and population dynamics of the willow tit *Parus montanus*. Ann Zool Fennici 20: 99-114
Orell M, Ojanen M (1986) Relation of fitness of female Great Tits to clutch size and number of broods. Ornis Fennica 63: 120-127
Perrins CM (1965) Population fluctuations and clutch-size in the great tit, *Parus major* L. J Anim Ecol 44: 695-706
Pettifor RA, Perrins CM, McCleery RH (1988) Individual optimization of clutch size in great tits. Nature 336: 160-162
Reid WV (1987) The cost of reproduction in the glaucous-winged gull. Oecologia (Berl.) 74: 458-467
Røskaft E (1985) The effect of enlarged brood size on the future reproductive potential of the rook. J Anim Ecol 54: 255-260
Slagsvold T (1984) Clutch size variation in birds in relation to nest predation: on cost of reproduction. J Anim Ecol 53: 945-953
Smith HG, Källander H, Nilsson J-Å (1987) Effect of experimentally altered brood size on frequency and timing of second clutches in the great tit. The Auk 104: 700-706
Smith HG, Källander H, Nilsson J-Å (1989) The trade-off between offspring number and quality in the great tit *Parus major*. J Anim Ecol 58: 383-401
Smith JNM (1981) Does high fecundity reduce survival in song sparrows? Evolution 35: 1142-1148

Stearns SC (1976) Life-history tactics: review of the ideas. Quart Rev Biol 51: 3-47
Svensson L (1970) Identification guide to European Passerines. Stockholm
Tinbergen JM (1987) Costs of reproduction in the great tit: interseasonal costs associated with brood size. Ardea 75: 111-122
Tuomi J, Hakala T, Haukioja E (1983) Alternative concepts of reproductive effort, costs of reproduction, and selection in life-history evolution. Amer Zool 23: 25-34
Williams GC (1966) Adaptation and natural selection. Princeton Univ Press, Princeton

COSTS AND BENEFITS OF REPRODUCTION OF THE COLLARED FLYCATCHER, *FICEDULA ALBICOLLIS*

János TÖRÖK and László TOTH
Department of Systematic Zoology and Ecology, Eötvös University, H-1088 Budapest, Puskin u. 3., Hungary

The cost of reproduction hypothesis has been the focus of many recent avian studies (DeSteven 1980, Nur 1984a, b, Slagsvold 1984, Gustafsson and Sutherland 1988, Korpimaki 1988, Orell and Koivula 1988, Pettifor et al. 1988, Smith et al. 1988, 1989). Reproductive cost can be expressed as a direct reduction in reproductive output of the birds, for example, a decreased number of fledged young, and/or a reduction in the future reproductive output, or reduced fertility of the parents in subsequent years. It is generally accepted (Nur 1984a,b, Gustafsson and Sutherland 1988, Pettifor et al. 1988) that data from experimentally modified broods are more useful for the estimation of trade-off functions between clutch size and other fitness components, than data from natural broods. However, using an experimental approach to gather data from different populations of the same species, could lead researchers to draw different conclusions. For example, Smith et al. (1988, 1989) found evidence for reproductive costs in a Swedish population of the Great Tit, while Pettifor et al. (1988) refuted the possibility of trade-offs between clutch size and survival rates, or future fecundity, of both parents and young in the Wytham population of the same species. These results demonstrate that our knowledge of the timing and action of costs in different populations or species, is scant.

The aim of this study was to identify the costs associated with reproduction in the Collared Flycatcher, *Ficedula albicollis*. First we used correlative data from natural broods, but since the individual optimization of clutch size can mask the costs (Nur 1989), we performed a manipulation experiment.

In addition we measured parental feeding frequency, which may be an important factor in shaping both nestling and parental survival (Nur 1984c).

METHODS

The Collared Flycatcher is a small, single-brooded, migratory passerine species, which is a common breeding bird of Central-European deciduous forests. The breeding phenology of this species was investigated as a part of studies on hole-nesting birds in Hungary, from 1982 to 1988. Data from 1020 complete clutches were obtained from a 20 ha study plot situated in a middle-aged oak forest near Budapest. 280 nest-boxes were checked regularly during the

NATO ASI Series, Vol. G 24
Population Biology of Passerine Birds
Edited by J. Blondel et al.

breeding season, and clutch size, number of hatched young, number of fledged young, and body mass of 13 day-old nestlings, were recorded. Nestling weight was measured with a Pesola spring balance to an accuracy of 0.1 g in 293 broods. All young and about 80% of adult females were ringed each year. We used the recapture rate as a measure of local survival for both parents and young. This measure underestimates the real survival for both young and adults because of unknown emigration and unknown number of non breeders (Perrins 1965, Nur 1984b, Nur and Clobert 1987).

A manipulation experiment was carried out in a 32 ha plot with 380 nest-boxes in 1988 and 1989. Over this period the broods of 219 Collared Flycatchers were manipulated while 347 broods were used as controls. Broods of 6 or 7 nestlings, hatching on the same day, were randomly chosen as enlarged, reduced or control broods. When nestlings were two days old, 2-3 ($n=151$) or 4-5 ($n=68$) nestlings were transferred between two experimental nests. There is no reason to suspect that this procedure in itself, increased the mortality of chicks. Both parents were caught with traps at the boxes and weighed on the day of hatching or the following day, and 12 days later. All nestlings and about 90% of the females from experimental broods were ringed, in both years.

Feeding frequencies at 49 experimental and 55 control nests were automatically recorded over two years. The recorder consisted of a Light Emitting Diode (LED) producing a beam which was broken by the birds when they entered or left a box. The time interval between successive interruptions was stored in an 8 kilobyte C-MOS memory, with a capacity of 4000 data entries. We were able to determine the number of visits and the time of the first and last feeding visits of the day, but it was impossible to separate the activity of each sex. Feeding frequency was recorded for two 24h periods beginning with day 10. In the analysis we used the mean number of visits over the two days. SPSS statistical package (Nie et al. 1975) was used throughout.

RESULTS

Natural broods

Clutch size of the Collared Flycatcher varied between 1 and 9 eggs (Tab. I). We only used data from nests containing 4-8 eggs ($n=982$) in the analyses, because clutches of 1-3 were usually found at the end of the breeding season, while clutches of 9 were probably the result of intraspecific nest parasitism. The two most common clutch sizes were 6 and 7.

The number of hatched young increased with clutch size ($Y = 0.810X - 0.276$, $r^2=0.125$, $p<0.0001$, $n=982$). There was a significant linear relationship between the number of fledged young and brood size (Fig. 1a, $Y = 0.790X + 0.049$, $r^2=0.201$, $p<0.0001$, $n=834$). On the other hand there was no significant difference in fledging success, when comparing different brood

Table I. Frequency distribution of clutch size of the Collared Flycatcher in an oak forest, Hungary during seven years.

Year	Clutch size											
	1	2	3	4	5	6	7	8	9	mean	SD	n
1982	-	-	2	10	24	47	11	2	-	5.6	0.96	96
1983	-	-	-	2	18	54	55	4	-	6.3	0.80	133
1984	-	-	-	9	30	46	67	8	2	6.3	1.04	162
1985	1	1	1	8	14	29	24	2	-	5.8	1.26	80
1986	-	-	6	33	82	75	20	5	-	5.4	1.01	221
1987	-	2	4	8	24	80	84	7	-	6.2	1.02	209
1988	-	-	-	2	11	60	43	3	-	6.3	0.74	119
Total	1	3	13	72	203	391	304	31	2	6.0	1.05	1020

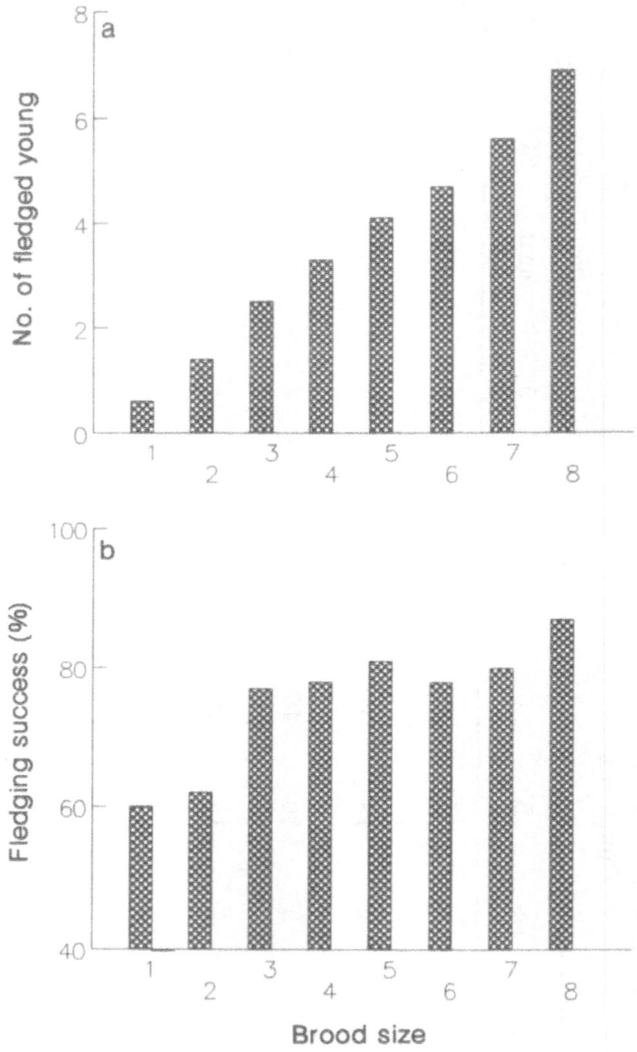

Fig. 1. Number of fledged young (a) and fledging success (b) as a function of brood size in the Collared Flycatcher during 1982-88; (n=5, 20, 35, 136, 197, 289, 141, 11 for the eight brood size categories, respectively).

sizes (no. of fledged per no. of hatched young), which showed that the mortality rate in large natural broods did not increase (Fig. 1b, Kruskal-Wallis one-way analysis of variance $X^2(7)=4.01$, $p=0.78$). Similarly, the mean nestling weight was not related to brood size (ANOVA $F=0.453$, $p=0.868$, NS). The recapture rate of nestlings during 1982-88 did not show any consistent trend with brood size (Tab. II). However, broods of 5-7 produced more recruits than other brood size categories. Local survival of adult females was independent of the number of reared young (Tab. II).

Table II. Recaptures of adult and nestling female Collared Flycatchers in relation to brood size. Data are pooled for seven years.

	Brood size							
	1	2	3	4	5	6	7	8
Adult females								
Ringed (n)	11	14	28	117	161	285	164	10
Recaptured (n)	3	5	9	26	36	65	40	2
Percent recaptured	27.3	35.7	32.1	22.2	22.4	22.8	24.4	20.0
Nestlings								
Ringed (n)	4	15	51	209	335	573	314	42
Recaptured (n)	0	0	2	7	21	39	22	2
Percent recaptured	0	0	3.9	3.4	6.3	6.8	7.0	4.8

Manipulated broods

In 1988 a significant difference was found between the number of young produced in the control and the extremely reduced brood size category (Fig. 2a, ANOVA $F=8.372$, $p<0.00001$). In 1989 the number of fledged young was smaller both in reduced and enlarged broods compared to the control, but the difference was only significant between the extremely enlarged and control broods (ANOVA $F=6.056$, $p<0.0001$).

The average fledging success differred considerably between the two years (Fig. 2b). In 1989 bad weather (heavy rainfalls during the parental care period) caused a much higher mortality among the nestlings. In both years the proportion of young surviving until fledging was related to brood size (Fig. 2b, Kruskal-Wallis ANOVA 1988: $X^2=36.579$, $p=0.0088$; 1989: $X^2=27.735$, $p=0.00001$).

Mean nestling body weight decreased with increasing brood size in both years (Fig. 3). The extremely enlarged broods differred significantly from the controls in both years (ANOVA 1988 $F=15.257$, $p<0.00001$; 1989 $F=8.996$, $p<0.00001$). In 1989 nestlings were lighter in each brood size category than in 1988 (Fig. 3).

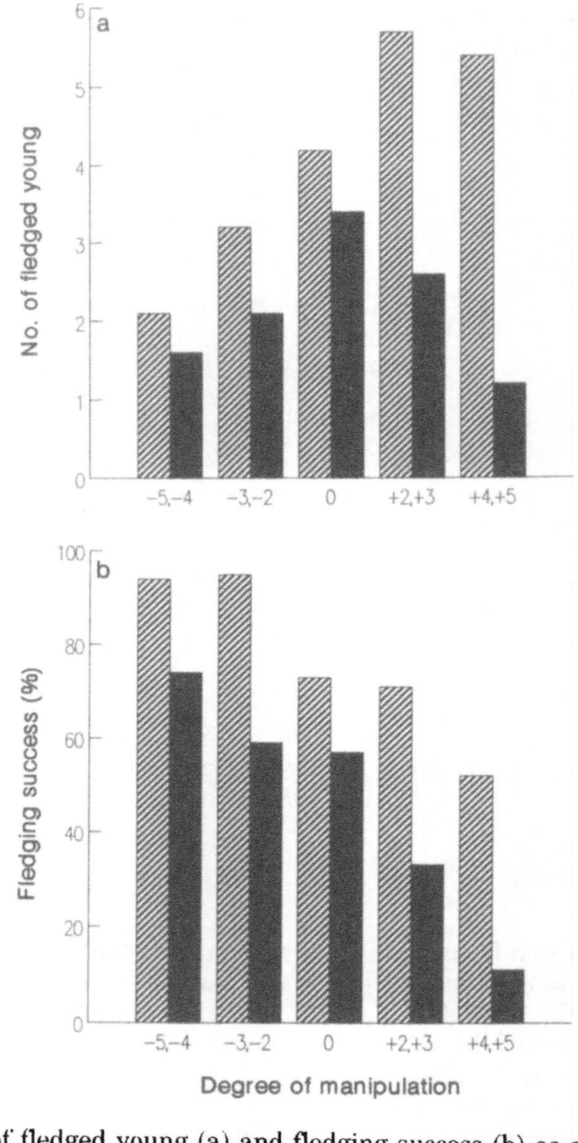

Fig. 2. Number of fledged young (a) and fledging success (b) as a function of the manipulated brood size of the Collared Flycatcher, in 1988 (hatched column) and 1989 (filled column); (n=31, 79, 347, 72, 37 for the five brood size categories, respectively).

Of the 211 females ringed in 1988, 77 were recaptured in 1989 (Tab. III). The recapture rate of females was independent of brood size ($X^2=1.26$, NS). The number of young recaptured seemed at first sight too small to make it possible to test for a manipulation effect. In fact the proportions (Table III) did

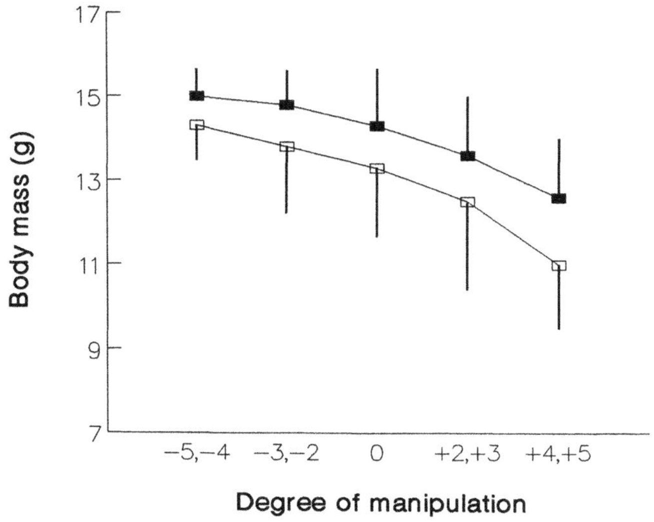

Fig. 3. Weight (mean, S.D.) of nestlings of the Collared Flycatcher at the age of 13 days in relation to manipulated brood size category in 1988 (filled square) and 1989 (empty square); (n=29, 71, 178, 65, 33 for the five brood size categories, respectively).

Table III. Recaptures of adult and nestling female Collared Flycatchers in relation to brood size manipulation.

	Degree of manipulation				
	-5,-4	-3,-2	0	+2,+3	+4,+5
Adult females					
Banded (n)	22	36	91	38	24
Recaptured (n)	7	15	30	14	11
Percent recaptured	31.8	41.7	33.0	36.8	45.8
Nestlings					
Banded (n)	16	64	169	106	51
Recaptured (n)	2	6	10	4	1
Percent recaptured	12.5	9.4	5.9	3.8	2.0

not even depart from homogeneity (G^2 test, $X^2(4)=4.999$, $p=0.287$). However, the degree of manipulation explained 98.6% of this G^2 statistic, and was significant (Logistic regression on degree of manipulation, G^2 test, X^2 $(1)=4.931$, $p=0.026$). Thus, the subsequent recapture rates of birds marked as young decreased as brood size increased.

Feeding frequency and weight loss of parents

The feeding frequency of individual broods (pooled data for two years) increased with respect to both natural and manipulated brood size. A stepwise multiple regression analysis using brood size and (brood size)1/2 as independent variables entered only the latter variable in both cases (natural broods: $Y = 337.9\ X1/2 - 139.1$, $r^2=0.590$, $p<0.0001$; manipulated broods: $Y = 255.8\ X1/2 + 46.0$, $r^2=0.592$, $p<0.0001$). This means that the number of feeding visits tends to saturate. Parents attending reduced broods fed their young significantly less often than those rearing non manipulated broods (Fig. 4, ANOVA 1988: $F=30.31$, $df=4$, $p<0.00001$; 1989: $F=4.24$, $df=4$, $p=0.007$) but in enlarged broods the amount of feeding visits remained at a similar level as that of the controls. Feeding visits per nestling declined as a second order

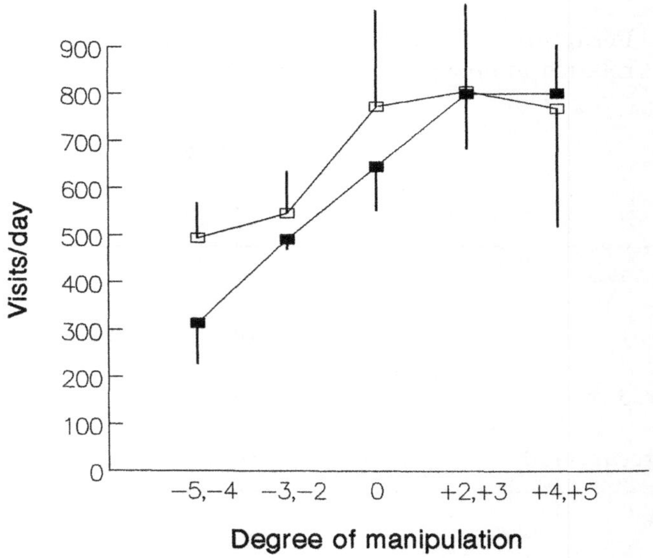

Fig. 4. Brood feeding frequency (mean, S.D.) in relation to brood size manipulation of the Collared Flycatcher in 1988 (filled square) and 1989 (empty square). (n=7, 15, 30, 15, 12 for the five brood size categories, respectively).

function of brood size in manipulated broods ($Y = 2.68X^2 - 46.76X + 292.4$, $r^2=0.733$ $p<0.00001$), while in case of natural broods a linear regression gave a better fit ($Y = -10.51X + 179.5$, $r^2=0.364$, $p<0.0001$). In both years only the extremely reduced broods differred significantly from the controls (Fig. 5, ANOVA 1988: $F=13.11$, $df=4$, $p<0.00001$; 1989: $F=10.36$, $df=4$, $p<0.00001$).

Weight loss of 80 female flycatchers measured in 1989 showed a nonsignificant trend of increase with brood size (Tab. IV). In 1988 ($n=83$) this trend was not clear; however the weight changes of females rearing reduced broods were smaller than those of females, which reared natural sized broods (ANOVA $F=4.566$, $p<0.002$). Males lost less weight than females, but their weight change did not show any tendency with brood size. Some authors assume that weight loss is an appropriate measure of the cost of feeding young (Nur 1984c, Orell and Koivula 1988, Smith et al. 1988). There was no evidence, from our results, that parents which fed their young more often suffered a higher weight loss. There were no significant correlations between brood feeding frequency and weight loss of females (1988: $r=0.022$, NS, $n=20$; 1989: $r=0.326$, NS, $n=24$) nor between nestling feeding frequency and weight loss of females (1988: $r=-0.096$, NS, $n=20$; 1989: $r=-0.203$, NS, $n=24$).

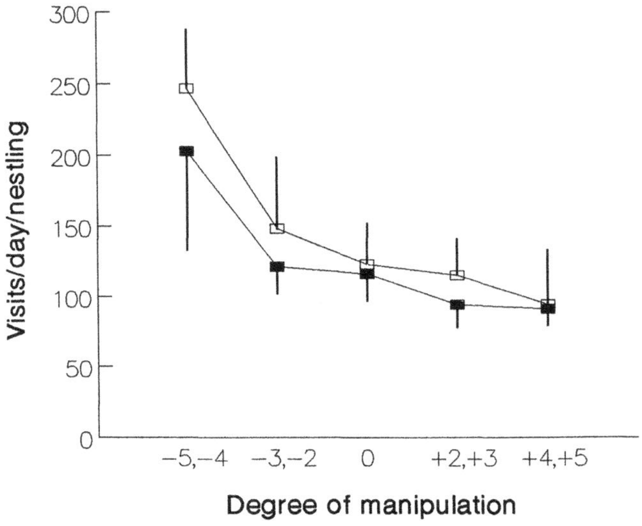

Fig. 5. Feeding frequency per nestling (mean, S.D.) in relation to brood size manipulation of the Collared Flycatcher in 1988 (filled square) and 1989 (empty square); ($n=7$, 15, 30, 15, 12 for the five brood size categories, respectively).

Table IV. Body mass loss of Collared Flycatcher females during the parental care period in relation to brood size in 1988 and 1989.

		Degree of manipulation				
		-5,-4	-3,-2	0	+2,+3	+4,+5
1988	mean (g)	0.09	0.10	0.19	0.15	0.15
	S.D.	0.07	0.07	0.10	0.09	0.08
	n	11	16	26	16	14
1989	mean (g)	0.10	0.16	0.20	0.23	0.28
	S.D.	0.07	0.06	0.10	0.21	0.26
	n	6	20	34	12	8

DISCUSSION

Several studies have reported an independence of clutch-size and of adult survival (see references in Orell and Koivula 1988). However, at least as many other studies have found evidence for clutch-size-dependence in adult survival (Askenmo 1979, Nur 1984a, Tinbergen et al. 1985, Orell and Koivula 1988). Data from unmanipulated broods of Collared Flycatcher showed that females rearing different numbers of young survived equally well. The number of young flycatchers produced till fledging was positively correlated with brood size, which is self-evident. However, young from different sized broods were equally well provisioned. Thus, the proportion of recruits in the next generation was independent of brood size. These results do not support the cost hypothesis.

In our study, as in that of Gustafsson and Sutherland (1988) in Gotland, Collared Flycatchers, nestling body weight was not related to brood size in natural broods, but it decreased with an increasing number of young in manipulated broods. Several studies have reported similar results for experimentally enlarged broods in other bird species (see Smith et al. 1989). The spring of 1989 was rainy and relatively cold during the period of parental care, thus starvation of young in large broods was more pronounced compared to the previous year. Orell and Koivula (1988) supposed that in years when foraging conditions were worse, Willow Tits decreased their food delivery rate, resulting in under-nourished young. We were able to measure accurately the feeding visits in Collared Flycatchers, and found a surprising result. In 1989, in spite of high feeding rates, nestling body weight was much lower in each category than in the previous year, when a lower feeding frequency was observed. Moreover, in large natural broods parents fed their young less often than parents with fewer young, but the nestling weight did not decrease in large broods. We must conclude that the relationship between the number of feeding visits and nestling weight is complex and future studies should consider factors such as e.g. environmental effects on foraging conditions, food availability, self-feeding behaviour and the number and size of delivered food items, as well as their nutritional quality.

The weight change of parents during the nestling period is considered as an estimate of feeding efforts (Nur 1984c, Jones 1987, Orell and Koivula 1988, Slagsvold and Lifjeld 1988, Smith et al. 1988). Alternatively, Norberg (1981) suggested that this weight change indicates an adaptive strategy of birds to reduce their flying cost while feeding their nestlings. In 1989 we found an increasing trend in the changes of body mass of female Collared Flycatchers with increasing brood size, although not significant. The increased amount of feeding visits did not increase the weight loss of females. These findings support Norberg's adaptation hypothesis. However, more data are needed to distinguish between the two hypotheses mentioned above.

Our results showed that Collared Flycatchers were unable to raise their feeding frequency above an average of 800 visits per day, even in large broods. A similar result was obtained by Smith et al. in Great Tits. However in Blue Tits (Nur 1984c) and Swallows (Jones 1987) an increased food delivery rate has been recorded, without an upper limit. Smith et al. (1988) suggested that the maximal feeding frequency depended not only on the physiological ability of parents but also on the amount of food items in the environment. Without more data on Collared Flycatchers it is impossible to test these ideas, but a third explanation is also apparent. The Collared Flycatcher is a long-distance migrant, which starts to migrate within a few weeks after breeding. Birds should keep themselves in good condition to endure the stress of migration. Therefore parents with large broods may not have worked so hard. Presumably this may be one reason why parental survival remained unaffected by both naturally and experimentally modified brood sizes. This upper limit in feeding activity resulted in lighter nestlings with lower survival rates until fledging in enlarged broods. The proportion of recruits also tended to be smaller from enlarged broods.

In the light of these data, it seems probable that the cost of reproduction in Collared Flycatcher can reduce both the quality and survival of the young.

ACKNOWLEDGEMENTS

We thank E. Pásztor, G. Meszéna and J. L. Martin for comments on an earlier draft of this paper, J.D. Lebreton for some statistical calculation and É. Ludvig, Gy. Kékesi and L. Vanicsek for help with the field work. R. Cowie significantly improved the English. Discussions in the Population Biology Working Group of Eötvös University helped to clarify the ideas.

We also thank Dan Wertheimer who designed the counters. This study was supported by the Hungarian Academy of Sciences and the Pilis Park Forestry.

REFERENCES

Askenmo C (1979) Reproductive effort and return rate of male pied flycatchers. Am Nat 114: 253-254

DeSteven D (1980) Clutch size, breeding success, and parental survival in the tree swallow (*Iridoprocne bicolor*). Evolution 34: 278-291

Gustafsson L, Sutherland WJ (1988) The costs of reproduction in the collared flycatcher *Ficedula albicollis*. Nature 335: 813-815

Jones G (1987) Parent-offspring resource allocation in Swallows during nestling rearing: an experimental study. Ardea 75: 145-168

Korpimaki E (1988) Costs of reproduction and success of manipulated broods under varying food conditions in Tengmalm's owl. J Anim Ecol 57: 311-323

Nie NH, Hull CH, Jenkins JG, Steinbrenner K, Bent DH (1975) SPSS: Statistical Package for the Social Sciences. McGraw-Hill, New York

Norberg RA (1981) Temporary weight decrease in breeding birds may result in more fledged young. Am Nat 118: 838-850

Nur N (1984a) The consequences of brood size for breeding blue tits: I. Adult survival, weight change and the cost of reproduction. J Anim Ecol 53: 479-496

Nur N (1984b) The consequences of brood size for breeding blue tits: II. Nestling weight, offspring survival and optimal brood size. J Anim Ecol 53: 497-517

Nur N (1984c) Feeding frequencies of nestling blue tits (*Parus caeruleus*): costs, benefits and a model of optimal feeding frequency. Oecologia (Berl.) 65: 125-137

Nur N (1989) The cost of reproduction in birds: Evaluating the evidence from manipulative and non-manipulative studies. In: Abstracts of the 74th Annual ESA Meeting (Toronto). p 216-217

Nur N, Clobert J (1987) Measuring Darwinian fitness in birds: a field guide. In: Proceeding of the 21th International Ornithological Congress (Ottawa). University of Ottawa Press, p 1121-1130

Orell M, Koivula K (1988) Cost of reproduction: parental survival and production of recruits in the Willow Tit *Parus montanus*. Oecologia (Berl.) 77: 423-432

Perrins CM (1965) Population fluctuations and clutch size in the great tit *Parus major* L.. J Anim Ecol 34: 601-647

Pettifor RA, Perrins CM, McCleery RH (1988) Individual optimization of great tits. Nature 336: 160-162

Slagsvold T (1984) Clutch size variation in birds in relation to nest predation: on the cost of reproduction. J Anim Ecol 53: 945-953

Slagsvold T, Lifjeld JT (1988) Ultimate adjustment of clutch size to parental feeding capacity in a passerine bird. Ecology 69: 1918-1922

Smith HG, Källander H, Fontell K, Ljungström M (1988) Feeding frequency and parental division of labour in the double-brooded great tit *Parus major*. Effects of manipulating brood size. Behav Ecol Sociobiol 22: 47-453

Smith HG, Källander H, Nilsson J-Å (1989) The trade-off between offspring number and quality in the great tit *Parus major*. J Anim Ecol 58: 383-401

Tinbergen JM, Balen JH van, Eck HM van (1985) Density dependent survival in an isolated great tit population: Kluyvers data reanalysed. Ardea 73: 38-48

BEHAVIOURAL ASPECTS OF REPRODUCTION

POLYGYNY AND BREEDING SUCCESS OF PIED FLYCATCHERS NESTING IN NATURAL CAVITIES

Rauno ALATALO[1], Allan CARLSON[2] and Arne LUNDBERG[3]

[1] Department of Biology, University of Jyväskyla, Yliopistonkatu 9, SF-40100 Jyväskylä, Finland
[2] Department of Wildlife Ecology, The Swedish University of Agricultural Sciences, S-750 07 Uppsala, Sweden
[3] Department of Zoology, Uppsala University, Box 561, S-751 22 Uppsala, Sweden

Many hole-nesting passerine birds accept, or even prefer, nest boxes. Hence most of our knowledge about the breeding biology, population dynamics, life-history evolution and so forth of such species comes from nest box studies. This, in some cases, might lead to erroneous conclusions. For example, breeding density in nest box areas may often be much higher than in the natural situation (e.g. von Haartman, 1971), possibly leading to unnatural density dependent effects. Furthermore, the use of nest boxes may reduce predation risks (e.g. Nilsson 1975, 1984 a,b, Møller 1989), while the routine procedure of cleaning boxes after each breeding season may reduce the number of parasites, thus affecting nest site choice, mate choice and reproductive success (Møller 1989).

In the Pied Flycatcher *Ficedula hypoleuca* many studies have dealt with breeding success, how females choose mates and/or the polyterritorial behaviour of males (e.g. von Haartman 1951, Alatalo et al. 1984, 1986, Slagsvold 1986, Slagsvold and Lifjeld 1986). All these studies with one exception (Alatalo and Lundberg 1984) have made use of nest boxes.

Here we examine mating behaviour and reproductive performance of Pied Flycatchers nesting in natural cavities, and compare the result with data from separate nest box breeding populations. We also focus on male polyterritoriality and the reproductive success of females of different mating status to examine the question whether natural data are consistent with nest box data.

METHODS

The study was carried out from 1983 to 1985 in a deciduous forest (Andersby, 60°09'N/17°49'E) north of Uppsala, central Sweden. The total study area consisted of 58.9 ha of forest. In the first year (1983), however, we only searched 38.8 ha of the area. The wood has been used in different ways, mainly for grazing by cattle and horses. This ceased in the 1920's when the open parts became overgrown. To recreate former land use structure, 31.2 ha of the forest (=14,1 ha of the area searched in 1983) was thinned in 1967/68 (Hytteborn

NATO ASI Series, Vol. G 24
Population Biology of Passerine Birds
Edited by J. Blondel et al.

1975), parts of which are now grazed by cattle while the rest (26.8 ha) is untouched. The area became a nature reserve in 1987. No nest boxes exist anywhere near this "natural" area, and our own nest box areas were located about 30 km south of it. We checked that birds inhabiting natural holes were not different from those breeding in boxes (e.g. mean plumage score of males according to Drost's (1936) scale: natural cavities: 3.5 $\pm$ SD 1.0, n=31, boxes: 3.4 $\pm$ 1.2, n=250, z=0.35, P>0.10).

In 1983 we searched the area every second day in May and twice a week in June, in 1984 every day in May and twice a week in June, and in 1985 every day in May. A total of 31 Pied Flycatcher males were individually colour ringed (8, 17 and 6 in each year). If not caught, they were recognized by plumage patterns of the back and/or by the shape of the white forehead patch. Difficult males, in particular if suspected to be polyterritorial, were carefully checked several times. This allowed us to determine the mating status of most males and females in the area.

We found a total of 105 nest holes used for breeding or defended by the unmated flycatcher males. In nests used for breeding we checked the start of egg laying (n=84), clutch size (n=74) and fledging success (n=84). In depredated nests we recorded whether the female, the clutch or nestlings were taken by looking for feathers and down in the cavity or at the entrance. The nest holes were inspected using a dentist's mirror and for each hole we measured features like size, height, tree species, tree condition etc., and we also classified nests according to their origin (see also Alatalo et al. 1988).

RESULTS

Breeding data

The mean breeding density of Pied Flycatchers in this forest was 0.6 pairs/ha although in unmanaged parts the density was slightly above one pair/ha. This could be compared with the mean of 1.8 pairs/ha (range 0.4 - 3.1) found in our nest box areas in deciduous forest (Alatalo et al. 1985). Thus, with boxes in excess we, on average, obtained threefold breeding densities.

We found laying date to be rather similar in natural cavities and nest boxes (Tab. I). On average, the median and mean laying dates were three and 2.8 days, respectively, earlier in boxes than in natural cavities (Tab. I, Alatalo et al. 1988). Primary females laid slightly earlier than monogamous females (25.6, n=7 versus 26.4, n=65; 1=1 May) while secondary females were considerably later (35.8, n=9).

In two of the years clutch size was slightly larger in boxes than in natural cavities, while in one it was smaller, and overall mean clutch size in natural sites was 6.45 while it was 6.62 for nest box areas in the same years (t=1.63, P>0.10; Tab. I, Alatalo et al. 1988). Since the median laying date differred between artificial and natural cavities, and the seasonal decrease in clutch size

Table I. Mean laying time, clutch size and fledging success for Pied Flycatchers breeding in natural cavities and boxes 1983-85. For laying date 1 = 1 May. • = $P<0.05$, •• = $P<0.01$, ••• = $P<0.001$; t-test for clutch size and Mann-Whitney U-test for laying date and fledged young.

	Laying date mean/N	Clutch size mean ± SD/N	Fledged young mean ± SD/N	% predation
1983				
cavities	26.7/21	6.45 ± 0.69/20	4.67 ± 2.97/21	23.8
	•••	ns	ns	
boxes	22.6/97	6.67 ± 0.72/97	5.85 ± 1.67/97	2.1
1984				
cavities	26.8/35	6.60 ± 0.86/30	4.13 ± 2.98/35	34.3
	ns	ns	ns	
boxes	27.9/49	6.51 ± 0.96/49	4.53 ± 2.63/49	18.4
1985				
cavities	30.0/28	6.25 ± 0.61/24	2.68 ± 3.18/28	57.1
	ns	•	•••	
boxes	29.6/20	6.63 ± 0.67/48	6.22 ± 0.87/48	0.0
total				
cavities	27.8/84	6.45 ± 0.74/74	3.79 ± 3.12/84	39.3
	•••	ns	•••	
boxes	25.0/166	6.62 ± 0.77/194	5.61 ± 1.92/194	5.7

is about 0.06 eggs/day (Alatalo et al. 1984), the small clutch size difference can be attributed to the slight difference in timing of laying. For primary females clutch size was 6.50 ± SD 0.84 (n=6) and for monogamous 6.63 ± 0.56 (n=57), while it was 5.50 ± 0.53 (n=8) for secondary females. The smaller clutch size of secondary females was not only due to their later breeding; monogamous females laying at the same time as secondary females on average laid 6.11 ± 0.33 eggs (n=9; t-test, $P=0.012$).

Pairs in natural cavities produced significantly fewer fledglings (3.79) than did pairs breeding in boxes (5.61) in the same years (Tab. I), and the main reason for this difference was the higher rate of predation in natural cavities (39.3 % versus 5.7 %; Tab. I). Predation reduced the fledging success of females of all mating categories; for primary females by 28.5 %, for monogamous by 38.4 % and for secondary by 44.4 %. However, among successful nests there was no difference in number of fledged young between natural cavities and boxes (monogamous females: cavities: 6.35 ± SD 0.70, n=40; boxes: 6.21 ± SD 1.01, n=171). If comparing number of fledglings produced by secondary females compared to simultaneous monogamous females in all natural nests (2.67 ± SD

2.55, n=9 versus 3.33 ± 3.20, n=9), secondary females managed to raise 80 % of that of monogamous females (U-test, P=0.63). If comparing only nests that gave rise to fledglings the proportion remained the same, but in this case the difference was significant (4.80 ± 0.45, n=5 versus 6.00 ± 0.71, n=5; P=0.012).

Predation and territory quality

Predation rate in natural cavities varied between years, and the predators probably were mainly mustelids. In 27 instances we knew the time of predation, and in 21 cases it occurred during laying or incubation. Also, out of nine events of predation during incubation, the female was taken by the predator in six cases, and might have been killed also in some of the others.

We compared characteristics of nests that became depredated with those that were not. First, if we consider the origin of the nest hole, those made by the Lesser Spotted Woodpecker *Dendrocopos minor* had the highest predation rate (72.8 %, n=11), Great Spotted Woodpecker *D. major* intermediate (41.4 %, n=17) while "fallen branch" holes were the safest (30.9 % predation rate, n=55). Second, if we consider properties of the nest hole and the nest tree, we found tree height, nest site height, angle of opening and distance from nestcup to entrance to be the most important factors in relation to predation (Tab. II). Thus, high nests in tall trees, with a nestcup far from the entrance and with an entrance hole pointing upwards were safer than nests with the opposite characteristics. That tree height entered as a significant variable is mainly due to the fact that many nests located in stumps of dead trees became predated. A discriminant analysis between depredated and successful nests showed that a model including tree height (coefficient = +1.21), angle of opening (+0.67) and tree circumference at 2 m (-0,77) correctly classified 78.6 % of all nests as predated or not. In reality this means that safe holes are those in high living trees, which are thin for their height, and where the nest hole has arisen by a fallen branch. By using these three variables we calculated a standardized canonical discriminant function coefficient, with a mean of zero and a SD of one, as a general measure of nest hole quality. It should be kept in mind that in the Pied Flycatcher the territory mainly consists of the nest hole (von Haartman, 1956). Then, if comparing territory/nest hole quality for sites occupied by females of different mating status, or by unmated males, we found a trend for the first nest of polygynous males to have the best territory quality scores whereas secondary nests and secondary territories (where the male failed to attract a female) had lower scores. A tendency in the same direction was observed when comparing simultaneous monogamous and secondary nests. The same pattern was also found for entrance size (mean of height and width), where monogamous and primary females occupied holes with smaller openings (x=35.9 mm) than did secondary females and unmated males (x=40.6 mm). Though all the differences are not statistically significant, the results clearly show that second females of polygynous males cannot compensate for their secondary status by having higher quality territories, as predicted by polygyny threshold models.

Table II. Nest site properties in relation to predation

Character	Successful mean ± SD(N)	Predated mean ± SD(N)	P t-test
Tree height (m)	15.6 ± 4.9(48)	11.8 ± 5.9(33)	<0.01 □
Angle of opening	6.4° ± 17(49)	-3.0° ± 8.5(32)	<0.01 □
Hole height (m)	5.6 ± 2.5(42)	4.4 ± 1.9(32)	<0.05
Distance to entrance (mm)	161 ± 66(45)	131 ± 54(32)	<0.05
Tree circumference at 2 m (cm)	130 ± 49(51)	118 ± 60(33)	ns □

□ Variables included in the discriminant function

Male behaviour

In the best studied year, i.e. 1983, 67 % of all males took up a second distant territory after having attracted a first female. This figure is the same as the estimate given from nest box studies in Finland by von Haartman (1956).

On average, among mated males (n=72), 12.5 % attracted more than one female while 8.9 % of all males (n=79) remained unmated. These figures could be compared with those we have obtained from nest box areas; in deciduous forest the proportion attracting more than one female ranged between 6.8 % and 11.3 % and in coniferous between 6.3 % and 39.4 %, the higher values referring to situations where boxes were in excess (Alatalo and Lundberg, 1984).

The median distance between the two nests of polygynous males was 230 m (N=9, range 120 - >300 m) while the corresponding figure for the distance between the two territories of polyterritorial males was 250 m (n=17, range 80 - 540 m). These distances are longer than those found by us for flycatchers breeding in nest box areas (m d=150 m, n=27; Alatalo and Lundberg, 1984), and thus in that respect nest box data probably underestimate natural polyterritorial distances.

DISCUSSION

In natural deciduous forests nest holes probably are plentiful though presumably many are of poor quality, as reflected by the lower breeding densities of Pied Flycatcher as compared to nest box areas. This study shows that laying dates and clutch size of the Pied Flycatcher are rather similar in natural cavities and nest boxes, while more fledglings are produced per nest in boxes. Previous studies on natural cavities have found that clutch size is often smaller in natural cavities than in boxes (Karlsson and Nilsson 1977, Nilsson 1984a, Gustafsson and Nilsson 1985, Slagsvold 1987). One explanation is that clutch size is reduced with decreasing cavity size, and that natural cavities are smaller than standard nestboxes. However, we found no difference in clutch size between natural cavities and boxes and, as reported earlier (Alatalo et al. 1988), no relationship existed between cavity size and clutch size. Moreover, many natural cavities in our area were larger than normal boxes (Alatalo et al. 1988).

The reason for the lower fledging success in natural cavities was predation. However, predation in boxes can be high also and probably even higher than in natural holes if the predator learns that boxes contain prey and remembers the position of boxes (e.g. Dunn 1977, Sonerud 1985). Risk of predation was related to properties of the nest hole and the nest tree. Holes low down were less safe than holes high up, especially if they were located in tree stumps. Also, if the entrance was less visible from the ground, predation rates were lower, suggesting that the main predators were ground living (probably *Mustela* spp.).

The components of reproductive success of females of different mating status were similar in the natural situation compared to boxes. Thus, clutch size was slightly higher for monogamous than for primary female while it was significantly lower for secondary females. The same holds for breeding success, and among un-predated nests secondary females produced significantly fewer offspring (80 %) than concurrently laying monogamous females. This reduction in fledging success for secondary females was not so much due to a higher predation rate as to starvation of some of the nestlings.

In relation to polyterritorial polygyny, this study suggests that scarcity of nest holes does not explain why males take up distant rather than nearby secondary territories . Nest holes are often clumped, so that there would be ample opportunities to attract additional females to nearer nest sites. It also seems unlikely that males take up secondary territories to be able to retain their female and quickly restart their breeding in cases of predation of the first nest (Slagsvold and Lifjeld 1986), since in many cases of predation the female was killed and there are thus few females to retain. Therefore, at the moment, the most viable hypotheses for explaining polyterritorial polygyny are the deception (Alatalo et al. 1981) and the female-female aggression (Breiehagen and Slagsvold 1988) hypotheses. The first one suggests that males take up distant secondary territories to hide the fact that they are already-mated while the latter implies that males are forced by females to move long distances because females are aggressive towards each other. This observational study cannot unambigously be used in favour of one or the other of these hypotheses. However, the data on timing of breeding and nest clumping may suggest that female-female aggression is a less likely mechanism for the spacing out of territories. First, secondary females start egg-laying approximately 10 days after the primary female, that is, when the first female is probably incubating and can hardly afford to search for and chase secondary females. Secondly, female-female aggression cannot hinder males in attracting secondary females very close to another pairs' nest in many cases. Clearly, however, secondary Pied Flycatcher females in the natural situation raise fewer offspring than concurrent monogamous females, and they cannot compensate their status by acquiring better territory quality, though good holes are available late in the season also.

ACKNOWLEDGEMENTS

We thank A.P. Møller and S. Ulfstrand for valuable comments on the manuscript. The study was supported by grants from the Swedish Natural Research Council (to AL).

REFERENCES

Alatalo RV, Lundberg A (1984) Polyterritorial polygyny in the pied flycatcher *Ficedula hypoleuca* - evidence for the deception hypothesis. Ann Zool Fennici 21 : 217-228

Alatalo RV, Carlson A, Lundberg A (1988) Nest cavity size and clutch size of pied flycatcher *Ficedula hypoleuca* breeding in natural tree-holes. Ornis Scand 19: 317-319

Alatalo RV, Lundberg A, Glynn C (1986) Female pied flycatchers choose territory quality and not male characteristics. Nature 323: 152-153

Alatalo RV, Lundberg A, Ståhlbrandt K (1984) Female mate choice in the Pied Flycatcher *Ficedula hypoleuca*. Behav Ecol Sociobiol 14: 253-261

Alatalo RV, Lundberg A, Ulfstrand S (1985) Habitat selection in the Pied Flycatcher *Ficedula hypoleuca*. In: Cody M (ed) Habitat selection in birds. Academic Press, New York, p 59-83

Alatalo RV, Carlson A, Lundberg A, Ulfstrand S (1981) The conflict between male polygamy and female monogamy: the case of the Pied Flycatcher *Ficedula hypoleuca*. Am Nat 117: 738-753

Breiehagen T, Slagsvold T (1988) Male polyterritoriality and female-female aggression in pied flycatchers *Ficedula hypoleuca*. Anim Behav 36: 604-605

Drost R (1936) Ueber das Brutkleid männlicher Trauerfliegenfänger, *Muscicapa hypoleuca*. Vogelzug 6: 179-186

Dunn EK (1977) Predation by Weasels (*Mustela nivalis*) on breeding tits (*Parus* spp.) in relation to the density of tits and rodents. J Anim Ecol 46: 633-652

Gustafsson L, Nilsson SG (1985) Clutch size and breeding success of Pied and Collared Flycatcher *Ficedula* spp. in nest-boxes of different sizes. Ibis 127: 380-385

Haartman L von (1951) Successive polygamy. Behav 3: 256-274

Haartman L von (1956) Territory in the Pied Flycatcher (*Ficedula hypoleuca*). Ibis 98: 460-475

Haartman L von (1971) Population dynamics. In: Farner DS, King JR (eds) Avian Biology. Academic Press, New York, Vol I, p 391-459

Hytteborn H (1975) Deciduous woodland at Andersby, Eastern Sweden. Above-ground tree and shrub production. Acta Phytogeogr Suec 61: 1-96

Karlsson J, Nilsson SG (1977) The influence of nest box area on clutch size in some hole-nesting passerines. Ibis 119: 207-211

Møller AP (1989) Parasites, predators and nest boxes: facts and artefacts in nest box studies of birds? Oikos 56: 421-423

Nilsson SG (1975) Clutch size and breeding success of birds in nest boxes and natural cavities. (In Swedish with English summary) Vår Fågelv 34: 207-211

Nilsson SG (1984a) Clutch size and breeding success of the Pied Flycatcher *Ficedula hypoleuca* in natural tree-holes. Ibis 126: 407-410

Nilsson SG (1984b) The evolution of nest-site selection among hole-nesting birds: the importance of nest predation and competition. Ornis Scand 15: 167-175

Slagsvold T (1986) Nest site settlement by the Pied Flycatcher: does the female choose her mate for the quality of his house or himself? Ornis Scand 17: 210-220

Slagsvold T (1987) Nest site preference and clutch size in the Pied Flycatcher *Ficedula hypoleuca*. Ornis Scand 18: 189-197

Slagsvold T, Lifjeld JT (1986) Mate retention and male polyterritoriality in the pied flycatcher *Ficedula hypoleuca*. Behav Ecol Sociobiol 19: 25-30

Sonerud GA (1985) Nest hole shift in Tengmalm's owl *Aegolius funereus* as defence against nest predation involving long-term memory in the predator. J Anim Ecol 54: 179-192

POLYGYNY AND NEST PREDATION IN THE ROCK PIPIT DO FEMALES TRADE MALE ASSISTANCE AGAINST SAFETY ?

Conny ASKENMO and Raimo NEERGAARD

Department of Zoology, Section of Animal Ecology, University of Göteborg, P.O. Box 25059, S-400 31 Göteborg, Sweden

Mating systems in territorial birds have been discussed in terms of female mate choice since the polygyny-threshold model appeared (Verner and Wilson 1966, Orians 1969). Females are assumed to choose optimally on the basis of some important male characteristics or the amount of resources controlled by territorial males. Males of high "quality" or those living in rich territories may attract several females while other less well supplied males remain unmated. Recently the idea of free choice has been questioned; preferred options may not always be obtainable because of conflicts of interest between males and females (Alatalo et al. 1981, Davies and Houston 1986, Wesolowski 1987, Davies 1989).

Few studies have been successful in unraveling the detailed basis for female choice (but see Pleszcynska 1978, Burley 1981). While the significance of sexually selected male characters remains controversial on theoretical grounds (Kirkpatrick 1987) many studies have convincingly shown that females take territorial quality or male parental capability into consideration (Searcy 1982). Age-related male characters sometimes influence attractiveness because they correlate with territory quality (Searcy 1979, Alatalo et al. 1986).

Nest predation is a potential danger with considerable impact on reproductive output. The role of predation in mate selection and mating systems is poorly understood, but circumstantial evidence suggests that females evaluate predation risks when choosing breeding sites. Females abandon breeding sites after nest predation (Furrer 1979, Wunderlee 1984) and prefer vegetation that provides good protection against nest predators (Holm 1973, but see Lenington 1980); they may also choose aggressive males to ensure good protection of the nest site (Dickinson and Lein 1987). Birds also avoid breeding close to nests of predators (Geer 1978, Reynolds 1979).

A female can reduce the risk of nest predation in at least two ways: by choosing a male with safe nest sites in his territory, or by avoiding areas with high predator density. The first option is the only one available when predator populations are evenly distributed over the breeding habitat. When predator density is lower in some places those refuges should be preferred, with possible consequences for the mating system if one sex values space more, and defends larger exclusive areas against intruders of the same sex.

This is a presentation of results from a study of polygyny frequency and nest predation in two populations of Rock Pipits *Anthus spinoletta*. We will show that polygyny frequency varies locally, and present evidence suggesting a preference for breeding sites with few predators. We also discuss the possibility that female preference for safe breeding sites coupled with intense competition

NATO ASI Series, Vol. G 24
Population Biology of Passerine Birds
Edited by J. Blondel et al.

among males for a limited number of territories may favour a polygynous mating system.

METHODS

The Rock Pipit is, in southern Scandinavia, a mono-territorial migrant passerine living close to the shoreline; we studied the mating system of this species on two islands off the coast of southwestern Sweden. The population on the island of Nidingen has been studied since 1981; the study was expanded in 1984 when the population on the island of Malön was included. Malön is the largest island (approx. 1 km^2) and lies close to the mainland, while Nidingen is less than 0.2 km^2 and is situated about six kilometres from the coast.

Pipit territories are usually lined up along the shore, and interior parts of larger islands, like Malön, are never used by the pipits. Because of this one-dimensional spacing pattern of territories, we did not use densities to describe population levels. Instead we calculated the number of resident pipits per 100 m shoreline; since we never observed territorial pipits in some stretches of the shore on Malön, we excluded these sectors when calculating the length of the available shore line.

A large fraction of the pipits returning in spring were already colour-banded; unbanded birds were captured and banded shortly after arrival in wire-cage traps placed at feeding sites, or later by presentation of mounts inside traps in combination with playback of song; some birds were caught in traps placed over nests with large nestlings. Each season a small number of birds were unbanded or wore metal rings only; mostly they could be identified within the season due to restricted movements inside a territory. Nestlings were banded with metal rings, except a few broods which escaped notice before fledging; from 1988 onwards many nestlings were colour-banded.

Pipits were sexed on the basis of morphology or behaviour. Male Rock Pipits have longer wings (Larsson and Strandvik 1986), and most pipits were caught and measured at least once. The sex of resident individuals was checked during the breeding season by means of their behaviour. Males perform a conspicuous flight display and show special postures inside the territory; females build the nest and incubate alone.

Nests were found in different ways; by watching nest-building females, by flushing incubating females or watching them from a distance when returning to the nest after a foraging trip, or by watching parents from a blind when feeding nestlings. Wooden sticks or stones found close to the nest were used to mark their location.

Pair-bonds were mostly determined on the basis of one or more of the following critera: observations of a male consorting with a female during the prelaying period, observations of a male giving alarm calls close to a female's nest, or observations of a male feeding nestlings or fledglings. Most pair-bonds were easily determined, but in some cases we had no good information that tied a female to a particular male; we then considered her mated with the owner of the territory in which she nested. Some of these females may have been assigned to the wrong male, especially on Nidingen, because of the dense territorial system and lack of detailed knowledge about territorial borders.

Breeding success was recorded by repeated nest inspections. Nestlings were counted and banded at an age of 8-12 days, and nests were inspected for dead young after fledging. Robbed nests were easy to identify because predators usually pulled out the nest cup. Some nestlings probably escaped the predator when nests were robbed in late stages, because old nestlings leave the nest quickly when disturbed. We recorded total failure for all nests with signs of predation unless we later observed fledged young or saw parents with food or heard them giving alarm calls; in this case we assigned these nests a fledgling production of one, or the number actually observed. Nest predation rates were calculated as the proportion of nests destroyed by predators before fledging. This method underestimates the true predation rate but should be adequate for an analysis of predation pressure differences between populations.

Natal dispersal tendencies were analysed by means of local recruitment rates. We calculated two different measures for each population: (1) the proportion of fledglings from each cohort which returned and settled either on Malön or Nidingen, (2) the proportion of recruits each year that were born on the island where they settled. These rates mainly reflect site fidelity of first year birds because most recruits (219 out of 236) were identified as settlers in their first breeding season. Some of the seventeen recruits identified later in life may have bred elsewhere in previous years, and some first-year males may have lived as nonterritorial floaters. Nevertheless, in our calculations of recruitment rates we assumed that these seventeen birds, already in their first season, were stationary at the place where they later were identified, but belonged to the small group of unidentified birds present in most season. Independent data on movements of identified birds support the assumption that breeding dispersal is uncommon (see below).

We used the SAS computer package (SAS Institute Inc. 1985) for storing and handling of data and for statistical analysis. Statistical tests based on ranks were used throughout the analysis, but for descriptive purposes we used means and standard deviations.

RESULTS

Population density

Variation in habitat suitability should produce a patchy distribution of populations (Fretwell and Lucas 1969), with higher densities in more suitable habitats. Our data suggest that Nidingen is more attractive than Malön; population levels were higher on Nidingen in four out of five years with complete cover of both populations (Table I), and the number of resident pipits per 100 m shoreline was significantly higher on Nidingen (Table I and II).

Table I. Population levels in different years. All males (also unmated) claiming territories, and all females making breeding attempts or consorting with territorial males (nest not found) are included. Individuals observed only temporarily (floaters) are excluded. Calculations of number of birds per 100 m shoreline are based on shore sectors used for breeding at least once during the study.

	Total stationary population		Number of birds per 100 m shoreline	
Year	Nidingen	Malön	Nidingen	Malön
1981	117	-	4.3	-
1982	99	-	3.7	-
1983	104	-	3.9	-
1984[1]	103	58	3.8	1.8
1985	54	80	2.0	1.3
1986	78	74	2.9	1.2
1987	60	53	2.2	0.85
1988	73	58	2.7	0.94
1989	85	55	3.1	0.89

[1]. In 1984 the study area on Malön covered approximately half of the shoreline normally occupied by territorial pipits.

Sex-ratio

Sex-ratios in resident populations should be related to habitat suitability if one sex has greater freedom to choose among breeding habitats (Fretwell and Calver 1969, Davies 1989). A surplus of females in high density areas would indicate that female fitness is less depressed by density, and suggests that mate sharing does not necessarily discourage females from settling.

Females always outnumbered males in both populations, except in two seasons with even sex-ratio on Malön, and the male deficit was larger on Nidingen (Table II). We never recorded a surplus of males, although a small number of the territorial males were usually unmated. Furthermore, the sex-ratio among residents probably underestimated the basic population sex-ratio, because five males removed during the prelaying season in 1987 were replaced by four new males, suggesting that there was a pool of nonterritorial male floaters.

Table II. Population characteristics expressed as averages (±SD) of annual estimates; for Nidingen separate statistics are given for 1981-89 (recruitment 1982-89) and for 1984-89 (recruitment 1986-89) when both populations were studied.

	Nidingen all years		Malön		Nidingen later years
Number of birds per 100 m shoreline	2.2±0.79	**	1.2±0.34	**	2.8±0.65
Sex-ratio, males/female	0.80±0.06	*	0.90±0.08	(*)	0.81±0.07
Average number of mates, all males	1.27±0.09	*	1.13±0.11	(*)	1.25±0.10
mated males	1.34±0.13	*	1.19±0.08	(*)	1.31±0.12
Proportion unmated among territorial males	0.05±0.03	NS	0.05±0.04	NS	0.04±0.03
Nest predation rate	0.08±0.06	**	0.30±0.09	*	0.10±0.07
Proportion fledglings settled on Malön or Nidingen	-		0.099±0.02	*	0.20±0.07
Proportion birds locally born among recruits	0.71±0.14	*	0.34±0.12	*	0.80±0.07

Significance levels calculated by means of Wilcoxon rank-sum test:
** $p<0.01$, * $p<0.05$, (*) $p<0.10$, NS $p>0.1$.

Mating success

Several males became polygynous in both populations, since more females than males settled, and some territorial males remained unmated; the number of males with more than one female varied between seasons; on Nidingen we recorded 6-16 males (14-44%), and on Malön 3-5 males (8-21%). Most of them had two females but a few trigynists were recorded (Fig. 1) The average mating success was higher on Nidingen, although the difference was not fully significant over the six years when both populations were studied (Table II). This is a necessary consequence of the larger surplus of females on Nidingen, because we found no firm evidence of unmated females. The lower mating success on Malön was not due to a higher frequency of unmated males but the harem size of mated males was smaller (Table II).

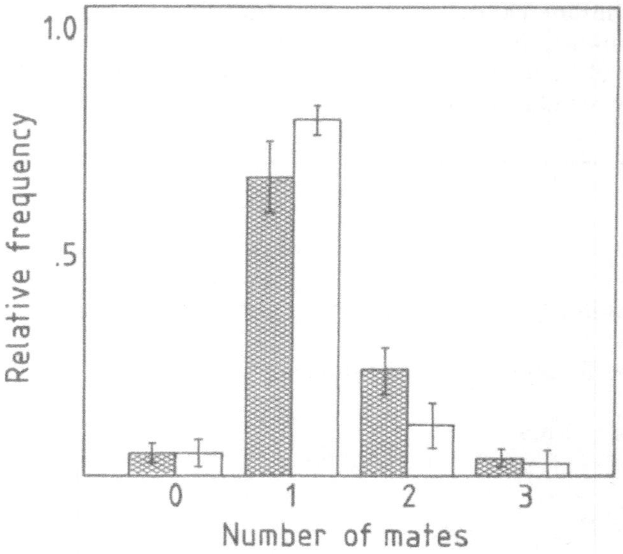

Fig. 1. Distribution of males in relation to number of mates. Average frequency over nine seasons for Nidingen (filled bars) and six seasons for Malön (empty bars). Vertical lines represent ±2SE.

Nest predation

Rock Pipit nests are easily accessible for predators when placed in grass or herb vegetation, but some nests are built below boulders or drift-wood and are better protected. We have little information about the importance of different predators since we never saw a predator in action. Corvids are probably the most important avian predators but gulls and some waders, like the Turnstone *Arenaria interpres* and the Oystercatcher *Haematopus ostralegus* may have destroyed some nests. A few pairs of Hooded Crow *Corvus corone* bred on Malön every year and since 1988 also one pair of Magpie *Pica pica*. Flocks of non-breeding crows visit Malön regularly. Avian predators are uncommon on Nidingen during the breeding season because of the isolated position. Small flocks of crows (usually less than five) visit the island now and then, but stay only for a few hours; no corvids were found breeding during this study.

In a few nests we found remains of nestlings or females clearly indicating a mammalian predator. Mink *Mustela vison* have been observed on Malön several times and smaller mustelids may occur. Occasional mink have been observed on Nidingen but no other predatory mammal.

Not surprisingly pipit nests were robbed more frequently on Malön; predation was more severe on Malön in all six years when both populations were studied, and the average predation rate was three times higher than on Nidingen (Table II).

Breeding success

Females tended to do better on Nidingen because of the lower predation rate (Fig. 2), although the difference was significant only in 1987. It is possible that also other conditions for raising young, for instance the food supply, were superior on Nidingen, making this island more attractive. We tested this possibility by examining the number of fledglings produced from successful nests, i.e. nests that were not robbed. There is no indication of better conditions on Nidingen; fledgling production from nests that escaped predation was similar, except for 1984 when production was significantly lower on Nidingen (Fig. 2). This year many nests on Nidingen were destroyed because of flooding during a severe storm.

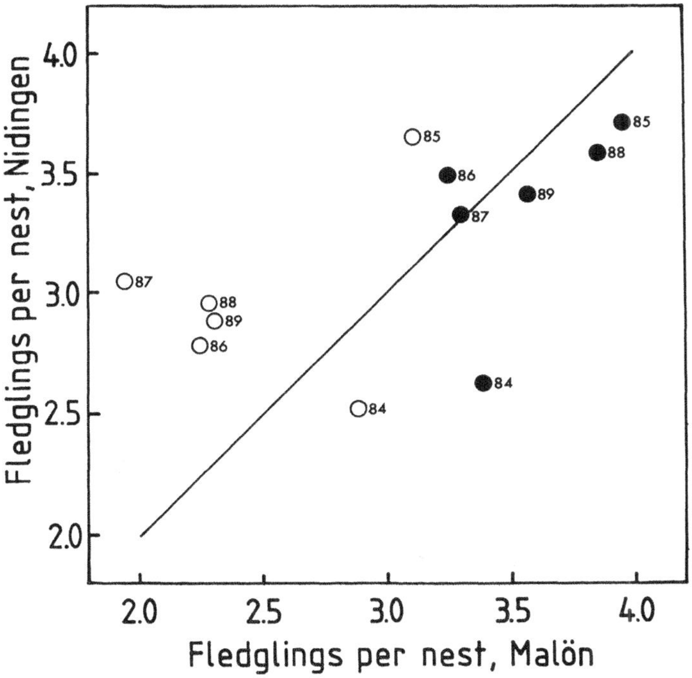

Fig 2. Relation between estimates of fledgling production per nest from Nidingen and Malön during six seasons (figures refer to year); circles (o) denote average production based on all nests, including those destroyed by predators; dots (•) show corresponding estimates for successful nests; line denotes equal production. Significant difference between Nidingen and Malön significant in two cases: 1984 for successful nests, and 1987 for estimates based on all nests ($p<0.05$ and $p<0.01$, Wilcoxon rank-sum test).

Dispersal

By definition inhabitants of attractive areas should be inclined to stay, and birds born in such areas should try to remain there if possible. One would therefore predict less dispersal from Nidingen.

We found no significant difference between the two populations in turnover rate, i.e. the proportion of recruits in the stationary population. Breeding dispersal between the two populations was uncommon; three adult females moved from Malön to Nidingen and one in the opposite direction; none of the males bred on both islands, but one of them held a territory on Nidingen, late in the season, after having lost his nest on Malön because of predation. Judging from this small sample, adult birds tend to move away from a breeding site with high predation rate towards safer sites, in spite of a higher density there.

Fledglings from Nidingen were more often recruited as settlers either on Nidingen or on Malön. Out of the 685 fledglings known to be born on Nidingen in 1985-88 a total of 133 (19%) settled on either Malön or Nidingen, while only 35 (10%) out of 352 fledglings from Malön were found as settlers somewhere on the islands ($X^2=14.7$, $p<0.001$). The probability of recruitment varied between seasons, but was significantly higher for birds born on Nidingen (Table II). Furthermore, recruits on Nidingen were more often locally born; 82% of 143 recruits on Nidingen were born there, while only 34% of 93 recruits on Malön were born on this island ($X^2=52{,}4$ $p<0.001$). The proportion of locally born birds among recruits, calculated annually, was significantly higher on Nidingen (Table II), although the variation was considerable (Fig. 3). Local offspring, both males and females, were more common on Nidingen in all four years from which recruitment data were available from both populations (Fig. 3).

Local recruitment from the cohorts born in 1985-88 was similar for females and males. On Malön 17 out of 53 (32%) female recruits, and 15 out of 40 (38%) male recruits were locally born; the corresponding figures for Nidingen are 63 out of 76 (83%) female recruits, and 54 out of 67 (81%) male recruits; neither of these differences are statisticallly significant. Local recruitment on Nidingen was significantly lower for females than for males in earlier years (cohorts born in 1981-84): 39 out of 73 (53%) female recruits, and 49 out of 68 (72%) male recruits were locally born ($X^2=4.4$ $p<0.05$). This result suggests that the notion of less dispersive females, proposed by Greenwood and Harvey (1982) for birds with a resource-defence mating system, does not hold in a short-term perspective. Apparently, conditions favouring high local recruitment do not necessarily occur simultaneously in males and females.

DISCUSSION

Polygyny was apparently a consequence of a male deficit on both islands. This is in agreement with experimental results showing that polygyny can be induced by removal of males (Smith et al. 1982), and confirms the importance

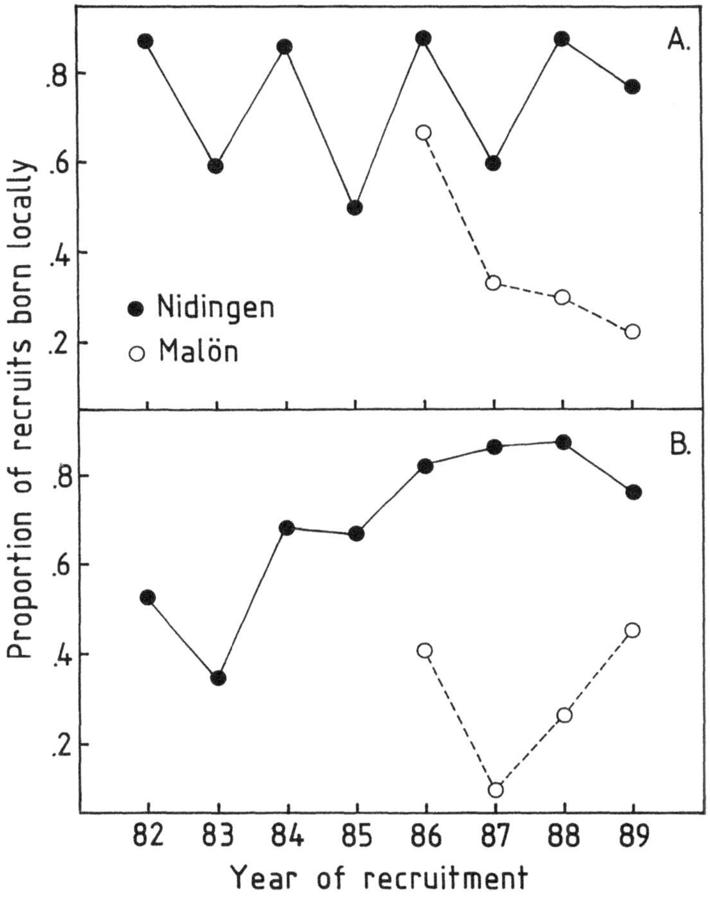

Fig 3. Proportion of locally born male (A) and female (B) recruits in different years; circles (○) refer to Malön, and dots (●) to Nidingen.

of the "operational sex ratio" for mating systems (Emlen and Oring 1977). The basic problem then is to understand why there is a shortage of territorial males, despite the presence of nonterritorial floaters indicated by replacements of removed males. We suggest that intra-sexual competition and a conflict of interest between the sexes play important roles; our data indicate that females compete for safe nest sites, which makes predation safe areas attractive also for males competing for access to mates. Hence, areas, like Nidingen, with low risk of nest predation should be densely populated, but why should the sex-ratio become more female-biased in these areas ?

One possible reason might be that males and females react differentially to intruders. If mating success is correlated with territory size (McLean and Seastedt 1979, Davies and Lundberg 1984), resistance against intruders of the same sex may be stronger in males, because they may pursue an area-maximizing strategy (Hixon 1987). Males with larger territories should be able to secure more mates and sire more offspring. Female fitness, on the other hand, is not likely to be linearly related to territory size unless she adopts a polyandrous strategy with multiple clutches cared for by different males; this option is not available for Rock Pipit females because males do not incubate. Without this option a territory over a certain size should be of no value to the owner, because the marginal benefit, in terms of brood survival, does not balance the increasing defence cost. Hence, females should be less inclined to spend time repelling intruders; the optimal female strategy during the pre-laying period might be to spend less time fighting than males but more time foraging in order to prepare nutritionally for the demanding time of egg formation and incubation. Hence, selection may favour highly aggressive males and more tolerant females with possible consequence on the sex-ratio. Time-budgets indicate that Rock Pipit females are less aggressive and forage more than males, suggesting that tolerance against intruders might be different (unpubl. data).

In attractive areas with high population densities males should be more unwilling than females to reduce territory size if male fitness is more severely affected by decreasing territory size. An asymmetry like this might prevent males from matching the number of females in high density areas, because some males would be able to defend several female home-ranges (Davies and Lundberg 1984, Wesolowski 1987).

The cost of nest predation is obvious, but females may reduce this cost by producing repeat clutches. Still, reproductive effort will be larger and may reduce female survival; furthermore, a delay due to nest predation may be critical for double-brooded species like the Rock Pipit, if time or energy constraints prevent laying of a second clutch after a repeat clutch. There are also costs associated with harem membership; females may face a choice between a safe nest site with reduced male assistance and an unsafe site with full assistance. Joining a harem will be an optimal strategy only if costs due to mate sharing do not exceed the benefits obtained by settling in a predation safe-habitat. Excluding predation effects, there is a weak tendency of higher reproductive success on Malön suggesting there are such costs (Fig. 2); the lower density on Malön may also improve breeding success. However, if predation losses are taken into account, females breeding on Nidingen tended to do better, in spite of possible costs associated with more frequent polygyny and high density (Fig. 2). Similarily, Dhondt (1987) found polygynous Blue Tits *Parus caeruleus* only in subplots with the highest reproductive success.

The polygyny-threshold model does not unequivocally predict higher average fitness in better territories, but such a relation would be more likely when territories occur in discrete, uniform patches, with great difference in quality between the patches (Altman et al. 1977, Wittenberger 1979, Davies 1989). This pattern partly applies to the situation described here, because of the clear difference in predation rate between the islands. However, the fact that unmated and monogamous males were recorded on both islands suggests that there was also a local variation in quality. This variation may have caused an overlap between the two islands in the quality of options available to females, as suggested by the occurrence of polygynous males on both island; some good territories on Malön may have been a better choice than poor territories on Nidingen, in spite of the higher risk of predation.

As only two populations were studied we do not know if the relationship between polygyny frequency and predation pressure is general. Data from a pipit population breeding on the mainland coast close to our populations fit into this pattern; nest predation rate was around 50%, i.e. higher than on Malön, and polygyny was never recorded (Elfström 1988).

The natal dispersal pattern suggests that young pipits are more inclined to settle in the natal area when it is predation-safe. One alternative explanation for the high recruitment rate of locally born individuals on Nidingen is that the isolated position of this island constrains natal dispersal. This is not very likely because many juveniles born on Nidingen have been observed as visitors on Malön during the summer, and dispersion would be possible during the spring migration.

In conclusion, the results of this study indicate that female Rock Pipits trade mate exclusiveness against security; but we must learn more about the costs of polygyny to females and about dispersal constraints before we can conclude that females choose optimally according to the polygyny-threshold model; there might be an important element of conflict between males and females involved in the processes shaping the mating system of Rock Pipits.

ACKNOWLEDGEMENTS

We thank all those people, too many to be mentioned here, who worked for us on Nidingen as staff members of the bird observatory. We are especially grateful to Uno Unger for his dedicated work both in the field and at the desk, and to Tommy Järås and Jonas Wikström for field assistance. We thank Richard Cowie and an anonymous referee for suggesting many improvements to the manuscript. This research was supported by grants to CA from the Swedish Natural Science Research Council and to RN from "Wilhelm och Martina Lundgrens vetenskapsfond" and "Elis Wides fond".

REFERENCES

Alatalo RV, Carlsson A, Lundberg A, Ulfstrand S (1981) The conflict between male polygamy and female monogamy: the case of the Pied Flycatcher *Ficedula hypoleuca*. Am Nat 117:738-753

Alatalo RV, Lundberg A, Glynn C (1986) Female Pied Flycatchers choose territory quality and not male characteristics. Nature (Lond.) 323:152-153

Altman SA, Wagner SS, Lenington S (1977) Two models for the evolution of polygyny. Behav Ecol Sociobiol 2:397-410

Burley N (1981) Mate choice by multiple criteria in a monogamous species. Am Nat 117:515-528

Davies NB (1989) Sexual conflict and the polygamy threshold. Anim Behav 38:226-234

Davies NB, Houston AI (1986) Reproductive success of Dunnocks, *Prunella modularis*, in a variable mating system. II. Conflicts of interest among breeding adults. J Anim Ecol 55:139-154

Davies NB, Lundberg A (1984) Food distribution and a variable mating system in the Dunnock *Prunella modularis*. J Anim Ecol 53:895-912

Dickinson TE, Lein MR (1987) Territory dynamics and pattern of female recruitment in Red-winged Blackbirds (*Agelaius phoeniceus*). Can J Zool, 65:465-471

Dhondt AA, (1987) Polygynous Blue Tits and monogamous Great Tits: Does the polygyny-threshold model hold? Am Nat 129:213-220

Emlen ST, Oring LW (1977) Ecology, sexual selection and the evolution of mating systems. Science, N.Y., 197:215-223

Elfström T (1988) Meadow Pipit *Anthus pratensis* L. and Rock pipit *A. spinoletta littoralis* Brehm. In: Andersson S (ed) Birds in the agricultural landscape. Vår Fågelvärld, suppl no 12, SOF, Stockholm, p 267-276. (Swedish with English summary)

Fretwell SD, Lucas HL Jr (1969) On territorial behaviour and other factors influencing habitat distribution in birds. I. Theoretical developement. Acta Biotheoretica 19:16-30

Fretwell SD, Calver JS (1969) On territorial behaviour and other factors influencing habitat distribution in birds. II. Sex ratio variation in the Dickcissel (*Spiza americana* Gmel). Acta Biotheoretica 19: 37-44

Furrer RK (1979) Brutplatzverschiebungen bei der Wacholderdrossel *Turdus pilaris* nach Nestverlust. J Orn 120: 86-93

Geer TA (1978) Effects of nesting Sparrowhawks on nesting tits. Condor 80: 419-422

Greenwood PJ, Harvey PH (1982) The natal and breeding dispersal of birds. Ann Rev Ecol Syst, 13: 1-21

Hixon MA (1987) Territory area as a determinant of mating systems. Amer Zool 27: 229-247

Holm CH (1973) Breeding, sex ratios, territoriality, and reproductive success in the Red-winged Blackbird (*Agelaius phoeniceus*). Ecology, 54: 356-365

Kirkpatrick M (1987) Sexual selection by female choice in polygamous animals. Ann Rev Ecol Syst 18:43-70

Larsson L, Strandvik P (1986) Wing length as a character for sexing Rock Pipits *Anthus spinoletta littoralis*. Ringing and Migration 7:71-74

Lenington S (1980) Female choice and polygyny in Red-winged Blackbirds. Anim Behav 28: 347-361

MacLean SF, Seastedt TR (1979) Avian territoriality: sufficient resources or interference competition. Am Nat 114: 308-312

Orians GH (1969) On the evolution of mating systems in birds and mammmals. Am Nat 103: 589-603

Pleszcynska WK (1978) Microgeographic prediction of polygyny in the Lark Bunting. Science (N.Y.) 201:935-937

Reynolds TD (1979) The impact of Loggerhead Shrikes on nesting birds in a sagebrush environment. Auk 96:789-800

SAS Institute Inc. (1985) SAS/STAT Guide for Personal Computers, Version 6 Edition. Cary, NC., p.378

Searcy WA (1979) Male characteristics and pairing success in Red-winged Blackbirds. Auk 96:353-363

Searcy WA (1982) The evolutionary effects of mate selection. Ann Rev Ecol Syst 13:57-85

Smith JNM, Yom-Tov Y, Moses R (1982) Polygyny, male parental care, and sex ratio in Song Sparrows: an experimental study. Auk 99:555-564

Verner J, Willson MF (1966) The influence of habitats on mating systems of North American passerine birds. Ecology 47:143-147

Wesolowski T, (1987) Polygyny in three temperate forest Passerines (with a critical re-evaluation of hypotheses for the evolution of polygyny). Acta Orn 23:273-302

Wittenberger JF, (1979) The evolution of mating systems in birds and mammals. In: Marler P, Vandenbergh J (eds) Handbook of Behavioral Neurobiology. Vol. 3. Plenum Press, New York. p 271-349

Wunderlee JM Jr (1984) Mate switching and a seasonal increase in polygyny in the Bananaquit. Behaviour 88:123-144

INCUBATION PATTERNS OF BLUE TITS (*PARUS CAERULEUS*)

Richard J. COWIE [1] and John A. NOVAK [2]

[1] School of Pure and Applied Biology, University of Wales, College of Cardiff, P.O. Box 915, Cardiff, CF1 3TL, Wales U.K.
[2] Department of Biology, Colgate University, Hamilton, New York 13346, U.S.A.

Our work on the breeding ecology of tits in suburban environments (Cowie and Hinsley 1987) has indicated that a major factor contributing to the poor breeding success of Blue Tits (*Parus caeruleus*) in this habitat, is the high percentage of eggs (32%) that fail to hatch.

We believe that this may be due to the females having to leave their eggs for abnormally long periods, during incubation, to find sufficient food. This, combined with the rapidly increasing risk of lethal chilling of embryos below a certain threshold of nest attentiveness (Webb 1987), may explain the poor hatching success.

In order to test this idea we have developed a nest-box activity recorder, which not only records the incubation pattern of the female, but can also distinguish visits made to the nest by the male. This is important because in common with other *Parus* species, only the female Blue Tit incubates the eggs, but she is often fed by the male who visits the box with food, as well as feeding her when she leaves the nest.

In this paper we report a preliminary investigation in deciduous woodland, using this apparatus, in which we have recorded the incubation pattern of Blue Tits in a natural woodland environment.

Our results provide an interesting comparison with similar data collected for other *Parus* species such as the Great Tit (*P. major*) (Royama 1966, Haftorn 1981); the Willow Tit (*P. montanus*) (Haftorn 1979); the Coal Tit (*P. ater*) (Haftorn 1984) and the Marsh Tit (*P. palustris*) (Nilsson and Smith 1988).

METHODS

Data were collected from Blue Tits breeding in an area of deciduous woodland in the Vale of Glamorgan to the West of Cardiff, (Long. 2°21' W, Lat. 51°26' N).

Prior to the breeding season, nestboxes constructed out of concrete (see Cowie and Hinsley 1987 for details), were installed at a density of four boxes per hectare. A 20mm length of plastic piping, with an internal diameter of 32mm and an external diameter of 36mm, was inserted into the hole of the nestboxes. This "dummy" insert was packed securely into the larger hole using

NATO ASI Series, Vol. G 24
Population Biology of Passerine Birds
Edited by J. Blondel et al.

a non-setting mastic. A perch connected to a microswitch was fitted on the outside of the box, so that the top of the perch was 5mm below the bottom of the hole. The plan was that the birds would trigger the perch on entering the box, but not on leaving.

When the nesting season had begun, and females had commenced egg-laying, the females were caught while roosting in the box at night. They were fitted with a colour ring to which was glued a piece of folded tin-plate, approximately 4 x 4 x 1.5mm. Once ringed, the females were weighed and placed back on the eggs.

One or two days before the estimated start of incubation, the nest-box was fitted with an "activity recorder". This had three connections to the nestbox. First a coil, which, in conjunction with the circuitry, acted as a metal detector. This coil replaced the dummy plastic insert in the nestbox hole. When the female, with the ferrous metal attached to her ring, passed through the hole, it caused a frequency shift in the oscillation of the coil, which was detected by the recorder. The second sensor was connected directly to the perch microswitch, and the third, to a thermistor fitted through a small hole in the side of the box.

The activity recorder was similar in principle to that described by Royama (1959), but recorded data directly to an electronic memory unit. The recorder worked in the following way. When the perch alone was activated it was registered as a visit by the male. When the coil detected metal within 10 s of the perch being released it registered the female as entering the box. When the coil alone was triggered, but was not preceded by the perch, it registered the female as leaving the box. Every thirty seconds the following information was recorded to the memory unit: (1) whether the female was in, or out of the box; (2) the number of male visits made in the preceding 30 seconds (maximum=3); and (3) the box temperature, in degrees Celsius, to the nearest integer. Every 48 hours the memory units were exchanged and the data downloaded to a BBC microcomputer.

Throughout the data collection period, the operation of the activity recorders was frequently monitored by direct observation. Activation of the perch and the coil were signalled by different tones in an earpiece which was connected to the recorder via a 30m lead. Two anomalies in the data were identified, which were later removed during the data analysis. These were: (1) when the females left the box, they sometimes hopped on the perch, which was thus registered as a male visit; all "male visits" coinciding with female exits were therefore disregarded; (2) males would frequently hop on the perch both on entering and leaving a box; identifying these occurrences was made difficult by the fact that the males would sometimes spend several minutes in the box before leaving; we therefore decided to disregard all male records which occurred within 300 s of the perch first being triggered.

Throughout the results section and discussion we have used the terminology established by Haftorn in his series of papers on incubation patterns in other *Parus* species (Haftorn 1979, 1981, 1984).

RESULTS

Data were collected from seven Blue Tit nests between 20th May and 10th June 1986. The timing of key events in the breeding cycle, the periods over which data were collected, and the mean daily box temperature throughout these periods, are shown in Table I. When means are referred to in this section, unless stated otherwise, they are calculated by averaging the data for all seven nests.

Table 1. Summary of breeding statistics from the seven nests studied. Columns are, from left to right: date of first egg (day 1 = April 1st); number of eggs; date incubation started; dates between which data were collected; date of hatching; and the mean, standard error, and range of daily temperatures, from 0600-1900h, throughout the data collection periods. All the data were collected in 1986.

	Date	No.	Date	Dates	Date	Daily box temp. °C		
Box	Egg1	Eggs	Inc.	Data	Hatch	mean	s.e.	range
b03	50	7	56	61-69	70	13.2	0.52	11.3-16.0
b04	40	11	57	58-68	69	13.0	0.43	11.3-16.1
b07	40	11	52	53-61	65	13.0	0.22	12.0-13.8
b18	39	10	50	50-58	63	13.0	0.46	10.4-14.6
b20	42	16	58	58-70	71	13.3	0.41	11.6-16.6
b30	44	12	56	60-68	69	12.1	0.59	9.9-16.6
b31	43	9	52	57-64	65	12.9	0.61	10.9-16.1

Pattern of male activity

On average, male Blue Tits visited their females at the nest every 42.6 mins (s.e. = 1.91, n = 7). The underlying frequency distributions were highly skewed, however, (see Fig. 1), and before performing any statistical test they were transformed by taking their natural logarithm. The resulting distributions were tested for normality using the Shapiro-Wilk test (Shapiro and Wilk 1965).

Differences in the interval between male visits, at individual nests, were tested using a one-way analysis of variance. There were no significant differences. However, this is likely to result from the high amount of variation involved, especially day-to-day variation in the number of visits at each nest (mean coeff. variation = 60%). The pattern of visiting also varied throughout the day (Fig. 2). It increased dramatically in the early morning, peaked around 07.00h, and thereafter declined throughout the rest of the day.

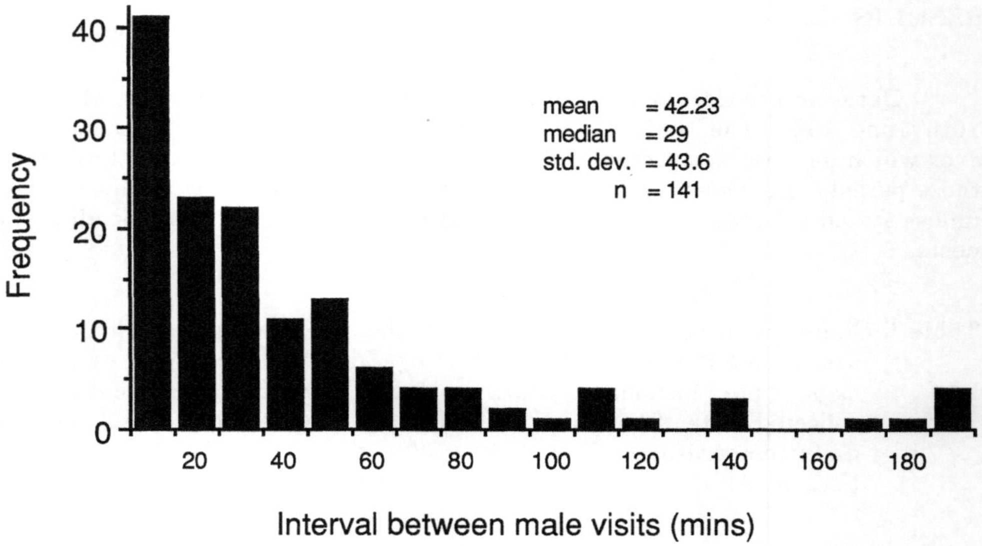

Fig. 1. An example frequency distribution of the intervals between male visits. Data from box 18, for the period 20-28th May 1986.

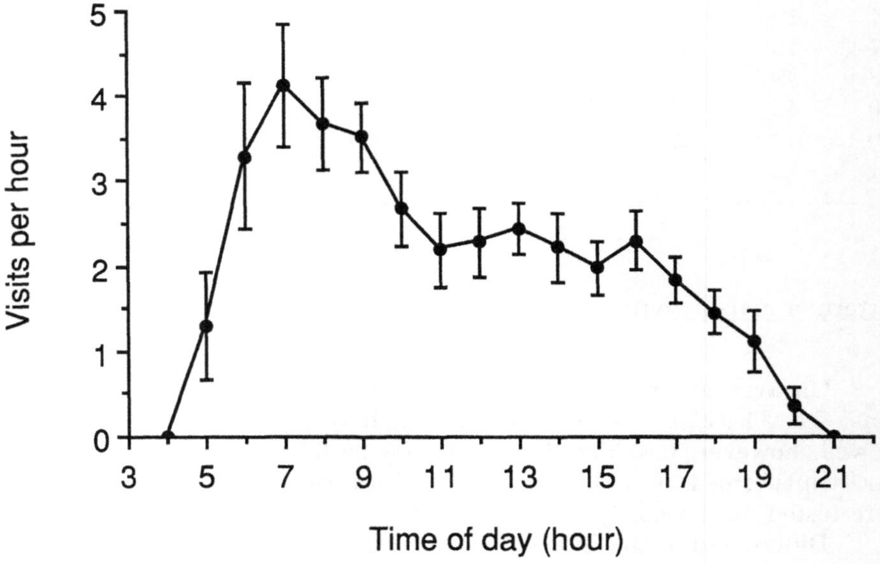

Fig. 2. Mean number of male visits in relation to time of day. (Means for seven nests, bars represent standard errors).

In order to investigate the causes of this variation, we performed a multiple regression on the (log) interval between male visits; with hour-of-the-day, and nest-box temperature as the two independent variables. The results of this analysis are shown in Table II. Overall, the amount of variation in the male inter-visit-interval that was explained by these two variables was rather small, but the relationship with time-of-day was significant at four out of the seven nests, and with temperature at two of them. Unfortunately, box temperature and time of day were highly correlated at all the nests, but in six out of the seven analyses the majority of variation explained by the regression analysis came from the time-of-day.

Table II. Multiple regression analysis of the interval between male visits in relation to time-of-day (hours) and nestbox temperature. (ns = not significant, * = $p<0.05$, ** = $p<0.01$, *** = $p<0.001$).

Box	Hour	Temp.	$\%r^2$	n
b03	*	ns	8.1	109
b04	ns	ns	5.0	121
b07	*	ns	12.0	95
b18	***	**	25.5	137
b20	ns	ns	2.8	90
b30	**	ns	15.2	164
b31	ns	**	21.7	112

Female incubation rhythm

The females' active day began, on average, 102 mins after sunrise (s.e. = 6 mins, n = 76); lasted for 845 minutes (s.e. 13 mins, n = 7); and ended 28 minutes before sunset (s.e. = 12 mins, n = 7).

Figure 3 shows the pattern of attentiveness, throughout the day, averaged over all seven nests. There was an initial decline in attentiveness once the females' day began, but between 06.00 and 19.00, the birds spent approximately 82% of the time incubating their eggs. Thereafter, the proportion increased until by 21.00h all the birds were again roosting in their boxes.

As the degree of attentiveness remains fairly constant between 06.00 and 19.00h, we have restricted the rest of our analysis of the incubation rhythm to this period.

As previous studies (Haftorn 1979, 1981, 1984) have shown, the incubation pattern of tits consists of relatively long periods of incubation, interspersed with short periods away from the nest. The Blue Tit is no exception, and incubation follows a very similar pattern to that recorded in the Great Tit (Haftorn 1981). Figure 4 depicts frequency distributions of all the

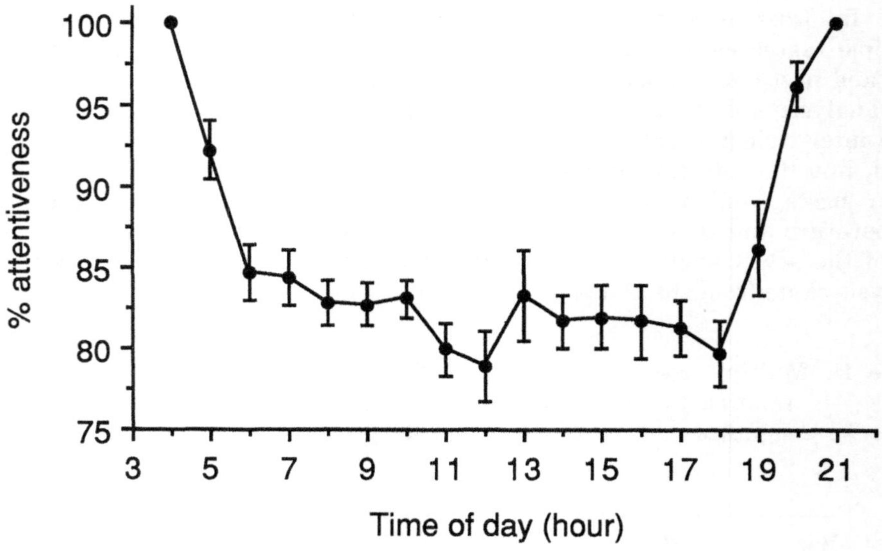

Fig. 3. The relationship between female attentiveness and time of day. (Means for seven nests, bars represent standard errors).

periods-on and periods-off at one nest (b04), over a ten day period. The frequency distribution of period-offs is highly skewed to the right, and, although there appears to be a smaller degree of variation, in six out of the seven nests the coefficient of variation was greater for periods-off than for periods-on.

Averaging over all seven nests the mean period-on was 29.13 mins (s.e. = 2.65, n = 7) and the mean period-off was 6.32 mins (s.e. = 0.88, n = 7). As before, both periods-on and periods-off were normalized using a log transformation, before carrying out any statistical tests. One-way analysis of variance indicated that there were highly significant differences between nests for the duration of both periods-on ($F = 54.53$, d.f. = 6,1472, $p < 0.001$) and periods-off ($F = 56.92$, d.f. = 6,1527, $p < 0.001$).

Multiple regression analysis was again used to examine the relationship between incubation parameters, and various independent variables such as time-of-day (hour), nest-box temperature, and the number of male visits. As well as log transformations for the periods on and off, the percentage attentiveness was normalized using an arcsine transformation. The results are shown in Table III.

Attentiveness seems to be little affected by the three variables tested, and only two of the overall regression analyses were significant. Surprisingly, the number of male visits per hourly period, had no effect on attentiveness at any of the seven nests. At only one nest (b20) was attentiveness related to time of day, and the variable having the greatest influence was temperature. As might be expected, this relationship was negative: the birds spent less time incubating when the temperature was higher. Although temperature and time-of-day are themselves highly correlated, temperature appears to have a greater

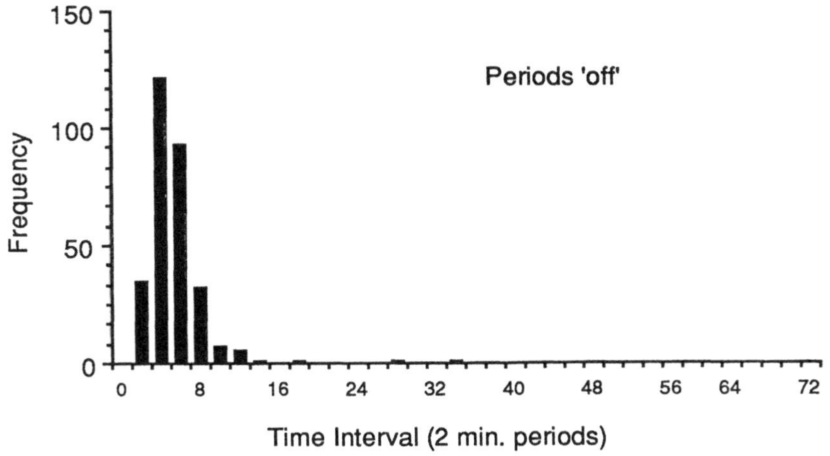

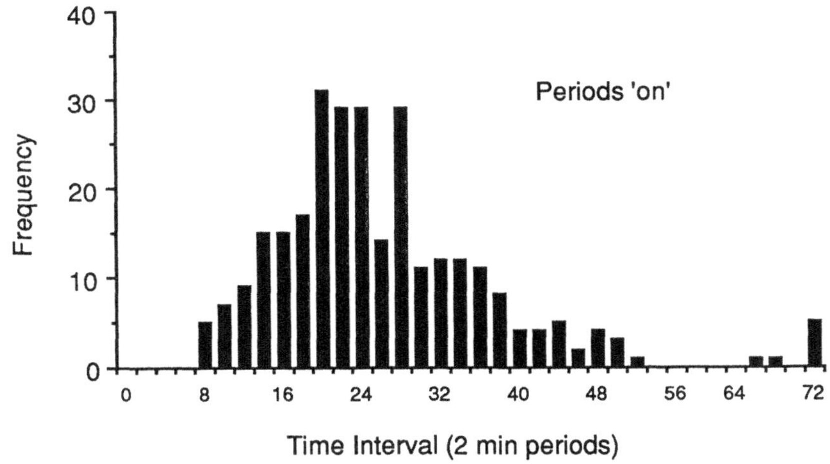

Fig. 4. An example frequency distribution of female period-off and period-on. Data from box 04 for the period 28 May - 7 June 1986.

influence than hour. In general, however, the relationship between temperature and attentiveness is rather weak, perhaps because of the relatively small range of temperatures experienced throughout the study period. Interestingly, one of the clearest relationships between the two variables was found at box 30, where the daily range of temperatures was greatest (Table I).

Multiple regression analyses of periods-off were significant at six out of the seven boxes (Table III), although it is difficult to determine whether time or temperature exerted the major influence. In general, periods-off tended to increase with both time-of-day and temperature.

Table III. Summary of multiple regression analysis on percentage attentiveness, periods-on and periods-off. Independent variables are number of male visits; time-of-day (hour); and box temperature. Males did not visit the box during periods-off and male visits are not included in this analysis. (ns=not significant; *=$p<0.05$, **=$p<0.01$, ***=$p<0.001$)

	(asin) % attentiveness					(log) periods-on					(log) periods-off			
Box	male	hour	temp	$\%r^2$	n	male	hour	temp	$\%r^2$	n	hour	temp	$\%r^2$	n
b03	ns	ns	-*	7.2 ns	97	***	***	-***	33.1 ***	162	ns	*	7.1 **	161
b04	ns	ns	ns	1.1 ns	137	***	**	ns	13.4 ***	252	***	***	17.6 ***	252
b07	ns	ns	-*	17.4 ***	153	**	ns	ns	3.8 *	228	ns	**	7.1 ***	230
b18	ns	ns	ns	2.5 ns	101	***	*	ns	16.5 ***	130	**	ns	7.3 **	128
b20	ns	-**	ns	6.4 *	158	*	-**	-*	13.4 ***	194	ns	-*	2.5 ns	203
b30	ns	ns	-*	4.6 ns	114	***	***	-***	39.4 ***	178	*	**	13.3 ***	178
b31	ns	ns	ns	3.9 ns	99	***	***	ns	17.5 ***	200	*	ns	4.6 **	201

Periods-on were tested against hour, box temperature and the number of male visits made during each incubation bout. The overall regression analysis was significant at each box, and in all cases there was a significant relationship with the number of male visits. An example of this relationship is shown in Figure 5. Time-of-day had a significant effect at six out of the seven boxes (Fig. 6) and at three boxes there was a significant relationship with temperature. As might be expected, periods-on tended to decrease in length as the nestbox became warmer.

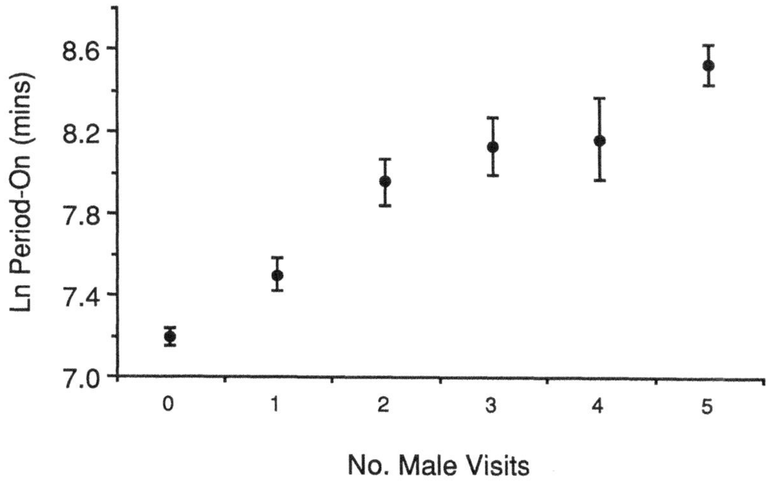

Fig. 5. The relationship between the number of male visits and the length of a period-on (data from box 30).

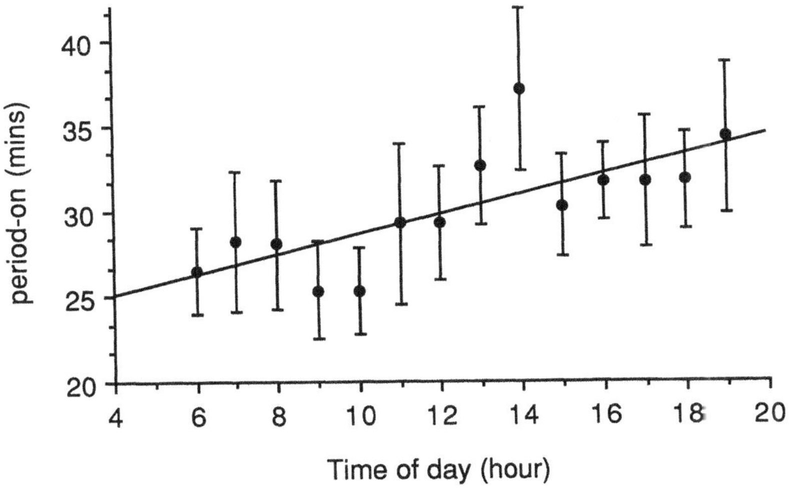

Fig. 6. The relationship between the length of period-on and the time of day (means for seven nests, bars represent standard errors).

In most of the analyses summarized in Table III, only a small proportion of the variance is explained. In general, the number of male visits, the time-of-day, and box temperature had more influence on periods-on (4%-39%) than on attentiveness (1%-17%) or periods-off (2%-17%).

Variation between boxes

Variations in incubation variables, such as rate of male visiting, attentiveness, periods-on, and periods-off, were tested against various "breeding" variables such as date of first egg, number of eggs, and days to hatching. Spearman Rank Correlation coefficients were non-significant for all the relationships tested.

DISCUSSION

The pattern of incubation at Blue Tit nests is very similar to that recorded for other species (Table IV). The active day of an incubating Blue Tit begins 102 minutes after sunrise, which is close to the time (98 mins) recorded by Haftorn (1981) for Great Tits breeding in Northern Norway. Despite the longer day-length in these Northern latitudes, the length of the active day is surprisingly similar (Blue Tits = 845 mins; Great Tits = 812 mins; $t=1.79$, $df= 10.9$, $p=0.10$). This is as a result of a highly significant difference in the time at which they go to roost (Welsh Blue Tits: 28 mins before sunset; compared to 262 mins for Norwegian Great Tits ($t=11.47$, $df=9.6$, $p<0.001$).

Table IV. A comparison of incubation behaviour of five *Parus* species, showing mean periods-on, periods-off and attentiveness.

Species	Authority	n	p.-on (min.)	p.-off (min.)	attent. %
Coal tit	(Haftorn 1984)	(5)	35.4	4.23	89
Willow tit	(Haftorn 1979)	(2)	20.9	6.45	76
Blue tit	(this study)	(7)	29.13	6.32	82
Great tit	(Haftorn 1981)	(11)	28.49	10.42	73
Marsh tit	(Nilsson and Smith 1988)		-	-	78

During this period, the total amount of incubation (attentiveness) appeared to be little affected by variables such as the rate of incubation feeding by the male, the time of day, or the temperature. At most nests, however, the length of a period-on was positively related to the time of day, and the number of male visits (Fig. 5). This relationship with the number of male visits is difficult to interpret. The most plausible explanation, is that the extra food delivered by the male enables the female to spend longer in the box, but it is difficult to understand why this is not reflected by a relationship between male feeding and attentiveness. For example, in an experimental manipulation which involved providing incubating Pied Flycatchers (*Ficedula hypoleuca*) with extra food (Smith et al. 1989), females spent more time on the nest and shorter periods away from it. An alternative explanation is that there is no causal relationship between the number of feeds a female receives and the length of a period-on. If the males visit at a constant rate, then a female will receive more male visits during a period-on, which is, for some other reason, longer. However, at the majority of boxes the mean rate of male visiting decreased throughout the day (Fig. 2), whilst the mean length of periods-on increased (Fig. 6). If the length of a period-on is independent of the number of male visits, we should therefore expect a negative relationship between the number of visits and the length of a period on. The fact that the relationship is positive suggests that they are causally related. It is also significant that the rate of male visiting is highest in the early morning, when the energetic demands on the female are likely to be greatest, having spent the night in the box with no opportunity to feed and when the temperature tends to be lowest. Smith et al (1989) have shown that, in the Pied Flycatcher, female nutritional state affects the rate of male incubation feeding.

Obviously, one major drawback of our study is that we have only measured the number of male visits to the nest. Although in 93% of the observed instances, this was accompanied by courtship feeding, the male also regularly feeds the female during her periods-off. For the Great Tit, Royama (1966) recorded rates of courtship feeding, during periods-off, as high as once per minute. Since a Blue Tit spends, on average, 152 minutes per day in periods-off, and on average we recorded feeds once every 43 minutes during an average daily period of 693 mins spent on the nest, this suggests that only about 10% of the incubation feeding occurs at the nest. Variation in the rate of courtship feeding during periods-off, may well explain why we failed to find any correlation between the rate of male visits to the nest and attentiveness.

ACKNOWLEDGEMENTS

Rob Tucker designed the nestbox activity recorders, constructed them, and wrote the software for down-loading the data. Without his help, this project would not have been possible. The Glamorgan Naturalists Trust, kindly allowed us access to their Coed Garnllwyd reserve. This research was, in part, supported by Colgate Damon Grant 30-350-18.

REFERENCES

Cowie RJ, Hinsley SA (1987) Breeding success of blue tits and great tits in suburban gardens. Ardea 75: 81-90

Haftorn S (1979) Incubation and regulation of egg temperature in the willow tit (*Parus montanus*). Ornis Scand 10: 220-234

Haftorn S (1981) Incubation rhythm in the great tit (*Parus major*). Fauna norv Ser C Cinclus 4: 9-26

Haftorn S (1984) The behaviour of an incubating female coal tit (*Parus ater*) in relation to experimental regulation of nest temperature. Fauna Norv Ser C Cinclus 7: 12-20

Lack D (1971) Ecological Isolation in Birds. Blackwell, Oxford

Nilsson J, Smith HG (1988) Incubation feeding as a male tactic for early hatching. Anim Behav 36: 641-647

Royama T (1959) Test of an automatic nest-recorder. Brit Birds 52: 295-302

Royama T (1966) A re-interpretation of courtship feeding. Bird Study 13: 116-129

Shapiro SS, Wilk MB (1965) An analysis of variance test for normality (complete samples). Biometrika 52: 591

Smith HG, Källander H, Hultman J, Sanzen B (1989) Female nutritional state affects the rate of male incubation feeding in the pied flycatcher *Ficedula hypoleuca*. Behav Ecol Sociobiol 24: 417-420

Webb DR (1987) Thermal tolerance of avian embryos: a review. Condor 89: 874-898

HELPING AT THE NEST IN EUROPEAN BEE-EATERS: WHO HELPS AND WHY?

C.M. LESSELLS

Dept. of Animal and Plant Sciences, Univ. of Sheffield, Sheffield S10 2TN, UK
and
Station Biologique de la Tour du Valat, Le Sambuc, 13200 Arles, France.

Avian cooperative breeding has been regarded as an ideal system in which to investigate the role of kin selection in the evolution of altruistic behaviour (Brown 1987). However, many studies of helping at the nest have been carried out on communally breeding species in which the costs and benefits of helping behaviour are confounded with joint exploitation or defense of a territory. In contrast, helping behaviour in European Bee-eaters *Merops apiaster* is not associated with territoriality, allowing the relative roles of individual (direct) and kin (indirect) selection to be more readily assessed. This paper first provides an overview of helping behaviour and second concentrates on the types of individuals who become helpers and the selection pressures that favour helping by them.

EUROPEAN BEE-EATERS

The European Bee-eater is a colonially breeding migratory species. The birds arrive in the breeding colony in early May and have departed again by the end of August. The interval from the first egg being laid until the start of hatching (which is highly asynchronous; Lessells and Avery 1989) is about 28 days, and from hatching to the completion of fledging about 32 days. Clutches consist of 4 - 7 eggs laid at 2 day intervals in a nest burrow. The nest chamber is at the end of a tunnel a meter or more in length, built in a sand or earth bank or in level ground. Some nest burrows are re-used in consecutive years, but the majority of pairs excavate burrows anew each year. Unlike in the closely related White-fronted Bee-eater *Merops bullockoides* (Hegner et al. 1982; Hegner and Emlen 1987), there is no feeding territory and, although relatives sometimes nest in close proximity, individuals do not roost in other pairs' burrows. Even during the nestling period, visits to other nests are rare, and birds almost never enter the nests of other birds at this time unless they have become helpers.

NATO ASI Series, Vol. G 24
Population Biology of Passerine Birds
Edited by J. Blondel et al.

STUDY METHODS

European Bee-eaters breeding in a colony of about 100 pairs in the Camargue, southern France, have been studied since 1983. Fully-grown birds, caught by mist-netting and trapping at the nest, and chicks from between a third and a half of the broods each year have been marked with metal leg rings. Fully-grown birds were also marked with acrylic paint on the tail. This allows individual identification in the field but, because the tail is moulted after the end of each breeding season, birds must be caught each year to establish their identity. As a result, about a quarter of the breeding birds, including a disproportionate number of those who fail early in the breeding season, are not identified each year. The sex of fully-grown birds was established in the hand on the basis of plumage characteristics and wing length (Lessells and Ovenden 1989) and confirmed from behavioural observations of courtship feeding and copulation. One year old ("juvenile") birds can be reliably distinguished from older birds ("adults") in the hand on the basis of plumage (Lessells and Krebs 1989). Breeding birds were identified as early as possible each season. Since 1986, laying dates, hatching dates and clutch sizes have been monitored at all nests in the colony using an industrial endoscope and a sub-sample of about 20 nests watched for an hour daily, before and during laying. Since 1984, about 40 nests have been watched for an hour daily for the first 23 days after hatching. These observations provide detailed information on the behaviour of helpers, and hatching date information for 1984 and 1985. Additional information on helpers at other nests has been collected opportunistically.

HELPING BEHAVIOUR

About a fifth of the nests with chicks have helpers (19%; n=239). The majority of these have a single helper (75%; n=53), some have two helpers (19%), and occasionally nests may have three (4%) or four (2%) helpers. Helpers contribute to provisioning the brood, but unlike in White-fronted Bee-eaters (Emlen and Wrege 1988), they do not help with incubation. They may start helping at any stage during the nestling period, and continue to provision chicks after they have fledged. Helped broods receive more prey items and the breeding male, but not female, at helped nests delivers food at a lower rate than at unhelped nests (Table I). These differences in provisioning rate appear to have consequences for the survival of helped broods and their parents. Brood size at fledging is higher in helped broods, and a higher proportion of the breeding birds that are helped return to the colony in the following year than of those birds who raise their chicks unaided (Table I). Ideally, experimental manipulations are needed to show that helping causes these differences (Brown et al. 1982), but these are impractical because of the sample sizes required. However, pairs which are subsequently helped do not lay larger clutches or breed earlier than those who are not (Table I). This suggests that there are not quality differences between birds who are or are not helped.

Table I. Provisioning rate and reproductive success of helped and unhelped broods.

	Unhelped[1] Broods	Helped[1] Broods	P[2]
Feeding rate[3]:			
Total (including helpers)	284	322	<0.05
Males	155	119	<0.05
Females	129	121	ns
Brood size at fledging	4.76	5.27	0.035
% chicks returning	18.5	19.9	ns
% parents returning:			
Males	31.6	43.9	ns
Females	28.8	47.6	<0.05
Hatching date[4]	59.6	60.7	ns
Clutch size	5.96	6.05	ns

[1]: "Unhelped" broods were not helped for any of the first 23 days after hatching. "Helped" broods were helped for at least part of this period.
[2]: P values are given controlling for year effects. 3-way G tests were used to compare return rates of males and females. Other comparisons were made using 2-way ANOVAs.
[3]: Nests were watched for 1 hour daily for the first 23 days after hatching. Feeding rate is the number of items taken to the nest over these 23 h of observation.
[4]: Days after 30 April (*i.e.* 1 = 1 May).

Table II. Relationships between helpers and breeding birds.

	Sex of helper	
Relative(s) helped	Male	Female
Both parents	6	1
Father, not mother	1	
Father, possibly mother	1	
Mother, not father	4	
Mother, possibly father	2	
Brother	4	
Son	3	1

WHO ARE HELPERS?

Helpers tend to be close relatives of the breeding pair (Table II). The range of relatives helped is greater than in many other species with helpers at the nest and includes not only both or either of the parents, but also siblings and offspring. Because not all of the population are marked, there are also many helpers whose relationship to the breeding pair is unknown. This introduces biases and Table II therefore indicates the range, but not necessarily the true relative frequencies, of helpers with different kinds of relationship to the breeding pair.

One important question is whether helpers are ever non-relatives or relatives of low relatedness to the breeding pair. Such instances are difficult to detect, again because of the incompleteness of genealogies known from ringing. Three (6%) of the male helpers were not the parents, offspring or same-brood siblings of the breeding pair. In two cases they might have been full siblings of the breeding pair but from different broods. In the remaining case the maximum coefficient of relatedness between the helper and nestlings was 0.125. Although the possibility that a substantial number of helpers are unrelated to the pairs they help cannot be excluded, it can be shown that helpers preferentially help relatives: helpers known to have relatives (parents, offspring or same-brood siblings) are much more likely to help them (24/25 cases) than expected on the basis of a random choice of broods in the colony. Also, after nest failure (see below) birds are more likely to become helpers if they are known to have relatives in the colony, and rarely help the pair in the nearest nest to their own, so that the preferential choice of relatives is not based merely on nesting proximity.

The analysis of the relatedness of helpers is based on ringing histories and does not take into account reductions in relatedness due to extra-pair copulations (EPCs) and intra-specific nest parasitism (ISNP), behaviours which occur in the study colony and which have been shown by DNA fingerprinting (T.Burke pers. comm.) to lead to extra-pair paternity and maternity. Acoustic recognition of offspring by parents develops a few days before fledging (Lessells et al. in press) and cross-fostered nestlings subsequently help foster rather than natural relatives. Chicks resulting from EPCs or ISNP also presumably recognize the pairs who raised them as their parents, and this reduction in kinship must be taken into account in any quantitative cost-benefit analysis of helping.

Helpers differ from randomly chosen individuals in the population. The vast majority of helpers are male (93%; n=70). Male helpers comprise a slightly higher proportion of juveniles (68%; n=59) than occurs in the breeding population (55%; n=328) although this difference is not significant.

Helpers are not individuals who are unable to raise broods of their own. 85% (n=13) of adult male helpers and 100% (n=3) of adult female helpers whose breeding attempts have been monitored in previous years had successfully raised chicks. Similarly, many helpers attempt to breed in the season in which they later help: 47% (n=58) of male helpers are known to have

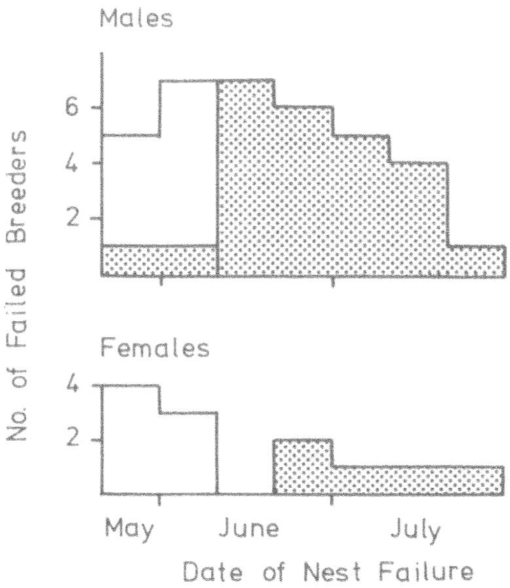

Fig. 1. Re-nesting and helping by failed breeders in relation to the date of nest failure. Unshaded: re-nested after nest failure. Shaded: helped after nest failure. Mann-Whitney U test, re-nesters vs helpers: males, $P<0.001$. Females, $P=0.001$.

made an unsuccessful breeding attempt earlier in the same season. This figure is an underestimate because many helpers are not already painted when they are first observed helping. 82% (n=33) of already painted male helpers have made an earlier unsuccessful breeding attempt. This estimate is open to potential biases in either direction, but indicates that a large proportion of helpers may be failed breeders. Not all failed breeders become helpers; some re-nest and the probability of helping rather than re-nesting depends on the date of nest failure (Fig. 1). Birds failing before early-mid June tend to re-nest, and those failing after this date to help. Some birds (45% of males, n=67; 81% of females, n=62) are recorded neither re-nesting nor helping after nest failure.

WHY DO MALES AND LATE-FAILED BREEDERS BECOME HELPERS?

It has been shown above that helpers are predominantly male and late-failed (as opposed to early-failed) breeders. There is also a tendency for male helpers to comprise a higher proportion of juveniles than male breeders. In order to explain why these types of individuals are over-represented as helpers it must be explained why the balance of selection pressures is tipped away from nesting or re-nesting in favour of helping. Helping might increase the future reproduction of helpers by increasing their chances of obtaining a feeding territory, nest site, mate or helper (Zahavi 1976; Woolfenden and Fitzpatrick 1984), but such individual benefits of helping behaviour are likely to be small in European Bee-eaters: there is no feeding territory, nesting burrows are rarely re-used, and helpers have never been observed to acquire one of the breeding pair as a mate in future seasons. Any explanation must therefore concentrate on a reduction in individual benefits from breeding or an increase in kin-mediated benefits from helping. Individual benefits from attempting to breed maybe low either because individuals are excluded from breeding or because they are less successful if they do breed. Kin-mediated benefits will be relatively important when close relatives are available as recipients of help or if the benefit of helping in terms of increased production of offspring by the breeding pair is high relative to the costs to the helper in increased mortality or decreased future reproduction. There are no data allowing this last possibility to be assessed, so the discussion below concentrates instead on exclusion from breeding, low reproductive success and the presence of relatives as possible explanations.

Exclusion from breeding

The possibility that individuals of a particular age or sex tend to be excluded from breeding can be assessed using the apparent frequency of non-breeding by individuals that are known to be alive and likely to breed in the study colony if they breed anywhere (Lessells and Krebs 1989; Table III). Juvenile females are recorded breeding in a slightly higher proportion of years than adult females, and juvenile males in a lower proportion of years than adult

Table III. Non-breeding in relation to age and sex[1].

	Males		Females	
	Not recorded breeding	Recorded breeding	Not recorded breeding	Recorded breeding
Juveniles	10	9	2	4
Adults	7	16	16	13
Total	17	25	18	17

[1] Data are included for birds known to have been alive and likely to breed in the study colony if they did breed. For juveniles this was birds ringed as nestlings and recorded breeding at least once as adults. For adults this was the intervening year(s) between first and last recorded breeding for birds first and last recorded breeding at least two years apart.

males, although neither difference is significant. Comparison of the apparent frequency of non-breeding by males and females (although complicated by potential sex biases in dispersal) reveals a non-significant tendency for males to be recorded breeding in a higher proportion of years than females, a trend opposite to that needed to explain helping by males. Thus exclusion from breeding is a possible explanation of helping by juvenile males as opposed to adult males, but not of helping by males as opposed to females.

Low breeding success

Differences in breeding success between juveniles and adults are reviewed by Lessells and Krebs (1989). Juveniles of either sex tend to breed later and have smaller clutches than adults, although these differences are only significant for females. Differences in clutch size are not translated into significant differences in brood size at fledging. The major difference between adults and juveniles is in the proportion of their fledged chicks which return to the colony in subsequent years; about 10% of the chicks of juveniles and 20% of the chicks of adults return. The difference is too large to be accounted for by the slightly later breeding of juveniles compared with adults. Because of strong assortative mating for age it is not possible to determine whether it is characteristics of the father or mother which are responsible for the difference in the return of fledged chicks, but regardless of the sex causing the difference these estimates represent the expected values for juveniles of each sex. There are no particular reasons to expect differences in the dispersal of chicks of juveniles and adults, so these differences may reflect differences in post-fledging survival of the chicks of juveniles and adults and could explain helping by juveniles as opposed to adults but not males as opposed to females.

In order to assess whether low breeding success will explain helping by late-failed breeders, seasonal variation in breeding success needs to be known. Ideally such an analysis should be based on pairs whose timing of breeding has been experimentally manipulated, but such manipulations are not practicable in European Bee-eaters. Observed brood size at fledging tends to peak early in the season, and the return rate of chicks shows a similar pattern (Fig. 2a and b), although neither relationship is significant. Combining these two measures (Fig. 2c) suggests that broods hatching at the end of the season produce only about half the number of recruits to the colony as those hatched at the time of peak breeding success. This low success of late-hatched broods may explain why late-failed breeders help rather than re-nest.

Presence of relatives

The likelihood of relatives being available in the colony as recipients of helping behaviour depends on the pattern of dispersal. Of birds ringed as

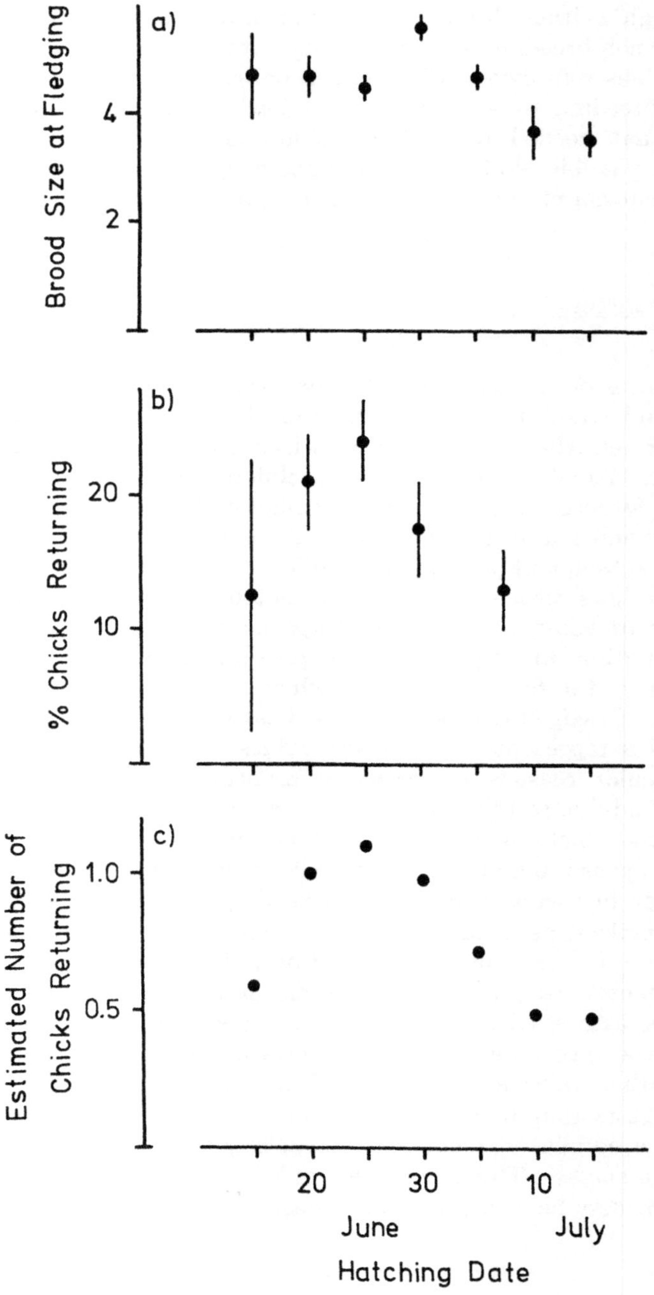

Fig. 2. Seasonal variation in reproductive success. (a) Brood size at fledging. (b) Return rate of nestlings. (c) Estimated number of chicks returning to the colony (based on values shown in (a) and (b)). Vertical bars are standard errors.

nestlings, about four times as many males as females are retrapped in subsequent years (105 males, 25 females). The sex ratio of the fully-grown birds caught is much closer to unity (511 males, 487 females) and once birds have started breeding in the colony, males and females are equally likely to return to the colony in future years. The excess of fully-grown males ringed as nestlings therefore appears to reflect a strong sex-bias in juvenile dispersal.

As a result of this, males should be more likely to have relatives breeding in the colony than females, and this may contribute to the predominance of males (but not juveniles) among helpers. However, the sex bias in dispersal does not explain all of that in helping; the sex bias in the presence of parents and siblings should be of the same magnitude as that in juvenile dispersal, but there should be no sex bias in the presence of offspring. If the bias in helping results only from the bias in the presence of relatives, it should be as, or less, extreme than the sex bias in juvenile dispersal. In fact, the reverse is true; 93% (n=70) of helpers are males, but only 81% (n=130) of nestlings returning to the colony are males. The residual excess of males among helpers could be explained by some males being excluded from breeding, males being more likely to survive nest failure, or males being more likely to help after nest failure when relatives are present.

The presence of relatives is unlikely to be the explanation of helping by late-failed (as opposed to early-failed) breeders.

ARE ALL HELPERS FAILED BREEDERS?

Over a half of male helpers are failed breeders, and it has been argued that these birds help rather than re-nest because of the low expected success of re-nesting attempts. Similarly, over 90% of helpers are male, and some, but not all, of this bias is explained by the sex bias in dispersal. A possible explanation for the residual excess of males among helpers is the exclusion of juvenile males from breeding, and this would also explain the tendency for a higher proportion of male helpers than male breeders to be juveniles. However, neither the excess of juveniles among male helpers nor the tendency for juvenile males apparently to breed less frequently than adult males are significant, and this leaves the question of whether all helpers are failed breeders. The way in which individuals are marked for field identification makes it difficult to demonstrate that individuals have not attempted to breed before becoming helpers; one male helper is known to fall into this category, so at least some helpers are non-breeders. Annual variation in the frequency of helping in relation to the nest failure rate confirms the connection between helping and nest-failure, but suggests that some helping may occur in the absence of nest failure (Fig. 3). However, the expected shape of the relationship is unknown so it cannot be used to estimate the frequency of helping by non-relatives.

Whether or not male helpers do comprise substantial numbers of non-breeders, the minimum estimate of the proportion of failed breeders among helpers is higher than in most other species with "helping regular in a moderate

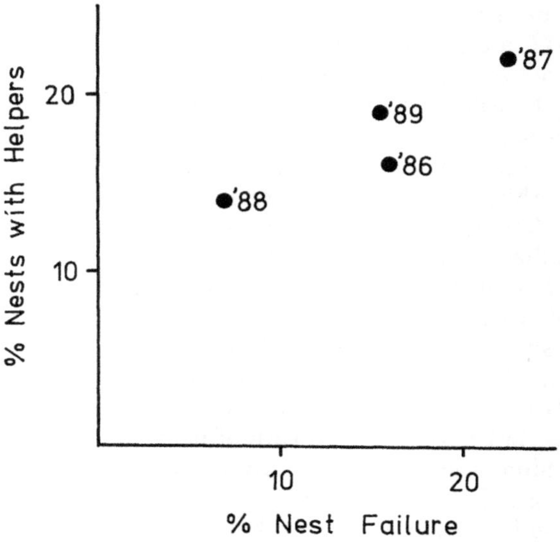

Fig. 3. Annual variation in the frequency of helping in relation to nest failure rate. The frequency of helping is based on nests watched for 1 h daily during the nestling period. Nest failure rates were calculated for all nesting attempts in which eggs were laid.

fraction of the species" including White-fronted Bee-eaters (Brown 1987; Emlen and Wrege 1988). Many of the species in this category of helping have a preponderance of males among helpers, and this is often explained by a shortage of females leading to males being excluded from breeding (e.g. Reyer 1980). Although European Bee-eaters are similar to these other species in having predominantly male helpers, the selection pressures involved are clearly different; at least half of the helpers are failed breeders, and dispersal rather than exclusion from breeding is the major explanation for helping mainly by males.

KIN SELECTION, INDIVIDUAL SELECTION OR NO SELECTION?

Although kin selection has been a primary interest in the study of helping behaviour in birds, two major challenges have been made regarding the importance of kin selection in the evolution of helping behaviour. The first points out that individual benefits of helping behaviour may account for its evolution, making kin selection arguments unnecessary (Zahavi 1976; Woolfenden and Fitzpatrick 1984). The second suggests that helping behaviour may be an epiphenomenon of parental feeding behaviour of no selective advantage *per se* (Jamieson and Craig 1987; Jamieson 1989). In European Bee-eaters the increased reproductive output of helped pairs and relatedness of helpers and recipients demonstrate that kin selection occurs by definition (Reyer 1980), but critics might argue that helping behaviour would have evolved even in the absence of this kin selection. However, helping behaviour is preferentially directed towards relatives to an extent that cannot be explained by spatial proximity. This allows the rejection of these alternative hypotheses because they cannot explain such kin favouritism. In communally breeding species, spatial proximity and kinship are often confounded to an extent that kin favouritism cannot be unequivocally demonstrated (Emlen and Wrege 1988, 1989). It is therefore colonially breeding species, such as the European Bee-eater, rather than communally breeding species, which provide the strongest evidence for the evolution of helping behaviour through kin selection (Brown 1987; Emlen and Wrege 1988, 1989).

ACKNOWLEDGEMENTS

I would like to thank Mark Avery, Nonie Coulthard and the many others who have helped with data collection over the years, Dr Luc Hoffmann for hospitality at the Station Biologique de la Tour du Valat, Monsieur Deville for access to the study site, the Centre de Recherches sur la Biologie des Populations d'Oiseaux for permission to ring and paint Bee-eaters, David Bryant, Finn Kjellberg, John Krebs, and Lulu Stader for comments on previous drafts of this paper, the British Ecological society, British Ornithologists' Union, Fondation Fyssen, Natural Environment Research Council, Tour du Valat and the University of Sheffield for financial support, and the Royal Society for a travel grant to attend the Evisa conference.

REFERENCES

Brown JL (1987) Helping and communal breeding in birds. Princeton University Press, Princeton

Brown JL, Brown ER, Brown SD and Dow DD (1982) Helpers: effects of experimental removal on reproductive success. Science, 215: 421-422

Emlen ST and Wrege PH (1988) The role of kinship in helping decisions among White-fronted Bee-eaters. Behav Ecol Sociobiol, 23: 305-315

Emlen ST and Wrege PH (1989) A test of alternate hypotheses for helping behavior in White-fronted Bee-eaters of Kenya. Behav Ecol Sociobiol, 25: 303-319

Hegner RE, Emlen ST and Demong NJ (1982) Spatial organization of the White-fronted Bee-eater. Nature 298: 264-266

Hegner RE and Emlen ST (1987) Territorial organization of the White-fronted Bee-eater in Kenya. Ethology 76: 189-222

Jamieson IG (1989) Behavioral heterochrony and the evolution of birds' helping at the nest: an unselected consequence of communal breeding? Amer Nat 133: 394-406

Jamieson IG and Craig JL (1987) Critique of helping in birds: a departure from functional explanations. Perspect Ethol 7: 79-98

Lessells CM and Avery Ml (1989) Hatching asynchrony in European Bee-eaters *Merops apiaster*. J Anim Ecol 58: 815-836

Lessells CM and Krebs JR (1989) Age and breeding performance of European Bee-eaters. Auk 106: 375-383

Lessells CM and Ovenden GN (1989) Heritability of wing length and weight in European Bee-eaters (*Merops apiaster*). Condor 91: 210-214

Lessells CM, Coulthard ND, Hodgson PJ and Krebs JR (in press) Chick recognition in European Bee-eaters: acoustic playback experiments. Anim Behav

Reyer H-U (1980) Flexible helper structure as an ecological adaptation in the Pied Kingfisher (*Ceryle rudis rudis* L.). Behav Ecol Sociobiol 6: 219-227

Woolfenden GE and Fritzpatrick JW (1984) The Florida Scrub Jay: demography of a cooperative-breeding bird. Princeton University Press, Princeton

Zahavi A (1976) Co-operative nesting in Eurasian birds. In Frith HJ and Calaby JH (eds). Proceedings of the 16th International Ornithological Congress, 685-693. Australian Academy of Sciences, Canberra

WHAT DETERMINES THE TIMING AND ORDER OF NEST-LEAVING IN THE MARSH TIT (*PARUS PALUSTRIS*)?

Jan-Åke NILSSON

Department of Ecology, Animal Ecology, University of Lund,
S-223 62 Lund, Sweden.

Despite several decades of intensive research on the breeding biology of birds, very little is known about the behaviour of parents and young during the nest-leaving phase. Factors responsible for the initiation of nest-leaving may have bearings upon a potential conflict between parents and their offspring. Certain nest-leaving behaviours have been assumed in the nest failure hypothesis as an explanation for asynchronous hatching (Clark and Wilson 1981).

During the breeding seasons of 1988 and 1989, I studied the nest-leaving behaviour and factors that determine nest-leaving in Marsh Tits (*Parus palustris*). The study centred around the following questions: (1) In which order do individual nestlings leave the nest? (2) What factors and which individual(s) initiate the nest-leaving process? (3) What factors determine the duration of the nest-leaving phase?

METHODS

All nestlings studied hatched in identical nestboxes within a study area of 64 km^2, 20 km east of Lund, southern Sweden. The area consists of small deciduous forests and groves interspersed with permanent pastures and agricultural fields. Most of the data presented below were collected during the breeding season of 1988, with some supplementary data from 1989.

Thirteen days after hatching, the nestlings were marked with an individual combination of colour-rings. To facilitate identification, I dyed the breast of all nestlings within a brood with a unique combination of different colours. At the same time, I measured body weight (to the nearest 0.1 g), wing length (to the nearest 0.5 mm) and tarsus length (to the nearest 0.05 mm). As an index of body condition, I used $\sqrt[3]{\text{weight}}$/tarsus length.

Using frequent nestbox inspections and by making video-recordings during nest-leaving, I ascertained the order and time when nestlings left the nestbox. From the video-recordings, I also obtained information on parental feeding frequencies and of their visual as well as auditory behaviour before and during nest-leaving of their offspring.

To calculate the nestling period, I used the time when 50% of the clutch had hatched as the starting point. When information on the exact hatching time of the mid-young was lacking, I used the observation closest to 50% hatchlings

NATO ASI Series, Vol. G 24
Population Biology of Passerine Birds
Edited by J. Blondel et al.

in the brood and converted this to the mid-point according to the hatching rate presented in Nilsson and Smith (1988a). The nestling period was taken to end when 50% of the brood had left the nestbox. When exact information was lacking, I assumed a constant rate of nest-leaving between observations. In cases where the nestbox was found empty but all young had been present during the previous inspection, I used the midpoint between the two inspections. I judge the uncertainty in the calculations of nest-leaving to be slight since each nestbox was inspected several times during the entire day.

Statistical analysis was performed with SYSTAT (Wilkinson 1987), and all statistical tests are two-tailed. Times are given in Swedish Summer Time = GMT + 2 hours.

RESULTS

Daily pattern and length of the nest-leaving process

Individual nestlings left their nestbox at nearly any time of the day, although rather few left around noon (Fig. 1). Most of the young fledged in the morning but as many as 35% left after noon.

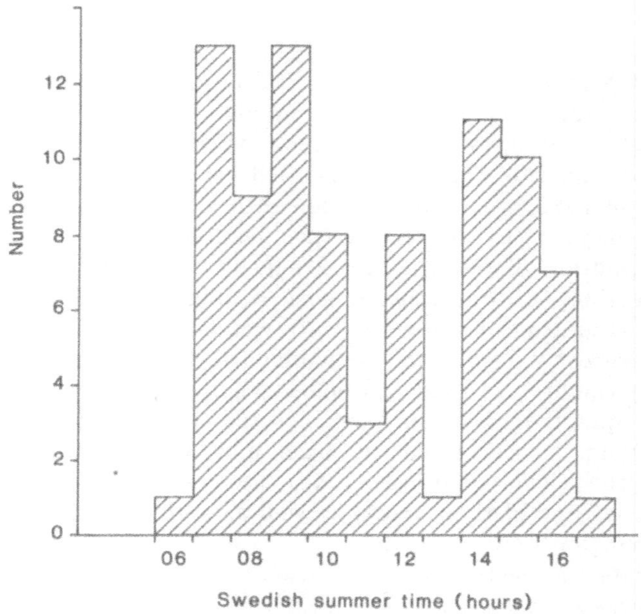

Fig. 1. The number of nestlings leaving the nest at different times of the day. Data from 1988 and 1989.

When both the first and the last nestling were observed leaving the nest, it was possible to measure the total duration of nest-leaving. This was known for 11 broods (Fig. 2). In 10 of these, all nestlings left the nest within 5 to 275 min (mean=115 min; S.D.=80.5). The only exception was a brood that took 1284 min (21 h 14 min) to fledge. One of the nestlings left the nest in the morning while the rest of the brood stayed for at least 6 h. The last young fledged early in the morning the next day.

To increase sample size, I included broods that I visited at least every third hour during the fledging period. I assumed fledging to have begun immediately after the first inspection with all young still present and to have ended immediately before the first time I found the nest empty. By this method, I overestimated the length of the fledging period but a further eight broods (Fig. 2) could be used to calculate the length of the fledging period (mean=201 min; S.D.=60.0; N=8).

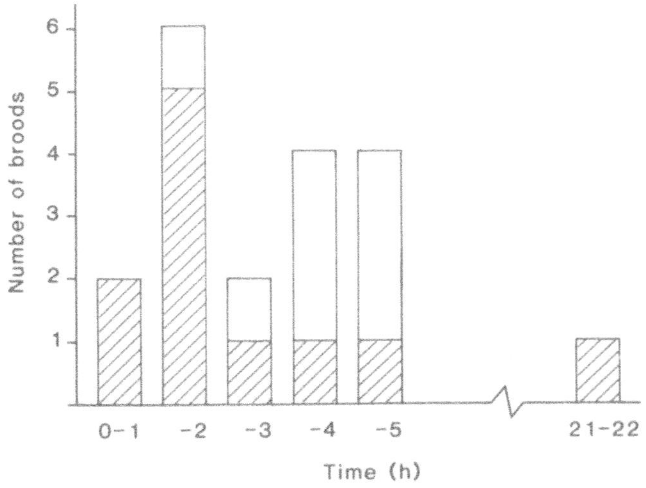

Fig. 2. The distribution of nest-leaving spans (time from first to last young to leave in each brood). Hatched bars represent exactly known and open bars assumed nest-leaving spans. See text for further explanation. Data from 1988 and 1989.

Nest-leaving order among the young

Fourteen nests were inspected when half (40-60%) of the brood had already left. The first nestlings to leave were significantly larger than those leaving the nest later (Tab. 1).

Table 1. Relationship between nest-leaving and body measurements. Mean deviation from brood mean for nestlings in the first half of the brood to leave the nest (early fledglings) and for those in the second half (late fledglings). Data from 1988. Tested with t-test.

	Early fledglings	Late fledglings	t_s	P
Tarsus length	+0.073 (49)	-0.073 (49)	2.54	<0.02
Weight	+0.275 (56)	-0.274 (56)	4.59	<0.001
Wing length	+1.223 (56)	-1.224 (56)	6.48	<0.001

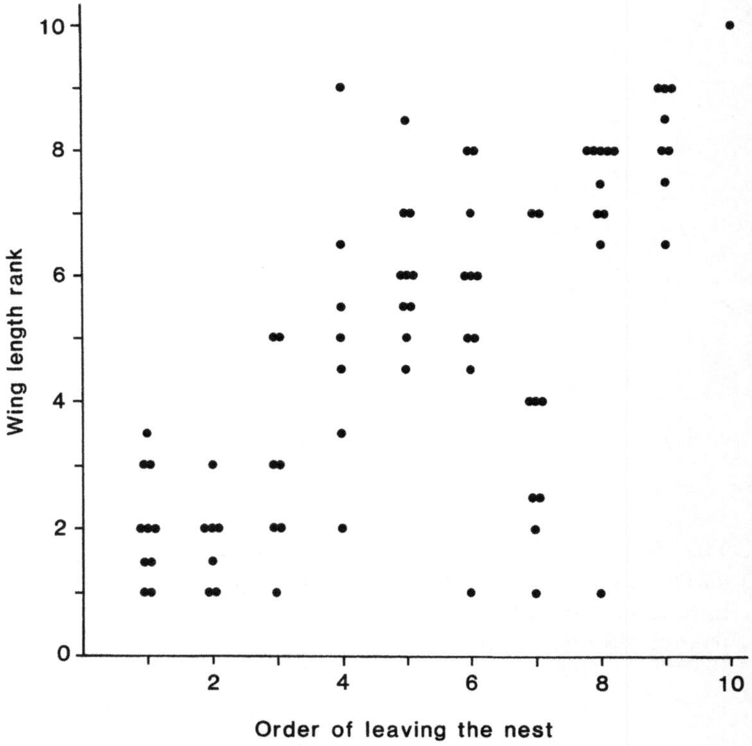

Fig. 3. Relationship between wing length rank and order of nest-leaving (rank 1 = the nestling with the longest wing in the brood). Data from 1988 and 1989. Tested with Spearman rank-order correlation: $r_s = 0.684$; $N = 79$; $P < 0.001$.

In 18 nests, I recorded the order in which all or some of the nestlings left the nest. By ranking the nestlings according to their size, from the largest to the smallest, I was able to correlate nestling size with order of nest-leaving. The best predictor of nest-leaving order was wing length (Fig. 3). Since wing length and weight of nestlings are positively correlated in the Marsh Tit (Nilsson 1989), weight (Spearman rank-order correlation, $r_s=0.505$; $N=79$; $P<0.001$) and body condition ($r_s=0.405$; $N=64$; $P<0.005$) also correlated positively with the order of nest-leaving, whereas tarsus length did not significantly ($r_s=0.209$; $N=64$; N.S.). The above relations could occur either if the largest young leave the nest first while the last young leave at random with respect to size or if the smallest young are the last to leave while the others leave at random. To investigate such effects, I separately calculated the correlation between wing length and fledging order of the first 5 nestlings that left the nest and that between wing length and fledging order for nestlings 6-10. Both correlations showed a significant positive relation (fledging order 1-5: $r_s=0.760$; $N=41$; $P<0.001$, fledging order 6-10: $r_s=0.606$; $N=38$; $P<0.001$) indicating that the nestlings with the longest wings leave first followed by nestlings of decreasing wing length; the last nestlings to leave are those with the shortest wing.

Weight, wing length and tarsus length of the largest and smallest nestling when they were 13 days old, the brood means of the three body measurements, the number of nestlings, and time of season were used as predictors of the total length of the nestling phase for a brood, in a stepwise multiple regression analysis. In both 1988 and 1989, only one parameter emerged as a significant predictor of the length of the nestling phase: wing length of the nestling with the largest wing in each brood ($P<0.001$; Fig. 4). This parameter accounted for 49.7% and 62.5% of the variation in nestling-time in 1988 and 1989, respectively.

Thus, the most long-winged nestling decides when to leave the nest, the others following in descending wing length order within 2-3 h.

The role of the parents during nest-leaving

One possible mechanism for parents to entice young to leave the nest is by calling. However, in the 42 nestlings observed leaving the nest during such circumstances that I would have heard the parents' calls, only 3 (7%) left when either one or both parents vocalized.

Another way in which parents may force their young to leave the nest is by decreasing the frequency of food provisioning. I have no indication of such a decrease prior to nest-leaving, if anything the feeding frequency increases (mean deviation between feeding frequency during the hour preceeding the start of nest-leaving and during the same hour but the day before: 9.25 feeds per hour; S.D.$=12.74$; $N=4$; N.S.). Changes in feeding frequency per young can nevertheless be responsible for the short duration of the nest-leaving period. This is because when the first half of the young had already left the nest, those remaining got significantly fewer feedings per nestling than during the early part of the nest-leaving phase (Fig. 5).

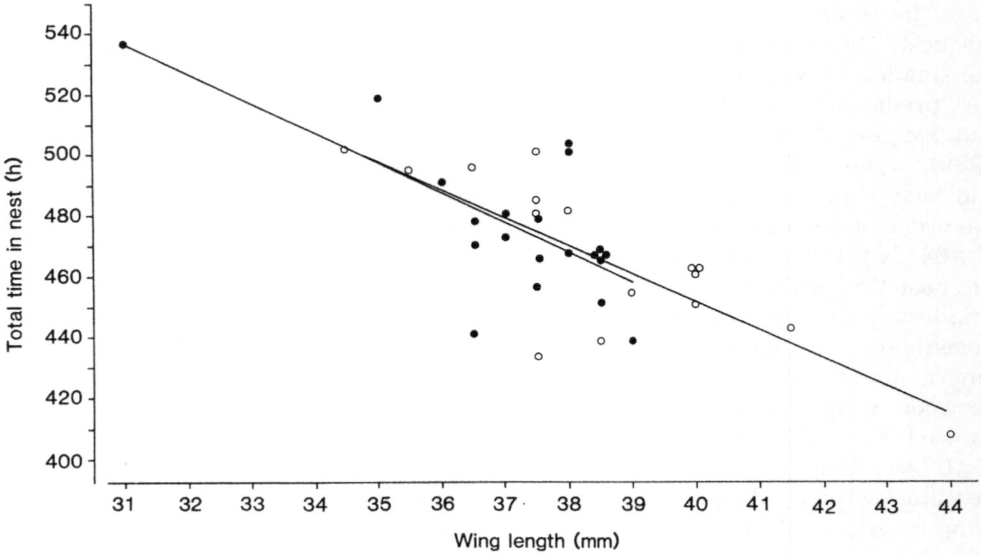

Fig. 4. Relationship between the total nestling period (mid-hatch to mid-fledging) and wing length of the most long-winged nestling in each brood, as measured on its 13th day, for 1988 (•) and 1989 (○). The regression equations, plotted as solid lines, are, 1988: $Y=835.3-9.66X$, and 1989: $Y=820.5-9.21X$.

DISCUSSION

The length of the nest-leaving period

Most of the Marsh Tit broods left the nest within 3 h, the shortest period being 5 min. A short nest-leaving period seems to be the general rule among tits (Gibb 1950; Hinde 1952; Perrins 1979). In some cases parents may even trick the last young out of the nest by presenting a caterpillar in the entrance-hole without delivering it (Walker 1972; Perrins 1979). Even though this behaviour was not observed in this study, other parental behaviours may force the last nestlings out of the nest. In the Marsh Tit, the last half of the brood to leave received less food per nestling than they did during the immediately preceding period. The main reason for this probably is that parents first satiate those young that have already fledged (Lemel 1989). Thus, the last nestlings may have to leave the nest in order to get enough food. This needs not, however, be a parental strategy to actively force the last young out of the nest. The feeding of already fledged young may serve to get them silent as quickly as possible (Lemel 1989). Thus, the decrease in number of feeding trips to the nest

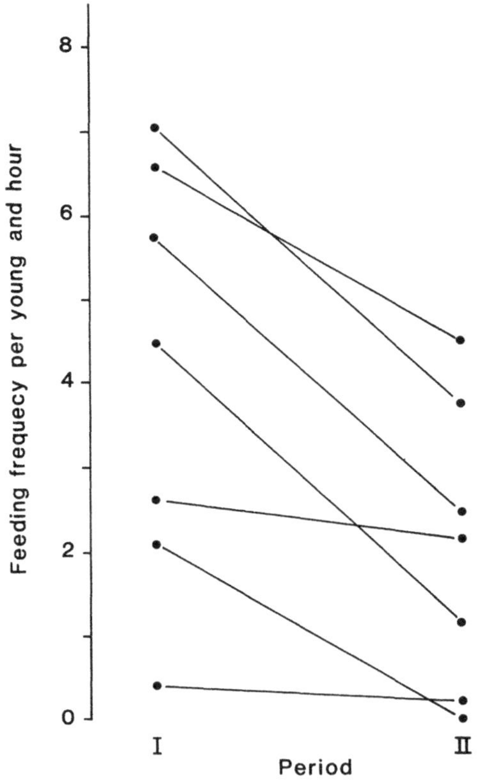

Fig. 5. Feeding frequency per nestling and hour for the same brood during two different periods. Period II: when the last half of the nestlings are leaving the nest; period I: the corresponding length of time to period II but just before the median young leaves the nest. In some cases period I includes feeding frequencies before the initiation of nest-leaving. Data from 1988 and 1989. Tested with Wilcoxon signed ranks test: N=7; P=0.018.

after fledging has begun may be an effect of the parents' efforts to avoid attracting predators to the surroundings of the nest-site.

Some of the broods in Fig. 2 where manipulated to increase hatching asynchrony (Nilsson in prep.). The time it took for all eggs in a clutch to hatch ranged from half a day to two days. This means that although a nestling is two days younger than its oldest sibling, i.e. the one likely to induce nest-leaving, all nestlings leave the nest at about the same time. Thus, the nestling period will be shorter for the last-hatched young, as also reported for Great Tits (*Parus major*) and Blue Tits (*P. caeruleus*) (Gibb 1950). The youngest nestlings may then fledge 1-2 days earlier than would be optimal for them. On the other hand, the oldest and therefore largest nestlings will stay longer than would be necessary according to their developmental stage. This means that they are capable of leaving the nest at least two days earlier, although probably in poorer condition, if the nest is attacked by a predator, as predicted by Clark and Wilson (1981).

The order of nest-leaving

Nestling Marsh Tits leave their nest in descending size order with the largest one leaving first. This is also true when the nest-leaving order is calculated separately for the first and second half of the brood. The fledging order probably reflects a dominance hierarchy among the nestlings. Nestlings seem to compete for the best begging positions in the nest and, at least when none of them is completely satiated, this position is usually occupied by the largest nestling (Greig-Smith 1985; Gottlander 1987). As the nestlings' mobility increases, in part due to wing development, the most advanced nestlings tend to jump or flutter against a feeding parent in order to monopolize the food brought (Bengtsson and Rydén 1981). The next step for the best-developed, dominant nestling is to sit in the entrance-hole, a position for which the nestlings compete (Hinde 1952), and wait for food-deliveries. When the wing is sufficiently developed, it flies out of the nest to meet an approaching parent.

Timing of the nest-leaving period

The timing of nest-leaving can be viewed as a conflict between parents and their offspring. Thus, the precise initiation of nest-leaving can be decided by all nestlings, one of them or the parents. In some studies, the nest-leaving of individual nestlings has been observed to follow calls made by the parents (Arnold and Arnold 1952; Hinde 1952). However, it is not known if the parents call to initiate nest-leaving or merely to try to get all the young out of the nest in a rather short time period. In the Marsh Tit, I could not discern any calls from the parents when the first young left the nest. The few nestlings that left the nest following calls from their parents where all late in the fledging order. Another way in which parents may entice the young out of the nest is by reducing their feeding frequency. As discussed above, a reduction in the feeding frequency per young may explain why all nestlings leave soon after the first one has left. However, in the few broods for which I have good data on feeding frequencies for some time before the initiation of nest-leaving, no such reduction in feeding frequency was evident.

A stable or increasing feeding frequency until fledging has been reported in several studies, e.g. Royama (1966), Morehouse and Brewer (1968), Leffelaar and Robertson (1986). However, also declining feeding frequencies close to nest-leaving have been observed; a general decline (Grundel 1987), a decline in one year but not in another (Gibb 1950) and a decline in some pairs but not in others, the decline often due to one of the parents ceasing to feed the nestlings altogether (Kluijver 1950). Thus, although not a general trend, parents might under certain circumstances reduce their feeding frequency, the effect of which may be that the young leave the nest.

In the Marsh Tit, the initiation of nest-leaving was related to the wing length of the young with the longest wing. Furthermore, it seems that, as soon as a chick has acquired a wing of a certain threshold length, it initiates the nest-leaving process of the brood. This would indicate that it is the most long-winged nestling that decides when the brood is going to leave the nest, in contrast to the most long-winged one leaving due to some sort of parental manipulation.

These results differ in one important aspect from those obtained by Lemel (1989) on nest-leaving in young Great Tits. In six broods studied, fledging order correlated negatively with weight. However, the first chick to leave was lighter and had a poorer body-condition index than the following ones. One important difference between the two species may help to explain the different results. Selection to be ready for dispersal as early as possible is very intense in the Marsh Tit. Those young that become independent early in the season have a much higher probability of getting established and of surviving until the next breeding season than late ones (Nilsson and Smith 1988b; Nilsson 1989). Thus, parents and offspring may have the same goal, i.e. to shorten the time until independence as much as possible. In this way the parent-offspring conflict, at least the conflict over time-investment, may be relaxed in the Marsh Tit (Nilsson and Smith 1985). Great Tits do not have the same rigid territorial system during the non-breeding season and are thus less dependent on early establishment for their survival (Gibb 1954; Nilsson and Smith 1988b). Furthermore, unlike Marsh Tits, Great Tits often lay a second clutch. Thus, the advantage to the parents of a rapid first breeding, i.e. a higher probability of having a second clutch, does not accrue to the young. This may lead to greater conflicts over time-investment between parents and their offspring than in the Marsh Tit. If parents and young disagree about when to leave the nest, parents may try to entice young to leave. The chick that is easiest for the parents to manipulate into leaving the nest is probably the hungriest one. When this one has left, the large, dominant nestlings also have to leave to preserve their priority of access to food brought by their parents.

If the first nestling leaves after attaining a certain wing length, broods in which food intake per nestling is low for some reason, will stay in the nest longer than nestlings in broods which are fed more adequately. In many brood manipulation experiments it has been shown that nestlings in enlarged broods are smaller than nestlings from natural or reduced broods (e.g. Nur 1984; Smith et al. 1989). Furthermore, Great Tit nestlings from enlarged broods stay one day longer in the nest than nestlings from reduced broods with the control brood nestlings having an intermediate nestling period (Smith et al. 1987). Such a time cost could be part of a reproductive cost of having a clutch size larger than the modal one.

ACKNOWLEDGEMENTS

I am grateful to Thomas Alerstam, J. H. van Balen, Hans Källander and J. L. Martin for comments on earlier drafts of the manuscript. The study was supported by grants from the Swedish Natural Science Research Council.

REFERENCES

Arnold GA, Arnold MA (1952) The nesting of a pair of blue tits. Br Birds 45: 175-180

Bengtsson H, Rydén O (1981) Development of parent-young interaction in asynchronously hatched broods of altricial birds. Z Tierpsychol 56: 255-272

Clark AB, Wilson DS (1981) Avian breeding adaptations: hatching asynchrony, brood reduction, and nest failure. Q Rev Biol 56: 253-277

Gibb J (1950) The breeding biology of the great and blue titmice. Ibis 92: 507-539

Gibb J (1954) Population changes of titmice, 1947-1951. Bird Study 1: 40-48

Gottlander K (1987) Parental feeding behaviour and sibling competition in the pied flycatcher *Ficedula hypoleuca*. Ornis Scand 18: 269-276

Greig-Smith P (1985) Weight differences, brood reduction, and sibling competition among nestling stonechats *Saxicola torquata* (*Aves*: *Turdidae*). J Zool, Lond (A) 205: 453-465

Grundel R (1987) Determinants of nestling feeding rates and parental investment in the mountain chickadee. Condor 89: 319-328

Hinde RA (1952) The behaviour of the great tit (*Parus major*) and some related species. Behaviour Suppl 2 p.134-135

Kluijver HN (1950) Daily routines of the great tit, *Parus m. major* L. Ardea 38: 99-135

Leffelaar D, Robertson RJ (1986) Equality of feeding roles and the maintenance of monogamy in tree swallows. Behav Ecol Sociobiol 18: 199-206

Lemel J (1989) Body-mass dependent fledging order in the great tit. Auk 106: 490-492

Morehouse EL, Brewer R (1968) Feeding of nestling and fledgling eastern kingbirds. Auk 85: 44-54

Nilsson J-Å (1989) Causes and consequences of natal dispersal in the Marsh Tit, *Parus palustris*. J Anim Ecol 58: 619-636

Nilsson J-Å, Smith HG (1985) Early fledgling mortality and the timing of juvenile dispersal in the Marsh Tit *Parus palustris*. Ornis Scand 16: 293-298

Nilsson J-Å, Smith HG (1988a) Incubation feeding as a male tactic for early hatching. Anim Behav 36: 641-647

Nilsson J-Å, Smith HG (1988b) Effects of dispersal date on winter flock establishment and social dominance in Marsh Tits *Parus palustris*. J Anim Ecol 57: 917-928

Nur N (1984) The consequences of brood size for breeding blue tits II. Nestling weight, offspring survival and optimal brood size. J Anim Ecol 53: 497-517

Perrins C (1979) British tits. Collins, London.

Royama T (1966) Factors governing feeding rate, food requirement and brood size of nestling great tits *Parus major*. Ibis 108: 313-347

Smith HG, Källander H, Nilsson J-Å (1987) Effect of experimentally altered brood size on frequency and timing of second clutches in the great tit. Auk 104: 700-706

Smith HG, Källander H, Nilsson J-Å (1989) The trade-off between offspring number and quality in the great tit *Parus major*. J Anim Ecol 58: 383-401

Walker JES (1972) Attempts at fledging of a runt great tit. Bird Study 19: 250-251

Wilkinson L (1987) SYSTAT The system for statistics. Evanston, IL, SYSTAT Inc.

TACTICS AND COUNTER-TACTICS OF SEXUALLY SELECTED INFANTICIDE IN TREE SWALLOWS

Raleigh J. ROBERTSON

Department of Biology, Queen's University, Kingston, Ontario, Canada K7L 3N6

Infanticide, the killing of dependent, conspecific young, has only recently been recognized as an adaptive behaviour which can be moulded by natural selection through some advantage which accrues to the infanticidal individual (Trivers 1972, Hrdy 1979). Sexually selected infanticide, in which reproductively mature individuals kill offspring of another individual in order to gain a breeding opportunity, is relatively common in mammals, but apparently rare in birds (Rohwer 1986). Mock (1984) suggests the paucity of sexually selected infanticide in birds is due to the prevalence of monogamy in this taxon. However, even monogamous species can face severe competition for breeding opportunities and hence strong sexual selection (Crook and Shields 1985, Robertson and Stutchbury 1988). Recent studies (e.g. Goldstein et al. 1986, Møller 1988a, Emlen et al. 1989) reveal that sexually selected infanticide may be more common in birds than previously realized, and infanticide and its threat may be an important factor in the evolution of both male and female reproductive strategies (Freed 1987, Møller 1988a).

The typical pattern of sexually selected infanticide involves a floater male which has either replaced or evicted a resident male. The replacement male kills the infants that were sired by the previous male and thereby acquires both a breeding territory and a female which is receptive much sooner than if she were to complete rearing her present brood (Hrdy 1979, Crook and Shields 1985, Freed 1986, Møller 1988a, Robertson and Stutchbury 1988).

Infanticide by replacement male birds poses a conflict of interests between the sexes. While the infanticidal male benefits by the earlier opportunity to sire his own offspring, the female must pay the cost of both the time and energy required for renesting. Consequently, natural selection should favour counter-tactics in females which function to prevent infanticide by males, especially in cases where females are able to raise young without male assistance. Such counter-tactics are apparently important in several species of mammals (Andelman 1987, Hiraiwa-Hasegawa 1988).

Sexually selected infanticide is most likely to evolve in those species which face limited opportunities for breeding. Species in which mates or nest sites are in short supply are likely candidates. Secondary cavity nesting birds, those which rely on the availability of cavities excavated by other species, often have large floater populations as demonstrated by the rapid response to provision of nest boxes (Holroyd 1975, Stutchbury and Robertson 1985, Freed 1986, Romagnano et al. 1986). Many of these are short-lived passerines which face strong selection to breed, but are often forced to facultatively delay breeding (Studd and Robertson 1985) due to a shortage of nest sites.

NATO ASI Series, Vol. G 24
Population Biology of Passerine Birds
Edited by J. Blondel et al.

Tree Swallows (*Tachycineta bicolor*) are small (21g), short-lived (Robertson and Rendell in press), secondary cavity nesters which compete intensively for breeding opportunities (Stutchbury and Robertson 1987). I predicted that Tree Swallows would be infanticidal because they meet the conditions of (a) limited breeding opportunities, (b) females renesting after failure, and (c) females not usually moving after nest failure (cf. Rohwer 1986). They also exhibit some of the features which Møller (1988a) argued would favour sexually selected infanticide, namely colonial breeding and asynchronous nesting, although in contrast to Møller's argument, they have only one brood per season. Aggression over cavity or nest box occupancy, frequent intrusions (Stutchbury and Robertson 1987), fights with injuries or even death (Lombardo 1986, Robertson et al. 1986), and usurpation of nest sites (Leffelaar and Robertson 1985) are all evidence of severe competition for breeding opportunities in Tree Swallows. The purpose of this study was first to determine whether infanticide by replacement males occurs in this species, second to determine the effect of nest stage on the male's decision of whether to adopt or kill the eggs/nestlings, and third to examine possible counter-tactics that are employed by females.

METHODS

These studies were conducted from 1985 until 1989 on populations of Tree Swallows breeding in grids of nest-boxes which have been in place since 1976 at the Queen's University Biological Station in southern Ontario, Canada. Each year, females and most males are captured using mist nets and nest-box traps (Stutchbury and Robertson 1986) prior to egg-laying, and given individually identifiable marks with acrylic paint on the wings and tail.

Since usurpations or death of a resident followed by replacement are relatively infrequent, and it is difficult to observe an instantaneous event like infanticide, an experimental approach was used. In order to create situations where infanticide would be most likely to occur if it is indeed part of the behavioural repertoire, resident males were permanently removed from certain experimental nests. Removals were done at various times of the laying, incubation, and nestling stages so that the behaviour of the female and of replacement males arriving at different stages of the nesting cycle could be examined. Nests were watched for the arrival of a replacement male, and the response of the female toward potential and actual replacements was monitored. Nests were checked daily and feeding rates of females, replacement males and control males were measured.

RESULTS

Infanticide by replacement males

Nestling stage replacements

Experiments were first designed to determine whether replacement male Tree Swallows are infanticidal. Resident males were removed from 15 nests which contained nestlings that were 1 or 2 days old. In 7 cases, a replacement male arrived, in all cases while nestlings were $\leq$ 10 days old. Five of these males killed the nestlings by pecking them while in the nest, or by removing live nestlings and dropping them outside the nest box. The other 2 replacement males were considered indifferent in that they neither harmed nor helped feed the nestlings. Three of the 5 infanticidal males renested, two with new females and one with the widowed female (Table I). These experiments and observations established that Tree Swallow males which replace a resident will kill nestlings and may thereby gain a breeding opportunity (Robertson and Stutchbury 1988).

Table I. Outcome of experimental removal of resident male Tree Swallows resulting in replacement males arriving during the nestling, incubation and egg-laying stages.

	Nestling	Incubation	Egg-laying
No replacement	8	3	4
Replacement	7(+1)	14(+1)	7
Indifferent	2	0	0
Adopt	0	3	7
Infanticide	5(+1)	11(+1)	0
No renest	2(+1)	1	-
Renest widowed female	1	9(+1)	-
Renest new female	2	1	-

Note: figures in brackets are for 2 removals done during egg-laying in which replacements arrived at a later stage of nesting.

Incubation stage replacements

Experiments were then designed to determine whether males which replace while there are eggs in the nest would destroy the eggs or otherwise disrupt the clutch. Removal of resident males was done at various stages of incubation to establish the role of the interval between arrival of the replacement male and hatching in determining whether the replacement male would adopt or kill the nestlings that had been sired by the previous male. Fourteen replacement males arrived following removal of males from 17 nests. Most (9/14) replacement males were seen to copulate with the resident female, and all attended the nest while incubation continued. None of the males disrupted the nest or eggs. Three of these males, all of which had arrived on 2 of incubation, adopted the nestlings and provided about 50% of the feeding and care as soon as the eggs hatched. However, 11 of the replacement males, which had arrived from day 2 to 14 of the 14 day incubation period, killed the nestlings soon after the eggs hatched (Table I). Ten of these males subsequently renested. These results indicate that the time of the replacement male's arrival is important in determining whether the male kills or adopts the brood.

Replacements during egg-laying

Male removals were next done during the egg-laying period to establish whether a replacement male that had some probability of paternity would adopt the entire brood. Seven replacement males arrived following removal of males from 11 nests. All seven of the replacement males adopted the brood and made approximately 50% of the feeding trips, similar to the control pairs. Six of these males had arrived prior to the day on which the last egg was laid, while 1 arrived on the day the clutch was completed (Table I). None of the adopting males renested since they were involved in rearing the adopted brood. Two additional males removed during egg-laying were not replaced until after the clutch was complete. One replacement (BGG3'-89), which arrived one day after clutch completion, is strongly suspected of having killed the nestlings soon after they hatched. Another replacement (HUD6-89), which did not arrive until the nestlings were 4 days old, attempted infanticide in that he pecked the nestlings causing gashes on the heads of two of them, but this male did not return after he was captured for marking. At another nest (BGD6-89), neighbouring males copulated with the widowed female and defended her nest site against intruders, but the neighbours neither disturbed nor adopted the nestlings.

Taken together, these results indicate that replacement males which arrive before the clutch is complete, such that they have a reasonable possibility of fertilizing at least one egg, usually adopt the nestlings. Replacements which arrive shortly after clutch completion, such that there may be some uncertainty regarding paternity, may either adopt or kill the nestlings. Replacements which arrive well after clutch completion, so there is no opportunity to fertilize any eggs, usually kill the nestlings (Fig. 1). Surprisingly, males which arrive after clutch completion do not disrupt the clutch but rather wait until the end of the 14 day incubation period and then kill the nestlings soon after the eggs hatch.

Female counter-tactics

Females are frequently aggressive toward the replacement male at the time he arrives. Indeed, at least one of the 2 males which were considered to be indifferent may have been prevented from killing the nestlings by the aggressive defense of the female. The initial aggressive response by the female appears to be similar to nest site defense against intruders by either sex of the

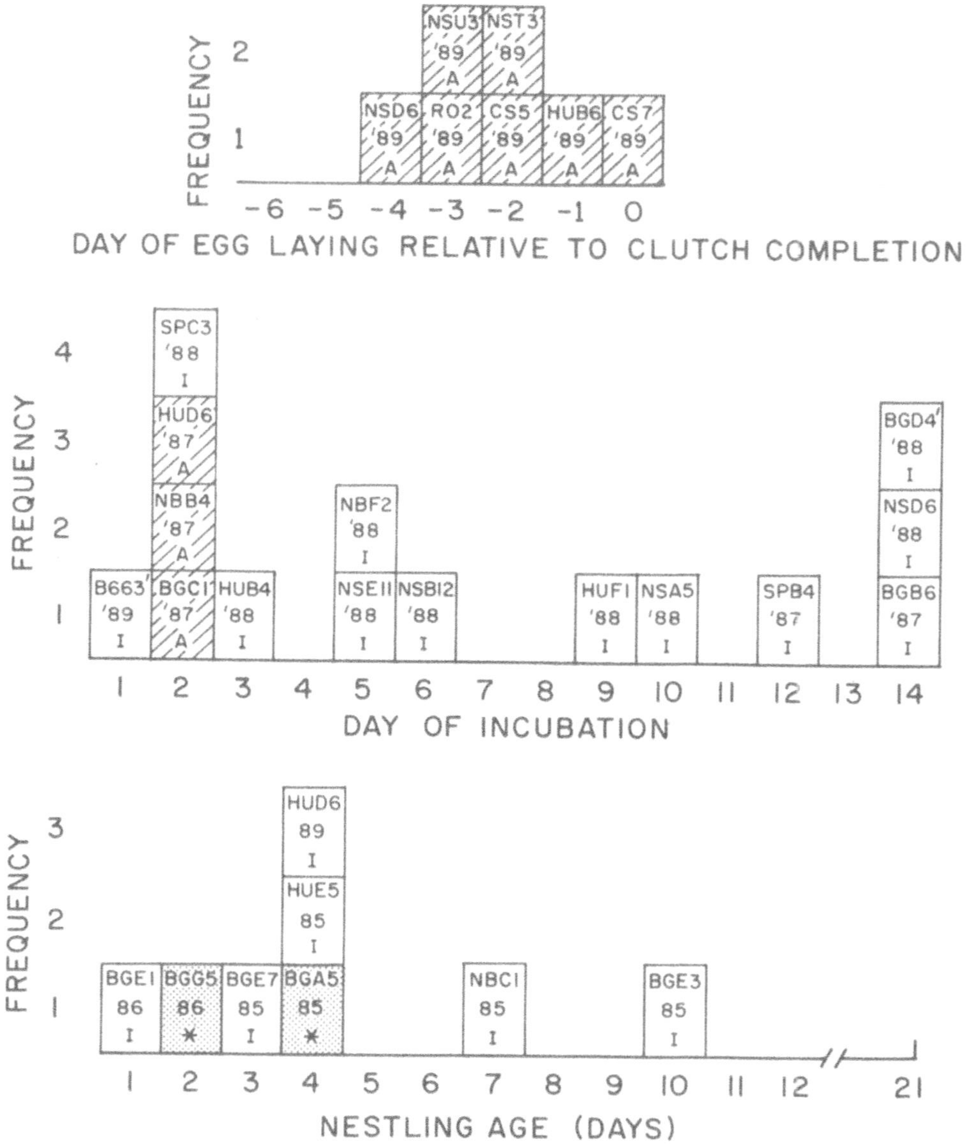

Fig. 1. Frequency distribution of arrival day of replacement males in relation to nest stage, showing the ultimate fate of the nest. Each box represents a nest with a replacement male and shows the nest identification, the year, and whether the replacement male adopted (A), was indifferent (*), or committed infanticide (I). Males which arrived during incubation did not commit infanticide until after the eggs hatched. (The last egg of the clutch is laid on day 0 of egg laying. Day 1 of incubation is the day following the day of clutch completion. Day 1 of the nestling period is the day of hatch).

resident pair. However, in the absence of the resident male, potential replacement males usually persist and engage the female in vigorous fights. Intermittent fighting between the resident female and the replacement male may continue for up to a day or more.

Most females are receptive and copulate with the replacement male once the aggressive response declines. This may occur within the first hour, or up to several days after the arrival of the replacement.

To determine whether copulation patterns of females with replacement males were different from those of control pairs, one grid of 18 boxes was watched regularly from mid-April to mid-May and copulations were recorded. Copulations seen at times other than during systematic watches were also recorded. Copulations occur most frequently on top of the nest-box or at a nearby perch, with the male hovering just above the perched female while giving a characteristic chatter. The male intermittently alights on the female's raised wings as she assumes a crouched position and lifts her tail, and cloacal contact is made. Each copulation bout involves from 1 to 10 or more cloacal contacts with the male hovering briefly above the female between each one. Copulation by mated pairs begins about 12 days prior to laying of the first egg and continues until the clutch is complete, with the frequency of copulations peaking 1 or 2 days before clutch initiation (Fig. 2). No copulations by control, mated pairs were seen after clutch completion. In contrast, females mating with a replacement male copulate even though they lay no additional eggs and continue to incubate the eggs, all or most of which had been fertilized by the previous male. Based on the number of cloacal contacts per copulation, the copulations with replacement males are similar in terms of female receptivity and male persistence to those between control pairs prior to clutch completion (Fig. 2).

DISCUSSION

Infanticide is clearly adaptive for males which replace a previous resident during the incubation or nestling stage. Floater males which do not breed have no reproductive output in the current year. In contrast, most replacing males which are infanticidal do obtain a breeding opportunity. Of 18 replacement males which killed nestlings, 14 renested (11 with the widowed female and 3 with a new female) and 11 of them fledged at least one young in that year.

Adoption is adaptive for the replacement male only if he arrives during egg-laying so there is some possibility of fertilizing at least one egg. Males which arrived after clutch completion but yet adopted the brood were apparently behaving in a maladaptive manner since they did not fledge any of their own offspring in the current year. In Tree Swallows, which have only a single brood, and in which pair bonds are not maintained between seasons, it seems unlikely that adopting in one season confers an advantage in the following season. Interestingly, adopting males, which at best shared a relatively small portion of the paternity of the brood, contributed as much

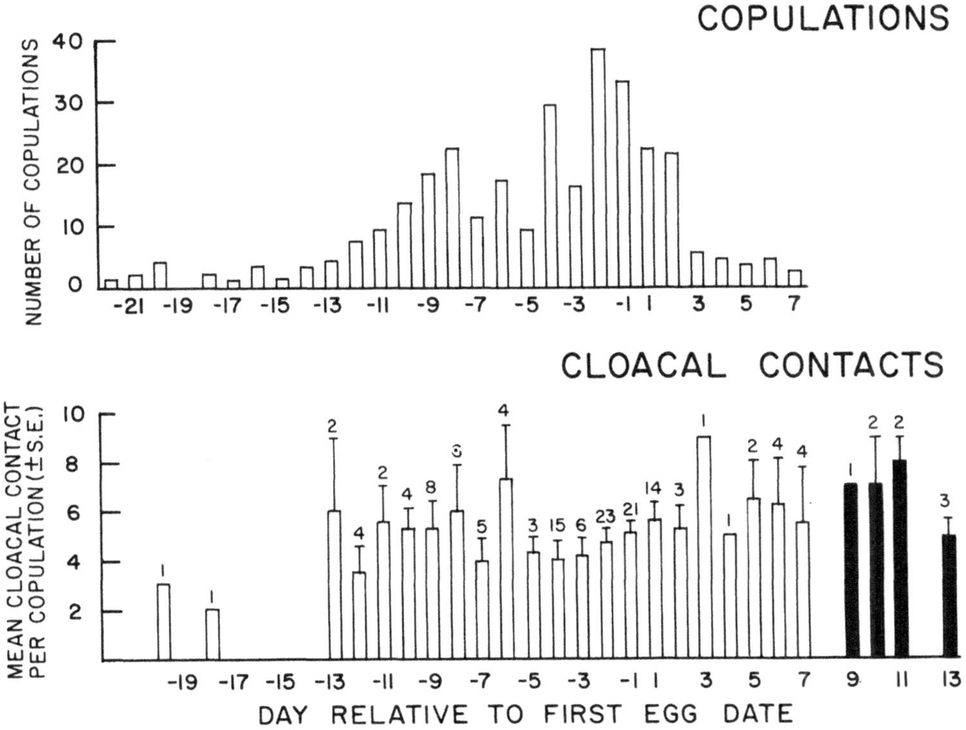

Fig. 2. Total number of copulations observed in 1988 during regular watches plus ad libitum sightings, and number of cloacal contacts per copulation in relation to first egg date. No copulations occurred for control pairs after their clutch was complete (day 7). Data for the number of cloacal contacts per copulation after day 7 (shaded bars) involve replacement males. Day 1 is the day on which the first egg was laid. Number of copulations on which the mean number of cloacal contacts per copulation is based is shown above each bar.

parental care in feeding nestlings as control males. This contrasts with Møller's (1988b) finding that male Barn Swallows (*Hirundo rustica*) adjusted their parental care in relation to their certainty of paternity. In Tree Swallows, males which adopted the broods in which they contributed little or no paternity may have been fooled into behaving in this manner by the female's prolonged sexual receptivity.

Gjershaug et al. (1989) suggest that widowed Pied Flycatchers (*Ficedula hypoleuca*) which copulate even though they have a full clutch or brood are engaging in a form of "marriage entrapment" as a means of obtaining male assistance in rearing their brood. Such male assistance would increase the female's reproductive success, but the prevention of infanticide may have an even more important influence on the evolution of prolonged copulation behaviour by widowed females.

Adoption by replacement males is always advantageous to the female, especially in this population where lone females never raise full broods and rarely fledge even partial broods. If the replacement male adopts, it not only increases the female's chances of fledging her current brood, but also saves her the time and energy that would be required to renest if the male were to kill the nestlings. The fact that females confronted with a replacement male become receptive and copulate at the same time they incubate, while control pairs do not copulate after clutch completion, suggests that the receptivity of the female may be functioning to manipulate the male's behaviour.

Male Tree Swallows which replace during egg-laying and early incubation could avoid the apparent error of caring for nestlings they did not sire if they destroyed any eggs that were in the nest at the time of, or shortly after, their arrival. Indeed, Freed (1987) argues that tropical House Wrens (*Troglodytes aedon*) evaluate paternity and kill eggs that appear within one fertility period of their arrival. Destruction of eggs would also allow the replacement male to avoid wasting the time of up to two weeks between his arrival and hatch. Perhaps Tree Swallows lack the ability to remove or destroy eggs without fouling the nest and rendering it unusable. However, the fact that renesting females occasionally build a new nest on top of a previously used one containing eggs or dead nestlings makes this explanation unlikely.

Infanticide by replacement males is not simply an artifact of our experimental manipulations. In 1987 and 1988 there were 4 cases where resident males were either preyed upon or evicted by a replacement male. In all cases, the replacement males, which arrived during incubation, killed the nestlings on the day that they hatched. Clearly, infanticide as a means of gaining a breeding opportunity is a normal component of the male Tree Swallow's reproductive strategy. Whether prolonged copulation by a female faced with a potentially infanticidal male is a tactic which functions to manipulate the male's behaviour remains to be firmly established.

This study demonstrates that strong competition for breeding opportunities, as occurs in many secondary cavity nesting species, has had an important influence in shaping the behaviour of male Tree Swallows and probably that of females as well. Infanticide, or the threat of infanticide, may shape behaviours such as nest guarding (Møller 1988a), copulation, dispersal after nest failure, response to intruders and neighbours, and male-female interactions. This influence may be more widespread than has previously been recognized and should be explored in other species of birds.

ACKNOWLEDGEMENTS

I thank Peter Dunn, Sarah Kalff, Scott Morrill, Wally Rendell, Bridget Stutchbury, Lisa Venier, and Larissa White for assistance in the field, the Natural Sciences and Engineering Research Council of Canada for financial support, Queen's University Biological Station for logistical support, and the School of Graduate Studies and Research at Queen's University for helping pay travel costs to the NATO workshop.

REFERENCES

Andelman SJ (1987) Evolution of concealed ovulation in vervet monkeys (*Cercopithicus aethiops*). American Naturalist 129: 785-799

Crook JR and Shields WM (1985) Sexually selected infanticide by adult male barn swallows. Animal Behaviour 33: 754-761

Emlen ST, Demong NJ, and Emlen DJ (1989) Experimental induction of infanticide in female Wattled Jacanas. Auk 106: 1-7

Freed LA (1986) Territory takeover and sexually selected infanticide in tropical house wrens. Behavioral Ecology and Sociobiology 19: 197-206

Freed LA (1987) Prospective infanticide and protection of paternity in tropical House Wrens. American Naturalist 130: 948-954

Gjershaug JO, Jarvi T, and Roskaft, E (1989) Marriage entrapment by "solitary" mothers: a study on male deception by female Pied Flycatchers. American Naturalist 133: 273-276.

Goldstein HM, Eisikovitz D and Yom-tov Y (1986) Infanticide in the Palestine Sunbird. Condor 88: 528-529

Hiraiwa-Hasegawa M (1988) Adaptive significance of infanticide in primates. Trends in Ecology and Evolution 3: 102-105

Holroyd GL (1975) Nest site availability as a factor limiting population size of swallows. Canadian Field Naturalist 89: 60-64

Hrdy SB (1979) Infanticide among animals: a review, classification, and examination of the implications for the reproductive strategies of females. Ethology and Sociobiology 1: 13-40

Leffelaar D and Robertson RJ (1985) Nest usurpation and female competition for breeding opportunities by Tree Swallows. Wilson Bulletin 97: 221-224

Lombardo MP (1986) A possible case of adult intraspecific killing in the Tree Swallow. Condor 88: 112

Mock DW (1984) Infanticide, siblicide, and avian nestling mortality. In: Hausfater G and Hrdy SB (eds) Infanticide: Comparative and Evolutionary Perspectives. Aldine, New York. p 3-30

Møller AP (1988a) Infanticidal and anti-infanticidal strategies in the swallow *Hirundo rustica*. Behavioral Ecology and Sociobiology 22: 365-371

Møller AP (1988b) Paternity and paternal care in the swallow, *Hirundo rustica*. Animal Behaviour 36: 996-1005

Robertson RJ, Gibbs HL, and Stutchbury BJ (1986) Spitefulness, altruism, and the cost of aggression: evidence against superterritoriality in Tree Swallows. Condor 88: 104-105

Robertson RJ and Rendell WB (in press) A comparison of the breeding ecology of a secondary cavity-nesting bird, the Tree Swallow (*Tachycineta bicolor*) breeding in nest-boxes and natural cavities. Canadian Journal of Zoology

Robertson RJ and Stutchbury BJ (1988) Experimental evidence for sexually selected infanticide in Tree Swallows. Animal Behaviour 36: 749-753

Rohwer S (1986) Selection for adoption versus infanticide by replacement mates in birds. Current Ornithology 3: 353-395

Romagnano L, Lombardo MP, Stouffer PC, and Power HW (1986) Suspected infanticide in the starling. Condor 88: 530-531

Studd MV and Robertson RJ (1985) Life span, competition, and delayed plumage maturation in male passerines: the breeding threshold hypothesis. American Naturalist 114: 439-446

Stutchbury BJ and Robertson RJ (1985) Floating populations of female tree swallows. Auk 102: 651-654

Stutchbury BJ and Robertson RJ (1986) A simple trap for catching birds in nest boxes. Journal of Field Ornithology 57: 64-65

Stutchbury BJ and Robertson RJ (1987) Behavioral tactics of subadult female floaters in the tree swallow. Behavioral Ecology and Sociobiology 20: 413-419

Trivers RL (1972) Parental investment and sexual selection. In: Campbell B.G. (Ed.) Sexual Selection and the Descent of Man 1871-1971. Aldine, Chicago, p. 136-179

BROOD DEFENCE IN GREAT TIT (*PARUS MAJOR*) PAIRS AND THE CONCORDE FALLACY

Wolfgang WINDT

Arbeitsgruppe für Verhaltensforschung, Ruhr-Universität Bochum,
Postfach 102148, D-4630 Bochum 1, West Germany

Theory predicts that an animal's decision of how much to invest into an ongoing project should depend only on future benefits and not on past costs (Dawkins and Carlisle 1976, Boucher 1977). A violation of this rule has become known as the Concorde fallacy. But past costs and future benefits are often closely correlated. This difficulty may be circumvented by uncoupling both, i.e. manipulating future benefits and leaving costs unchanged or vice versa (Robertson and Biermann 1979, Weatherhead 1982, Curio 1983, Carlisle 1985, Windt and Curio 1986).

Great Tits defend (actual investment) their young (future benefits) which they have produced (past costs) against a Pygmy Owl *Glaucidium passerinum*, known as a marauder of nest boxes. The purpose of this study was to examine the influence of past costs and future benefits on nest defence behaviour.

GENERAL METHODS

All experimental groups in an experiment were balanced for number of young and age of young as closely as possible. Nest defence tests were carried out alternatingly in the experimental groups to limit the interdependence of results over time.

To measure nest defence, a predator stimulus was presented 3 m in front of the nest box of the pair to be tested; the stimulus consisted of a caged Pygmy Owl and a multi-specific mobbing chorus emitted from a loudspeaker beside the owl. Response strength was measured as a risk index, unless indicated otherwise. The risk index increased linearly with both proximity to the owl and time spent near the owl (for more details see Windt and Curio 1986).

Statistical analysis was carried out with a two-tailed Mann Whitney U-test, unless indicated otherwise.

NATO ASI Series, Vol. G 24
Population Biology of Passerine Birds
Edited by J. Blondel et al.

RESULTS AND DISCUSSION

Manipulation of future benefits

In a first experiment in a study area in Bochum, I examined the influence of future benefits by making the number of young in first broods as different as possible: the number of young was reduced to 2 in randomly chosen broods at 10 days of age, and the young thus obtained were used to enlarge comparable broods.

There was no significant ($p > 0.05$) difference in nest defence between the two experimental groups (Fig 1a).

To make sure that no possible effect of the brood size manipulation went undetected, for example, by a higher feeding-of-young-activity of the parents in the enlarged group, thus preventing parents from reacting more strongly to the owl as was observed, in the following year feeders with mealworms were placed near the nest boxes of both experimental groups 3 days before a trial (Fig. 1b). Use of the feeders did not result in a significant difference in the defence of young between the experimental groups. The slightly stronger reaction in both groups than in the first experiment may be due to the feeders.

Do Great Tits ever adjust their response to a predator in relation to a change in brood size? A similar manipulation of number of young, this time in second broods near Wolfsburg, with no supplemental feeding, yielded a marked difference ($p < 0.001$) in brood defence between reduced and enlarged broods (Fig. 1c). This demonstrates that Great Tits are able to detect a change in brood size and adjust their investment in the ongoing brood.

The premise that more young are a greater benefit is not invariably correct (Pettifor, Perrins and McCleery 1988). Lindén (1987) and Smith (1988) reported similar brood size manipulation experiments with Great Tits, 6 days after hatching. They found that the number of local recruits next spring was greatest in an unmanipulated control group. Both reduced and enlarged broods resulted in fewer recruits: in enlarged broods because fledglings are of poor quality, in reduced broods because there are simply fewer fledglings. Given this result, the first mentioned experiment with only two experimental groups was repeated by adding a dummy manipulated control group (Fig 1d). Control broods were defended significantly ($p < 0.05$) more strongly than were enlarged or reduced broods. Very likely this is not only a mere effect of a disturbance caused by a change in number of young, because in second broods enlarged ones are defended much more strongly than reduced ones. Hence, defence was flexibly adjusted to the value of the brood as assessed by surviving recruits.

In a number of birds, among them Great Tits, weight of fledglings is a good predictor of survival (Murphy 1983, Westmoreland and Best 1987, Smith 1988 and in one of two years Nur 1984). In enlarged broods, one week after manipulation, the weight of fledglings was significantly ($p < 0.01$) lower than in the control or reduced group (reduced: 19.5 g, control: 20.7 g, enlarged: 14.6 g). These weight differences may induce differences in survival of broods of different size which in turn may result in control broods having the greatest value.

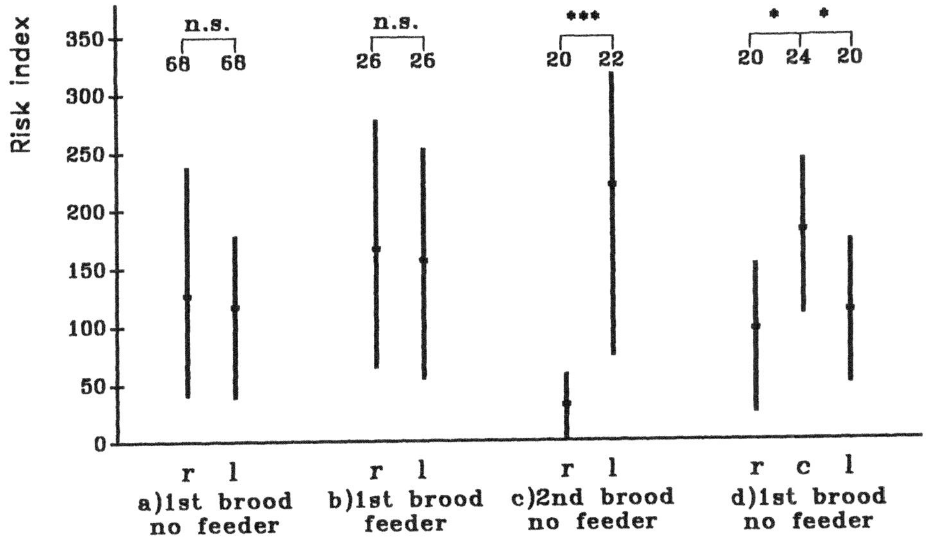

Fig. 1. Medians and interquartile ranges of risk index for defence of reduced (r), control (c) and enlarged (l) broods. Sample size indicated above ranges; U-test, two-tailed. For more details see text.
**, $P<0.01$; ***, $P<0.001$.

But there is still an open question: There were no differences in defence of reduced and enlarged number of young in first broods, but there were in second broods. What made enlarged second broods so valuable for the parents? The answer might be that enlarged second broods are of a similar size as unmanipulated first broods. Accordingly the parents might be able to feed them sufficiently. Why then are second broods so small? Even in first broods there is a significant difference in the number of young with the highest fledging success and the mean number of young found naturally (Perrins 1979). In second broods this difference might be even greater, because second broods are of low value (Curio and Regelmann 1982). Unfortunately, it was not yet possible to test the hypothesis that enlarged second broods are of higher value than enlarged first broods.

Manipulation of past costs

If past costs are manipulated theory predicts no effect on parental investment. An increase of past costs was achieved by tying together the primaries 8 and 9 on one side about 2.5 cm from distal with a nylon tie in both male and/or female Great Tits of a pair. This allowed the birds to survive but lowered their flight capacities for several weeks, as tested in an aviary (a comparable, but non-removable handicap was used by Lifjeld and Slagsvold 1988). For the experiment, Great Tits were captured about the 2nd day after hatching and some of them were handicapped in this manner and some not. They were recaptured about the 12th day. If there was a handicap it was removed and finally the bird was tested for brood defence on day 14. Because it was impossible to catch every bird that needed to be captured for a trial, there were four groups in the analysis: Firstly there was a group of randomly chosen birds that had never been captured. Second the control group, that was captured, but not handicapped. Third, the "released" birds were handicapped for 10 days but released before the test. Fourth, birds that were handicapped, but not recaptured before the test, were tested when still handicapped.

The handicap did not affect the weight of young significantly. Perhaps non or dummy manipulated partners compensated for eventually lower feeding frequencies by handicapped tits.

The comprehensive risk index cannot be applied for a comparison of all groups, because sitting in the same distance will probably not involve the same risk for still handicapped birds in comparison to others. The additionally used behavioural measures are: latency of approach to the test site as measured from the onset of a taped mobbing chorus, average distance and the minimum distance.

In the latency of approach there is no difference between uncaptured birds and control birds (Fig 2a). Hence, Great Tits that had been captured do not differ in their brood defence from an unmanipulated group. Released birds approach danger significantly later than control birds. This difference is not in the expected direction if Great Tits were behaving Concordian and thus were influenced by past costs.

The difference between control birds and released birds disappears in the average distance (Fig. 2b) and in the minimum distance (Fig 2c). That still handicapped birds have an extremely long minimum distance does not necessarily mean that they take a lower risk in this respect. An H-test does not reveal a difference between uncaptured, control and released birds concerning the risk index (Fig. 2d). So this experiment also fails to prove, that in Great Tits past costs influence the actual investment.

The difference in latency of approaching danger between control and released birds (Fig. 2a) may be due to the experience of the released birds that one of the observers in the test was regarded a dangerous predator that handicapped them for several days.

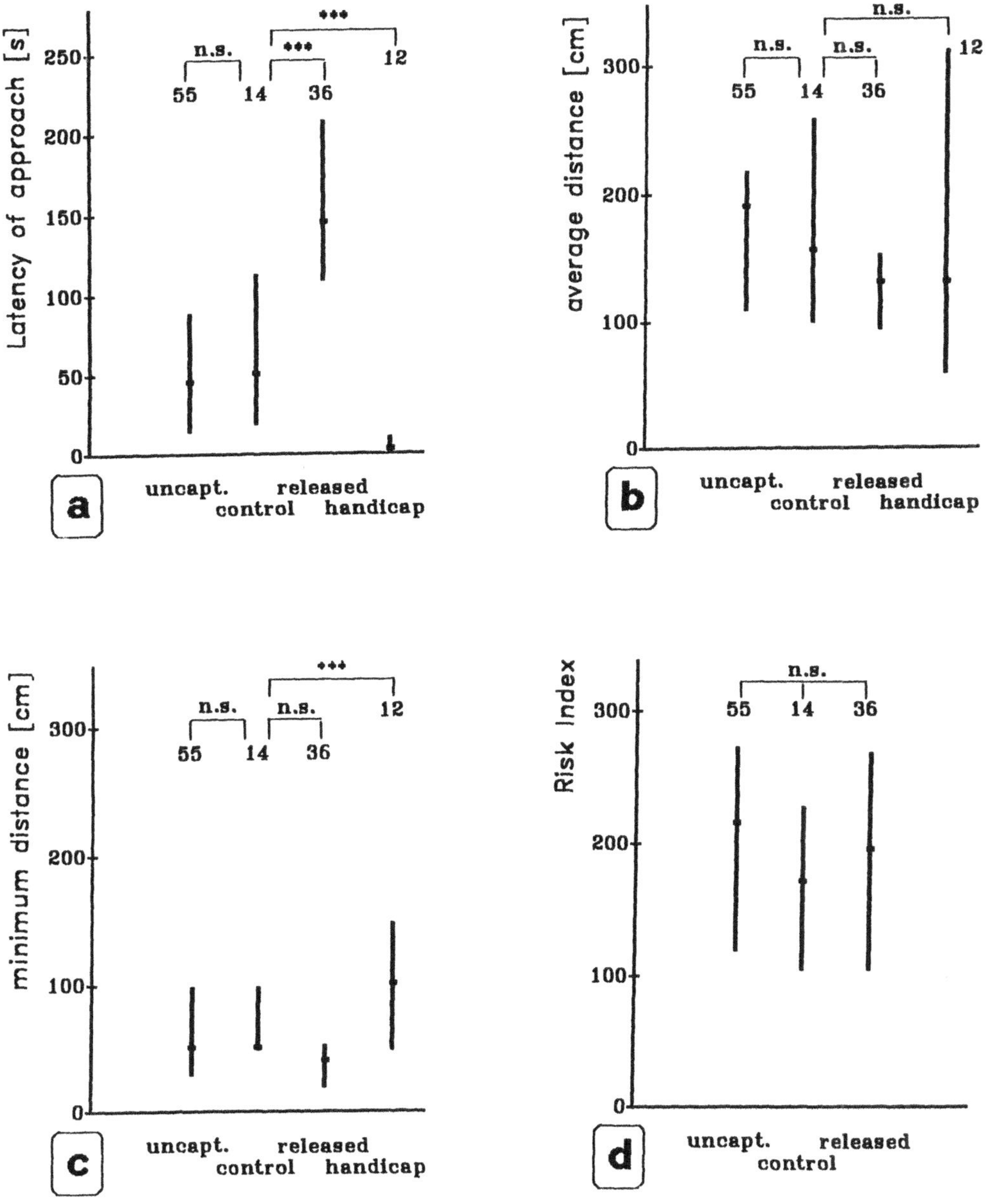

Fig. 2. Medians and interquartile ranges of (a) latency of approach, (b) average distance, (c) minimum distance and (d) risk index for uncaptured (uncapt.), captured but not handicapped (control), first handicapped and then released (released) and during test still handicapped (handicap) birds.
***, $P<0.001$.

If this explanation holds true, it is even more remarkable, that still handicapped birds have such a short latency of approach. Unfortunately this not intented experimental group is not a random sample of the population and this part of the experiment needs to be repeated properly. So this conclusion may only be drawn with reservation: If Great Tits use their handicap as a predictor for mortality in winter and thus for their residual reproductive value, this result may be a hint that even in a short-lived bird like a Great Tit life history may have some influence on reproductive effort.

ACKNOWLEDGEMENTS

I am extremely grateful to all who facilitated my work. There is not enough room here to name them all. Prof. Dr. E. Curio, the supervisor of this project, encouraged my work constantly by entering a lot of ideas and facilities. The Außenstelle für Populationsökologie, especially Dr. R. Berndt and Dr. W. Winkel helped me for my 2nd brood experiments. The work was funded by the DFG, grants Cu 4/28-2 and Cu 4/28-3.

REFERENCES

Carlisle TR (1985) Parental response to brood size in a cichlid fish. Animal behaviour 33: 234-238

Curio E (1983) Begehen Kohlmeisen den "Concorde-Fehler"? Journal für Ornithologie 124: 196-198

Curio E (1987) Brood defence in the Great Tit: The influence of age, number and quality of young. Ardea 75: 35-42

Curio E, Regelmann K (1982) Fortpflanzungswert und "Brutwert" der Kohlmeise (*Parus major*). Journal für Ornithologie 123: 237-257

Dawkins R, Carlisle TR (1976) Parental investment, mate desertion and a fallacy. Nature 262: 131-132

Lifjeld JT, Slagsvold T (1988) Effects of flight costs on the optimal diet: an experiment with pied flycatchers *Ficedula hypoleuca* feeding nestlings. Ornis Scandinavica 19: 111-118

Lindén M (1987) Clutch size and fitness in Great tits. Abstract ESEB Foundation Congress Basel

Murphy MT (1983) Clutch size in the eastern king bird: factors affecting nestling survival. The Auk 100: 326-334

Nur N (1984) The consequences of brood size for breeding Blue Tits II: nestling weight, offspring survival and optimal brood size. Journal of Animal Ecology 53: 497-517

Perrins CM (1979) British Tits. Collins, London, 213-214

Pettifor RA, Perrins CM, McCleery RH (1988) Individual optimization of clutch size in great tits. Nature 336: 160

Regelmann K, Curio E (1983) Determinants of brood defence in the Great Tit *Parus major* L. Behavioral Ecology and Sociobiology 13: 131-145
Robertson RJ, Biermann GC (1979) Parental investment strategies determined by expected benefits. Zeitschrift für Tierpsychologie 50: 124-128
Smith HG, Källander H, Nilsson JÅ (1988) The trade-off between offspring number and quality in great tits *Parus major*. In: Smith HG Reproductive costs and offspring quality: the evolution of clutch size in tits (*Parus*). Dissertation Lund University, Lund, Sweden
Weatherhead PJ (1982) Risk taking by red-winged blackbirds and the concorde-fallacy. Zeitschrift für Tierpsychologie, 60: 199-208
Westmoreland D, Best LB (1987) What limits mourning doves to a clutch of two eggs. Condor 89: 486-493
Windt W, Curio E (1986) Clutch defence in the Great Tit (*Parus major*) pairs and the Concorde Fallacy. Ethology 72: 236-242

RECRUITMENT, FLOATERS, IMMIGRANTS

SOCIAL ORGANIZATION OF WINTER FLOCKS OF WILLOW TITS *PARUS MONTANUS* IN A NORWEGIAN SUBALPINE BIRCH FOREST

Svein HAFTORN

University of Trondheim, The Museum, 7000 Trondheim, Norway

Much attention has recently been given to the study of titmouse winter flocks in different parts of the world and the literature on the topic is rapidly growing. In June 1987 I began studying a colour-marked population of the Willow Tit *Parus montanus* at Venabu, Ringebu in Central Norway. The aim was to describe the fate of each individual and its social relationship to other birds in the study area as thoroughly as possible, including flock formation and the vocal repertoire. One specific objective was trying to improve our understanding of the problem of flock size determination, a crucial issue discussed by several authors (e.g. Ekman 1979, 1989, Smith 1988, Nilsson 1988, Hogstad in print). To my knowledge this is the first study of titmouse winter flocks in the subalpine birch region.

STUDY AREA, MATERIALS AND METHODS

The habitat at Venabu (61°40'N) is subalpine birch *Betula pubescens* forest with some spruce *Picea abies* and pine *Pinus sylvestris* trees intermingled, lying about 900 m above sea level and close to the tree limit. The study area covers about 290 ha. The Willow Tit is the only tit species breeding in the area, except for a few Great Tits *P. major* which occasionally associate with the Willow Tits in winter.

Up to now I have restricted the study to pure field observations and recordings of bird vocalizations. Thus I have not disturbed the population by any kind of experimental intervention, except for catching the birds once to band, sex and age them. All birds were individually colour-marked. I caught the birds at 9 feeders spread out in the study area. Artificial feeding was not continuous, however, but usually confined to periods of short duration. Up to November 1989 the study is based upon 112 colour-marked Willow Tits. Recordings were made on a cassette recorder: Sony TC-D5 Pro, with an integrated parabolic reflector (Telinga Pro Type III). Analyses of vocalizations were made using Uniscan II spectral analyzer, with hardcopy digital spectrograms printed on Epson FX-85 printer.

NATO ASI Series, Vol. G 24
Population Biology of Passerine Birds
Edited by J. Blondel et al.

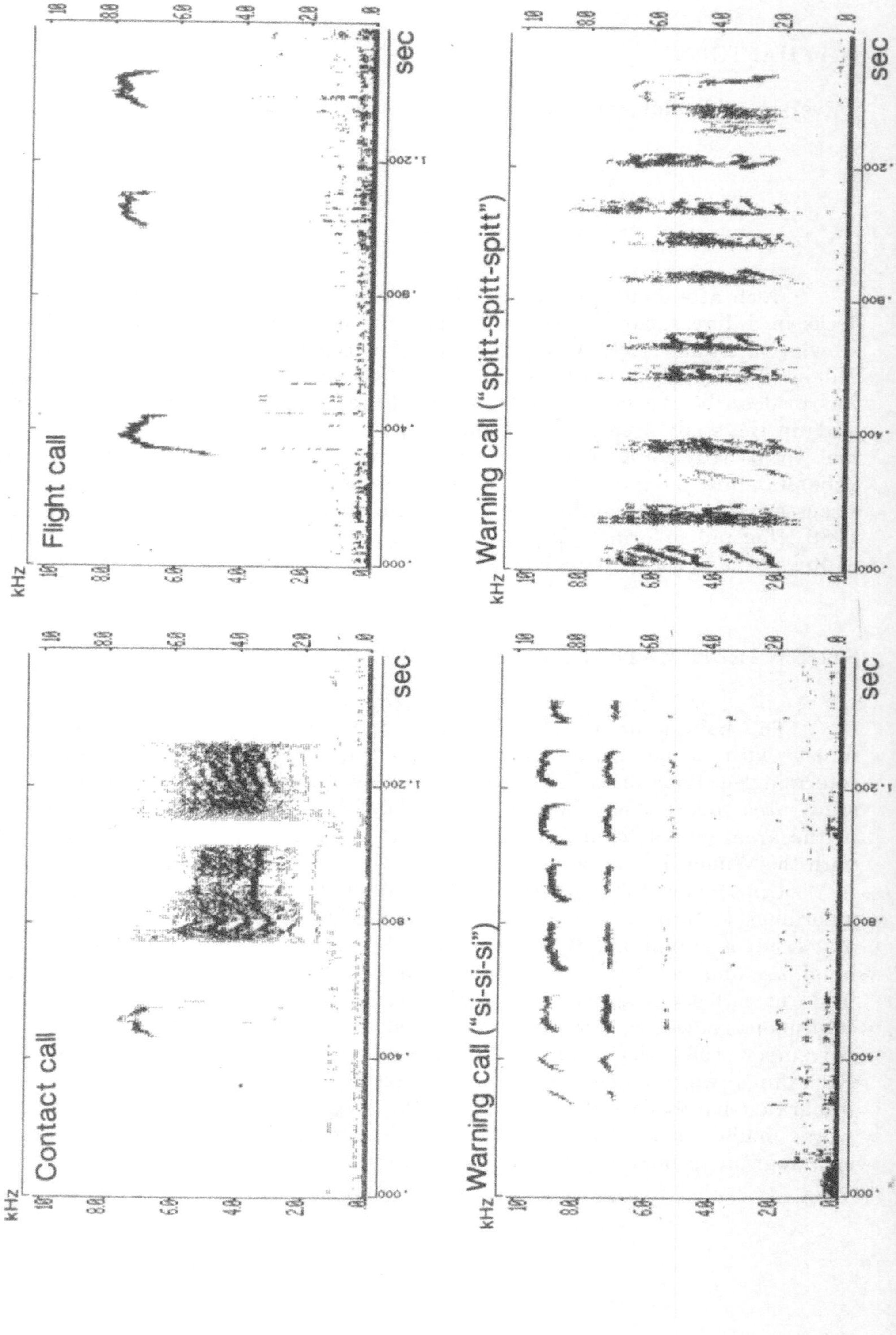
Contact call
Flight call
Warning call ("si-si-si")
Warning call ("spitt-spitt-spitt")
kHz
sec
.000
.400
.800
1.200

RESULTS AND DISCUSSION

After leaving the nest, the juveniles are dependent on their parents for a further period of about 2 weeks, which I define as the period of dependence. During this period they become independent gradually and start foraging by themselves.

In the latter part of the period the juveniles alternately beg intensively for food while tightly following the parents, or forage efficiently somewhat apart from the parents, often in the outskirts of the family flock which at this stage may be more or less spread out.

In contrast to general belief, the female parent is often alone when feeding the young in the latter part of the period of dependence. In spite of the fact that the male still is present in the territory and sometimes is seen together with his family he does not show the slightest tendency to feed the young. Interestingly, in such situations the young birds usually do not beg him for food.

In some broods all the siblings keep together until the parents stop feeding them entirely. However, in broods with marked asynchronous hatching the presumably oldest siblings may leave the family a few days ahead of the others. After the young have become independent of their parents they are not accompanied by them anymore, although the juveniles still may roam about in the territory for several days before they eventually disperse.

The first juvenile immigrants appear in the study area at a time when some of the local family flocks are still intact. The immigrants often temporarily intermingle with the intensely food-begging local juveniles, which may act aggressively toward the newcomers, without actually evicting them. The adults on the other hand do not seem to pay any attention to the strangers.

During the period of dependence the young birds develop their vocabulary, which is nearly complete before they leave the parents. The different calls seem to develop asynchronously, inasmuch as the non-aggressive calls develop before the aggressive ones. Thus the contact call *tsi tae tae*, flight call *sie sie* and warning calls *spitt spitt* and *sisisi* are fully developed relatively early in the period of dependence (Fig. 1), in contrast to the gargle, *pjae pjae* notes and song, which all contain elements of aggression (Fig. 2). It is noteworthy that the development of the latter utterings is postponed to the very end of the period of dependence, or immediately afterwards.

The vocabulary crystalizes out of the subsong, a long continuous series of notes which to begin with is completely unarticulated and babbling, apparently without any communicatory function. As time passes , however, the human listener starts recognizing species-specific calls integrated into the subsong (Fig. 3).

◄---

Fig. 1. Typical non-aggressive calls given by fledged Willow Tit juveniles still dependent on their mother. Recordings from the latter part of the period of dependence. Venabu 7-8 July 1989. Note incipient gargle immediately following the series of "spitt" warning calls (equivalent to the terminal elements in the complete gargle shown in Fig. 2).

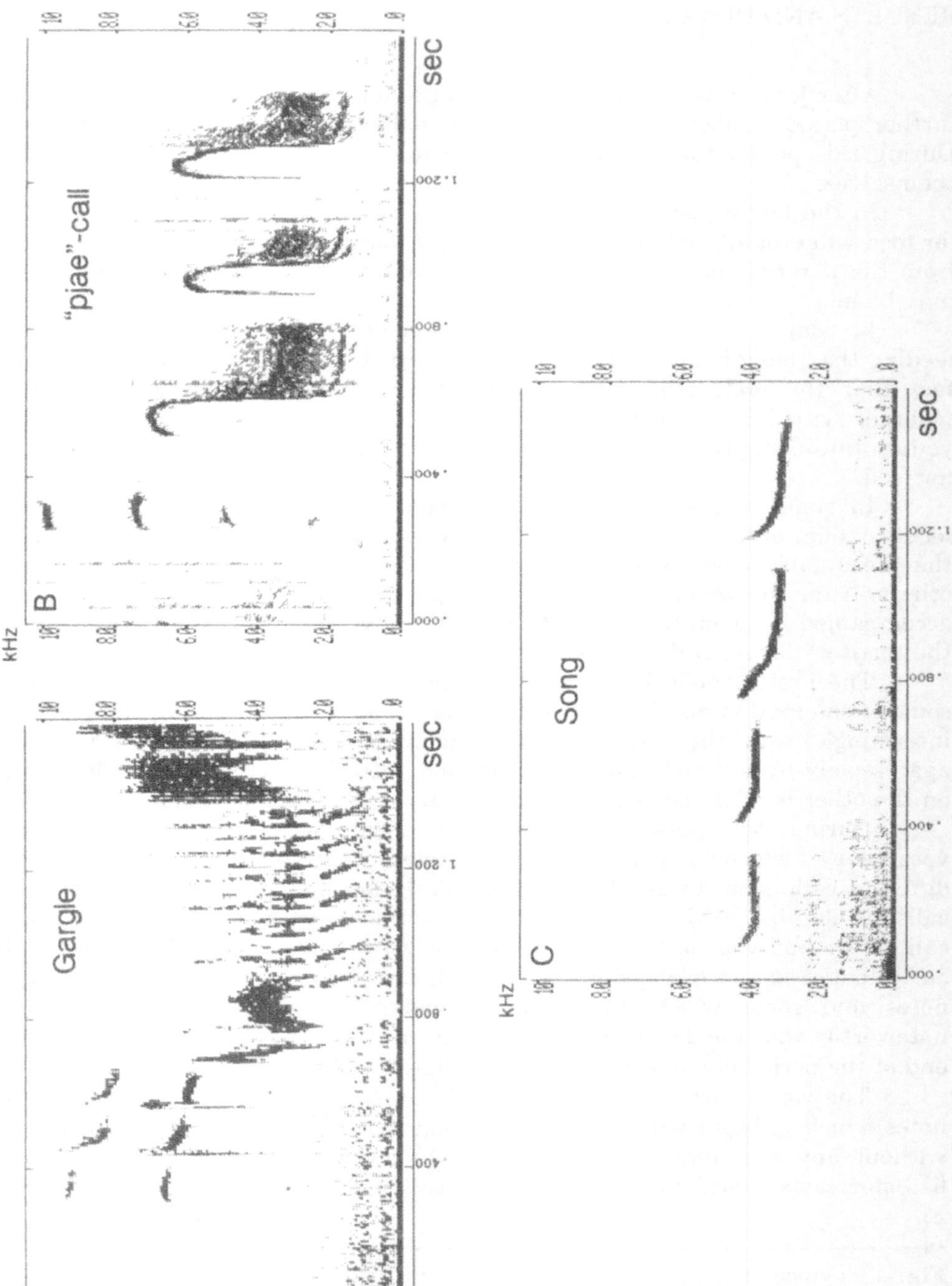
A
Gargle
B
"pjae"-call
C
Song
kHz
sec

That the non-aggressive calls develop ahead of the aggressive ones may be adaptive. Functional aggressive calls between siblings in a phase of life when they still are dependent on their parents would probably interfere with family coherence and possibly reduce their survival chances.

The social bond between the siblings is usually broken soon after the period of dependence, although some siblings may still keep together for a week or more. Finally they split up and associate with juveniles from other broods in small flocks which in my study area seldom exceed 10 individuals. During the phase of dispersal nearly all young hatched in the study area disappear, while the area in return is invaded by unbanded birds of unknown geographic origin.

During the next few weeks such flocks of young birds are frequently observed and are easily recognised by their restlessness and noisy performance with the typical food begging call *pipae* (Fig. 4) still being common in their well-developed vocabulary. Loud disputes with supplanting attacks and short-lasting grappling combats with gargling and other aggressive notes are commonly observed among the young. The function of such disputes, which often involve several individuals at the same time, probably serves to establish rank order, but may also be related to pairing. Some of the juveniles, presumably the males, spontaneously utter long series of soft gargling notes during foraging, even when they are completely alone (Fig. 5). Such intense gargling indicates a relatively high state of aggressiveness.

Roaming the wood, often in a more or less straight direction and at high speed, the juveniles pay no attention to existing territories. Adult territorial birds evidently recognize their status at once and do not prevent them from trespassing. I suspect that the *pipae*-call, primarily an ordinary food begging call, gets a secondary function after the young become independent. Obviously the call is not a demand for food anymore (cf. Smith 1980). Instead it probably serves as an appeasement call, a passport to territories. By giving this call the young birds announce their current status and probably preclude attacks from territorial adults. This call may help the juveniles in being accepted as flock members, as the following two examples suggest:

Two neighbouring established flocks were observed in an agonistic encounter at the border between the territories. I then noticed that a young bird at least twice passed from one side to the other without intervention from either part. The bird uttered *pipae*-calls.

In October 1988 a territorial old pair foraged alone. The mates were

Fig. 2. Typical vocalizations expressing state of aggressiveness, given by juvenile Willow Tits shortly after the period of dependence. A: gargle (combined with wing shivering) uttered during foraging by one juvenile toward another juvenile situated 20 cm apart. Venabu 3 August 1989. B: "Pjae"-calls are commonly heard among a variety of other calls during agonistic encounters between flocks. In this case the call came from an independent juvenile immigrant who attended a family flock of juveniles still dependent on their mother. Venabu 10 July 1989. C: Song notes uttered by a juvenile in a flock of other juveniles. Venabu 3 August 1989.

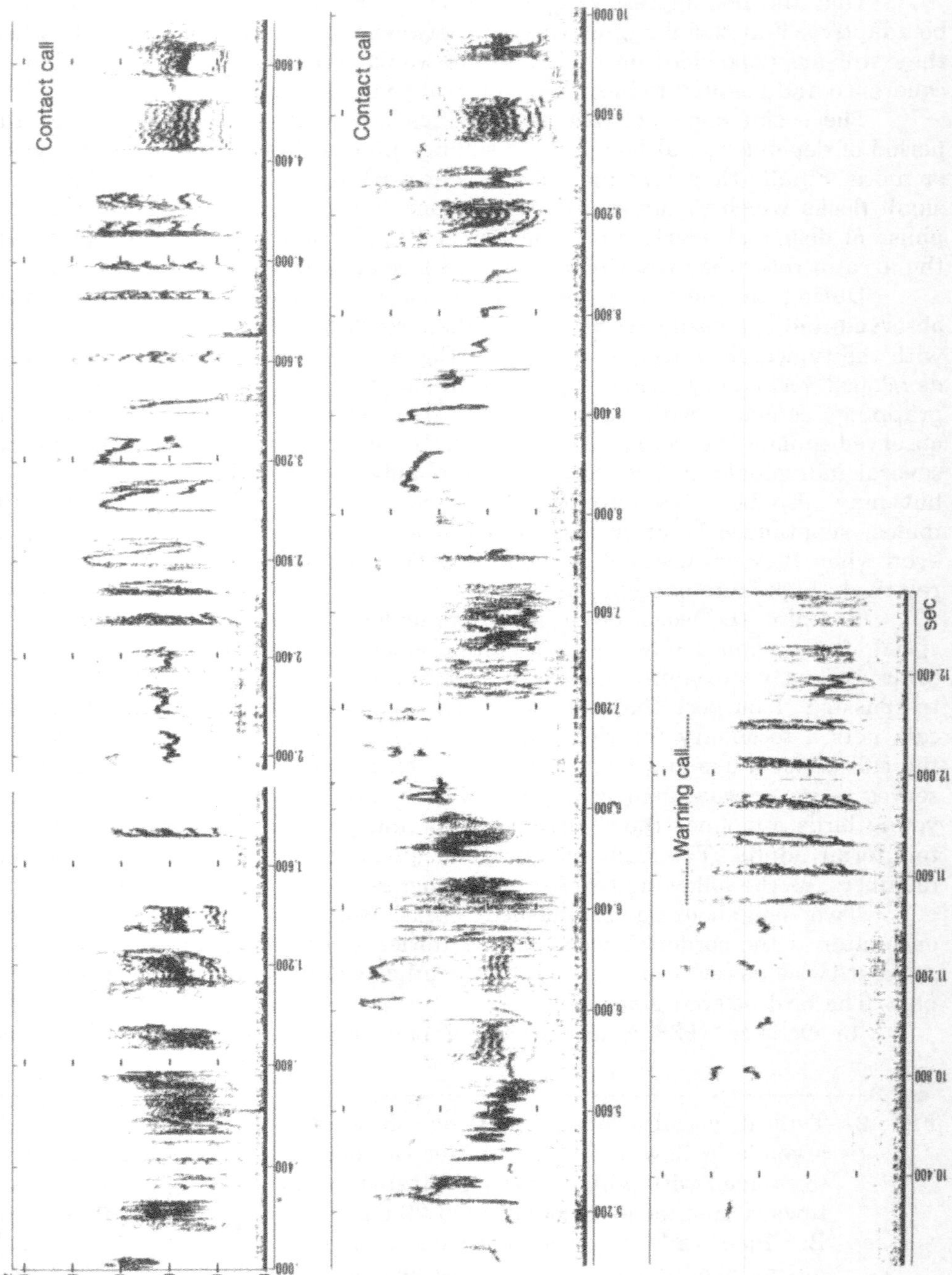

kHz
10
8.0
6.0
4.0
2.0
0
Contact call
Contact call
Warning call
sec
.000
.400
.800
1.200
1.600
2.000
2.400
2.800
3.200
3.600
4.000
4.400
4.800
5.200
5.600
6.000
6.400
6.800
7.200
7.600
8.000
8.400
8.800
9.200
9.600
10.000
10.400
10.800
11.200
11.600
12.000
12.400

intensively busy with food searching and partly food storing and they behaved very quietly, only uttering soft contact calls. Suddenly the female gave a typical *pipae* call immediately followed by the same sound from the male nearby. In the next moment two young birds, one male and one female, approached the pair, also giving *pipae* calls. The young birds then associated with the old pair without any signs of aggression. From then on all four birds foraged together in harmony and no more *pipae* calls were heard!

A few young have settled and found their final companionship for the winter already in late July, but for most individuals the dispersal period lasts to August-September.

The winter flocks of Willow Tits in my study area are relatively small and usually consist of 2-6 individuals only. In one case a "flock" consisted of a single bird only. An adult male lost his mate in early autumn and remained alone until the middle of January when a newly immigrated young female joined him and became his new mate.

As previously shown by other authors (e.g. Glase 1974, Ekman 1979, Smith 1984, 1988, Hogstad 1987) the adult territorial pair, having nested in the area at least once, usually forms the flock nucleus, around which the juvenile birds gather. The typical flock in my area consists of one adult pair and 1-3 unrelated juveniles, but deviations from this pattern are rather common. Thus, flocks with up to 4 adults occur, and some flocks contain either adults or juveniles only. This is similar to the Black-capped Chickadee *P. atricapillus* and Mountain Chickadee *P. gambeli* (D. A. McCallum, pers. comm.)

Several authors (Ekman 1979, Hogstad 1987, Smith 1984, 1988) emphasize that the flocks usually contain an even number of males and females. Also in my area the sexes are approximately equally represented in the flocks, but commonly the flock contains an extra individual of either sex. Thus flocks of 3 or 5 birds occur approximately at the same frequency as flocks of 4 birds (N=29).

Male-female bonds between juveniles may be established early in the summer, in fact before the pair attends a winter flock. I have seen courting displays by one juvenile already in July, shortly after it became independent.

In general, pairing may take place at any time of the year according to the circumstances. Thus, if one member of a pair dies during the winter, the surviving member may find a new mate either within the flock or outside. In one case an adult dominant male had 4 successive mates within one year. He lost his first mate in mid February 1988 and rapidly paired with a juvenile from his own winter flock. When she disappeared in December next winter the male mated with a young female, again from his own winter flock. This third mate was last seen on 13 January and already 3 days later the old male had got a new mate which this time was found in a neighbouring flock. This was in

Fig. 3. Segment (duration ca 13 sec) of a continuous series of babbling notes given by a foraging juvenile Willow Tit in the latter part of its period of dependence. Well-developed contact and warning calls crystallize out. Venabu 8 July 1989.

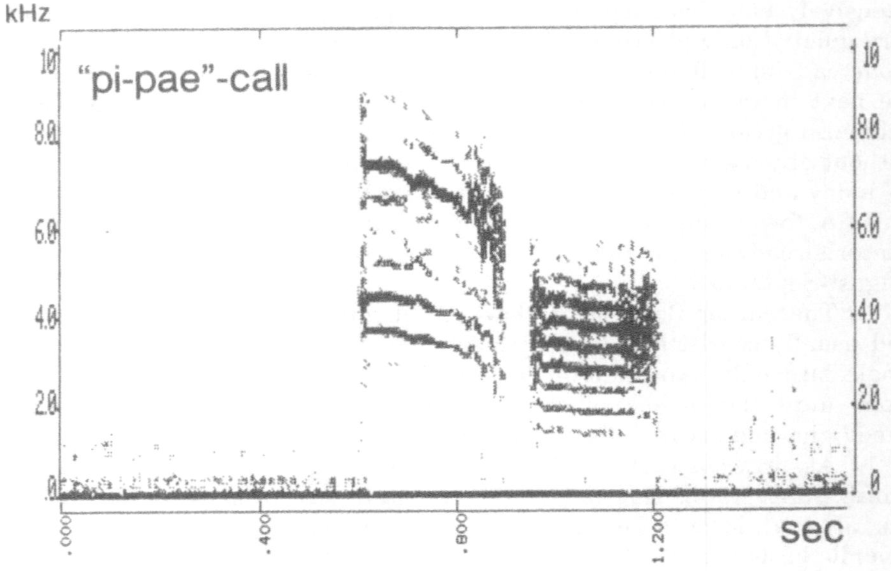

Fig. 4. Typical begging call uttered by a fledged juvenile Willow Tit still dependent on its mother. Venabu 7 July 1989.

fact his only chance to remate because his own flock at that time of year had been reduced to one member only, namely himself.

The bond between the different members of the flock varies in strength, as also has been found in other studies (e.g. Ekman 1979). Strongest is the bond between mates, but even an old territorial pair may intermittently separate in winter and forage independently of each other out of calling distance.

As commonly observed in other studies of the Willow Tit winter flocks often divide in subunits during foraging, and subsequently reunite later on (Haftorn 1958, Ulfstrand 1975, Hogstad 1988, Glase 1974, see also Odum 1942). Individuals which belong to this flock system are called regulars by Smith (1987). In addition to regulars there are also birds being more loosely attached to the system, termed floaters or switchers. I have in my study area several examples of this kind. Thus a young male suddenly appeared in the central study area in mid January 1989 and stayed within the territory of the local flock for the rest of the season. Though he attended the local flock from time to time, he more frequently accompanied a young female, which also periodically attended the local flock. This young female roamed about more in the study area than the young male, however, and was successively seen together with representatives from different flocks. I would say that they both belonged to the category of floaters, but the female was a typical switcher as well, implying that she associated with several flocks in succession. After all, according to my experience it is hard to make clear distinctions between regular

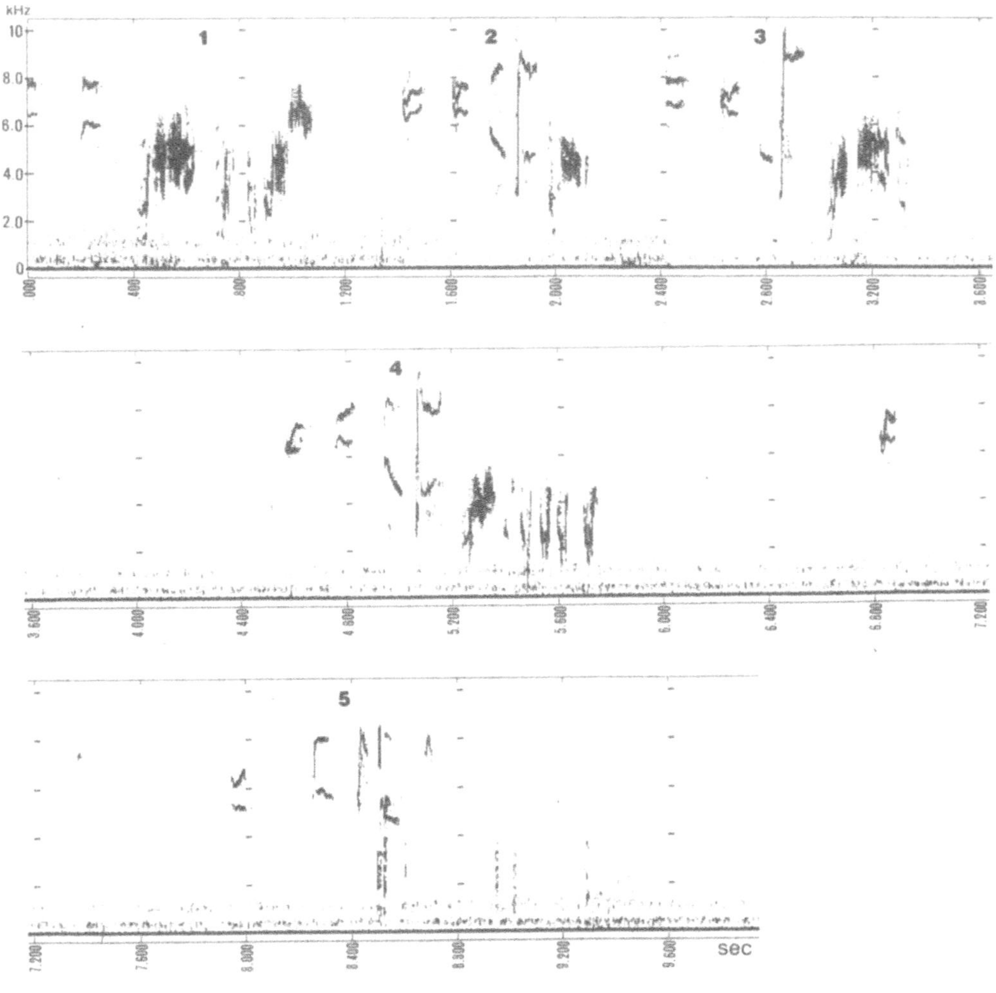

Fig. 5. Series of 5 gargles uttered spontaneously within 10 sec by an independent juvenile during foraging together with other juveniles. Venabu 3 August 1989.

flock members and winter floaters. Rather there is a gradation between two extremes.

Generally Willow Tits pair for life, and the mates keep together the year around. Nevertheless, divorces occur, even after successful breeding, as exemplified by the following two events. In the first case a one year old female left her mate after having raised 7 young. She joined a flock two territories apart from her earlier territory and was never seen together with her ex mate anymore.

The events of the second case are rather complicated. One flock consisted of 4 individuals, one old pair together with an adult male and a young female, which presumably also formed a pair. In spring the alpha male suddenly disappeared. The other male immediately took over the widow, whereupon they started breeding. The deserted number two female then paired with a young male in a neighbouring flock which besides this male only contained an old pair. The young pair raised 6 young successfully in an area where the winter ranges of the two flocks overlapped. Surprisingly the young pair divorced immediately after the breeding season, whereafter the birds associated with their former winter flocks, which in the meantime had changed composition due to mortality and adoption of newcomers. Similar sequences of re-pairing have been observed in the Mountain Chickadee (D. A. McCallum, pers. comm.).

Other studies of the Willow Tit in Scandinavia have shown that winter flocks stay in exclusive territories. Ekman (1979) even claimed that he could draw the boundaries within an error of only ± 20 m in his study area in a swedish coniferous lowland. In my subalpine study area it is impossible to define the territories with approximately the same accuracy. In fact, except for certain stretches, clear boundaries hardly exist and the territories overlap to some extent. The pattern seems intermediate between Ekman's findings and the situation in for example Desrochers et al.'s (1989) study area in Canada, where winter flocks of Black-capped Chickadees did not occupy exclusive or defended areas (see also Smith and Buskirk 1988).

Now I come to the crucial question on flock formation and the principle of flock size determination. Hogstad (1989) has recently provided evidence that flock size is limited by established juveniles in the flock. Based on experimental work at Budalen in Central Norway he concludes, first, that floaters were present in the area and, second, that resident juvenile males prevented them from attending intact flocks. The field observations on which these conclusions were based, were all done at feeding tables where the flock members aggregated (Hogstad, pers.comm.).

Also Ekman et al. (1981) and Nilsson (1988), working with the Willow Tit and the Marsh Tit *P. palustris*, respectively, seem to be of the opinion that newcomers are actively prevented from settling in winter flocks which are already "filled up".

Having observed flock formation during three successive seasons I have not been able to verify the hypothesis that the flock size is determined by the already established birds, whether young or adults. The overall behaviour of the birds during the critical period of flock formation has instead led me to the hypothesis that the decision whether or not an individual should be integrated in a flock is largely left to the bird itself. Experiences with the Mountain Chickadee point in the same direction (D. A. McCallum, pers. comm.)

My view is supported by field observations of which the following may serve as an example. Two young birds, one male and one female, were banded in the same territory in the middle of August 1988 and later became a pair. On 11 September they foraged together with an old territorial pair, with which they spent part of their time during the following month. Although a relationship evidently had been achieved between the two young birds at this time, they were also seen separately together with other birds. Thus, on 17 November I discovered the young female in a neighbouring territory together with the local flock, consisting of one old male and his young mate, and two young males. They all were busy foraging and the newcomer behaved as an ordinary member of the flock. Nevertheless, there was obviously a state of aggression directed toward the newcomer, especially from the established young female, which carried out supplanting attacks three times within one hour. Also the two young males and even the adult male supplanted the newcomer once each. However, the overall hostility exhibited was moderate and interrupted the newcomer's foraging time only for seconds. None of the regulars seriously tried to drive the newcomer away and she continued following the flock as long as I followed it. One month later, in the middle of December, the same young female was still seen periodically together with the same flock, but it was quite clear that the adherence was weak. Despite that, none of the regulars behaved aggressively toward her.

In the meantime the young female's future mate had frequently been visiting a feeder in a third territory, where he repeatedly had been attacked by members of the local flock. I do not know whether he actually tried to enter the flock or not. He was still on his own, roaming about. In December-January, however, he reallied with the young female and for the rest of the winter these two birds kept together, occupying parts of the former territory of the old pair mentioned above. This territory had just become vacant due to the death of the male.

The behaviour of this young couple and the interactions with conspecifics clearly show that hostility was not the reason why they did not associate permanently with the first old pair which they repeatedly had visited, and the events during the young female's short stays in the other flock indicate that she left this flock voluntarily.

In conclusion I would say that there is considerable evidence for the view that monitoring the quality of different territories and assessing the relationship to different home flocks are essential parts of dispersal. In search of flock settlement young birds by no means accept the first chance that comes along. The fact that they are frequently shifting companionship and are successively visiting different flocks strongly indicates that every single bird tries to maximize its fitness by testing and comparing available possibilities before making a decision on what to do, or by changing its decision when the social landscape changes. Interestingly, juveniles may repeatedly join and leave the same flock before they eventually settle.

Even after they have been accepted as regular flock members, the birds continue to be on the look-out for possibilities to increase their likelihood of successful reproduction. Thus, by interactions with neighbouring flocks and raids outside the territory they seem to be currently informed about the immediate status in the neighbourhood, especially with regard to bird density and food sources. Furthermore, they seem to be readily prepared for utilization of new chances which presumably increase their fitness. This is clearly indicated by the following example, which also illustrates another point of importance, namely that switching by an individual from one flock to another may dramatically alter the relationship with its former companions.

I refer to a series of interactions between two neighbouring flocks, which happened in the winter of 1988-89. One of the flocks contained 2 birds only, an old pair. The male in this flock suddenly disappeared in midwinter, and on 17

January (1989) the widowed female was seen together with the neighbouring flock which consisted of one old pair and a young male. The old female in this flock chased the widowed female several times, in contrast to the males which consistently remained neutral. Shortly thereafter the young male left his flock, paired with the widowed female and took over the vacant territory of her former mate. Interestingly, from now on the young male was not tolerated by his former companions anymore, and whenever they met, agonistic encounters were observed.

Removal experiments may also illustrate the birds' preparedness to react upon environmental changes. Thus in a Swedish study area Willow Tits entered vacated territories after removals (Ekman et al. 1981). At least some of the replacement tits were recruited from already established flocks in border territories, implying that regulars may leave their groups at any time during the autumn or winter if they presumably can improve their survival chances by moving to other places.

These examples, together with many other observations of local movements and interactions between birds during the non-breeding season, accentuate the fact that the population structure is characterized by an extensive flexibility, which *per se* is a clear warning against a rigid schematizing of the birds' social organization.

ACKNOWLEDGEMENTS

I want to thank D. A. McCallum and an anonymous referee for valuable comments on an earlier draft of the manuscript. The University of Trondheim has provided financial support.

REFERENCES

Desrochers A, Hannon SJ, Nordin KE (1988) Winter survival and territory acquisition in a northern population of Black-capped Chickadees. Auk 105: 727-736

Ekman J (1979) Coherence, composition and territories of winter-social groups of the Willow Tit *Parus montanus* and the Crested Tit *P. cristatus*. Ornis Scand 10: 56-68.

Ekman J (1989) Group size in dominance-structured populations. Ornis Scand 20: 86-88

Ekman J, Cederholm G, Askenmo C (1981) Spacing and survival in winter groups of Willow Tit *Parus montanus* and Crested Tit *P. cristatus* - a removal study. J Anim Ecol 50: 1-9

Glase JC (1974) Ecology of social organization in the Black-capped Chickadee. Living Bird 12 (1973): 235-267

Haftorn S (1958) Local movements and "total range" in *Parus cristatus* L. and *Parus montanus* Bald. (In Norwegian, English summary). Kgl Norske Vidensk Selsk Forh 30: 144-149

Hogstad O (1987) Social rank in winter flocks of Willow Tits *Parus montanus*. Ibis 129: 1-9

Hogstad O (1988) Advantages of social foraging of Willow Tits *Parus montanus*. Ibis 30: 275-283

Hogstad O (1989) The role of juvenile willow tits, *Parus montanus*, in the regulation of winter flock size: an experimental study. Anim Behav 38: 920-925

Nilsson JA (1988) Causes and consequences of dispersal in marsh tits: time as a fitness factor in establishment. Dissertation, Lund

Odum EP (1942) Annual cycle of the Black-capped Chickadee. Auk 59: 499-531

Smith CS, Buskirk J (1988) Winter territoriality and flock cohesion in the Black-capped Chickadee *Parus atricapillus*. Anim Behav 36: 466-476

Smith SM (1980) Demand behavior: A new interpretation of courtship feeding. Condor 82: 291-205

Smith SM (1984) Flock switching in chickadees: Why be a winter floater ? Am Nat 123: 81-98

Smith SM (1987) Responses of floaters to removal experiments of wintering chickadees. Behav Ecol Sociobiol 20: 363-367

Smith SM (1988) Social dynamics in wintering Black-capped Chickadees. Acta XIX Congressus Internationalis Ornithologici p. 2382-2390

Ulfstrand S (1975) Bird flocks in relation to vegetation diversification in a South Swedish coniferous plantation during winter. Oikos 26: 65-73

WINTER FLOATERS IN WILLOW TITS *PARUS MONTANUS* A MATTER OF CHOICE OR MAKING THE BEST OF A BAD SITUATION?

Olav HOGSTAD

Department of Zoology, University of Trondheim, N-7055 Dragvoll, Norway

Winter flocks of *Parus* species are formed during summer and early autumn as dispersing juvenile tits often settle with an adult mated pair that already occupy an exclusive year-round territory (see Hogstad 1989a). Such flocks often have a constant membership with individuals organized in a linear dominance hierarchy. As a consequence of adult occupancy of territories and social dominance, juveniles that become established as flock members generally become subordinates within each sex. Although flock formation usually takes place during July and August, movement of unsettled juveniles may continue into autumn (Smith 1976, 1984, Ekman 1979a,b, Weise and Meyer 1979, Ekman et al. 1981, Hogstad 1989b). These non-territorial birds, or winter floaters, can be characterized as individuals that wander among flocks or territories without settling. Since only flock members seem to achieve reproductive status the following spring (Ekman 1988, Hogstad 1989b), a crucial question arises: why be a winter floater? Is the wandering behaviour a strategy preferred by some birds, or are they forced to be floaters simply because the area is saturated by dominant conspecifics?

In the Black-capped Chickadee *Parus atricapillus*, Smith (1984, 1987) has suggested that winter floaters are unpaired juveniles that have chosen to become a floater visiting several flocks throughout the whole winter waiting for a vacant position among high-ranked, but not among low-ranked, flock members. When replacing a high-ranked resident, Black-capped Chickadee assume the rank position of the original vanished bird. In contrast, the Willow Tit *P. montanus* floaters apparently disappear in early winter (Ekman 1979a,b, Ekman et al. 1981), and during autumn dominant individuals are rapidly replaced by juvenile floaters that accept the lowest rank position of their sexes (Hogstad 1989b).

This paper reports the behaviour of floaters after experimental removal of flock members of different rank at different times during the autumn and winter to examine whether the floating system of the species is a strategy chosen by the birds.

METHODS AND MATERIAL

The study was performed from August 1988 to April 1989 in a 450 ha study area of subalpine mixed forest of pine and birch in Budal, 90 km south of Trondheim, Central Norway. Flock composition and dominance hierarchies of

NATO ASI Series, Vol. G 24
Population Biology of Passerine Birds
Edited by J. Blondel et al.

the Willow Tits in the area have been studied since 1979 (Hogstad 1987). Before the present study, all adults were already known and their territorial borders had been roughly determined. Feedertraps filled with sunflower seeds were placed near the centre of each of the 17 territories in the study area, and juvenile tits were caught during August and September, banded with unique colour combinations and sexed. The size and composition of the flocks within these territories were determined on the basis of coherence among individuals during the period. The hierarchical order of the individuals within each flock was determined by observing the outcomes (winners and losers) of agonistic encounters at feeders (see Hogstad 1987). About half (n=9) of the 17 flocks consisted of an adult mated pair and two juvenile pairs. The remaining flocks consisted mostly of five birds: the adult pair and three juveniles (n=3), or only one adult that was paired with a juvenile together with three (n=4) or four other (n=1) juveniles.

Twice a month between October and December and once in each of January and February, one bird from each of five to seven flocks was removed temporarily for 3-4 days (see Table I). The flocks from which birds were removed, were randomly determined among the 17 flocks at each removal date. The birds removed were selected on the basis of their dominance rank so that both high, middle and low-ranked birds were removed each time. However, because some individuals were trap-shy, this balanced design could not always be strictly followed. The removed birds were kept isolated from each other in small cages placed in an outdoor aviary near the study area. All these birds maintained or increased their body weight until they were released in their respective territories. Unfortunately five of the seven birds removed on 17 October and kept at the University of Trondheim died in captivity.

While the removed birds were in captivity, their territories were searched carefully for new birds not originally belonging to the flock. The flocks were visited two to six times a day, mostly at the feeders, but also when the flock members were gathered after I had used playback of Willow Tit song and call notes. All newcomers caught were individually colour-banded and sexed. If a newcomer was observed together with one or more resident flock members for more than one day, it was considered to be a floater that had "replaced" the removed bird. To determine the dominance rank and behaviour of the remaining original flock members and the floater, the social rank order within the flocks were redetermined. Behaviour and activity patterns of the seven floaters (five males and two females) that replaced the birds removed on 6 October were recorded during the first two days after they entered their respective flocks, and were compared with the activity patterns of the original juvenile flock residents during the same days. Activities recorded were foraging, vigilance, aggression (chasing, supplanting attack), preening and resting (see Hogstad 1988a) when the bird was in tree, in a bush <2m tall or on the ground. Behavioural data were collected on ten occasions for each of the seven floaters, and 44 and 34 occasions for male and female juvenile resident flock members, respectively. Chi square tests were used when examining statistical differences.

RESULTS

Replacements of removed birds

One to three days after removal of the birds, juvenile Willow Tits of the same sex as those removed were observed and caught together with the remaining flock members. All birds removed on 6 and 17 October and on 9 November were replaced, while four of six were replaced on 25 November and one of six was replaced on 10 December (Table I). None of the birds removed on 29 December or in January or February was replaced.

Table I. Date and number of days resident flock members were removed from their flocks during October 1988 - February 1989, their dominance positions in the flock (1=top, 5 or 6=lowest) and the number of earlier-ringed (old) or unringed (new) floaters that replaced the removed ones. Only one bird at a time was removed per flock.

Date removed	No. of days removed	No. of birds removed	Dominance rank of birds removed	Replacements Total	Old	New
6 Oct.	4	7	1, 2, 2, 4, 4, 5, 6	7	3	4
17 Oct.	4	7	2, 2, 3, 3, 4, 4, 5	7	2	5
9 Nov.	4	6	1, 3, 3, 4 ,5, 6	6	3	3
25 Nov.	4	6	1, 2, 2, 3, 3, 6	4	2	2
10 Dec.	3	6	1, 2, 3, 4, 4, 5	1		1
29 Dec.	3	6	1, 1, 3, 4, 4, 5	0		
16 Jan.	4	5	1, 1, 3, 4, 5	0		
17 Feb.	3	5	1, 1, 4, 5, 6	0		

None of the newcomers was a territorial member of another flock in the study area. The total of 25 replacements were made by 19 individuals. They were observed together with one or more resident flock members, both when attending a feeder and when foraging naturally. Seven of these birds were observed only during the two or three days they were in the territories where they had replaced a removed bird. Of the remaining 12 birds, five males became resident members of flocks joined on 18-20 October because the removed birds died in captivity. The other seven birds were observed two or three times (three birds) or four to seven times (four birds) after they had replaced removed birds or on other occasions. One of these birds later became a resident when one of the resident males disappeared on 27 November. The remaining six birds were observed for the last time in October or November, after they had stayed in the study area for 10-68 days (mean=35 days).

When the removed birds were released where captured three or four days earlier, they immediately regained their original rank; the inserted floaters left their respective flocks one or two days after the return of the residents. The withdrawal of the floaters was apparently a result of a marked increase in supplanting attacks and chasings of the floaters done by the resident juvenile males (not quantified in this study, but see Hogstad 1989b).

Of the 19 floaters only four had been caught and colour-ringed prior to the removal experiment. These four visited feeders close to the border in the southern part of the study area, two on 20 and two on 29 September. The floaters were generally silent and extremely difficult to observe. Nearly all were observed only when they attended a feeder together with residents. In contrast to residents, however, the floaters did not respond to playback of songs or calls.

The mean wing length of the floaters (males: 65.5mm; females: 62.4mm) did not differ from that of the juvenile flock members (males: 65.8mm; females: 62.4mm; Mann-Whitney U test, n.s.), indicating that the floaters were not smaller birds than were the residents.

Do the floaters maintain a home range?

Seven floaters were observed in the study area on at least two occasions over 10-68 days without ever becoming a resident. Three of these birds were observed at successively more northly locations and seemed to be moving slowly through the study area. The remaining four birds, two females and two males, were observed within an area of 100-150 ha that covered four to six territories. One of these birds, a juvenile male, became resident on 27 November when the alpha male disappeared from its territory. Prior to this, the juvenile male had been observed as a floater within an area covering six territories. This male survived the winter and was observed last in April the following year. Although the material is scanty, it indicates that floaters may keep within a restricted area for a relatively long period when visiting different flocks.

Dominance rank and behaviour of the floaters

When among flock residents, the floaters behaved submissively and were dominated by all flock members. When attending a feeder, they waited for the other birds to leave first or sought seeds on the ground. The floaters foraged more in bushes (23%) and on the ground (19%) and less in trees (58%) than did resident juveniles (ground 5%, bushes 4%, trees 91%; $P<0.001$ for males, $P<0.01$ for females). The floaters were more vigilant than the residents ($P<0.05$) and showed no aggressive behaviour (Table II). They also were significantly more vigilant and foraged less when a resident was close by than when a resident was farther away or absent ($P<0.01$; Table III).

Table II. Different activities (%) of floaters and juvenile residents in the removal period 6-10 October 1988.

	N	Foraging	Vigilant	Aggression	Preening	Resting
Floater males	50	34	66	0	0	0
Floater females	20	25	75	0	0	0
Resident males	44	41	43	9	5	2
Resid. females	34	47	50	0	3	0

Table III. Percentage of foraging and vigilance observations of floating males ($N=5$) and females ($N=2$) when a resident juvenile flock member was close ($<$5m) or farther away ($>$5m) or absent from the floaters. N denotes the total number of foraging and vigilant records.

	Close			Absent		
	N	Foraging	Vigilant	N	Foraging	Vigilant
Males	23	17	83	27	48	52
Females	6	-	100	14	36	64

DISCUSSION

In Willow Tits, where the preferred habitat becomes saturated long before all individuals have acquired a winter territory (Ekman 1988), juveniles benefit from being a territorial flock member because such membership gives them priority over territories when dominant adults disappear (Ekman 1988). Settlement in a territory also increases the probability of survival through the first winter (Ekman et al. 1981). Winter survival of Willow Tits is probably related to dominance rank in the flock (Koivula and Orell 1988, Hogstad unpubl. data) and rank among juveniles seems related to time of settlement (Nilsson and Smith 1988, Nilsson 1989 for Marsh Tits *P. palustris*). The importance of an early settlement is also indicated for Willow Tits. A study of skull ossification of juvenile males from five flocks revealed that the dominant of the males was the oldest in each flock (Sign-test, $P=0.05$; Hogstad unpubl.). That the floaters were not smaller than the juvenile flock members, strongly suggests that floaters are not second-rate birds, but simply might be birds that have been born too late to find a vacant place. During the ten years I have studied tits in the Budal area, nearly all birds settled where they first were caught. Only a small fraction of the resident flock members were observed in July or the beginning of August in a different place from where they settled.

This postponed settlement of a few birds might have either been a result of preclusion from existing flocks or because the birds were seeking a better rank position or a better quality territory.

The floater density of about one per territory, as observed in the present study, was probably higher than normal in the study area because of a warm summer and autumn and a presumed low juvenile mortality. In 1988 about 25% more juveniles were caught compared to previous years. However, irrespective of density, the lack of replacements after mid-December revealed that there were no floaters in the area during winter, either because of mortality or emigration. The same has been found earlier, for Willow, Crested *Parus cristatus* and Marsh Tits (Ekman 1979a, Ekman et al. 1981, Nilsson and Smith 1988, Hogstad 1989c). Some floaters, however, find a vacant position and become a resident flock member. That happened as late as 27 November for one of the birds in this study. The repeated loud calling of the widowed bird searching for its mate, probably attracts floaters and makes it easy for them to find out whether there is a vacancy or not, and to assess the mating status of the resident bird.

Floaters were observed attending flocks for brief periods, but like floater Black-capped Chickadees, they never settled permanently unless vacancies had appeared (see also Ekman et al. 1981, Hogstad 1989b). When associated with flocks, floaters were significantly more vigilant, lost foraging time, and obviously had lower intake rates of food than did the resident juvenile members. Floaters might, therefore, be forced to choose the lesser of two evils: to enlarge an existing flock and accept the lowest rank position, thereby reducing their probability of winter survival (Ekman 1987, Hogstad 1988b), or to remain floating while hoping to find a vacant position within a "filled" flock. However, since Willow Tit floaters apparently were prevented by resident juvenile flock members from joining "filled" flocks (Hogstad 1989b), and no floaters were observed to settle and enlarge such flocks, the most profitable option for the floaters is to seek a vacancy. Apparently, floaters either can stay within a restricted area and repeatedly check among a few known flocks or they penetrate larger areas while behaving more like a migrant. However, even if Willow Tit floaters choose to replace the most subordinate flock member, the chances of finding an opening in a flock before mid winter seem small because of a low winter mortality of residents before January (Hogstad 1989c). The floater behaviour is therefore not a strategy chosen by the birds. It seems more likely that the birds are making the best of a bad situation.

REFERENCES

Ekman J (1979a) Coherence, composition and territories of winter social groups of the willow tit *Parus montanus* and the crested tit *P. cristatus*. Ornis Scand 10: 56-68

Ekman J (1979b) Non-territorial willow tits *Parus montanus* de Selys-Longchamp in late summer and early autumn. Ornis Scand 10: 262-267

Ekman J (1987) Exposure and time use in willow tit flocks: the costs of subordination. Anim Behav 32: 508-514
Ekman J (1988) Subordination costs and group territoriality in wintering Willow Tits. Acta XIX Congr Internat Ornithol 1986: 2372-2381
Ekman J, Cederholm G, Askenmo C (1981) Spacing and survival in winter groups of willow tit *Parus montanus* and crested tit *P. cristatus* - a removal study. J Anim Ecol 50: 1-9
Hogstad O (1987) Social rank in winter flocks of willow tit *Parus montanus*. Ibis 129: 1-9
Hogstad O (1988a) The influence of energy stress on social organization and behaviour of Willow Tits *Parus montanus*. Fauna norv Ser C, Cinclus 11: 89-94
Hogstad O (1988b) Rank-related resource access in winter flocks of Willow Tit *Parus montanus*. Ornis Scand 19: 169-174
Hogstad O (1989a) Social organization and dominance behavior in some *Parus* species. Wilson Bull 101: 254-262
Hogstad O (1989b) The role of juvenile willow tits, *Parus montanus*, in the regulation of winter flock size: an experimental study. Anim Behav 38: 920-925
Hogstad O (1989c) Subordination in mixed-age bird flocks - a removal study. Ibis 131: 128-134
Koivula K, Orell M (1988) Social rank and winter survival in the willow tit *Parus montanus*. Ornis Fenn 65: 114-120
Nilsson JA (1989) Causes and consequences of natal dispersal in the Marsh Tit, *Parus palustris*. J Anim Ecol 58: 619-636
Nilsson JA, Smith H (1988) Effects of dispersal date on winter flock establishment and social dominance in marsh tits *Parus palustris*. J Anim Ecol 57: 917-928
Smith SM (1976) Ecological aspects of dominance hierarchies in Black-capped Chickadees. Auk 93: 95-107
Smith SM (1984) Flock switching in chickadees: why be a winter floater? Am Nat 123: 81-98
Smith SM (1987) Responses of floaters to removal experiments on wintering chickadees. Behav Ecol Sociobiol 20: 363-367
Weise CM, Meyer JR (1979) Juvenile dispersal and development of site-fidelity in the Black-capped Chickadee. Auk 96: 40-55

DIFFERENCES IN RECRUITMENT OF YOUNG BY IMMIGRANT AND RESIDENT GREAT TITS IN WYTHAM WOOD

Robin H. McCLEERY[1] and Jean CLOBERT[2]

[1] Department of Zoology, Edward Grey Institute of Field Ornithology, South Parks Road, OX1 3PS, Oxford, Great Britain
[2] Laboratoire d'Ecologie, URA 258, C.N.R.S., Ecole Normale Supérieure, 46 rue d'Ulm, 75230 Paris cédex, France

Throughout the long term study of Great and Blue Tit populations in Wytham wood, now in its 43rd year, it has been apparent that the Wytham population was not isolated from its surroundings. Great tits are known to breed in hedgerows and gardens near the wood, and there are other woodlands within 5-10 km though, being managed, they are thought to contain less nest sites than Wytham itself. If Wytham is optimum habitat for Great tits it might be thought that the wood would be entirely populated with birds originating there and that those breeding in the surroundings were the excess production from Wytham. However, it is clear that this is not the case, even allowing for the incomplete trapping of adults. In any given year about half the breeding adults are birds that are not known to have been raised in Wytham (Fig. 1). It will be noticed from this figure that the proportion of Wytham born first year breeders increased noticeably after 1976, when a program of replacing the original wooden boxes with a weasel-proof concrete pattern was completed. The suggestion is that losses to weasels were significantly affecting the structure of the population (Dunn 1976, McCleery and Perrins 1989). Analysis of the life tables shows that R_0, which should be close to unity for a population in equilibrium, is much less than 1, indicating that nests in Wytham are producing far too few recruits to account for the overall numbers in the wood. The most likely explanation seems to be that Wytham wood is a part of a larger population and is exchanging individuals with it. The only way to be certain of this is to study the surroundings for evidence of migration by Wytham birds, a project which is just beginning. However the purpose of this paper is not to study the dynamics of the whole population, but to investigate whether immigrant birds differ in certain respects from those that were raised in the wood.

In an earlier paper (Clobert et al. 1988) we noticed that the apparent survival rate was higher for immigrants, especially females, than for residents. Such a consistent difference between animals which differ only in whether they originate within or outside the study population is at first sight surprising. In this case the persistence of the difference throughout the study is of interest. If immigrants live longer but produce young at the same rate as residents then most Wytham birds will eventually be the descendants of an immigrant, so either immigrant characteristics are not strongly heritable or there is a

NATO ASI Series, Vol. G 24
Population Biology of Passerine Birds
Edited by J. Blondel et al.

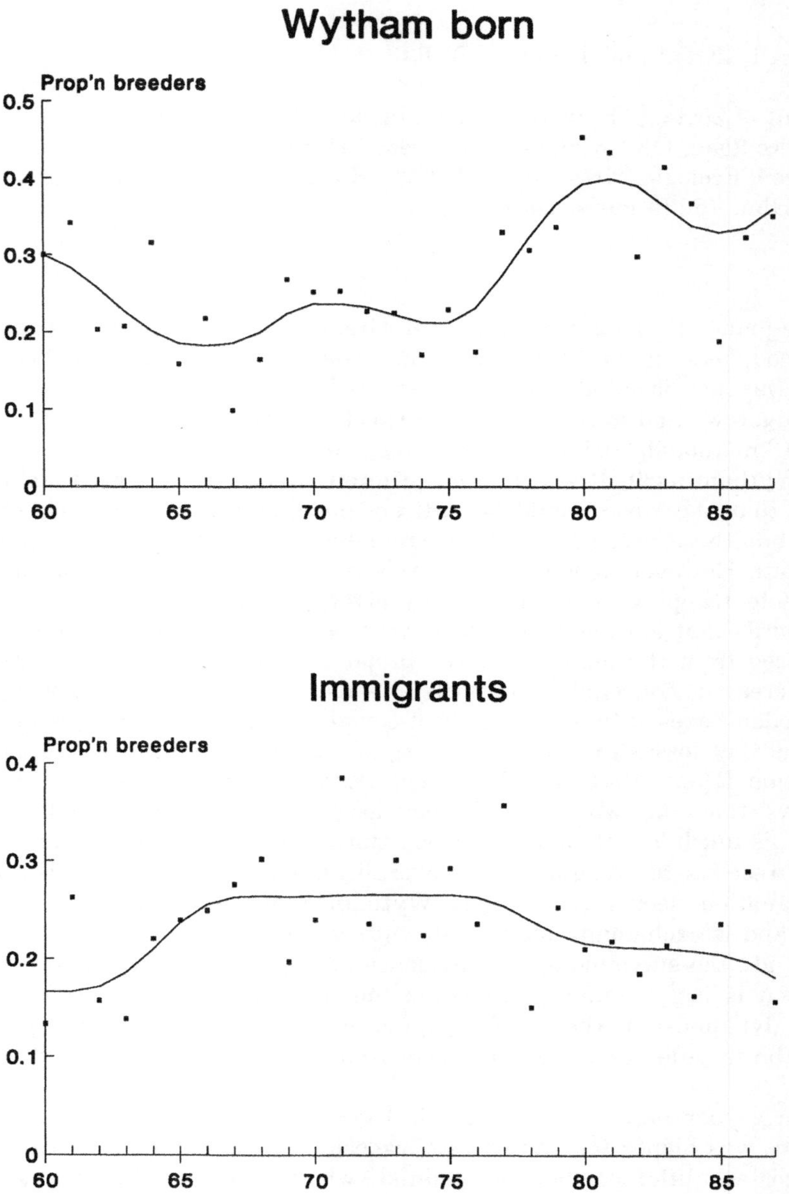

Fig. 1. Proportion of first breeders which are of (a) resident or (b) immigrant origin.

corresponding deficit in fecundity of immigrant birds. The purpose of this study is to examine the apparent fecundity of immigrants as opposed to residents to see if there is any substance in this suggestion.

METHODS: DATA COLLECTION

The data are the routinely collected information about Great Tit pairs breeding in nest boxes in Wytham wood, near Oxford. The nest round technique has been extensively reported elsewhere (Perrins 1965, 1979). It is known for each nest the date of the first egg, how many young were produced, and how many were subsequently recruited as breeders in the population. For the parents we can say how old they are, and whether they were themselves raised in Wytham.

Although the study started in 1947, the full extent of Wytham was not boxed until about 1958, and prior to 1964 few male breeders were caught. For this paper we have used breeding attempts in the seasons 1964-1987, with recruited young scored up to 1989. Nests which were subjected to manipulations, or where one or other parent was not trapped, have been dropped from the analysis. Overall this means that just over 50% of the recorded nests have been used in the analysis; in some years the figure is much lower due to extensive manipulations, and the success rate at trapping both parents has improved from about 40% (1964-74) to about 75% (1981-89).

METHODS: ANALYSIS

As a measure of breeding success we use the "local recruitment rate" of young because the contributions of one of us to survival and recruitment analysis (Clobert et al., this volume) were not practicable at the time we started the study. Local recruitment rate obviously underestimates the true survival rate as it ignores non-capture, emigration and non-breeding. In mitigation we would say that, where they have been estimated, differences between immigrants and residents in recapture rates are small (Clobert et al. 1988). Where possible we have checked biasses due to movement.

It has also been suggested that life time reproductive success or reproductive value in a season should be used as a measure of success. This may be the ideal measure, but in practise the reproductive output in a single season will be a good predictor of the "true" success provided there are no negative correlations between the fecundity and survival of individuals. As far as can be judged at present there is no convincing evidence for such trade-offs in the life histories of Wytham Great Tits (Pettifor et al., 1988), though the matter remains under review (see Nur 1988, and this volume).

To investigate the effects on the parental status (immigrant or resident) on the success of a nest we use a General Linear model with appropriate error distribution and link function as implemented in the program GLIM (Nag 1986, Aitken et al. 1989). This allows the addition and substraction of factors (in the analysis of variance sense) or variables (in a regression sense) to a linear model with constant coefficients. The change in deviance resulting from adding or subtracting components to the model gives a measure of the proportion of variation in the data explained by the variable of interest. As always in such exploratory analysis, especially when looking for a relatively weak signal against a good deal of statistical noise, the conclusions need to be interpreted with caution and, at best, provide clues which may guide further investigation.

RESULTS

Figure 2 shows the distribution of number of "surviving" young per nest. It is strongly skewed to the left, with most nests producing no survivors, some one, rather fewer 2 and so on. Analysis of factors governing the number of young produced per nest, when it has a distribution like this is best carried out by fitting a logarithmic model with Poisson error distribution. For such a model the change in deviance follows the Chi-Square distribution, so significance testing is straightforward.

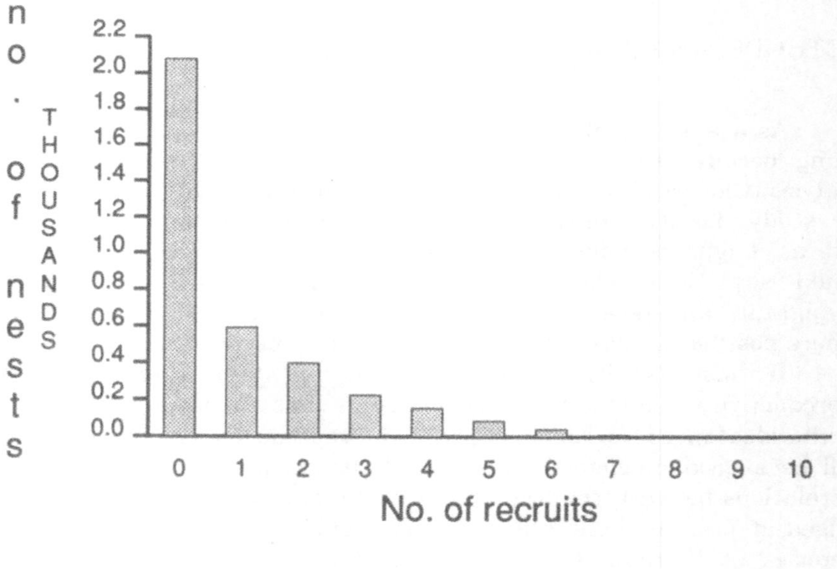

Fig. 2. Numbers of nests producing different numbers of recruited young. Recruits are defined as individuals which were recorded breeding in Wytham wood in any subsequent breeding season.

Preliminary analysis: age of the parents

Table Ia shows the results of fitting explanatory variables in the manner of step-wise multiple regression. Because the overall survival rates differ significantly between years it is necessary to fit a factor for the YEAR effect before looking at factors of interest and, as expected, it is highly significant. In view of previous work, we first examined the effects of the age of the parents. Number of survivors is not affected by female age but older males produce significantly more surviving young. There is no significant interaction term, so the male age effect is independent of the age of the female. This confirms earlier work on a smaller data set (Perrins and McCleery 1985).

Residents and immigrants

Fitting a factor for the resident/immigrant status of the male parent (MSTA) produced a significant change in deviance, and shows that male immigrants produce significantly fewer recruited young than residents (Table Ib). A factor for female parent's immigration status (FSTA) similarly produces a highly significant change in deviance and shows that nests where the mother is an immigrant produce significantly fewer recruited young. The MSTA.FSTA interaction term is not significant however; this shows that the male and female effects can be considered as additive on a logarithmic scale. We checked for interactions between immigration status and YEAR, which would indicate that the difference between immigrants and residents differed between years. The MSTA.YEAR interaction is not significant, indicating that the effect of male status is the same each year, but the FSTA.YEAR effect is significant, suggesting that for females the magnitude or direction of the difference between immigrant and resident female success differs from year to year. The coefficients for this factor are plotted against year in Figure 3.

Interactions between status and age

An obvious possibility would be that differences in status would be apparent in first-year birds but not in older ones. This is not straightforwardly the case (Table Ib): the interaction terms between male status and age (MSTA.MAGE) and female status and age (FSTA.FAGE) are not significant. However it turns out that the interaction between male status and female age (MSTA.FAGE) is significant ($p<0.05$); the coefficients of the model indicate that older females do slighly worse when mated to an immigrant male than when mated to a resident. The effect is weak, and could be due to a variety of causes which we have yet to investigate. It could be in particular an artefact resulting from the overdispersion in our data (var.> mean), which makes the Poisson distribution too stringent. There is no corresponding FSTA.MAGE interaction.

Table Ia. Effects of year, female age and male age on no. of recruits per nest. Linear model with Poisson errors and logarithmic link, using only pairs for which both males and females were known. The variable names are as follows: YEAR factor levels for different years. FAGE, MAGE factor levels for 1st year female and male parents (1) and older birds (2). FSTA, MSTA factor levels for parent immigration status, residents (1) or immigrants (2). Coefficients are given where significantly different from 0, and where they relate to a continuous variable, or one with only 2 levels. Thus in this table the coefficient for MAGE, a 2 level factor, is 0.16 for level 2 (older birds) meaning that older birds as a group produce more recruited young than 1st years.

Variable	change in deviance	d.f.	sig.	coeff.
+YEAR	545.4	24	**	
+FAGE	2.8	1	n.s.	
+MAGE	16.4	1	**	0.16
+MAGE.FAGE	3.3	1	n.s.	

Table Ib. Effects of male and female immigration status.
Poisson model as in Table Ia, adding status factors to the model YEAR+FAGE+MAGE+FAGE.MAGE. Variable labels as in Table Ia.

Variable	change in deviance	d.f.	sig.	coeff.
+MSTA	7.0	1	**	-0.10
+FSTA	11.9	1	**	-0.13
+FSTA.MSTA	0.8	1	ns	
+MSTA.MAGE	0.02	1	ns	
+FSTA.FAGE	0.8	1	ns	
+FSTA.MAGE	0.9	1	ns	
+MSTA.FAGE	4.9	1	*	-0.18
+FSTA.YEAR	65.5	23	**	
+MSTA.YEAR	24.8	23	ns	

Table Ic. Larger broods do better. Coefficients for no. ringed (RING) and its square (RIN2) added to the Poisson model:
YEAR + MSTA + FSTA + MAGE + FAGE + FSTA.YEAR

Variable	change in deviance	d.f.	sig.	coeff.
+RING	233.9	1	**	+0.56
+RIN2	41.3	1	**	-0.02

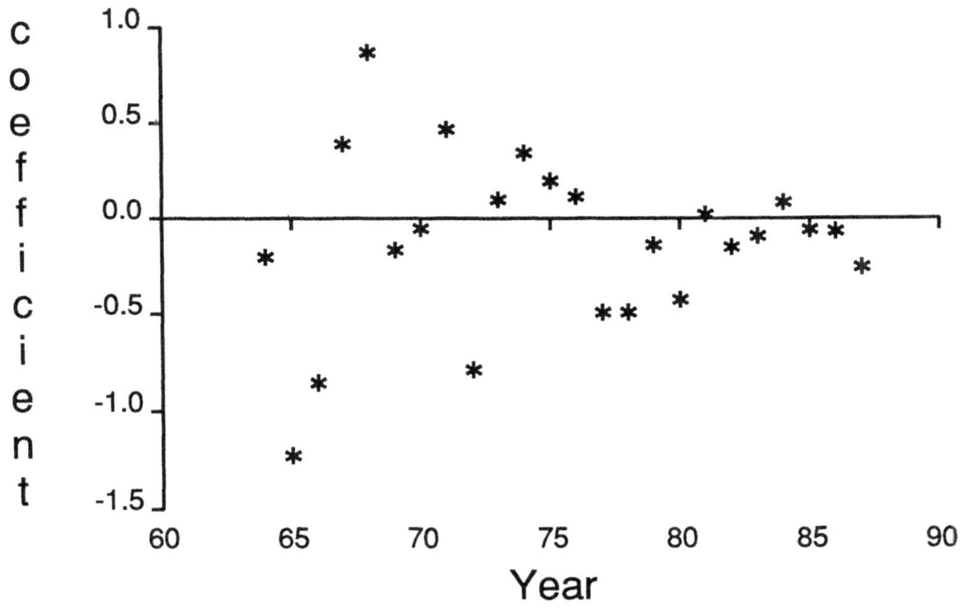

Fig. 3. Coefficients of the FSTA.YEAR interaction term in the model YEAR+MAGE+MSTA+FSTA.YEAR. In most years immigrant females do relatively worse, though in some they do better.

Initial conclusions: immigration status and recruitment

The observation of a difference in the number of recruited young produced by immigrants and residents is, in effect, the main result of the paper and the remaining analysis is a check against some spurious trivial causes and a search for clues as to how this effect might come about.The non-significance of the interaction terms between status and age imply that the relative fecundity disadvantage of immigrants is the same for birds of any age, and the non-interaction between the effects of status for males and females means that the effect applies to both parents independently of the other status. A complication is the significance of the female age/male status interaction, which must await further investigation.

Number of fledglings, or post-fledging survival?

It is clear from Table Ic that larger broods generally produce more survivors. Fitting variates for number ringed in the Poisson model gives a significant positive coefficient for the linear term in number ringed, though there is also a significant negative coefficient for the term in number ringed squared, indicating that the biggest broods do slightly less well. The obvious question is, do immigrant parents differ from residents in the brood size they produce?

Table II shows the results of fitting a linear model for broodsize (scored as number ringed). As before, between-year differences and age effects were examined first. There is a measurable effect of male age, and a slight effect for female age, in line with previous work (Perrins and McCleery 1985), but there is no sign that there is a difference between immigrants and residents in the number of young fledged.

Table IIa. Effects of female and male age on brood size. Linear model with normal errors.

Variable	F	$d.f._1$	$d.f._2$	sig.
+YEAR	17.9	24	2489	**
+FAGE	5.04	1	2488	*
+MAGE	9.75	1	2487	**
(+MAGE.FAGE	0.01	1	2486	ns)
-FAGE	1.90	1	2486	ns

Table IIb. Effects of immigration status on brood size after factoring out effects of year and age. Adding factors to the model: YEAR+FAGE+MAGE.

Variable	F	$d.f._1$	$d.f._2$	sig.
+FSTA	2.01	1	2485	ns
+MSTA	1.96	1	2484	ns
+FAGE.FSTA	2.23	1	2483	ns
+MAGE.MSTA	1.35	1	2482	ns

Another possibility is that the viability of young produced differs slightly because the young of immigrants are fledged slightly later than those of residents, since there is a general trend for offspring from earlier broods to have a higher survival rate (Perrins 1979). In Table III we examine the date of 1st egg using a linear model with normal errors. As expected there are strong between-year differences, and females lay significantly (about 2 days) earlier when they are more than one year old. In this analysis males also seem to have slightly earlier nests when older (about 1 day) and this effect is independent of the age of the female, since the interaction term between male and female age is not significant. There is no effect of male parents' immigration status on first egg date, but female immigrants do seem to lay slightly (about 1 day) later than female residents. This effect seems to be consistent across years, as the interaction term between female status and year is not significant. This result is thus consistent with the observation that female immigrants produce slightly fewer recruited young, but it is unclear whether it completely explains it.

Table III. Effects of male and female age and status on 1st egg date.

Variable	F	$d.f._1$	$d.f._2$	sig.	coeff.
+YEAR	39.45	24	2494	**	
+FAGE	60.69	1	2493	**	-2.02
+MAGE	10.55	1	2492	**	-0.96
+FAGE.MAGE	1.25	1	2491	ns	
+MSTA	0.30	1	2490	ns	
+FSTA	10.97	1	2490	**	0.95
+FSTA.YEAR	1.04	23	2467	ns	

Survival or dispersal?

The results obtained so far, combined with earlier work, seem to suggest that immigrant birds are slightly less successful at raising young which themselves recruit into the Wytham population. We have referred to this as "survival", but obviously it may be due to differential emigration. A simple explanation would be that immigrant parents are likely to have travelled further from their natal nest than residents and that they produce young who tend to move further away so that more of them move outside the study area. If the tendency to move is a gross difference, then nothing will be learned from comparing dispersal distances of individuals of different kinds which remain in the wood to breed, but it is at least worth checking the possibility that there are systematic differences between the offspring of immigrants and residents.

We examined this with a linear model for the logarithm of distance moved with normal errors. Females disperse slightly further than males, so this must be factored out at the outset, along with between-year effects. In view of the comments of van Noordwijk (1984) on measuring the heritability of movements we plotted the location of every nest together with an indication of the immigration status of the parents. Visual inspection did not reveal any pattern of settlement peculiar to immigrants so, although this has not been tested statistically, we believe any apparent differences between groups in movement would not be an artefact of the nest box positions.

The analysis of movement is presented in Table IV. As expected, the main effect for SEX is significant. The main effect for YEAR is not significant, but the YEAR.SEX interaction is, indicating variation over the years in the difference between the sexes. Neither female parents status, nor its interactions with year or sex of offspring are significant, so there seem no grounds for attributing the differences in recruitment rate due to female status to a gradation in dispersal leading to increased emigration from the study area. With male status the picture is more complicated. The MSTA.SEX interaction is significant in the absence of the YEAR.SEX interaction term, but not if the YEAR.SEX term is added. The three-way interaction SEX.MSTA.YEAR is significant, suggesting that in some years there are differences in the dispersal of male compared with female offspring of immigrant as opposed to resident male parents. The effects are rather slight, but we cannot be wholly confident that the dispersal of the offspring of immigrant versus resident males is always the same. On the other hand the effects do not seem sufficiently strong to wholly explain the differences in recruitment rate, and certainly do nothing to explain the differences in recruitment rate relative to female immigration status.

Table IVa. Movements of males and females: The SEX factor has levels male (1) and female (2), YEAR is a factor for year.

Variable	F	d.f.$_1$	d.f.$_2$	sig.	coeff.
+SEX	122.6	1	2031	**	0.39
+YEAR	1.42	25	1996	ns	
+SEX.YEAR	1.67	24	1972	*	

Table IVb. Effects of male and female immigration status on movement of offspring: adding factors to model: SEX + YEAR + SEX.YEAR

Variable	F	d.f.$_1$	d.f.$_2$	sig.
+MSTA	0.08	1	1971	ns
+FSTA	0.21	1	1970	ns
+SEX.MSTA	3.62	1	1969	ns
+SEX.FSTA	0.29	1	1968	ns
+SEX.MSTA.YEAR	1.31	48	1922	*

In 1989, for the first time, nest boxes were placed in gardens and hedges in the immediate surroundings of Wytham (Riddington and Gosler in prep.). Most of the adults caught at these boxes were unringed, indicating that they had not emigrated from Wytham. Among those that were of Wytham origin about half had come from nests where the male was an immigrant, and half from a nest where the female was an immigrant, but because the sample size is small it cannot presently be said whether this differs from the ratio of immigrants to residents breeding in Wytham this year (70% residents).

Effects of population density, winter food supply and temperature

We conclude from the analysis presented so far that there are differences in production of recruited young by immigrant males and females as opposed to resident males and females. The differences are not large (an immigrant will produce about 0.86 of a recruit for each one produced by a resident), but they seem to be robust and not due to any obvious artefact. In the remainder of the paper we look for clues as to how this process might come about, and compare the recruiting young with their parents.

In an earlier paper (Clobert et al. 1988) we found that a number of variables concerned with the overall population, the food supply and the weather had different influences on the survival rates of immigrants and residents (Table V). We examined the effects of the same variables on reproductive output of immigrants and residents, using a binomial model for survivorship per young, on the grounds that most variables might be expected to affect the chances of individual survival. Environmental variables were added in turn to the basic model including male and female immigrant/resident status and male age. For each variable we added the main effect to confirm that there was some influence on the population as a whole, and then the interaction terms respectively with male status and female status to see whether the variable had different effects on nests with immigrant or resident fathers or mothers. As we only had a complete set of environmental variables for 1964-1983, this analysis is on a slightly smaller data set than that reported in Table I.

In Table VIa we show the results of fitting variables related to population density. Negative coefficients show that in each case the success of nestlings was somewhat lower in years when the population density of Great Tits or Blue Tits was high. It does not seem to matter whether the Great Tit density is measured as population level, number of resident females or number of immigrant females.

In Table VIb we show the effects of fitting environmental variables. The food supply, scored as presence or absence of Beech mast in the winter after breeding, which would affect the survival rate of young (Perrins 1966), has the expected effects on the population as a whole but there is no evidence for a difference between residents and immigrants in their response.

Table V. Effects of external variables on the survival of adults of different immigration status as found by Clobert et al. (1988). + and - signs indicate significant positive and negative effects respectively, while the notation -/+ means that the effect was positive for 1st year birds and negative for older birds. Variable names are defined in Table VI.

	Male		Female	
	Res.	Imm.	Res.	Imm.
Density variables				
NRF	-	ns	-	-
NIF	-/+	-	-	-
GTD	-	-	-	-
BTD	-	-	ns	ns
Environmental variables				
Beech	ns	+/-	+	ns
Overall t	+	ns	ns	(+)
Summer t	ns	ns	ns	ns
Autumn t	ns	+/-	+/-	ns
Winter t	+	ns	-/+	ns
Spring t	ns	ns	ns	ns

The effect of overall winter temperature, i.e. mean daily temperature from October to March, is more interesting. In warmer years the population as a whole has more success at producing young, but the interaction terms suggest that immigrant parents produce relatively fewer young in warmer years, especially for female parents. In Table VIc this result is examined further by breaking down the overall temperature into periods of the year. The result for overall temperature seems to depend largely on the temperature in the winter period, but inspection of a model fitted with separate coefficients for each level of the female status factor suggests that following warm winters offspring of residents are relatively more likely to be recruited than after cold winters, while offspring of immigrants have about the same chance of recruitment independent of temperature; there is some evidence for a similar effect for males as well (Fig. 4 top). A similar, though barely significant, relationship is seen between female status and spring temperature, while after warm summers the offspring of immigrant females do relatively better (Fig. 4 bottom). The presence of these interaction effects may explain why the lower productivity of immigrant females varies between seasons, while that of males does not, though further analysis would be instructive.

Table VIa. Interactions between immigrant/resident status and density variables. Binomial model for offspring recruitment, adding variables to MSTA+FSTA+MAGE. External variables: NIF, NRF No. of Immigrant or Resident females respectively. BTD, GTD density of Blue or Great tits.

Variable Added	Main Effect	Interaction MSTA	Interaction FSTA
NRF	-0.002***	ns	ns
NIF	-0.013***	ns	ns
GTD	-0.003***	ns	ns
BTD	-0.002***	ns	ns

Table VIb. Interactions between immigrant/resident status and environment variables. Binomial model, adding variables to MSTA+FSTA+MAGE. External variables: BC Beech crop, BCC Beech crop in previous winter, OVERALLT mean temperature October to March.

Variable Added	Main Effect	Interaction MSTA	Interaction FSTA
BC	+0.192***	ns	ns
BCC	- 0.103***	ns	ns
OVERALLT	+0.05	-*	-***

Table VIc. Interactions between immigrant/resident status and temperature variables. Binomial model, adding variables to MSTA+FSTA+MAGE. Variable names: OVERALLT as Table VIb; other variables are mean daily temperatures in each three monthly period as follows: Summer June-August, Autumn September-November, Winter December-February, Spring March-May.

Variable Added	Main Effect	Interaction MSTA	Interaction FSTA
OVERALLT	+0.05	-*	-***
Summer t	ns	ns	+*
Autumn t	+0.09***	ns	ns
Winter t	ns	-*	-***
Spring t	ns	ns	-

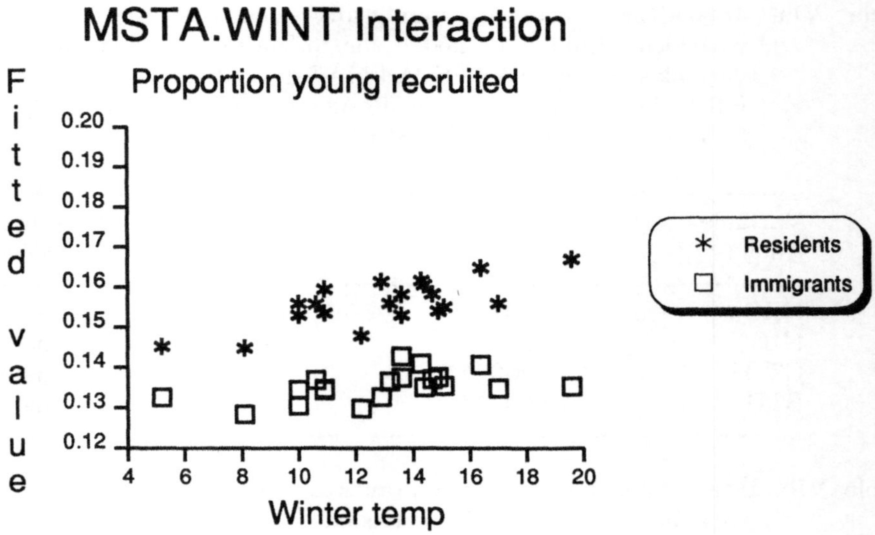

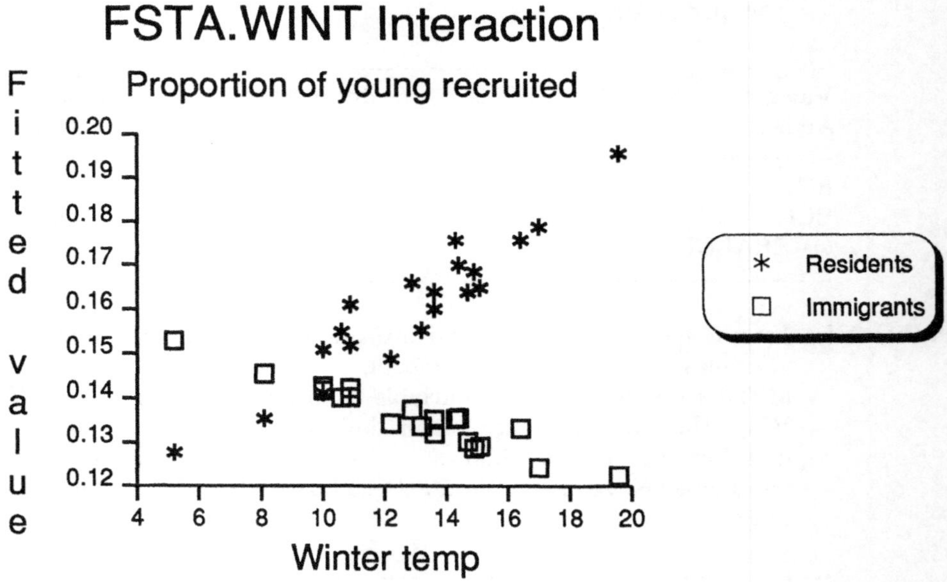

Fig. 4. Fitted values for the proportion of young which recruit from resident and immigrant males (a) and females(b) related to the mean temperature from December to February. The young of female residents survive better in warmer years than they do in colder ones, but the young of female immigrants do no better in warmer years.

DISCUSSION

Although we have identified certain respects in which immigrant birds differ from residents, we do not really understand the reasons for these differences. The results presented here need to be considered in relation to a similar analysis comparing the survival of immigrants and residents (Clobert et al. 1988).

The main conclusion is that birds that did not themselves originate in Wytham are marginally less successful at breeding than those that are Wytham born; the fact that their individual survival is no worse or slightly better suggests that this may be to do with the fact that they have moved to a different place to breed, rather than that they are in some way of lower quality. As far as we can see, the difference in recruitment of young arises from differences in the survival rate of the offspring after fledging, rather than from performance at the nest. Although we cannot completely exclude the possibility that the differences in recruitment are differences in emigration rather than survival, there is no definite evidence for this alternative view. Perhaps the most surprising result is the fact that there are no differences at all between immigrants and residents in the number of their offspring which recruit into the population, especially as most "residents" have "immigrants" among their immediate ancestors. It would be interesting to know whether immigrants differ in some genetic attributes from residents (cf. C.M. Perrins, concluding remarks), though to study this quantitatively would involve looking at grand-parental effects which would reduce even our data-set to a small number of cases. It is extremely interesting to note that the apparent differences in production of recruited young may balance the slight differential survival rate found previously by Clobert et al. (1988). This might be an example of ecological compensation of life history parameters as suggested by Sibly and Calow (1987), though we would be cautious in accepting this interpretation.

The difference in recruited young produced by immigrant and resident females differed strongly between years, though this effect was not apparent for males, leading to the obvious suggestion that population or environmental factors differing between years are responsible. Again this invites comparison with our previous analysis of the effects of environmental and population parameters on survival rates of immigrants and residents, summarised in Table V.

To compare the results of Table V and Table VI it needs to be understood that a significant interaction in Table VI represents a significant difference in the effects tabulated for male and female immigrants and residents in Table V. Thus the consistent negative signs for the effects of population variables in Table V agree fairly well with the significant negative signs for the main effects but lack of interaction with immigration status in Table VIa; the indication is that population density has a generally negative effect on the survival of adults and on the recruitment of young, but that these effects do not differ between immigrants and residents. Exceptions to this finding are the differences between the survival of immigrant and resident males relative to the numbers of immigrant or resident females in the previous season.

The effects of beech crop, probably a surrogate measure for late summer and winter food availablity, are clearly different for offspring compared with their parents. For adults we found that winter survival was strongly related to beech crop for first year male immigrants and for female residents, which are the least dominant in contests for food. For offspring, however, there was no difference in recruitment dependent on parental immigration status when beech crop and non beech crop years were compared.

Adult immigrants and residents have different survival rates relative to autumn temperatures but no such differences appeared for production of recruited offspring. On the other hand the positive effect of winter temperature on the survival of residents (except 1st year females) and neutral effect for immigrants seems to be similar to the effects on their offspring where the effect of winter temperature is, unexpectedly, that resident parents produce more recruits in warmer winters but that immigrants do not. This is a remarkable result, especially as we have been unable to find any relevant difference in the behaviour of resident parents which would account for the differences in the local survival rate of their young.

In summary, then, immigrant Great Tits produce slightly fewer recruits than Wytham born birds, but survive marginally better themselves. For females the differences in number of recruited young vary between years, apparently depending on the winter temperature following the breeding season, but do not seem to depend on the age of the parents. For males the effects are consistent between years, though apparently slightly dependent on the age of the mate.

The possibility that emigrants or immigrants might differ in certain respects from resident parts of the population is not a new idea, though relatively little detailed work has been done on it. Johnson and Gaines (1987) found that in prairie voles (*Microtus ochrogaster*) dispersing individuals (who were usually sub-adults) had a higher survival rate than adults remaining behind, and that both groups survived better than animals which attempted to disperse but were prevented from doing so. However the dispersers were moving into vacant habitat where higher survival may simply have been an effect of lower density. Our study may not be comparable in that we are comparing emigrants from an unknown population with individuals present at their destination rather than with residents staying behind in the source population.

Horn (1983), reviewing earlier work on dispersal suggested that differences might exist between different components of a population in the degree of emigration they enforced on their offspring in relation to the consequences of local recruitment for competition and inbreeding. Immigrants might, since their chance of survival to the next season is higher, benefit more from the dispersal of their young than residents. The applicability of this model here is uncertain, however, since we can not find any definite evidence that dispersal and local recruitment are related.

Although the precise meaning for the structure of the Wytham population is at present unclear, there do seem to be consistent differences in life-history parameters between birds which are raised in the wood and those which enter from the surroundings. We have partly characterised the way in which these differences correlate with population and environmental factors both internal and external to the population, but more information on the relationship between the Wytham population and that of the immediate surroundings might show us the significance of these differences for the processes governing the structure of both populations.

REFERENCES

Aitken M, Anderson D, Francis B, Hinde J (1989) Statistical modelling in GLIM. Clarendon Press, Oxford

Clobert J, Perrins CM, McCleery RH, Gosler AG (1988) Survival rate in the Great Tit *Parus major* in relation to sex, age and immigration status. Journal of Animal Ecology 57: 287-306

Dhondt AA, Olaerts G (1981) Variation in survival and dispersal with ringing date, as shown by recoveries of Belgian Great tits (*Parus major*). Ibis 23: 96-98

Dhondt AA (1979) Summer dispersal and survival of juvenile Great tits in southern Sweden. Oecologia 42: 139-157

Dunn EK (1976) Predation by weasels (*Mustela nivalis*) on breeding tits (*Parus* spp.) in relation to the density of tits and rodents. Journal of Animal Ecology 46: 633-652

Horn HS (1983) Some theories about dispersal. In: Swingland IR and Greenwood PJ (eds) The Ecology of Animal Movement. Clarendon Press, Oxford

Johnson M, Gaines M (1987) The selective basis for dispersal of prairie vole *Microtus ochrogaster*. Ecology 68: 684-694

McCleery RH, Perrins CM (1989) Evidence for the impact of predators on the population of great tits (*Parus major*) in Wytham wood. In: Perrins CM, Lebreton J-D, Hirons GJM (eds) Bird Population studies: relevance to Conservation and Management. Oxford University Press

NAG (1986) The Generalised Linear Interactive Modelling system. Release 3.77. The Royal Statistical Society, London

Nur N (1988) The cost of reproduction in birds: an examination of the evidence. Ardea 76: 155-168

Perrins CM (1966) The effects of beech crops on Great Tit populations and movements. British Birds 59: 419-32

Perrins CM (1965) Population fluctuations and clutch size in the Great Tit, *Parus major*. Journal of Animal Ecology 34: 601-47

Perrins CM (1979) British Tits. Collins, London

Perrins CM, McCleery RH (1985) The effect of age and pair bond on the breeding success of the Great Tit *Parus major*. Ibis 127: 306-315.

Pettifor RA, Perrins CM, McCleery RH (1988) Variation in clutch size in great tits: evidence for the individual optimisation hypothesis. Nature 336: 160-162

Sibly RM, Calow P (1987) Ecological compensation - a complication for testing life-history theory. Journal of Theoretical Biology 125: 177-186 van Noordwijk AJ (1984) Problems in the analysis of dispersal and a critique on its heritability in the Great Tit. Journal of Animal Ecology 53: 533-544.

RETURN RATES OF MALE PIED FLYCATCHERS : AN EXPERIMENTAL STUDY MANIPULATING BREEDING SUCCESS

Tore SLAGSVOLD and Jan T. LIFJELD

Zoological Museum, University of Oslo, Sarsgate 1, N-0562 Oslo 5, Norway

Fidelity to a former breeding area is strong in many bird species, and in particular in males, since previous experience may help in the competition for essential breeding resources, like a good territory and a good nest site. Because of such fidelity, return rates have repeatedly been used in analysing annual survival rates of adult male Pied Flycatchers *Ficedula hypoleuca*, e.g. when studying relationships with male plumage colour, and effects of brood size manipulation (Askenmo 1979, Järvi et al. 1987, Slagsvold and Lifjeld 1988, Alatalo and Lundberg 1989). A problem is that return rate is a mixture of survival and capture rate. Capture rate is not only dependent on the efficiency of observation within a study area, but on breeding dispersal. Thus, it has been questioned whether return rate can be used to estimate survival rate (Högstedt 1981; but see Curio and Regelmann 1982, p.242, for a reply). Here we present data for return rates of male Pied Flycatchers to study sites near Oslo, in southern Norway. Studies of other bird species have shown that an increase in dispersal rate may occur after a breeding failure (e.g. Harvey et al. 1979, Newton and Marquiss 1982, Pärt and Gustafsson 1989). We therefore asked if previous breeding success affects male return rate in our population. To answer this question, we compare male return rates for two study plots ; in the first plot we experimentally reduced breeding success by removing all females ; in the second plot we increased breeding success by releasing these additional females. Such experiments are necessary to ascertain whether an increased dispersal after a breeding failure is related to the failure itself, or to a qualitative difference between the birds involved and/or of their breeding territories. We therefore also analysed return rate in relation to various male characteristics such as age, body size, and plumage colour, to see if there was any relationship with male quality.

Male return rate may also be negatively influenced by reduced survival due to heavy parental investment made in previous season (e.g. Askenmo 1979). To answer this question, we compared return rates of males of different mating status (bachelors, monogamous, and polygamous males). We also compared male return rates in relation to number of young fledged, and hatching time of young.

NATO ASI Series, Vol. G 24
Population Biology of Passerine Birds
Edited by J. Blondel et al.

MATERIAL AND METHODS

A map of the study area (60° N, 11° E), with the plots Brenna, Tangen, and Sinober, is given by Slagsvold et al. (1988, Fig. 1). Brenna consists of deciduous forest, mainly birch *Betula verrucosa*, whereas Tangen is dominated by spruce *Picea abies* forest. Sinober consists of a mixture of coniferous (mainly spruce) and deciduous forest (mainly birch). The study plot from which females were removed (Brenna) is hereafter referred to as Oslo I; the data for the other two plots (Tangen and Sinober) on which females were released have been pooled and referred to as Oslo II. The distance between Oslo I and II is 3-4 km, and between Tangen and Sinober less than 1 km. Male return rate was 49% on both Tangen (N=135) and Sinober (N=158).

All males seen defending nestboxes were trapped and colour ringed each year (1985-1989) soon after arrival, and their body measurements recorded. The procedure was strictly similar on Oslo I and II. Plumage colour was scored according to Drost's (1936) scheme; males of score I-IV were regarded as blackish, those of score V-VII as brownish. Every female that settled on Oslo I was trapped soon after mating, and released on Oslo II (see Slagsvold et al. 1988, for description of female behaviour after release, and Stenmark et al. 1988, for female breeding success, during 1985 and 1986). However, late in the breeding season (4-17 June) some other females were trapped in a distant area and released onto Oslo I (see Slagsvold et al. 1988, Dale and Slagsvold 1990), where a few of the males there thus became late breeders.

In 1985 and 1986, empty nestboxes were available throughout the breeding season on Oslo I. After the main period of male arrival, all unoccupied nestboxes on this plot were closed on 25 May 1987, on 20 May 1988, and on 18 May 1989. No such nestbox manipulations were made on Oslo II.

Return rates were calculated on a year-to-year basis. Thus, males that were absent from a plot in one year but which re-appeared in a later year were not included in the returning birds group (these were few, however; see below). Data for males that disappeared immediately, or only a few days after trapping, have been excluded from the analyses, as have also those for males known to have died during the breeding season (although these males were classified as having returned if ringed in a previous year).

RESULTS

Differences between Oslo I and II

Rate of return was 31% (N=124) for Oslo I, and 49% (N=293) for Oslo II ($X^2=10.97$, $P<0.001$). Return rate was lower for Oslo I than for Oslo II in each study year, but was only significantly so in the first of the four seasons (1985/86). Overall, the values varied between 17 and 48% on Oslo I, and between 36 and 55% on Oslo II.

In 1987-1989 no nestboxes were available for males that returned after certain dates on Oslo I (see Material and Methods). This may have decreased the return rates for Oslo I in these three years. Altogether 11 returning males arrived after these critical dates on Oslo II. Even if these males are classified as non-returning, the return rate for Oslo II (45%) is still higher than that for Oslo I ($X^2=6.90$, $P<0.01$). Note also that the return rate for Oslo II (49%) is very high compared with values found for male Pied Flycatchers elsewhere in Europe (Table I; range 18-45%). Overall return rate for all these other localities is 31% (N=2568, Oslo I and II excluded), which is significantly different from the Oslo II value ($X^2=36.84$, $P<0.001$).

Table I. Return rates of male Pied Flycatchers in different parts of Europe

Country	Locality	%	N	Source
Norway	Oslo I	31	124	Present study
Norway	Oslo II	49	293	Present study
Norway	Trondheim	37	149	Slagsvold and Lifjeld 1988
Norway	Songli	31	107	Røskaft et al. 1986
Sweden	Ammarnäs	24	786	Røskaft et al. 1986
Sweden	Uppsala	18	179	Alatalo and Lundberg 1989
Sweden	Gothenburg	38	73	Askenmo 1979
Finland	Kristinestad	33	36	von Haartman 1960
Finland	Lemsjöholm	36	379	von Haartman 1960
Finland	Siuntio	29	34	Røskaft et al. 1986
Germany	Schlesien	20	51	Trettau and Merkel 1943
Germany	Hessen	35	37	Trettau 1952
Germany	Dresden	31	118	Creutz 1955
Germany	Harz	22	44	Røskaft et al. 1986
Germany	Lingen	40	359	Røskaft et al. 1986
Germany	Berlin	45	121	Curio 1959
England	Forest of Dean	39	95	Campbell 1955
Total		33	2985	

Similar proportions of first-year males were present on both study plots (41% and 39% on Oslo I and II, respectively). It was yearlings in particular that had a low return rate on Oslo I (Fig.1). Their return rate was only 16% (N=51), compared with 43% (N=115) on Oslo II ($X^2=10.81$, $P<0.002$). Return rate for older males was nonsignificantly lower on Oslo I than on Oslo II (41% versus 52%, $X^2=2.03$, NS). Hence, return rate was lower for yearlings than for older males on Oslo I ($X^2=7.96$, $P<0.01$), but not on Oslo II ($X^2=1.69$, NS). For older males, return rate was independent of age (Fig. 2). There was no significant difference in return rates for males classified as being at least two

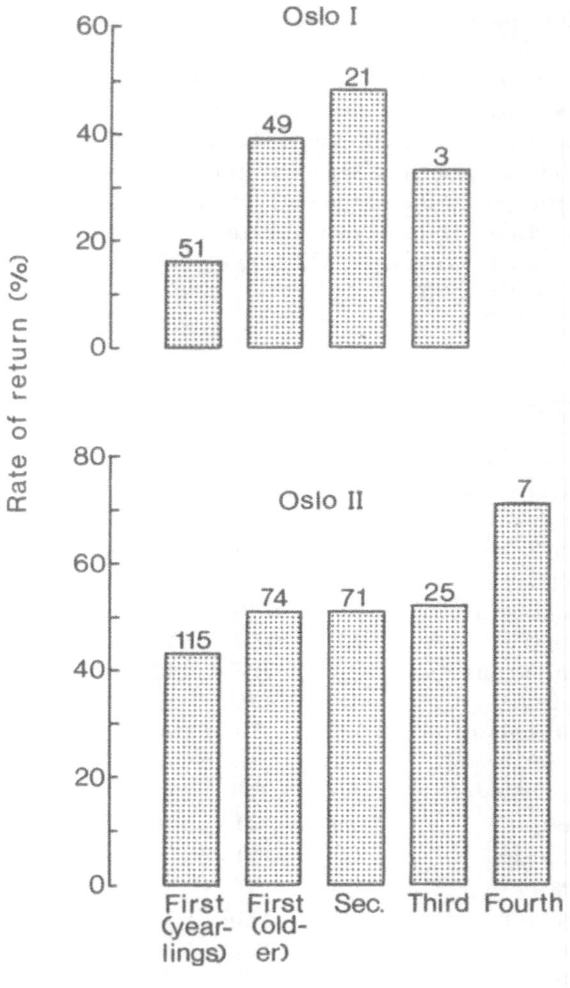

Fig. 1. Return rate of male Pied Flycatchers in relation to age and number of years present in the respective study plots. Sample size indicated above.

years old when they first settled in one of our study plots and those that had already been present previously (Fig. 1). The data for older males have therefore been pooled, as also the data for all study years, even though some individuals appear more than once in the analyses (see e.g. Askenmo 1979 for use of same method).

Four males (3%, N=124) that had defended nestboxes on Oslo I were found to have moved to Oslo II in a later year; no males moved from Oslo II to Oslo I. Two males present in one year on Oslo I were absent the following year but re-appeared in a later year. The corresponding number on Oslo II was four.

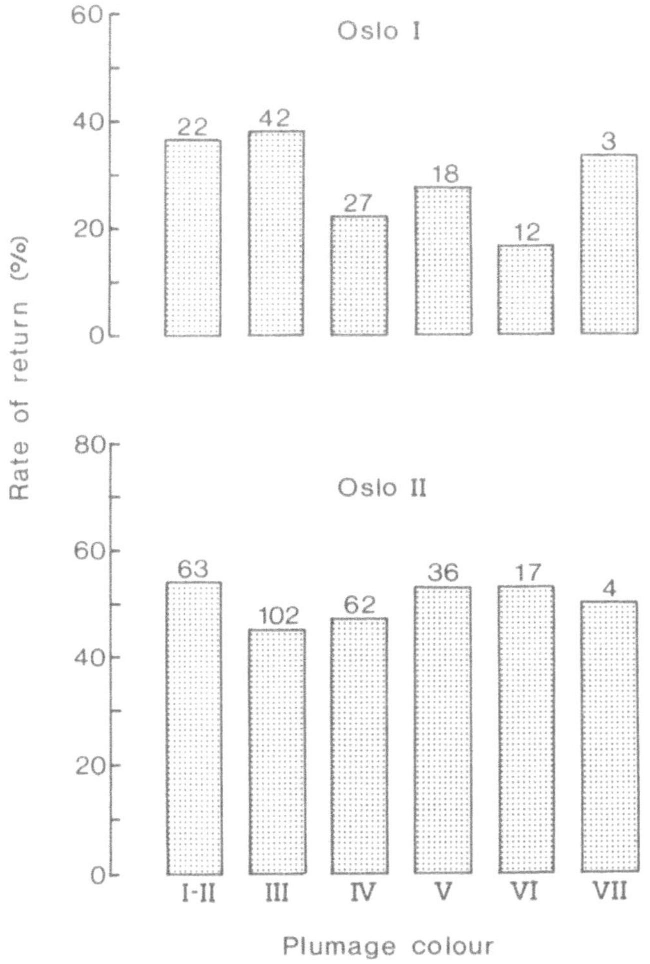

Fig. 2. Return rate of male Pied Flycatchers in relation to plumage colour. Sample sizes indicated above.

Male arrival time, body size, and plumage colour

There was no difference in mean arrival time on any plot between males that returned the following year and those that did not. Nor was any difference found between the two male groups in regard to mean body mass (recorded in early spring, soon after arrival), or to mean tarsus length (data not shown). Returning males on Oslo I had longer wings than non-returning males ($t=2.45$, 122 df, $P<0.05$), but this did not hold true when male age was taken into account, nor did it hold on Oslo II.

Earlier studies have yielded contrasting results as regards return rates of blackish and brownish males (Røskaft et al. 1986, Slagsvold and Lifjeld 1988). No significant correlation was found between return rate and plumage colour in the present study (Fig. 2). This held even when male age was taken into account : the return rates for blackish and brownish male yearlings on Oslo I were 20% (N=30) and 10% (N=21), respectively, and 38% (N=73) and 52% (N=42) for the same groups on Oslo II. The corresponding return rates for blackish and brownish older males on Oslo I were 39% (N=61) and 50% (N=12), and 52% (N=153) and 53% (N=15) for the same groups on Oslo II.

Mating success and breeding success

Males on Oslo I attracted 0-13 females each within a single breeding season. There was no difference in mean number of females attracted by males that returned and by those that did not (respective mean values 2.9, SD=2.2, N=38, and 2.6 females, SD=2.2, N=86).

Surprisingly, the return rate on Oslo I for the few males (N=35) that obtained a mate late in the season, that we did not remove, was no higher than that for the other males (Fig.3). Lack of difference in return rates of mated and unmated males was also found on Oslo II (Fig.2).

It may be argued that the males on Oslo I experienced the experimental removal of their mates as a kind of predation, and so were less willing to return. However, if we consider only the data for males on Oslo I that never attracted a female (*i.e.* those experiencing no "predation"), the return rate is also low (21%, N=14), and in fact lower than the overall value for the plot (31%, N=124), and lower than the value obtained for the unmated males on Oslo II (48%, N=23).

The return rate of polygamous males on Oslo II did not differ from that of other male groups (Fig. 3). Moreover, there was no significant difference in the mean number of young fledged for males that returned and those that did not return to Oslo II (respective mean values 6.0, SD=3.3, N=143, and 5.7 young, SD=3.6, N=150), even when male age was taken into account (data not shown).

Of males that attempted to breed late in the season on Oslo I, 25 succeeded in fledging at least one young. Surprisingly, return rate for these males was very low (20%; Fig. 4), in fact significantly lower than that for males that succeeded in breeding on Oslo II ($X^2=7.25$, $P<0.01$); the difference was significant for older males ($X^2=4.67$, $P<0.05$), but not for yearlings ($X^2=2.32$, NS), probably because of the small sample size for Oslo I (Fig. 4). Note that the return rate of those older males that did not succeed in fledging any young was high on Oslo I, in fact similar to the level of return rates recorded for Oslo II (Fig. 4). On Oslo II, return rate for males that did not succeed in fledging any young was nonsignificantly lower than that for successful males ($X^2=1.40$, NS).

One reason for the low return rate of breeding males on Oslo I may have been problems involved in rearing young late in the season, *i.e.* these males may have suffered increased mortality. This possibility was investigated by looking

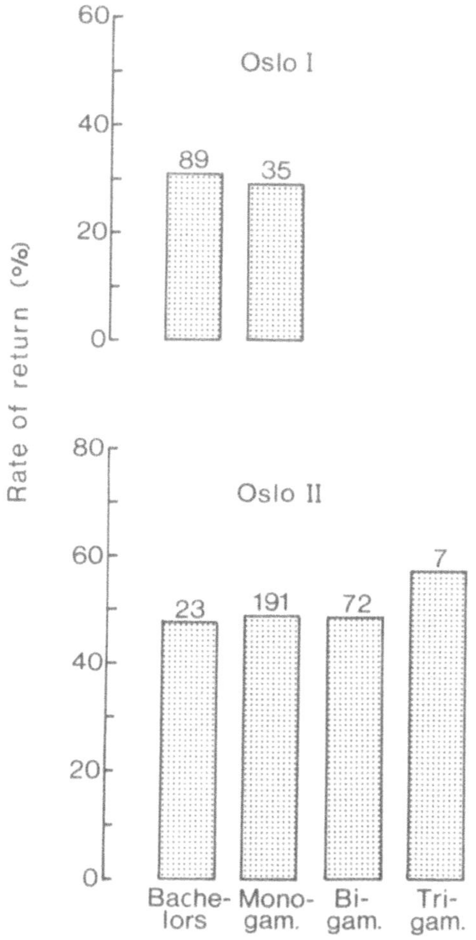

Fig. 3. Return rate of male Pied Flycatchers in relation to mating status. Sample sizes indicated above.

at return rate for Oslo II in relation to hatching time (Fig. 5). A tendency was found for males with late broods to have low return rate. Hatching time (first egg hatched) for males that bred on Oslo I was 29 June - 15 July. For males on Oslo II, the broods of which hatched after 25 June, return rate was only 30% (N=23). This is lower than that found for males with broods that hatched earlier on the same plot (51%, N=247), although not significantly so ($X^2=2.67$, NS). The 23 late breeding males on Oslo II cannot be considered low-quality males, because 7 were bigamously and 3 were trigamously mated.

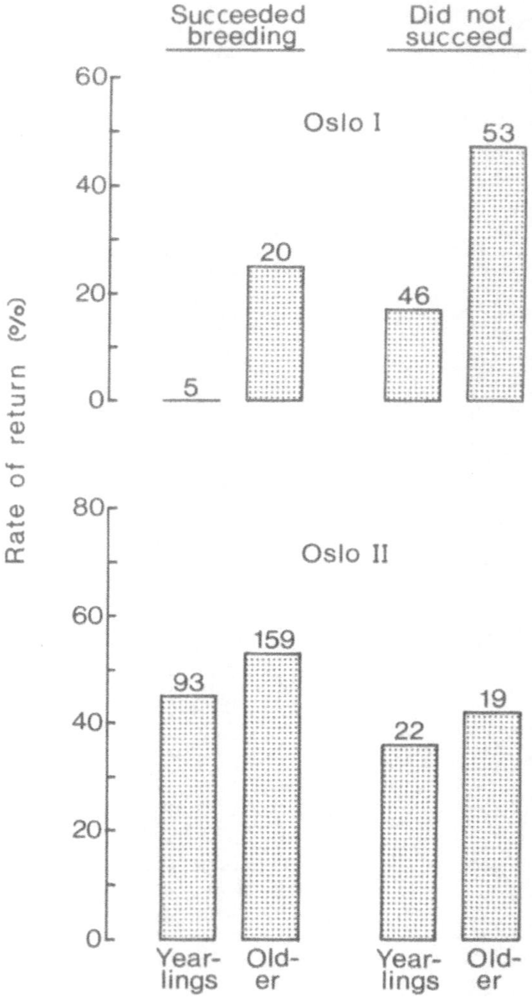

Fig. 4. Return rate of male Pied Flycatchers in relation to age and breeding success. Sample sizes indicated above.

DISCUSSION

Mortality or dispersal ?

The major result of the experiment was the difference in return rates for yearling males on Oslo I and II. This result could have been due to differences in (1) trapping efficiency, (2) survival rate, and/or (3) dispersal rate.

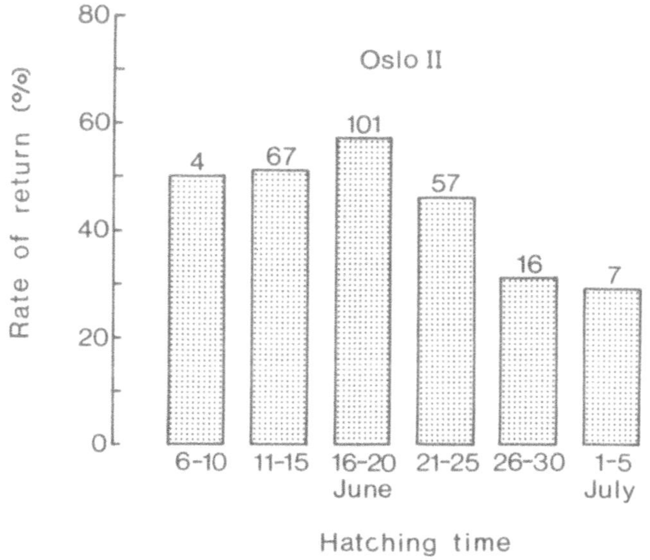

Fig. 5. Return rate of male Pied Flycatchers on Oslo II in relation to hatching time of the male's last brood hatched. Only broods of males for which at least one young fledged are included. Sample sizes indicated above.

The trapping procedure was the same on each plot. Most of our study area is surrounded by spruce forest, with few nesting holes available. The low number of males (only six) that reappeared after an entire year's absence, on both study sites, indicates that we recorded most of the males.

Because of low reproductive costs, it is unlikely that the low return rate on Oslo I represented high mortality of these particular males. Oslo I is situated only a few minutes flight away from Oslo II, and yearling males had a high return rate on Oslo II. If the predation rate on males was higher on Oslo I than on Oslo II we would expect low return rate also among older males on Oslo I, yet it was not much lower than on Oslo II.

We suggest instead that the low return rate of yearlings to Oslo I was due to dispersal. A high dispersal rate of first-year birds agrees with findings of other studies, both of the same (von Haartman 1949, Nyholm and Myhrberg 1983, Harvey et al. 1984), and other species (e.g. Drilling and Thompson 1988, Pärt and Gustafsson 1989).

Preferably, breeding success should have been manipulated in more than one study plot. In the present case the effects found from the treatment could have been due wholly to location effects, viz. the low return rate of Oslo I males may have been due to some other quality aspect of the area than female removal. This makes it difficult to analyse why young males did not like the Oslo I area. However, differences in return rates between the two study sites probably did not result from differences in habitat quality, since Oslo I included more deciduous woodland than Oslo II, and Pied Flycatchers seem to prefer to nest in deciduous woodland (Berndt and Winkel 1975, Alatalo et al. 1985).

It did not seem to be the breeding success of the individual male *per se* that was critical for the decision of whether to return or not, but the overall success of the local population. Late in the breeding season, when nests contained young, we often observed prospecting males repeatedly inspecting nest cavities from the entrance, but never feeding the young or trying to enter the nestbox. Perhaps males utilise the general presence of females and young to assess the value of a potential breeding area. This idea of assessment would explain low overall return rates on Oslo I, and high return rates of yearlings and unmated males on Oslo II. However, because of the experimental procedure, of removing females, we are not able to tell whether males on Oslo I responded to the low breeding success of their neighbours, or the high "predation" rate of females. We recommended a study in which the nest content is experimentally removed but not the female. A problem is however that most females would leave if breeding fails (Slagsvold and Lifjeld 1986).

The return rate on Oslo II was probably not much lower than the true survival rate. It is the highest recorded in Europe so far, and the rates were also high for both yearlings and unmated males. Earlier studies of Pied Flycatchers have indicated that return rate of males that have bred twice or more in a study area does reflect survival rate, and that it may be around 50% (von Haartman 1951, Nyholm and Myhrberg 1983). In our study area, not only return rates of previous breeders, but those of all older males may serve as an estimate of survival rate provided that all returning birds are captured ; this seemed to hold true for both localities studied. On the other hand, return rates of yearlings should be treated with caution. Probably, low male return rates recorded for many localities (Table I) will have been due to local, short-distance dispersal (cf. von Haartman 1951), in addition to a failure to identify returning males.

Cost of reproduction

The experimental addition of females to Oslo II resulted in an increase in the number of young fledged per male. However, this did not lead to any reduction in subsequent male return rate. Thus, rearing of many young, and polygynous mating of several males on Oslo II, did not seem to have involved any increase in parental costs in terms of lowering survival. The only indication of a cost of reproduction was the low return rate of males that reared young late in the breeding season. Male Pied Flycatchers start moulting in late June and in July (Ojanen and Orell 1982, Lifjeld and Slagsvold 1988), and it is possible that late breeders suffered from an overlap between moult and feeding of young.

We could find no relationship between male plumage colour and return rate. Our data therefore lend no support to the idea that blackish male invest more heavily in breeding than brownish-coloured males (Järvi et al. 1987), nor to the idea that such brownish males are in some respects phenotypically low-quality birds (cf. Slagsvold and Lifjeld 1988).

ACKNOWLEDGEMENTS

We thank B. A. Bjerke, T. Breiehagen, S. Dale, H. Lampe, H. Rinden, G. Stenmark, G.-P. Saetre, and C. Aas for assistance in the field, and K. Lessells and P.A. Tallantire for comments on the manuscript.

REFERENCES

Alatalo RV and Lundberg A (1989) Clutch size of the Pied Flycatcher *Ficedula hypoleuca* - an experiment. Ornis Fenn 66: 15-23

Alatalo RV, Lundberg A and Ulfstrand S (1985) Habitat selection in the Pied Flycatcher *Ficedula hypoleuca.* in: Cody ML (ed) Habitat selection in birds. Academic Press, London, p. 59-83

Askenmo C (1979) Reproductive effort and return rate of male Pied Flycatchers. Am Nat 114: 748-753

Berndt R and Winkel W (1975) Gibt es beim Trauerschnäpper *Ficedula hypoleuca* eine Prägung auf den Biotop des Geburtsortes? J Orn 116: 195-201

Campbell B (1955) A population of Pied Flycatchers (*Muscicapa hypoleuca*). Proc Int orn Congr 11 (1954): 428-434

Creutz G (1955) Der Trauerschnäpper (*Muscicapa hypoleuca* (Pallas)). Eine Populationsstudie. J Orn 96: 241-326

Curio E (1959) Beiträge zur Populationsökologie des Trauerschnäppers (*Ficedula hypoleuca* Pallas). Zool Jahrb 87: 185-230

Curio E and Regelmann K (1982) Forpflanzungswert und "Brutwert" der Kohlmeise (*Parus major*). J Orn. 123: 237-257

Dale S and Slagsvold T (1990) Random settlement of female Pied Flycatchers, *Ficedula hypoleuca.* Anim Behav 39, in press

Drilling NE and Thompson CF (1988) Natal and breeding dispersal in House Wrens (*Troglodytes aedon*). Auk 105: 480-491

Drost R (1936) Uber das Brutkleid männlicher Trauerfliegenfänger, *Muscicapa hypoleuca.* Vogelzug 6: 179-186

Haartman L von (1949) Der Trauerfliegenschnäpper. I. Ortstreue und Rassenbildung. Acta Zool Fenn 56: 1-105

Haartman L von (1951) Der Trauerfliegenschnäpper. II. Populationsprobleme. Acta Zool Fenn 67: 1-57

Haartman L von (1960) The Ortstreue of the Pied Flycatcher. Proc Int orn Congr 12 (1958): 226-273

Harvey PH, Greenwood PJ and Perrins CM (1979) Breeding area fidelity of the Great Tit (*Parus major*). J Anim Ecol 48: 305-313

Harvey PH, Greenwood PJ, Campbell B and Stenning MJ (1984) Breeding dispersal of the Pied Flycatcher (*Ficedula hypoleuca*) J Anim Ecol 53: 727-736

Högstedt G (1981) Should there be a positive or negative correlation between survival of adults in a bird population and their clutch size? Am Nat 118: 568-571

Järvi T, Røskaft E, Bakken M and Zumsteg B (1987) Evolution of variation in male secondary sexual characteristics: a test of eight hypotheses applied to Pied Flycatchers. Behav Ecol Sociobiol 20: 161-169

Lifjeld JT and Slagsvold T (1988) Mate fidelity of renesting Pied Flycatchers *Ficedula hypoleuca* in relation to characteristics of the pair mates. Behav Ecol Sociobiol 22: 117-123

Newton I and Marquiss M (1982) Fidelity to breeding area and mate in Sparrowhawk *Accipiter nisus*. J Anim Ecol 51: 327-341

Nyholm NEI and Myhrberg HE (1983) Breeding area fidelity of the Pied Flycatcher *Ficedula hypoleuca* at Ammarnäs, Swedish Lapland. Ornis Fenn 60: 22-27

Ojanen M and Orell M (1982) Onset of moult among breeding Pied Flycatchers (*Ficedula hypoleuca*) in Northern Finland. Vogelwarte 31: 445-451

Pärt T and Gustafsson L (1989) Breeding dispersal in the Collared Flycatcher (*Ficedula albicollis*): possible causes and reproductive consequences. J Anim Ecol 58: 305-320

Røskaft E, Järvi T, Nyholm NEI, Virolainen M, Winkel W and Zang H (1986) Geographic variation in secondary sexual plumage colour characteristics of the male Pied Flycatcher. Ornis Scand 17: 293-298

Slagsvold T and Lifjeld JT (1986) Mate retention and male polyterritoriality in the Pied Flycatcher *Ficedula hypoleuca*. Behav Ecol Sociobiol 19: 25-30

Slagsvold T and Lifjeld JT (1988). Plumage colour and sexual selection in the Pied Flycatcher *Ficedula hypoleuca*. Anim Behav 36: 395-407

Slagsvold T and Lifjeld JT, Stenmark G and Breiehagen T (1988) On the cost of searching for a mate in female Pied Flycatchers *Ficedula hypoleuca*. Anim Behav 36: 433-442

Stenmark G, Slagsvold T and Lifjeld JT (1988) Polygyny in the Pied Flycatcher, *Ficedula hypoleuca*: a test of the deception hypothesis. Anim Behav 36: 1646-1657

Trettau W (1952) Planberingung des Trauerfliegenschnäppers (*Muscicapa hypoleuca*) in Hessen. Vogelwarte 16: 89-95

Trettau W and Merkel F (1943) Ergebnisse einer Planberingung des Trauerfliengenfängers (*Muscicapa hypoleuca* Pallas) in Schlesien. Vogelzug 14: 77-90

WINTER REPLACEMENT RATES OF HIGH-RANKED CHICKADEES VARY WITH FLOATER DENSITY

Susan M. SMITH

Department of Biological Sciences, Mount Holyoke College, South Hadley, MA 01075, U.S.A.

Many parids show some version of an alternating social system, living in territorial breeding pairs during the spring and summer,then forming some sort of nonbreeding flocks during the rest of the year. Yet even these nonbreeding groups often turn out, on close examination, to be made up of varying numbers of pairs. Indeed, Nakamura (1975) suggested that the basic unit in parid social systems was the pair, with winter flocks being simply relatively stable aggregations of pairs that have come to live together for the nonbreeding season (see Haftorn, this volume). While this has since been challenged (e.g. Saitou 1978), more and more evidence has appeared suggesting that it is true, at least for several parid species. One of these is the Black-capped Chickadee (*Parus atricapillus*). Three main sorts of evidence support this idea. One is generated by observations of the behaviour of intact, free-living flocks; Ficken et al. (1981) have shown convincingly that individual Black-capped Chickadees associate with their own mates considerably more than they do with any other member of their flock. Another stems from the initial composition of flocks. At least in my own study area, chickadee flocks, when first formed, typically consist of even numbers of birds with even sex ratios, suggesting that pair formation may actually be part of flock formation (Smith 1984). A third sort of evidence comes from what happens when a bird vanishes from a winter flock. At least under some circumstances, high-ranked chickadees that disappear during the winter get replaced rapidly, sometimes within a matter of hours (Smith 1984, 1987).

Exactly who replaces these high-ranked birds? One might imagine a kind of domino effect occurring, for example, if an alpha male died, the beta male might break his pair bond and move up to replace him, after which the third-ranked male would move up to the beta position, and so on down the hierarchy. Evidently, however, this seldom, if ever, occurs. Instead, what has happened for many years in my study area is that the replacing bird is a winter floater, that typically settles into the flock and pairs with the mate of the vanished bird, leaving all other pair bonds in the flock intact (Smith 1984). Rarely, a high-ranked bird may be replaced by "complex substitution". Here, if a member of the alpha pair disappears, it is replaced by a member of the beta pair, after which a floater inserts into the beta pair: a process involving the breaking of one additional pair-bond.

My study area obviously has had an unusually high density of winter floaters, and earlier I predicted that, in situations with fewer floaters available for replacement, the domino effect described above might well occur (Smith 1987). This seemed to be supported by removal experiments done on other

NATO ASI Series, Vol. G 24
Population Biology of Passerine Birds
Edited by J. Blondel et al.

parids, where it was reported that beta males took over in flocks from which alpha males were removed (e.g. Dixon 1965, Hogstad 1987). Recently, a sudden, unexpected drop in winter floater density in my own study area has allowed me to test this prediction. I report here on the replacement process after this drop occurred.

STUDY AREA AND METHODS

The study area, approximately 35 ha of residential area, old field, and mixed second growth woodland near Mount Holyoke College, South Hadley, Massachusetts, has been described earlier (Smith 1984). In most winters, 4 or 5 winter chickadee flocks inhabit this area. Chickadees have been colour-banded there since the autumn of 1979, with each bird receiving a unique band combination for individual recognition. Birds were aged by the tail feather method of Meigs et al. (1983), and sexed using wing length and observing subsequent behaviour. Census and dominance data have been taken at least once weekly during each winter, and breeding pairs and their approximate territory boundaries recorded each summer.

Beginning on 31 August 1987, data were taken in two winter flocks for calculating association indices, after the method used by Ficken et al. (1981). I took data only on what they called "nearest associate observations": whenever more than two chickadees were feeding within 3 m of each other, the two closest individuals were recorded as being nearest associates. Association indices could then be calculated by the formula:

$$\text{Association index} = 2h/(a + b)$$

where:

h = number of samples containing both individuals together,
a = number of samples involving individual A, and
b = number of samples involving individual B.

Association data were taken through 31 January 1988.

RESULTS

Over the seven-winter period from autumn 1980 through spring 1987, the proportion of floaters in my study area varied from 18.9% to 27.3%, with an average proportion of 23.1% (Table I). However, in the next two years, the proportion of floaters dropped dramatically, to 10% or less, despite no decrease in the local number of flock regulars (Table I). Prior to this drop, the high density of floaters generally meant that floaters were available to move in whenever a high-ranked regular disappeared; as mentioned above, the replacement typically took place in a matter of hours. After the drop, however, conditions became different. Floaters still maintained relatively restricted home ranges, which included only three to five flock ranges. Therefore no single floater ranged over my entire study area (indeed, most had ranges that included

Table I. Study area winter floaters (M = male, F = female).

Year	No. of Flock Regulars	No. of Banded Floaters M	F	Total	% Floaters
1980-81	42	6	8	14	25.0
1981-82	40	7	8	15	27.3
1982-83	44	7	7	14	24.1
1983-84	52	8	6	14	21.2
1984-85	30	4	3	7	18.9
1985-86	48	7	6	13	21.3
1986-87	48	9	6	15	23.8
1987-88	54	3	3	6	10.0
1988-89	48	3	2	5	9.4

some combination of central and peripheral flocks). Hence situations could arise where high-ranked openings occurred with no floater available to fill them.

During the first seven winters, 25 floaters inserted into my flocks after high-ranked regulars had disappeared; all 25 bred locally the following spring. The number of successful insertions varied from 2 to 6 per winter, (average: 3.6). Moreover, over this same seven-winter period, only one suitable position remained unfilled by floaters (Smith 1987).

By contrast, over the next two winters, only one floater replaced a high-ranked regular. This figure is slightly misleading, as there were no high-ranked deaths in any of my central flocks in the winter of 1988-89. However, in 1987-88, there were a total of six: the alpha males vanished from three different flocks, as did two highest-ranked and one second-ranked female (Table II).

Table II. High ranked deaths and their consequences: 1987-88.

Bird	Rank*	Date vanished	Rank of replacement	When replaced	Source
ARBG	1m(6)	early Nov.	floater	early Nov.	floater
ABRR	1m(12)	mid-November	3m	Feb.-Mar.	same flock
AKBK	1m(6)	late December	2m	Feb.-Mar.	diff. flock
ARGK	2f(12)	mid-October	4f	Feb.-Mar.	same flock
AGOY	1f(6)	late December	2f	Feb.-Mar.	same flock
AOBO	1f(6)	late December	2f	Feb.-Mar.	same flock

* 1m(6) means top-ranked male in a six-bird flock;
2f(12) means second-ranked female in a twelve-bird flock.

All six of these birds disappeared before the end of December. Yet only one, the first high-ranked male to go, was replaced by a winter floater. This insertion was typical of those occurring in past winters, with the floater settling into the flock as the new alpha male within at least a day and a half of my last sighting of the original alpha male. One of the most obvious indications that a floater has inserted is how closely it sticks to its new mate (typically that of the bird it has just replaced); this was clearly the case here as well. By contrast, however, the other five positions, including two other alpha males, remained apparently unfilled for many weeks (Table II). Two of the five happened to be in flocks for which I was gathering association data. Flock 1 lost its alpha male (ABRR), and Flock 2 lost its top-ranked female.

Table III shows the association indices for the members of Flock 1 prior to ABRR's disappearance. Three clear pairings stand out: the top-ranked pair of ABRR with ARYY, which had bred together for the last three years; AROO and ARYB, which had bred together the previous summer; and ARGR and ARKB, two young birds which would breed together for the first time the following summer.

Table IV shows the association indices for the same flock after its alpha male (ABRR) had vanished. The widowed bird (ARYY) clearly did not associate particularly strongly with any other chickadee during this period. Moreover, the intact pair of AROO and ARYB continued to associate together as strongly as before. Yet the following spring, AROO deserted his mate to form a pair with the widowed top-ranked female (ARYY), leaving his former mate to find a new male. Clearly, therefore, the alpha male's position in this flock remained unfilled for at least two and a half months.

Similar behaviour occurred in Flock 2, which lost its top-ranked female (AOBO) in mid-December. Before her disappearance, the flock's three pairs associated strongly together, with the two young birds (AYBY and AYYR) showing the weakest of the three pair indices. After AOBO vanished, her mate (AORY) did not show very strong association with any other flock member, and the intact pair of AYRG and AYYB associated together just as strongly as before: their association index was 0.52 before AOBO's death, and 0.59 afterward. Yet the following spring, AYYB deserted her mate AYRG and paired with AORY, the alpha male whose mate had vanished more than a month and a half previously.

Similar processes occurred for the filling of the other three high-ranked positions listed in Table II. The source of the replacements is interesting. Of the three females, one was from a low-ranked pair in the same flock, and had lost her mate earlier. Both of the other two (including the one described above) broke intact pair bonds at the end of the winter to move up into the vacant high-ranked position. One of these was paired with a male who, once deserted, then moved to an adjacent flock and replaced the alpha male. Another male was replaced by a floater, and the third, from Flock 1, has been described above. Hence the six high-ranked deaths resulted in three broken pair-bonds, all many weeks later; and only one insertion by a floater.

Looking at the three broken pair-bonds, no evidence of a "domino effect" could be seen. Of the resulting three deserted mates, one managed to insert into a high-ranked opening elsewhere, and the other two paired with lower-ranked

Table III. Association indices in Flock 1 before the alpha male vanished: 31 August - 16 November 1987.

	ABRR	ARRG	AROO	ARGR	ARYR	ARYY	ARYB	AROY	ARKB
ABRR (13)[a]	X	.10(1)[b]	.18(3)	.08(1)	.09(1)	.46(6)	.07(1)	--	--
ARRG (8)		X	.14(2)	--	.12(1)	.10(1)	.08(1)	.13(1)	.12(1)
AROO (21)			X	.06(1)	.13(2)	.06(1)	.63(12)	--	--
ARGR (13)				X	.09(1)	.15(2)	.07(1)	.10(1)	.55(6)
ARYR (9)					X	.09(1)	--	.25(2)	.11(1)
ARYY (13)						X	.07(1)	.10(1)	--
ARYB (17)							X	.08(1)	--
AROY (7)								X	.13(1)
ARKB (9)									X

[a] Total number of times each bird was counted as a nearest associate.
[b] Number of nearest associate observations for each pair.
Birds are listed in descending order of rank, beginning with the alpha male.

Table IV. Association Indices in Flock 1 after the alpha Male had vanished: 17 November 1987 - 31 January 1988.

	ARRG	AROO	ARGR	ARYR	ARYY	ARYB	AROY	ARKB
ARRG (8)	X	.08(1)	.09(1)	.13(1)	--	--	.14(2)	.10(1)
AROO (16)		X	.13(2)	.08(1)	.08(1)	.58(9)	--	.14(2)
ARGR (14)			X	.09(1)	.17(2)	--	.08(1)	.52(7)
ARYR (8)				X	.22(2)	.17(1)	.21(2)	.10(1)
ARYY (10)					X	.24(3)	.10(1)	.09(1)
ARYB (15)						X	.15(2)	--
AROY (11)							X	.08(1)
ARKB (13)								X

Numbers in brackets as in Table III.

regulars each of whose mate had died earlier in the winter. All three subsequently bred in the study area.

Finally, dominance data indicate that the widowed birds whose mates were unreplaced for several weeks nevertheless managed to maintain their place in the hierarchy with no loss of rank.

DISCUSSION

The data from this study confirm that Black-capped Chickadees in my eastern study area, like those in the midwest (Ficken et al. 1981), associate in pairs within winter flocks. However, they do not support my prediction that, where winter floaters are unavailable, high-ranked positions will rapidly be filled by slightly lower-ranked flock members, causing a domino effect of several broken pair bonds. The key word here seems to be "rapidly". During the seven previous winters, the replacement process was very rapid indeed, leading me to expect it would be fairly fast even in situations with no available floaters--clearly an incorrect assumption. Moreover, a maximum of only one pair bond was broken for each replacement: hardly the domino effect I had pictured.

It is interesting that all six "widowed" birds, i.e. the mates of the high-ranked birds that vanished, maintained their original rank after their mates' deaths. While this is not surprising for the males, it is perhaps a bit unexpected for the females, whose rank has been suggested to be simply a reflection of that of her mate (e.g. Withiam et al., in press). All three females survived the summer, and they and their mates were the top-ranked pairs in their flocks the following winter. This clearly indicates that a female's rank can, once established, be maintained alone for many weeks.

Several reports exist of removal experiments where the alpha male is removed from a flock and the beta male takes over, e.g. Dixon (1965) for Mountain Chickadees (*Parus gambeli*) , and Hogstad (1987) for Willow Tits (*P. montanus*) . Such reports could, in fact, be somewhat misleading. Naturally, once the alpha male is removed, the next-highest ranked flock member becomes, by default, the top-ranked bird. But this does not mean that he has actually replaced the alpha bird. To find this out, one must study the behaviour of the alpha male's mate. If no bird associates closely with her, then it seems clear that the position has not been filled. The beta male may pair with her later, but until this happens, it is incorrect to say that he has fully become the flock's new alpha male.

It is difficult to explain why the first position was filled so very quickly by a floater, after which no other positions were filled until early spring. A few floaters were, in fact, available in some areas, and at least one did try to associate with a widowed alpha female, but to no avail. Possibly some minimum floater density is necessary for the rapid replacement typical of my study area in past years. Further research is necessary to establish exactly what this critical level might be.

REFERENCES

Dixon KL (1965) Dominance-subordination relationships in Mountain Chickadees. Condor 67: 291-299

Ficken MS, Witkin SR and Weise CM (1981) Associations among members of a black-capped chickadee flock. Behav Ecol Sociobiol 8: 245-249

Hogstad O (1987) It is expensive to be dominant. Auk 104: 333-336

Meigs JB, Smith DC and J Van Buskirk (1983) Age determination of Black-capped Chickadees. J Field Ornith 54: 283-286

Nakamura T (1975) A study of *Paridae* community in Japan. III. Ecological separation in social structure and distribution. Miscellaneous Report of the Yamashina Institute of Ornithology, 7: 603-636

Saitou T (1978) Ecological study of social organization in the Great Tit, *Parus major* L. I. Basic structure of the winter flocks. Japanese Journal of Ecology, 28: 199-214

Smith SM (1984) Flock switching in chickadees: why by a winter floater? Am Nat, 123: 81-98

Smith SM (1987) Responses of floaters to removal experiments on wintering chickadees. Behav Ecol Sociobiol 20: 363-367

Withiam ML, Lemmon D and Barkan CPL (in press) Pair-bonds in wintering Black-capped Chickadees: benefits to females. Auk, in press

HOW DOES TRANSLOCATION AFFECT THE SUBSEQUENT DISTRIBUTION OF BREEDING PAIRS IN A POPULATION OF PIED FLYCATCHERS (*FICEDULA HYPOLEUCA*) ?

Wolfgang WINKEL and Doris WINKEL

Aussenstation Braunschweig für Populationsökologie des Instituts für Vogelforschung "Vogelwarte Helgoland", Weddel, Bauernstr. 14, D-3302 Cremlingen, FR Germany

For many years nesting Pied Flycatchers (*Ficedula hypoleuca*) have been individually investigated in nestbox study areas near Braunschweig and Lingen in the northern part of the FR Germany. During the years 1985 to 1988 we intervened in the process of settlement of Pied Flycatchers in some of these areas through translocations. The aims of this experimental study were:

a) to examine the site fidelity of the translocated birds and analyse possible contributing factors resulting from their previous history,

b) to investigate the dispersion dynamics of the Pied Flycatcher in relation to the territories made vacant by translocations, and

c) to throw light on the question of non breeding adults in the Pied Flycatcher.

MATERIAL AND METHODS

The Pied Flycatchers were captured in deciduous wood habitats of the Braunschweig region when inspecting empty nestboxes before beginning to breed. This was made possible by means of front panels with a wire trap fitted behind the entrance hole. In addition to the numbered ring, the individuals were fitted with a coloured one. Sexes were usually identified by plumage characteristics. In 1985 and 1986 the caught Pied Flycatchers were placed into a cardboard container and within 2 hours taken by truck to a point of release up to 40 km away, where the birds were released at once. In 1987 the experimental birds were kept initially for 3 to 10 days in an aviary and were then transported to a point of release within 39 km from their capture area. In 1988, after being kept in an aviary for a period of 2 to 5 days, 69 Pied Flycatchers were transported during the hours of darkness to the Lingen area (pine and larch forest) - about 250 km away - and were released there at dawn.

From 1985 to 1988 a total of 505 Pied Flycatchers (324 males and 181 females) were translocated in this manner. No bird was displaced more than once in any year. 84 % of all Pied Flycatchers were transported to a nestbox study area (in the majority of cases another deciduous wood habitat), 16 % to localities where no nestboxes were available (Fig. 1). - Statistical methods used follow Sachs (1969).

NATO ASI Series, Vol. G 24
Population Biology of Passerine Birds
Edited by J. Blondel et al.

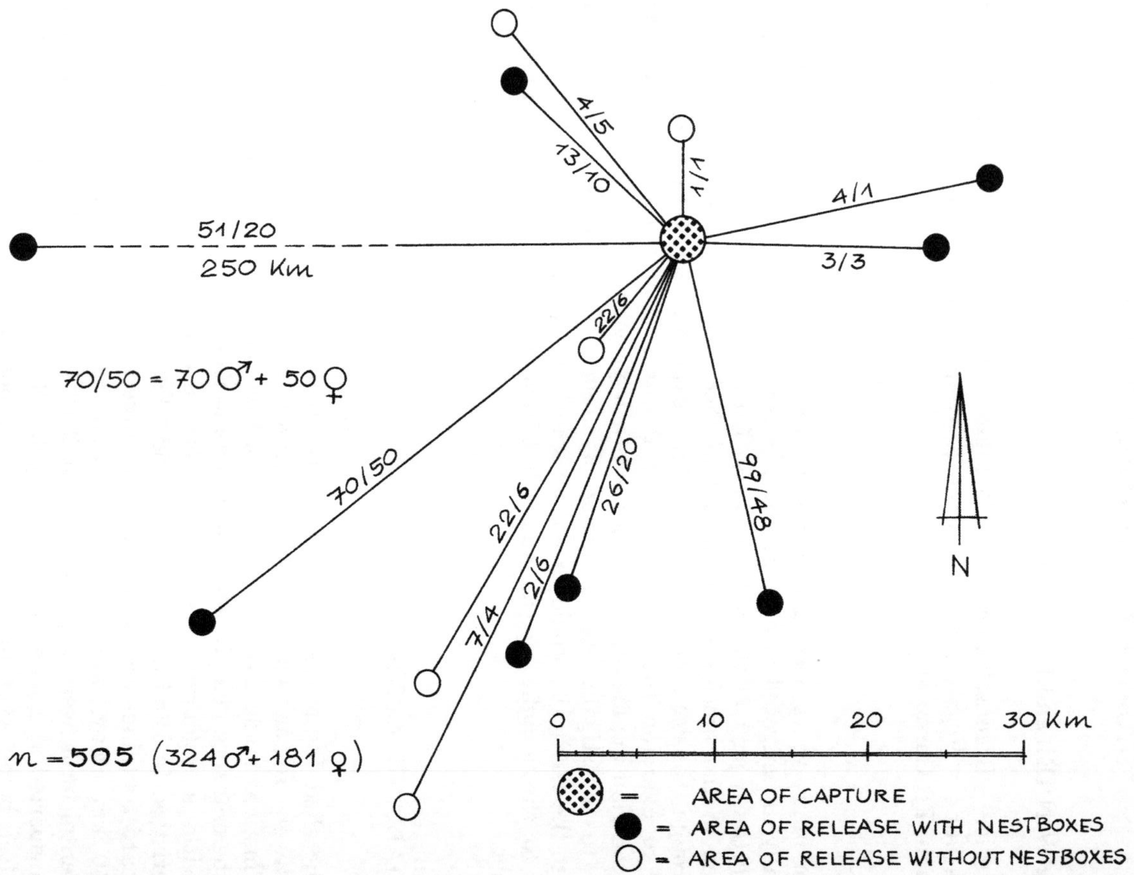

Fig. 1. General data of the translocation experiments.

In every year all breeding individuals in the areas of capture as well as in the release areas were examined with regard to their rings. By this method we could establish for instance what proportion of transported birds had returned for breeding and what proportion had settled in their area of release.

RESULTS AND DISCUSSION

Translocations to release areas without nestboxes

On release in an area without nestboxes the rate of return to the area of capture (based on recaptures in the year of translocation) was 46.8 %, however on release in a study area with nestboxes the rate was significantly lower (30.5 % ; $X^2=7.33$, $p<0.01$). The return of translocated Pied Flycatchers therefore depends on the presence or absence of nestboxes in the area of release. For the following detailed analysis we used only the data for translocations into nestbox areas.

Translocations into nestbox areas

Results did not differ significantly between the various release areas. Also the duration of captivity in an aviary had no discernable influence on the reaction of the translocated birds. Therefore the interpretation of the experiments may be applied to the whole data set. According to the previous history of the translocated birds, we subdivided the translocations into the following four groups :

Group 1 : Translocation of more-than-one-year old Pied Flycatchers which had already bred in the capture area at least once.

Group 2 : Translocation of one-year old Pied Flycatchers born in the capture area.

Group 3 : Translocation of more-than-one-year old Pied Flycatchers born in the capture area but without evidence of having bred there.

Group 4 : Translocation of Pied Flycatchers, to whom no criterion of group 1 - 3 applies, as a rule unringed individuals.

Group 1 : The diagram (Fig. 2) indicates that 63 % of translocated males and 88 % of females bred in the year of translocation in the area of capture or release. The majority homed to the capture area. Males seem to have less

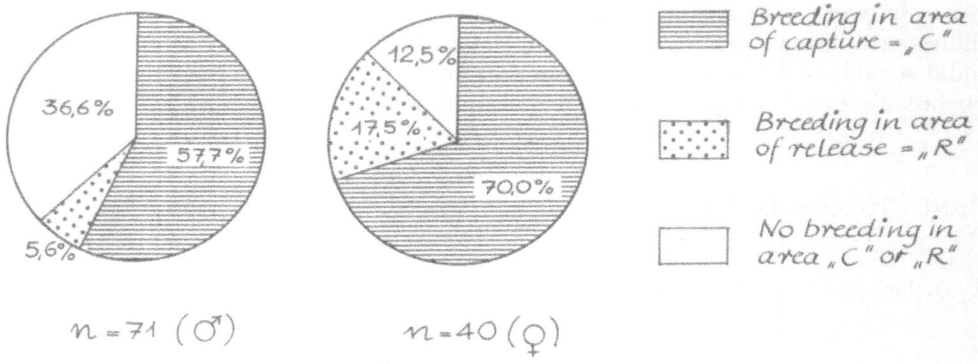

Fig. 2. Results of the translocations of more-than-one-year old Pied Flycatcher which had already bred in the capture area at least once (=Group 1).

tendency to return. This interpretation, however, is not correct, for the following reason : we found that the actual proportion of home-comers to the capture area in Group 1 is actually still much greater, especially in cases of males, than the figures based on evidence of breeding show. This is confirmed by catching attempts from vacant nestboxes in the main capture area "Schnäbel" until the time of egg laying or breeding in the years 1985, 1986 and 1987 : of 21 translocated males and 5 females , of whom we could find no evidence of breeding in the experimental year , we could identify 12 males and 1 female as non-breeding home-comers. No pairing of these birds took place in this particular year. This applies to males mainly, for the returning females apparently had no difficulty in finding a partner in spite of the delay caused by translocation. If one adds the number of unpaired returning birds to the number of returning breeding birds of the main capture area we have the following situation (Table I) : from the translocated males 78.2 % returned, while of the translocated females only 64.5 % were found to be home-comers (the difference is, however, not significant). This result corresponds with our findings from years without translocation experiments showing a higher breeding place fidelity of males (Winkel 1982). There are no indications that breeding success in year n influences the response to translocation in year n + 1.

Group 2 : In this group no significant difference between male and female Pied Flycatchers could be found. Therefore the findings for sexes were combined (Fig. 3). Only 46 % of translocated birds bred in the respective year in the area of capture or release, of these by far the majority bred in the release area, in contrast to Group 1. Though Pied Flycatchers have a strong bonding to their place of birth, which according to our findings results from an imprint-like process (Berndt and Winkel 1979, 1980), they are under certain conditions flexible in their choice of breeding place. This makes sense biologically, for juveniles frequently cannot prevail against intra - or interspecific competition in their place of birth. If translocated juvenile Pied Flycatchers find vacant

Table I. Data of translocation out of the capture area "Schnäbel" (Group 1-4), home-comers without breeding record are included.

Group	Sex	Returned to the area of capture = "C"	Breeding in the area of release = "R"	No evidence in area "C" or "R"
1	♂	43 = 78.2%	3 = 5.5%	9 = 16.4%
	♀	20 = 64.5%	7 = 22.6%	4 = 12.9%
2	♂ + ♀	6 = 19.4%	14 = 45.2%	11 = 35.5%
3	♂	19 = 76.0%	0 = 0%	6 = 24.0%
	♀	2 = 18.2%	5 = 45.5%	4 = 36.4%
4	♂	26 = 27.1%	23 = 24.0%	47 = 49.0%
	♀	3 = 4.6%	27 = 41.5%	35 = 53.8%

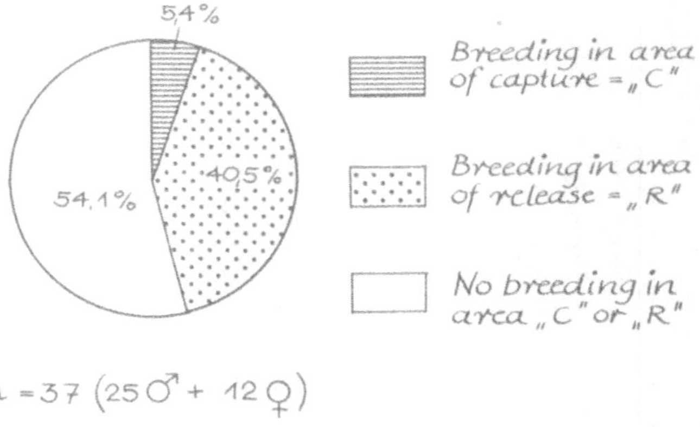

Fig. 3. Results of the translocations of one-year old Pied Flycatchers born in the capture area (=Group 2).

nestholes, it would generally be opportune to remain there for breeding rather than to return to the area of capture with uncertain expectation of obtaining a nesthole.

The data from 1985-1987, which include the number of individuals who returned but were only recorded as non-breeders, are given in Table I. The predominance of breeding in the release area (45 %) in comparison to returning to the capture area (19 %) is almost the same as shown in Fig. 3.

Group 3 : In this group 53 % of translocated males and 62 % of females were noted as breeding birds in the area of capture or release - males almost exclusively in the capture area, females in the release area (Fig. 4). The difference between the sexes is significant ($X^2=15.15$, $p<0.001$). The males of Group 3 reacted therefore like the birds of Group 1. This suggests that at least some of the males in Group 3 were living in the area of capture the year before as non-breeding adults, which could be confirmed through our data from 6 males. Their presence as surplus breeding stock apparently produces as strong a bond to the respective area as would result from actual breeding. The females of Group 3, however, returned in much smaller numbers. This is another indication that persistence as a non-breeder in the area occurs more frequently with males than with females. We assume that at least some of the females of Group 3 had bred elsewhere in the year prior to translocation (we have one example for this in our data set), which would explain their weaker bonding to the area of capture. The values for the translocation experiments out of the capture area "Schnäbel", in which the home-comers without breeding evidence are included (Table I) demonstrate the same difference between the sexes as shown by the data presented in Fig. 4.

Group 4 : Of these Pied Flycatchers only 44 % of males and 46 % of females could be found as breeding birds in the area of capture or release.

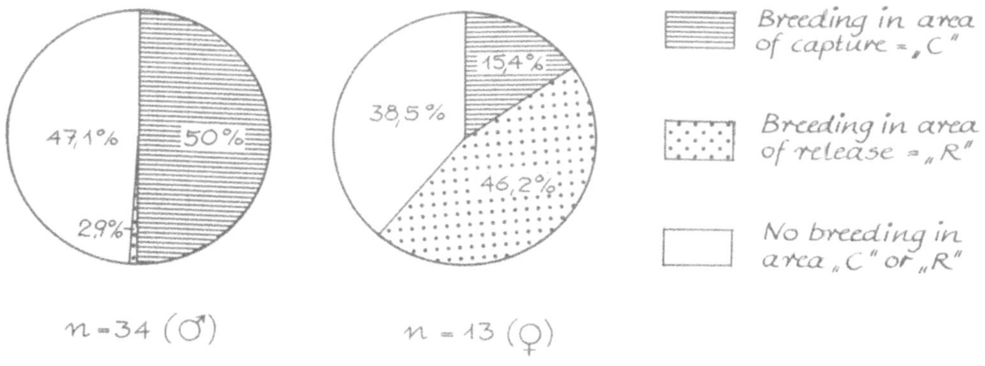

Fig. 4. Results of the translocations of more-than-one-year old Pied Flycatchers born in the capture area but without evidence of having bred there (= Group 3).

Females bred in the release area much more frequently than males (Fig. 5). The difference between the sexes is significant ($X^2 = 10.19$, $p < 0.001$). The fact that males of Group 4 returned more frequently to the area of capture than females might be the result of many translocated males being at least 2 years old and living as non-breeders in the area of capture during the year before translocation, whilst among the translocated females relatively more one-year olds were to be found (see also the findings of Group 3). This assumption is supported by our data for ringed birds in Group 4. From the ringed males 29 were at least two years old and only 8 had an age of one year whilst among females only 8 were at least two years old and 9 had an age of one year. Besides this, four males of Group 4 but no females could be found in an earlier

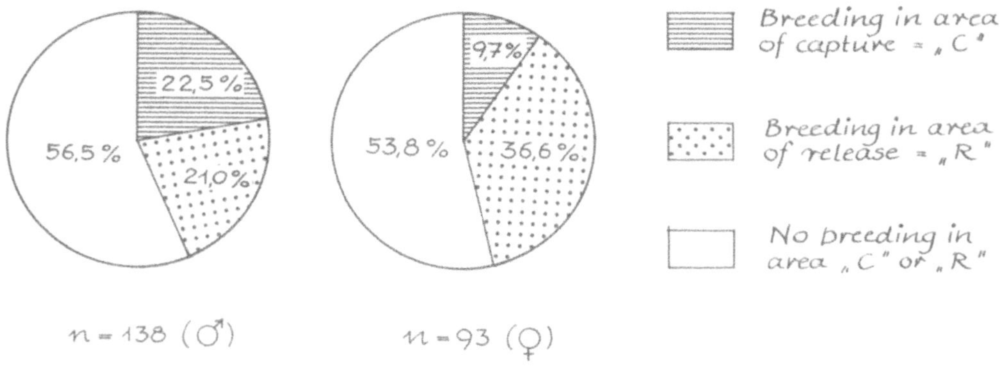

Fig. 5. Results of the translocations of Pied Flycatchers to whom no criterion of Group 1 - 3 applies, as a rule unringed individuals (=Group 4).

year as a non-breeder in the capture area. The data which include home-comers without breeding records show also clearly the difference between male and female behaviour (Table I).

Summing up the results of Group 1 - 4, it can be said that site fidelity behaviour of Pied Flycatchers would primarily be determined by the previous history of birds. In both sexes the bond to the capture site was reinforced above all through former breeding in this area rather than through the process of being born there. But a strong bonding to the area was also produced through the presence as surplus breeding stock (in our study plots the persistence as a non-breeder apparently occurs more frequently with males than with females).

Individual differences in the response to translocation

To demonstrate the individual differences in the reaction, we have selected four examples. They concern Pied Flycatchers which, in at least 3 successive years, were each first recorded in the capture area and then translocated. There was no relationship between response to translocation and breeding success of the previous year.

Example 1 (Fig. 6, left) : This male was translocated in the year 1985, 1986, 1987 and 1988 to different release areas (R1, R3, R2, and R8 respectively) ; the bird always returned to the capture area. Example 2 (Fig. 6, right) : This male reacted in 1986 and 1987 by returning, but in 1988 it remained in the release area. Example 3 (Fig. 7, left) : This female returned to the capture area in 1985, remained in the release area in 1986 and returned once more to the capture area in 1987. Example 4 (Fig. 7, right) : This male reacted in the year 1985 and 1986 by remaining in the release area but returned to the capture area in 1987 though it was translocated to the same release area as in the previous year, where it raised a brood in 1986.

From these examples it becomes clear, that the response to translocation does not only depend on the bird's previous history but also on the specific conditions of the release area. For translocated male presence of vacant nestholes in the immediate vicinity of the point of release would be of special significance, whilst for females especially the meeting with males presenting a nesthole would be important. This may explain deviations from the "normal" behaviour in respect to the urge to return to their place of origin.

Influence of translocations on the breeding population

From 1985 to 1988 the number of individually recorded Pied Flycatchers fluctuated in the capture area "Schnäbel" between 85 and 92 for males and between 57 and 81 for females (Table II, columns VII + VIII). There were therefore in some cases more than twice the number of individuals recorded than the number of breeding pairs might lead us to expect : the number of

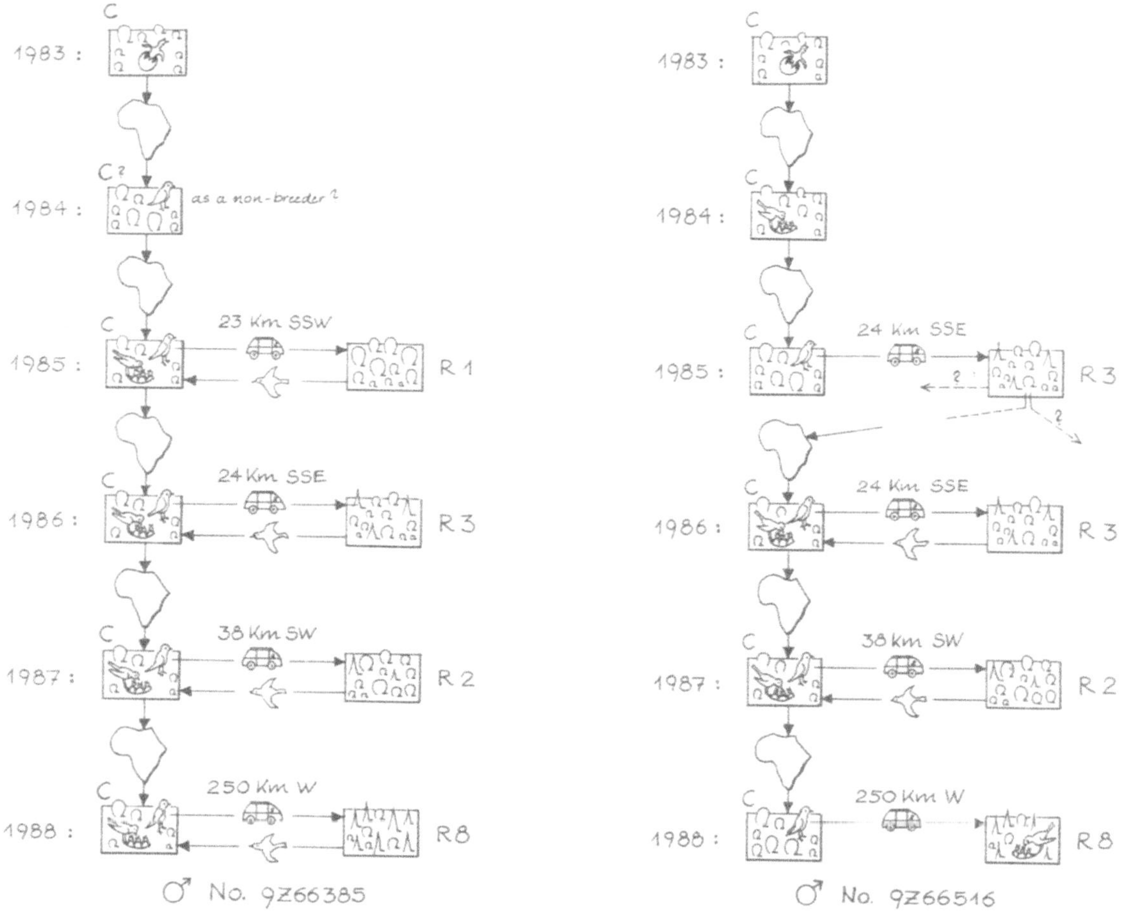

Fig. 6. Illustration of individual differences in the reaction of translocated Pied Flycatchers (examples 1 and 2).

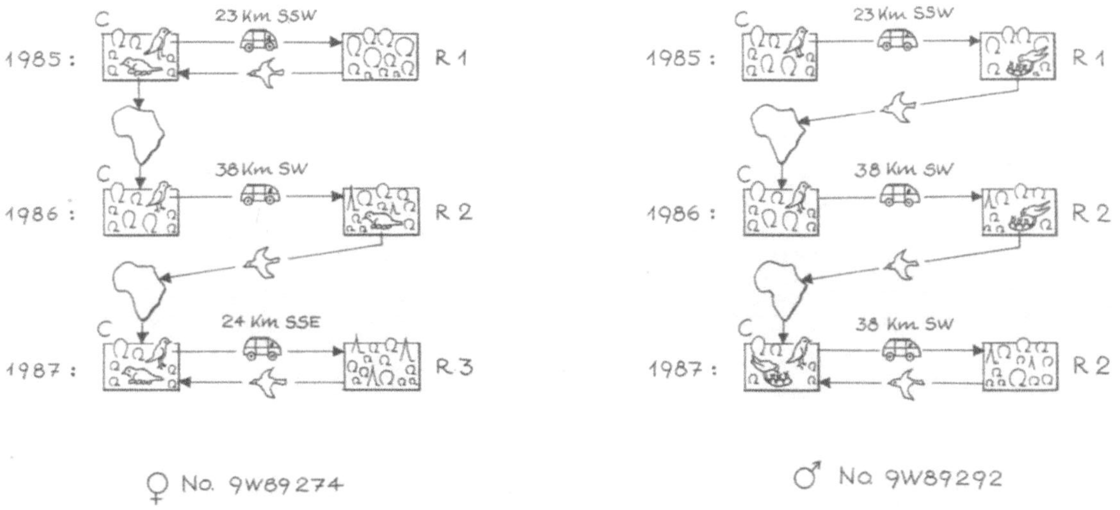

Fig. 7. Illustration of individual differences in the reaction of translocated Pied Flycatchers (examples 3 and 4).

Table II. Number of individually recorded Pied Flycatchers in the capture area "Schnäbel" (Oak and Hornbeam forest, 8.6 ha) in the translocation years 1985 - 1988.
Numbers in the second column = days and months (15.5 = May 15).

Year	Capture from vacant nestboxes	Birds caught in the capture area, transported, but no evidence of return		Non-breeders in the capture area		Breeding birds in the capture area		Total number of recorded individuals	
		I	II	III	IV	V	VI	VII	VIII
		♂	♀	♂	♀	♂	♀	♂	♀
1985	4. - 15.5 daily in addition: 21., 24., 28., 31.5, 5.6	33	35	22	4	31**	35	86	74
1886	28.4 - 9.5 daily	44	47	10	0	31	34	85	81
1987	27.4 - 2.5 daily in addition: 9., 17., 24.5	40	16	12	0	40	41	92	57
1988	29.4 - 2.5 daily (all individuals kept in an aviary and released at 3.5)	35	15	*	*	50**	58	85	73

* In this year only the breeding Pied Flycatchers were recorded individually. ** Including one instance of bigyny.

recorded broods varied from 34 to 58 (Columns V + VI) with an average of 42. In spite of the removal of a considerable number of individuals (Columns I + II) the number of recorded broods reflects the usual dispersion dynamics of Pied Flycatchers in this study area : for 1978-1984 on average 46 broods were recorded (concerning fluctuations and long term trends in the Braunschweig region see Winkel 1989). It may therefore be assumed that territories falling vacant on account of translocations would be taken over very promptly by new individuals, especially one-year old and unringed Pied Flycatchers (W. and D. Winkel, in preparation).

ACKNOWLEDGEMENTS

In 1985 and 1986 the translocation experiments were supported with great dedication by the late Dr. Rudolf Berndt and Mrs. Margrit Frantzen. Mrs. Frantzen also cared for the Pied Flycatchers during their temporary stay in an aviary during the years 1987 and 1988. We are very grateful to both also for stimulating exchange of views. Frank Hammersley kindly advised us in styling of the English text.

REFERENCES

Berndt R, Winkel W (1979) Verfrachtungs-Experimente zur Frage der Geburtsortsprägung beim Trauerschnäpper (*Ficedula hypoleuca*). J Orn 120 : 41-53

Berndt R, Winkel W (1980) Field Experiments of Imprinting to the Birthplace in the Pied Flycatcher *Ficedula hypoleuca*. Acta XVII Congr Internat Ornithologici Berlin 1978 : 851-854

Sachs L (1969) Statistische Auswertungsmethoden. 2. Aufl. Springer Verlag, Berlin Heidelberg N.Y.

Winkel W (1982) Zum Ortstreue-Verhalten des Trauerschnäppers (*Ficedula hypoleuca*) im westlichen Randbereich seines mitteleuropäischen Verbreitungsgebietes. J Orn 123 : 155-173

Winkel W (1989) Langfristige Bestandsentwicklung von Kohlmeise (*Parus major*) und Trauerschnäpper (*Ficedula hypoleuca*) : Ergebnisse aus Niedersachsen. J Orn 130 : 335-343

CONCLUSIONS

CONCLUDING REMARKS: DISPERSAL AND GENE FLOW

Christopher M. PERRINS

Department of Zoology, Edward Grey Institute of Field Ornithology,
South Parks Road, OX1 3PS Oxford, Great Britain

I wish to restrict my comments largely to one general area of the points which have been discussed during the conference, namely the extent to which birds can evolve different characteristics in different habitats. This problem has several inter-connected aspects concerning the rate of possible gene-flow between such habitats, the quality of individual birds and the associated problems of dispersal.

Local Adaptation

Looking first at the problems of local adaptation, several speakers, in particular Isenmann and Zandt, showed that, sometimes, local habitats differ in various respects, especially in the timing of food availability during the breeding season. Since it is important for the parent birds to be able to find large quantities of caterpillars for their nestlings, one might expect that the birds would shift their breeding season or alter their clutch-size in order to match these local variations in food supply. This, at least at times, they do not seem to do. The authors' explanation was that the birds dispersed freely between these habitats with the result that the concomitant gene-flow made it impossible for local populations, breeding in a range of different woodland types each with a different pattern of food availability, to evolve characteristics suitable for one particular habitat alone.

This view was criticised by a number of questioners, in part at least because we have no evidence to support such a suggestion. Nevertheless, I suggest this does not mean that it may not be true. Many years ago I suggested (Perrins and Moss 1975) that one of the reasons why the clutch-size of Great Tits in oak woodland in Britain seemed to be lower than that which was most productive (in the sense of Charnov and Krebs 1974) could be that such an effect was in operation. In Britain today, only a very small proportion of the countryside is forested; a large proportion of the tits nest in poor woodland, hedgerows or gardens (I shall from now on refer to all these poor habitats as gardens for the sake of brevity). The few studies that have been made in gardens show that nesting success tends to be low there, sometimes it is very poor indeed compared with nesting success in woodland (Perrins 1965). Although no studies of most productive clutch-sizes have been made in these gardens, it seems reasonable to assume that they would be lower than in woodland. However, many tits move between woodland and gardens. Hence we

NATO ASI Series, Vol. G 24
Population Biology of Passerine Birds
Edited by J. Blondel et al.

may be looking at a situation where natural selection would favour a larger clutch in woodland, but a smaller one in gardens; the observed clutch falls in between, being somehow a compromise between the most-productive clutch-size in each of the two habitats and, presumably, the numbers of birds raised in each habitat (Fig. 1). (It is necessary to stress that I am talking about genotypic effects; van Noordwijk et al. (1980) have shown strong heritable components in clutch-size. However, there is also much phenotypic variation in clutch-size; birds in poor habitats do lay smaller clutches than those in good woodland, but the extent of this variation in clutch-size in relation to habitat seems insufficient to account for different optima in different habitats or to explain the mismatch between these optima and the observed clutch-sizes).

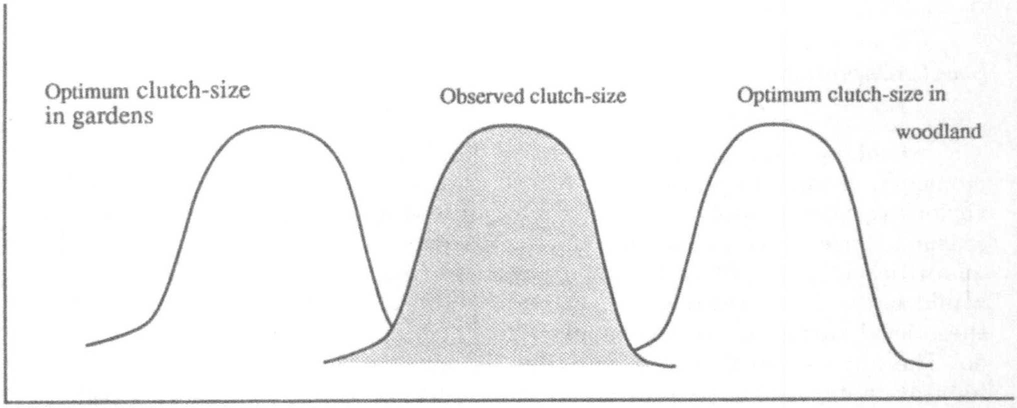

Fig. 1. The relationship between the most productive clutch-size in different habitats and the observed clutch-size of the population. The latter may not fall mid-way between the other two; its position will depend on the numbers breeding in the two habitats and the degree of dispersion.

The suggestion that the observed clutch-size is a result of differing selection pressures in woodlands and gardens has a number of implications. First, it implies that clutch-size must have become smaller over the last two millennia or so as man progressively reduced the proportion of wooded country in Britain. This suggestion can be made safely, since there is little chance of anyone being able to disprove it! However, there are other, more testable, implications as follows:

1. Since I argue that there is selection for a larger clutch in woodland and a smaller one in gardens, one might expect to find a slightly smaller clutch in immigrants to Wytham than the residents (since, on average, immigrants will come from gardens). This difference will be very small since, as argued above, both populations are intermixing freely, the immigrants will be of similar

genetic background to the Wytham birds. Hence any differences will result from selection for young in smaller than average broods in gardens and larger than average ones in woodland in the most recent generation. McCleery and Clobert (this volume) find no difference in brood-size between immigrant and resident females, after correcting for age and between year effects; McCleery (pers. comm.) confirms that the same is true of clutch-size.

2. The difference between observed and most productive clutch-sizes should be smaller in areas where the proportion of natural habitats is large such as, perhaps, may still be the case in some areas of northern Scandinavia. There, a very high proportion of the birds may still be nesting in natural habitat and there, the "forest genes" will not be markedly "diluted" by birds from man-made habitats where natural selection is favouring a smaller clutch. Hence the match between observed and most-productive clutch-size should be closer than in Britain.

3. It follows that, if the most productive clutch-size is larger than that observed in woodland then, in gardens, in contrast to woodland, the observed clutch ought to be larger than those which are the most productive. This could be checked by making an extensive study in gardens.

Dispersal

The degree to which a population can build up local adaptations is obviously going to depend on the scale of dispersal within that population and the size of the blocks of habitats. In the case of the tits we know that, in most of western Europe at least, the natural habitats are nowadays very fragmentary and many birds move from one patch to another. For example, at Oxford, we work in Wytham Woods, a block of woodland some 230 ha in extent; this is a large wood for southern Britain today. Yet about 50% of the Great Tits which breed in the boxes were not raised in those boxes. Since we believe that almost all the birds nesting within the wood do use the boxes, this implies that these birds are immigrants, coming from outside the wood. Similarly, in studies of life-time reproductive success, McCleery and Perrins (1989) showed that, in their life-times, individual Great Tits breeding in the wood produced fewer than one offspring which itself survived to breed in the wood; they should, of course, raise two young which survive to breed in a stable population. So, either our study area is a poor one for tits (which we do not believe to be the case) or, more probably, this discrepancy arises because there are other surviving offspring, but these emigrate from the wood. If this is the case, then roughly half the surviving young emigrate and roughly half the breeding population are immigrants. There is little opportunity for the evolution of characteristics specific to Wytham under such circumstances.

It would be interesting to know about other aspects of dispersal. For example, many studies (e.g. Kluyver 1951, Perrins 1965) have shown that, within a study area, more young from early first broods are found breeding in later years than is the case with young from later first broods. Strictly speaking we know nothing about the young which "disappear". The assumption has often been made that they die (Perrins 1965). However, because Kluyver (1971) was not convinced about this, he termed the observed survival "local survival", with the aim of stressing that the missing birds could either have died or dispersed elsewhere.

It is a common observation in studies of tits that early-hatched young show higher local survival. However, when Kluyver (1951) and Dhondt and Hublé (1968) studied the recovery rates of ringed tits from the national ringing schemes of the Netherlands and Belgium, they found that there was no correlation between the dates of ringing of nestlings (which they took to be a measure of their hatching dates) and their subsequent survival. Further, Dhondt and Hublé studied movement between small woods in the vicinity of Ghent and also showed that more late- than early-hatched young switched woods. Dhondt and Olaerts (1981) refined this, showing that the late-hatched young which survived and emigrated tended to be from early second broods; late-hatched young from first broods had a poorer survival rate than early ones. The simplest explanation that fits such findings is that not all later-hatched fledglings have a higher post-fledging mortality than the early-hatched ones, but, rather, that some of these late-hatched birds are more likely to emigrate.

Bird "Quality"

What are the emigrants doing? Having moved, do they stand a good chance of settling to breed somewhere and, if so, are they successful? There is some reason to suppose that these emigrants are not all fully able to compete with fit birds from early broods. We know for example that birds which fledge at low weights tend to remain at low weights for the rest of their lives (Garnett 1981) and that late-hatched birds rapidly go to the bottom of the peck order in flocks (Garnett 1981). Kluyver (1971) showed that young from second broods only had good local survival when competition from early-hatched young was removed.

This raises a number of questions about the background of the birds which are immigrants. Are they in any way inferior to residents? Are the immigrants to Wytham late-hatched emigrants from elsewhere? In what ways do they differ from Wytham-born birds? McCleery and Clobert (this volume) have shown differences in survival rates and reproductive success between immigrants and residents in Wytham. Gosler (1990) has produced some information relevant to these questions. He has shown that late-hatched young tend to have more un-moulted greater coverts than early-hatched young. This seems to be some indication of their "quality" during the post-juvenile moult and thereafter since those birds with more un-moulted coverts have lower fat scores during the winter. However, Wytham immigrant breeders do not have more un-moulted coverts than the residents, so there is no reason to suppose that these immigrants are of poor quality.

So, perhaps the key question is: "Why do some birds emigrate"? Do they do so because they are inferior to those that stay behind (though not so inferior that they die)? For example, it would be possible for later-hatched Wytham young to emigrate because they are of too low quality to obtain a territory (they are inferior for environmental, not genetic reasons). They might then breed successfully in gardens and some of their offspring might be of high enough quality to become immigrants to Wytham. Hence environmental factors might affect not only the chances of survival of the individuals, but also for those that do survive, the place where they breed.

So, if the emigrants are in any way inferior (other than by virtue of having a lower survival rate from fledging to breeding which we cannot measure, but which the analyses of ringing recoveries suggest is not true) what can we say about them? There could be another way of marrying these two conflicting findings. Most ornithologists have tended to select as their study sites areas of good habitat; this is not surprising, they tend to be pleasant places to work and the study birds are plentiful in them. However, it is possible that, by doing so, we are seeing only part of the story. For example, the observation that many of the late-hatched fledglings leave study areas is strictly only known in most areas for good habitats (the areas discussed by Dhondt and Hublé (loc. cit.) are probably rather poor in this sense since they are mostly town parks). Are the findings of ringing schemes explained by the fact that more of the late-hatched fledglings, having been "forced" out of good habitats, move to sub-optimal habitats? If many of the emigrants move into marginal habitats and raise few young, then, in a sense, they may best be considered genetically dead.

Productivity of different habitats

This raises another crucial question about which we know very little and this is the relative productivity of different habitats. Can all those populations which breed in poor habitats such as gardens maintain themselves or are they dependent on receiving a continuing influx from more productive habitats such as woodland? Certainly in poor garden habitats, not only do the parent tits lose most of their young, but those that are raised to fledging are of such poor weight that, at least in comparison with similar fledglings in woodland, they would stand very little chance of surviving (Perrins 1965). Is it possible that we have a situation where the garden populations are not able to maintain themselves without immigration and the immigrants tend to be the later ones from woodland. If this were the case, then two things might follow. First if the later-raised young are of poorer quality, then the poor breeding success in gardens may stem not only from poor quality habitats but also poor quality birds. Second, if woodlands are producing a surplus of birds which are helping to stock the gardens, then they must be producing larger numbers of survivors - and hence larger numbers of emigrants - than had previously been thought.

We could thus have the following situation. There is selection for a different clutch-size in gardens and in woodland, but the gardens are unable to evolve small enough clutches to raise enough healthy young which would enable them to maintain themselves. The woodlands continue to overproduce and provide the excess which maintain the gardens populations.

In summary, we begin to know a great deal about what happens in individual study areas, but we need to know much more about how the birds cope with a patch-work of different habitats, how much they can adapt to these and the reasons why some individuals move and others do not and the implications of such movement to the birds future success.

References

Charnov EL, Krebs JR (1974) On clutch-size and fitness. Ibis 116: 217-219

Dhondt AA, Hublé J (1968) Fledging date and sex in relation to dispersal in young Great Tits. Bird Study 15: 127-134

Dhondt AA, Olaerts G (1981) Variations in survival and dispersal with ringing date as shown by recoveries of Belgian Great Tits, *Parus major*. Ibis 123: 96-98

Garnett MC (1981) Body size, its heritability and influence on juvenile survival among Great Tits, *Parus major*. Ibis 123: 31-41

Gosler AG (1990) On the use of greater covert moult and pectoral muscle as measures of condition in passerines with data for the Great Tit *Parus major*. Bird Study 37: in press

Kluyver HN (1951) The population ecology of the Great Tit, *Parus m. major* L. Ardea 39: 1-135

Kluyver HN (1971) Regulation of numbers in populations of the Great Tit (*Parus m. major*). Proc Adv Study Inst Dynamics Numbers Popul (Oosterbeek 1970): 507-523

McCleery RH, Perrins CM (1989) Great Tit. In Newton I (ed) Lifetime Reproduction in Birds. Academic Press, p 35-53

Noordwijk AJ van, Balen JH van, Scharloo W (1980) Heritability of Ecologically important Traits in the Great Tit. Ardea 68: 193-203

Perrins CM (1965) Population fluctuations and clutch-size in the Great Tit *Parus major* L. J Anim Ecol 34: 601-647

Perrins CM and Moss D (1975) Reproductive rates in the Great Tit. J Anim Ecol 44: 695-706

CONCLUDING REMARKS: UNDERSTANDING POPULATIONS

Arie J. van NOORDWIJK

Zoologisches Institut, Rheinsprung 9, CH4051 Basel, Switzerland

In an attempt to weave some threads through the fabrics of observation and experimentation collected in these proceedings, it is useful to start with the basic question of what we hope to achieve eventually with our studies. Why would or why should society be interested in paying some biologists to watch birds ?

Not all of us are aware that the roots of the population studies, as we perform them, go back to applied research rather than to basic research. In the early 1930's Kluyver started to ring all Great Tits in a particular study area to study population level processes as the sum of what "happens" to all the individuals. He chose Great Tits, because he wanted to assess the importance of insectivorous birds in keeping the numbers of so-called pest species at an acceptable level. Moreover, Great Tits are big enough to be observed as individuals, numerous enough to collect reasonable sample sizes and because they are resident, it is possible to study the whole year cycle. These reasons for studying tits and similar birds are still valid.

The reasons for wanting to gain a better understanding of population dynamics are, if anything, more pressing than ever. Mankind imposes tremendous changes on the environment of virtually all species throughout the world. The high numbers of local and global extinctions - which is, after all, the same as a population density going to zero - are but one manifestation of this process.

Different approaches

In general, there is nothing that is more practical than a good theory. Wherever our studies are aimed at developing good theory, which is generally applicable by definition, one needs no additional justification. This does raise the question what sort of theory we are looking for. The studies reported on in these proceedings, have shifted away from trying to answer the question of what determines population size. Instead, the main question seems to have become why individuals are and do as they are and do, in other words to describe and explain the various aspects of our species. In all this, there is an emphasis on the evolutionary dimensions of these properties. What then is the connection between the properties of organisms, whose explanation we seek, and population dynamics ?

The honest answer is that we don't really know. What we do know, is that our attempts in explaining population densities always run into questions

NATO ASI Series, Vol. G 24
Population Biology of Passerine Birds
Edited by J. Blondel et al.

concerning these properties of individuals. It is, however, difficult to turn the process around and therefore we are using many different approaches to our studies. There is fortunately a growing tendency to combine several approaches in many of the individual studies. It is also the value of conferences like this one to learn about the results from other approaches and to draw conclusions from across studies. For example, in a growing number of studies the variation among individuals in clutch size is studied, which is a topic that can neither be classified as being at the population level, nor at the individual level. At the same time, it is impossible to study everything at once, there are simply far too many variables. What we want to do, is to find out which variables are the quantitatively most important ones, so that the others may be ignored. It is a priori unlikely that the same variables will be important in all or even in a majority of cases.

In an attempt to classify the main approaches to learn about population dynamics, I see seven types of approach, although pure forms are rare. I would characterize and caricature them as follows:

1) <u>The classical approach</u>; a description of average reproduction and average survival and the variation in space and in time therein. In a second step one attempts to correlate environmental parameters with the observed variation.

2) <u>The energetic approach</u>; using energy as a general currency, descriptions are made at the population level, or at the individual level. The former is more typical in systems ecology, the latter lends itself very well to experimental manipulations of individuals. However, it is usually still assumed that all individuals are intrinsically equal.

3) <u>The micro-environment approach</u>; similar to the classical approach, but using the environmental variation at the level of individuals, rather than at the level of populations. This approach is still rare, but shouldn't remain so, if we start using our increased computer power.

4) <u>The genetic approach</u>; when there is genetic variation, no further explanations are needed, natural selection will do it for you. When there is no genetic variation, one has to explain its absence.

5) <u>The decision making/optimisation approach</u>; birds know what is good for their fitness and fly around calculating what to do next. If we measure the costs and the benefits of each possible action, we know what each individual will do under all conceivable circumstances. The population dynamics will follow from the consequences of the decisions taken by individuals.

6) <u>The reaction norm approach</u>; it is a combination of 3) and 4) where genes are seen as programs that "process" information from the environment.

7) <u>The individual variation approach</u>; it is a combination of 2), 3) and 5) where one assumes that individuals are different for genetic reasons, for reasons of developmental history and for reasons of having different micro-environments. Hence, individuals "ought" to do different things. One may describe population fluctuations, by identifying the marginal recruits, the marginal survivors and the marginally produced offspring.

In all of these approaches one can find combinations of descriptive field studies, field experiments and theoretical studies. There are (too) few studies where one attempts to work in a similar way on the ecology of adjacent trophical levels, be they prey, predators or parasites. When these adjacent trophic levels are considered, they are usually treated as extrinsic, static quantities.

What was new at this conference ?

In part, we make progress by using more sophisticated electronic or molecular gadgetry to take up old problems with new methods. In part we make progress by collecting new critical data and performing critical experiments. There is little need to repeat the table of contents here. There are, however, clearly also advances made through changes in emphasis. The two major changes in our thinking that I see, are the increased interest in variation among individuals and in the connection between observable traits and decision making processes in the individual animal.

The variation among individuals may be caused by genetics, by the individuals history or by the current micro-habitat. In this area, we are about to progress from simple division of variance components to either genes or environmental factors to a view in which the genes constitute a program which processes information from the environment. To what extent it will be possible to collect good data on this issue is still unclear, but as a framework to treat various sources of variation among individuals it will certainly be valuable.

We are used of thinking about the properties of organisms in terms of traits - clutch size, body size, bill shape, flight speed are such traits - that each have a certain value in a certain case. This dissection of organisms into a set of trait values is a tool that is most pronounced among systematists and among geneticists, but by no means restricted to these subdisciplines. What are these traits, how do these values come about ? In some cases they can be seen as a decision or a series of decisions. The date of first egg laying, can easily be seen as a decision to produce a first egg, clutch size as a series of decisions to lay one more egg. Considering the decision(s) to go and hunt for one more caterpillar and whether to feed it to which nestling rather than eat it yourself, is a useful way of studying our organisms.

Combining these two elements, one may ask the question to what extent the variation among individuals is caused by differences in the decision rules, and to what extent the outcome of the same rule is different, because the input was different. Which brings us back to reaction norms. Let me make three additional remarks. 1) Evolution of these decision rules requires genetic variation in them. 2) Much of the variation in decision rules will not be genetic, but will be a consequence of recent history, i.e. the effect of learning of the individual. 3) There might well be geographic variation in these decision rules, which are the result of local adaptation. Gene flow may then be important in the maintenance of genetic variation for such decision rules.

Understanding population density

Eventually, our understanding of individuals and their properties at the population level should lead us back to population density. What we want are mechanistic models, that will allow us to make conditional predictions. To say that, if food does a1 and weather does b2, then population size will decrease by 50 %, or that with a2 and b1 it will increase by 100 %, even if those estimates have a 20 % error, is still much more useful than to be certain that next years population size will be between 50 % and 200 % of this year's. So we want to relate our understanding to environmental conditions and to identify the key-factors in the environment.

The interrelationships between patterns and processes are depicted in the figure. The body of our understanding lies in the processes, with a dominance batch of experimentation. Along one axis, we have the wings of spatial and temporal patterns, Along the other axis we have the continuum of time scale, ranging from the brainwork of decision making, to the ultimate explanations at the tail-end (but these explanations should rise above the level of tales). Finally, we have the two legs of observation and manipulation, that allow our beast to stand on firm ground. It is our job to ensure that none of these parts proliferates at the cost of others. A functioning bird needs a well proportioned and integrated body (of theory).

The in the great tit study
of bird populations
first the egg
then the chick
EN
VI
RON
MENT
food distribution
TREE
PRO CESS
EXPERIMENTS
long term monitoring
geographic patterns morphology
models
observations
ultimate tale
EPC

SUBJECT INDEX

NATO ASI Series G

Vol. 1: **Numerical Taxonomy.** Edited by J. Felsenstein. 644 pages. 1983. (out of print)

Vol. 2: **Immunotoxicology.** Edited by P.W. Mullen. 161 pages. 1984.

Vol. 3: **In Vitro Effects of Mineral Dusts.** Edited by E.G. Beck and J. Bignon. 548 pages. 1985.

Vol. 4: **Environmental Impact Assessment, Technology Assessment, and Risk Analysis.** Edited by V.T. Covello, J.L. Mumpower, P.J.M. Stallen, and V.R.R. Uppuluri. 1068 pages. 1985.

Vol. 5: **Genetic Differentiation and Dispersal in Plants.** Edited by P. Jacquard, G. Heim, and J. Antonovics. 452 pages. 1985.

Vol. 6: **Chemistry of Multiphase Atmospheric Systems.** Edited by W. Jaeschke. 773 pages. 1986.

Vol. 7: **The Role of Freshwater Outflow in Coastal Marine Ecosystems.** Edited by S. Skreslet. 453 pages. 1986.

Vol. 8: **Stratospheric Ozone Reduction, Solar Ultraviolet Radiation and Plant Life.** Edited by R.C. Worrest and M.M. Caldwell. 374 pages. 1986.

Vol. 9: **Strategies and Advanced Techniques for Marine Pollution Studies: Mediterranean Sea.** Edited by C.S. Giam and H.J.-M. Dou. 475 pages. 1986.

Vol. 10: **Urban Runoff Pollution.** Edited by H.C. Torno, J. Marsalek, and M. Desbordes. 893 pages. 1986.

Vol. 11: **Pest Control: Operations and Systems Analysis in Fruit Fly Management.** Edited by M. Mangel, J.R. Carey, and R.E. Plant. 465 pages. 1986.

Vol. 12: **Mediterranean Marine Avifauna: Population Studies and Conservation.** Edited by MEDMARAVIS and X. Monbailliu. 535 pages. 1986.

Vol. 13: **Taxonomy of Porifera from the N.E. Atlantic and Mediterranean Sea.** Edited by J. Vacelet and N. Boury-Esnault. 332 pages. 1987.

Vol. 14: **Developments in Numerical Ecology.** Edited by P. Legendre and L. Legendre. 585 pages. 1987.

Vol. 15: **Plant Response to Stress. Functional Analysis in Mediterranean Ecosystems.** Edited by J.D. Tenhunen, F.M. Catarino, O.L. Lange, and W.C. Oechel. 668 pages. 1987.

Vol. 16: **Effects of Atmospheric Pollutants on Forests, Wetlands and Agricultural Ecosystems.** Edited by T.C. Hutchinson and K.M. Meema. 652 pages. 1987.

Vol. 17: **Intelligence and Evolutionary Biology.** Edited by H.J. Jerison and I. Jerison. 481 pages. 1988.

Vol. 18: **Safety Assurance for Environmental Introductions of Genetically-Engineered Organisms.** Edited by J. Fiksel and V.T. Covello. 282 pages. 1988.

Vol. 19: **Environmental Stress in Plants. Biochemical and Physiological Mechanisms.** Edited by J.H. Cherry. 369 pages. 1989.

Vol. 20: **Behavioural Mechanisms of Food Selection.** Edited by R.N. Hughes. 886 pages. 1990.

Vol. 21: **Health Related Effects of Phyllosilicates.** Edited by J. Bignon. 462 pages. 1990.

NATO ASI Series G

Vol. 22: **Evolutionary Biogeography of the Marine Algae of the North Atlantic.** Edited by D. J. Garbary and G. R. South. 439 pages. 1990.

Vol. 23: **Metal Speciation in the Environment.** Edited by J. A. C. Broekaert, Ş. Güçer, and F. Adams. 655 pages. 1990.

Vol. 24: **Population Biology of Passerine Birds. An Integrated Approach.** Edited by J. Blondel, A. Gosler, J.-D. Lebreton, and R. McCleery. 513 pages. 1990.